T0282452

STATISTICS FOR FOOD SCIENTISTS

STATISTICS FOR FOOD SCIENTISTS

Making Sense of the Numbers

FRANK ROSSI
VIKTOR MIRTCHEV
Kraft Foods, Illinois, USA

Academic Press is an imprint of Elsevier
125 London Wall, London EC2Y 5AS, UK
525 B Street, Suite 1800, San Diego, CA 92101-4495, USA
225 Wyman Street, Waltham, MA 02451, USA
The Boulevard, Langford Lane, Kidlington, Oxford OX5 1GB, UK

Front and back cover design created by Kremena Rouhliadkova.

ISBN: 978-0-12-417179-4

British Library Cataloguing-in-Publication Data
A catalogue record for this book is available from the British Library

Library of Congress Cataloging-in-Publication Data
A catalog record for this book is available from the Library of Congress

For information on all Academic Press publications
visit our website at http://store.elsevier.com/

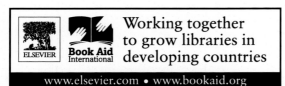

Publisher: Nikki Levy
Acquisition Editor: Patricia Osborn
Editorial Project Manager: Jaclyn Truesdell
Production Project Manager: Caroline Johnson
Designer: Greg Harris

Typeset by TNQ Books and Journals
www.tnq.co.in

Printed and bound in the United States of America

CONTENTS

ACKNOWLEDGMENTS

From Viktor and Frank:

A project of this nature is not the work of just two people. Without the help of our friends, co-workers, business partners, and our families, we would have never been able to write this book. We have learned a lot about what analytical skills are desired from interacting with folks from different walks of life. Our intent with the book was to share a framework for problem solving rather than presenting a bag of tricks, and hopefully a comprehensive approach that would help busy people do well.

The folks at Rutgers University's Department for Continuing Professional Education asked us to create and teach a statistics workshop for food scientists several years ago. The interactive course we developed, "Making Sense of the Numbers," which we have now delivered for a number of years, has served as the basis for this book. We are thankful to the organizers and the participants for allowing us to develop and instruct the course and to "practice-write" the chapters of this book.

Just as important is the assistance and advice that we received from our friends and colleagues at Kraft Foods and Mondelēz International. Rob Glennon was an early reader of our text and provided critical feedback that makes this book more useful and readable. His experiences, strong grasp of reality, wonderful anecdotes, and the invaluable lessons he shared over the years have helped make the book far more readable and engaging. In addition to Rob, the statisticians who we have worked closely with over the years have taught us much and helped us to provide great examples. These statisticians include Keith Eberhardt, Sandra Echols, Addagatla Babu, Jeff Stagg, Gael Le Barzic, Fernando Rosa, Maria Teresa Lopez, Anja Schleppe, Marty Matzinger, Eric Lagergren, Boris Polsky, Yvonne Wang, Stanislav Zakharkin, Trevor Duguid Farrant, and Kamaljit Sagoo. In these pages, you will find the fruits of their work with us over the years. Shamsheer Mahammad is a talented Rheologist at Kraft Foods who helped us greatly with measurements from and advice about from the Brookfield viscometer. Any errors about that measurement system are ours alone even after his patient advice.

Two additional people played a very large role who we would like to specially thank. Donka Mirtcheva provided critical feedback to every chapter, and the book is better for her suggestions and creative ways in framing issues and addressing challenges. Kremena Rouhliadkova created the book cover design that is original, attractive, and enticing to potential readers, and this alone should increase the book's popularity.

And, above all, we would like to thank our families for letting us progress on this writing adventure.

From Viktor:

To my precious wife, Silvia Mirtchev, thank you for your encouragement to write this book. I could not have done it without your unconditional support and understanding; or the love of our children, Maxim and Viano. You are a treasure and without your comfort and cheer over the years and during countless several late nights, while I was writing my heart out, I would have never been able to complete this project.

From Frank:

To my wife, Aylesa Singley, thank you for your continued inspiration and encouragement throughout this project and my career. I'm sure I could not have done this without you!

CHAPTER 1

Introduction

"Maria accepted the job offer, and she will be joining your team in two weeks!" Joan had spent 7 months on a quest to find a food scientist, and this short voicemail from Human Resources put an end to his efforts. He was so excited that he stepped out of his office to share the great news with Steve, a senior food scientist, working for him. Steve had also enjoyed meeting Maria and shared Joan's high opinion of her from the recent university job fair.

Maria, a recent graduate holding a Master's degree in food science, is a bright, young professional who is not only excited, she has already posted the good news on Facebook and Twitter and has texted all of her friends about her decision to join the Ultimate BBQ company.

"Help is on the way," said Joan, jokingly, to Steve. "Maria has accepted, and her strong critical thinking and analytical skills will help us improve the quality of our BBQ sauce."

Steve: "BBQ!" BBQ sauce has been one of the company's core products, and it has been around for over 100 years. Sadly, in the most recent years, it has lost market share due to product quality degradation, a trend that the company hopes to reverse quickly to maintain a market leading position. "Will Maria be equipped for such a big first project?"

Joan: "I agree with you Steve, and that is why I would like to ask you to be Maria's manager. Help her understand the complex process of making BBQ sauce, the similarities between our pilot plant and plant production, and introduce her to the rest of the team. As you know 'it takes a village' to make good BBQ."

"There have been a lot of recent complaints indicating that our BBQ sauce is not thick enough. Our consumers care deeply about our product. They have given us a second chance through the gift of feedback, and now we owe it to them. Steve, I cannot think of a better person than you not only to pass on a lot of BBQ sauce institutional knowledge to Maria, but also to help her to focus on the areas that need the most attention. You have worked closely with James in Marketing Research to understand the specifics of the consumers' feedback from the BBQ users' focus group sessions."

Statistics for Food Scientists
http://dx.doi.org/10.1016/B978-0-12-417179-4.00001-9

Steve: "I would love to help Maria. Since this will be a highly analytical project, I'll introduce her to Chad in Analytical Sciences, Louise in Quality, and the other subject matter experts that she will need to work closely with. I'll also organize a visit to both of our BBQ plants after she is situated in the Research and Development center."

In the spirit of good planning and to better prepare for a successful relationship with her, Steve decides to venture into the virtual space for traces of Maria. He revisits her LinkedIn profile and learns from it that Maria relishes challenges, never gives up, and most of all she enjoys working on teams. She seems to have some data analysis skills that Steve hopes he can learn from. At the same time, her profile suggests that she is interested in learning more about process capability analysis, which has been an area of focus for Steve in recent years. As Steve prepares to weave in the relationship between product concepts and analytical approaches, he realizes that he will need the help of others to enable Maria to see deeply into what is best for BBQ sauce.

Steve wants to help Maria try new approaches and statistical practices. Only after she sees some successes will she add them into her repertoire of statistical tools. These will enable her to develop critical practices to shape thought processes and transform the way scientists approach everyday quality and product development challenges. These practices will be relevant to many challenges that food scientists face, and Steve wants Maria to have a successful start into her corporate career. Now, onto Day 1...

CHAPTER 2

Descriptive Statistics and Graphical Analysis

One of the first objectives for Maria as a new employee is to develop an understanding of the BBQ sauce manufacturing process. Fortunately, the research and development center where she works is part of a larger facility that includes the main BBQ sauce production facility for her company. Steve arranges for her to visit the production facility in her second week. Lisa, the plant manager, will give her a tour.

Maria arrives at the facility a few minutes before the meeting is scheduled. Though only a 5-min walk from the main office building, she feels that she has entered another world. As she waits for Lisa to meet her, she can see the busy activity through the glass walls of the waiting room—bottles of sauce rattling down lines, forklifts moving pallets of materials, uniformed workers in hairnets, earplugs, and safety glasses bustling about. What a change from the quiet research labs where she spent her first week at her desk and in meetings!

When Lisa arrives, Maria is surprised to find that she appears to be quite young for someone who has responsibilities for a manufacturing facility. Short, trim, and completely unadorned, Lisa displays a confident manner.

"We are very happy to have you on board Maria," Lisa says as she vigorously shakes Maria's hand. "We were so excited when you accepted the position with us. Like you, I started in research right out of school. I never would have thought I would manage production one day."

"I see you are already wearing your safety shoes and lab coat. You will also need to remove all of your jewelry before we enter the plant," Lisa says as she guides Maria to a room with lockers and a dispensary for the gear required for entering the facility. She gives a hairnet and a set of earplugs to Maria and runs down a list of safety rules for the tour. "Keep your hands away from most everything and stay within the yellow lines on the floor and you will be fine," Lisa concludes.

When she enters the production floor, Maria is surprised at the noise level. Behind the glass in the waiting room, she saw the high level of activity, but she was unprepared for the accompanying noise.

Statistics for Food Scientists
http://dx.doi.org/10.1016/B978-0-12-417179-4.00002-0

"We do a lot of shouting around here," Lisa says as they start walking. "With the noise and the earplugs there is no other way!"

The next hour is a blur for Maria, as she encounters the many steps of BBQ sauce production. Ingredients arrive at a receiving dock and are moved by forklift to an internal storage facility within the plant. They are moved again to the manufacturing lines as they are needed. The sauce production starts with all ingredients mixed in an extremely large vat. Mixing continues as the temperature on the vat is increased. When the vat reaches a specific temperature, it is evacuated. The sauce moves through pipes to be filled into bottles. Then the bottles are capped. Although there are a number of production lines, they look similar, if not identical. The process seems straightforward enough, but Maria suspects that it may be more complicated than it appears.

Maria sees that on each line workers periodically pull bottles from the production line, and mark the date and time on the label. The bottles are placed on a table near the operator station.

Maria points and asks Lisa, "What is going on there?"

"The operators pull samples periodically to monitor product quality," Lisa responds. "Once an hour, the technician from the Quality Lab collects the bottles from all of the production lines and measures the viscosity. It is our most important quality measure. Consumers expect the sauce to cling to the meat as it cooks."

At the end of the tour, Lisa brings her to a stop and proudly exclaims, "I have saved the best for last! We just finished installation of our newest filling line a few weeks back. This is state-of-the-art technology that can fill at nearly twice the rate of our other lines. The viscosity should also be more consistent."

Maria looks on as Lisa points out the differences between the existing and new equipment. But one thing she sees again is the operator periodically pulling bottles from the line and marking the date and time, as on the other lines. Following Maria's eyes, Lisa comments, "We measure viscosity on the new line too. We collect samples as frequently as the other lines, but since the output from the line is much greater we may need to change our sampling strategy for this line."

As the tour concludes, Lisa mentions to Maria, "I hear that you have some data analysis skills from your university training. Can I have you look at the production data from the new and existing lines to verify the performance improvement on the new line?"

Maria eagerly agrees, "I did have terrific experience learning about statistics and data analysis working with my graduate advisor, Dr Wang. I'd be happy to take a look at the data."

Lisa agrees to organize and email the production data to Maria. "I'll take a look at the data and set up a time to discuss my findings," Maria promises.

Within a day, Maria receives an email from Lisa with a spreadsheet attached. The text of the email reads as follows:

> *It was nice meeting you yesterday. …I have included viscosity data from the new line, line 6, and the best performing line from our existing production lines, line 4, for two days of production last week. Please let me know what you find.*

The first few rows of attached spreadsheet of data look like this:

Date and Time	Line 4 Viscosity	Date and Time	Line 6 Viscosity
4/28/2013 8:06	4776	4/28/2013 8:05	3746
4/28/2013 8:14	4318	4/28/2013 8:13	4424
4/28/2013 8:23	4363	4/28/2013 8:22	4284
4/28/2013 8:31	4447	4/28/2013 8:30	4241
4/28/2013 8:39	3832	4/28/2013 8:38	4132
4/28/2013 8:48	4426	4/28/2013 8:47	4189
4/28/2013 8:56	4516	4/28/2013 8:55	4650
4/28/2013 9:04	5085	4/28/2013 9:03	4303
4/28/2013 9:13	4585	4/28/2013 9:12	4252

There are 116 data points for each of the two lines, each data point with a time stamp. For both lines, the data appear to come in roughly 8–9 min intervals.

Maria recalls from her work with Dr Wang a number of summary statistics that can be used to characterize the production data for lines 4 and 6. Having brought to the office most of her college textbooks, she locates and opens the introductory statistics textbook she used in Dr Wang's class. The book describes a list of summary statistics and includes illustrations. The list comprises two groups: measures of the center and measures of the spread of a set of numbers.

Three measures of the center are the mean, median, and mode:

1. Mean = This is the mathematical average; add up all of the values and divide by the number of observations.
2. Median = The median is the middle value of a set of numbers arranged in order from smallest to largest. This is the 50th percentile.
3. Mode = This is the most frequently occurring value in a dataset.

There is an illustration in the book that shows each of these measures for a given dataset. The illustration is a histogram, a graph depicting the distribution of the data values. Histograms group the data into a set of bins representing a subset within the range of the data. The frequency or percentage of occurrences in each bin is represented on the vertical axis. The illustration below identifies the three measures, which happen in this case to be close to one another.

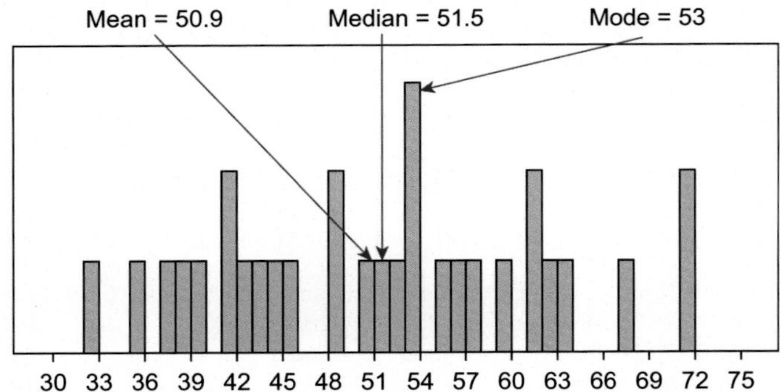

Further reading describes the advantages and disadvantages of the measures. While the mean uses all of the individual values in its calculation, it can be influenced by extremely large or small values. The median and mode are not influenced by extreme values since they do not use each individual value. The mode may not be unique, and it may not reflect the center of the distribution very well.

Four measures of the spread are the variance, standard deviation, range, and interquartile range:

1. Variance $= \frac{\sum_{i=0}^{n}(x_i - \bar{x})^2}{(n-1)}$
2. The standard deviation is the square root of the variance.
3. Range = (largest value – smallest value).
4. Interquartile range (IQR) = the middle value of the second half of the data—the middle of the first half of the data when the data is arranged in order from smallest to largest. In other words, it can be represented as follows:

IQR = (the value at the 75th percentile – value at the 25th percentile)

These measures of spread for the data in the illustration above are as follows: the variance is 112.36, the standard deviation is 10.6, the range is 39 (71–32), and the interquartile range is 17.75 (59.5–41.75).

These measures, too, have advantages and disadvantages. Like the mean, the variance and standard deviation use all of the individual values in their calculation, but they can be influenced by extreme values. The range is very easy to understand but does not use each value and is affected by extreme high or low values. The interquartile range ignores half of the data but is not affected by extreme values.

Maria had used the data analysis functions in Microsoft Excel to create summary statistics in Dr Wang's class. She recalls that she had installed the Analysis ToolPak add-in for Excel to accomplish this and will need to do so in preparation for creating descriptive statistics for the viscosity data. Dr Wang was a strong believer in looking at data in graphs before performing any statistical analyses. One graph she had used previously is a histogram like the one illustrated in her book. Comparing histograms for each of the product lines should help in understanding if they perform differently. Histograms are also created using the Analysis ToolPak in Microsoft Excel.

Maria arranges the histograms she creates in Excel to compare the lines. They do indeed look to be symmetric and bell-shaped. However, she quickly notes that Excel chose to use different bins for each line, so the scale is different on each graph.

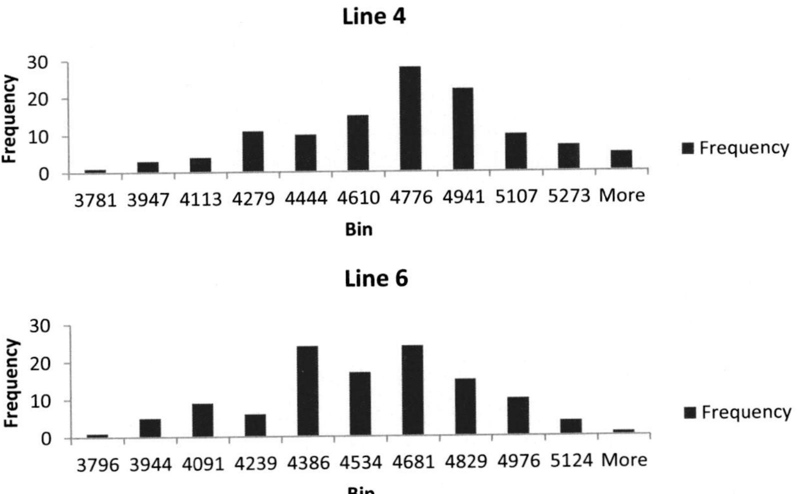

This makes it difficult to compare the lines. It is not apparent if the variability is similar for each line, or if they have similar centers. Excel is not making it easy for her to look at her data in the way that makes most sense to her. Looking back at the pop up menu from which she created the histograms, she sees that she has the ability to enter in the bin range, and with a bit of effort, she comes up with a range that represents the data from both lines reasonably well.

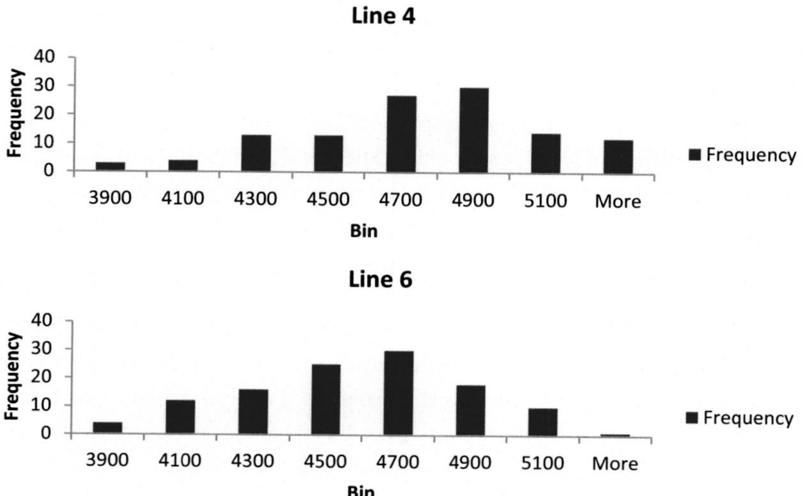

When Maria arranges the histograms to compare the lines, it does appear that the two lines perform differently. The center for line 6 seems somewhat lower in viscosity. The variability in line 6 seems to be smaller but not to a great degree.

It occurs to Maria that a different plot would indicate if there is some sort of increasing or decreasing trend in the data as it was collected over time. She uses Excel to graph the data in time sequence to see if there is an increasing or decreasing trend over time. She has difficulty using the date and time column for the horizontal axis and realizes that creating the plot without this will use the data point sequence number on the horizontal axis. This is sufficient for her purposes, since the data points from both lines are similarly spaced in time. Her graphs have viscosity on the vertical axis and time point sequence on the horizontal axis.

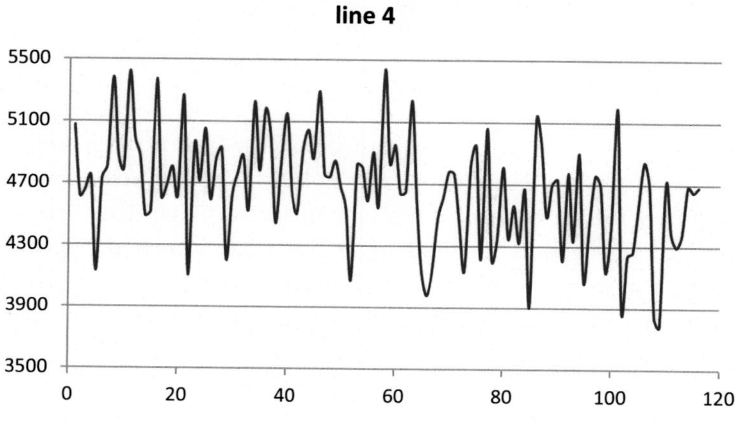

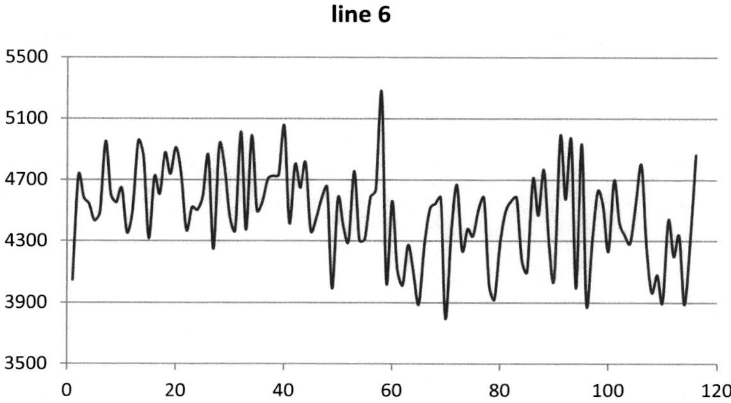

Maria does not observe an increasing or decreasing trend when the data is displayed in sequence, but the graphs do reinforce her earlier observation from the histograms that both the center and the variation in line 6 are smaller.

Maria has no problem creating the summary statistics for the two lines. The output from Excel not only includes the measures of the center and spread that she just read about, but it also gives her a number of additional statistics measures, as shown in a table that looks like the following:

Viscosity Line 4

Mean	4662.86
Standard error	32.95
Median	4686.60
Mode	#N/A
Standard deviation	354.90
Sample variance	125,955.02
Kurtosis	−0.09
Skewness	−0.20
Range	1657.19
Minimum	3781.39
Maximum	5438.59
Sum	540,891.86
Count	116

Viscosity Line 6	
Mean	4480.46
Standard error	28.02
Median	4505.23
Mode	#N/A
Standard deviation	301.75
Sample variance	91,052.26
Kurtosis	−0.35
Skewness	−0.10
Range	1475.20
Minimum	3796.18
Maximum	5271.38
Sum	519,733.58
Count	116

Maria digs back into her statistics textbook again to re-familiarize herself with the additional measures included in the Excel output. For both lines, the mode is listed as N/A. She thinks this is because all of the data values are unique. She recalls that the variance is simply the square of the standard deviation. She is not sure what the sum of the values may tell her. She learns that the standard error is the standard deviation divided by the square root of the sample size, and it is an estimate of the standard deviation of the mean, though she is not sure how that may be useful to her. And while the range is listed in the output, she is disappointed that the interquartile range is not.

Two new terms for her are skewness and kurtosis. Checking back in her statistics text, she learns that these describe the shape of the distribution of the data.

Skewness quantifies the extent to which the data is not symmetric—zero indicates symmetry, a negative number indicates a skew to the left side, and a positive number indicates a skew to the right side.

Kurtosis indicates the flatness of the distribution of the data compared to a bell shape—numbers above zero indicate a flatter distribution, and numbers below zero indicate a more peaked distribution.

For the viscosity data the skewness and kurtosis values are quite small, indicating that the distributions of the viscosity data from both lines can reasonably be considered symmetric and bell-shaped.

Maria arranges her descriptive statistics summaries and histograms in a report to present to Lisa. Dr Wang always stressed making things clear and concise when putting a report together. Maria decides to remove some of the descriptive statistics that she does not think will be useful when speaking to Lisa, and she arranges the table of these statistics for a more direct comparison of the production lines. She also formats the values so that there are not too many values after the decimal point, which contain no information.

	Viscosity Line 4	**Viscosity Line 6**
Mean	4663	4480
Median	4687	4505
Standard deviation	355	302
Kurtosis	−0.09	−0.35
Skewness	−0.20	−0.10
Range	1657	1475
Minimum	3781	3796
Maximum	5439	5271

Maria feels confident that she has information that Lisa will find interesting. She sets up a meeting to discuss the results with Lisa for the following day.

"Take a look at the analyses that I created for you on the two lines," Maria says as she spreads her analysis sheets on the conference room table. "It seems clear that the two lines are operating differently."

Lisa examines the sheets before her. She looks puzzled, as she is sure that the new line is performing better. After a short time she comments:

"The means and medians for the two lines look to be pretty close; I'm not sure I agree that the lines are performing that differently in that respect. The mean for line 4 is 4663; for line 6, it is 4480. There does not seem to be much of a difference given the scale of the data. And the standard deviation of 302 for line 6 is definitely smaller than the 355 for line 4. But maybe I'm seeing these results as I want them to be. Is there a way to make a definitive statement about differences between the lines?"

Maria realizes that she could have done more in analyzing this data. She recalls from her work with Dr Wang that she can more definitively compare the lines and tells Lisa that she will perform some additional analyses to get an answer.

CHAPTER 3

Hypothesis Testing

Lisa is right! As a plant manager, she ought to be able to demonstrate that the investment made in a new line is a good investment. Alternatively, if the new asset brings no additional, or only marginal, benefit why should she upgrade the rest of the lines? Lisa needs more than a summary of line performances. She should be able to defend the case with confidence that the asset investment has merits. She ought to be able to clearly answer the question: Is there a significant difference between line 4 and line 6 viscosities?

Maria learned about statistical hypothesis testing in one of Dr Wang's classes. By using a statistical hypothesis test, she can test the hypothesis that line 4 and line 6 perform similarly in terms of finished product viscosity. Thus, she will be able to address Lisa's concerns if there are viscosity differences between the two lines.

Since Maria already has the data files, she decides to run a hypothesis test to answer Lisa's questions. She looks at the options in the data analysis add-in and is a bit uncertain which method is most appropriate. In Excel, she can choose the following:

1. *t*-test for two samples assuming equal variances
2. *t*-test for two samples assuming unequal variances
3. *t*-test paired for two-sample means
4. *F*-test for two-sample variances

She remembers the discussion of paired data in her class—the two production lines run independently, so test 3 can be eliminated. Since the difference between the remaining two *t*-tests hinges on the assumption of equal or unequal variances, it seems best to test the variances first, and then choose the appropriate *t*-test to use.

The test of the sample variances answers the question: Is there a difference in viscosity variability between the lines? It is easy to perform

Statistics for Food Scientists
http://dx.doi.org/10.1016/B978-0-12-417179-4.00003-2

this test in Excel; simply select the data points for each of the lines and elect the "*F*-test for two sample variances."

F-Test Two Sample for Variances

	Line 4	Line 6
Mean	4663	4480
Variance	125,955	91,052
Observations	116	116
df	115	115
F	1.38	
$P(F \leq f)$ one-tail	0.04	
F critical one-tail	1.36	

Maria recalls that the *p*-value (listed in the output as $P(F \leq f)$) is an important part of the output for her to focus on, and that small values indicate that the variances of the two lines are different. Dr Wang had used probabilities of 0.05 as a cutoff value. The *p*-value of 0.04 in the output is smaller than 0.05, so she can conclude that there is a statistically significant difference in the viscosity variance between line 4 and line 6. Since the variance is smaller for line 6, the data indicate that line 6 delivers a more consistent product. This finding is what Lisa had expected, so Maria is excited to deliver the news.

The *t*-test for sample means answers the question: Is there a statistical difference in the average viscosity between lines 4 and 6? Since the variances are different for the lines, Maria chooses test 2 that assumes unequal variances.

t-Test: Two Sample Assuming Unequal Variances

	Line 4	Line 6
Mean	4663	4480
Variance	125,955	91,052
Observations	116	116
Hypothesized mean difference	0	
df	224	
t stat	4.22	
$P(T \leq t)$ one-tail	0.00	
t critical one-tail	1.65	
$P(T \leq t)$ two-tail	0.00	
t critical two-tail	1.97	

Again, Maria looks at the *p*-value listed as $P(T \leq t)$ as one of the important output pieces for her to focus on. She sees that there are two *p*-values in the output: one-tail and two-tail. She does not recall the difference between the one- and two-tail tests, but because both values are very small, the conclusion is the same; the average viscosity of the lines is statistically significantly different. The average of line 6 is smaller, so Lisa will be alarmed as a more viscous product is desirable.

Maria is not confident that she has used the right statistical analyses. Did she read the results correctly? She has to be sure about her findings before she reaches out to Lisa. She pulls out a textbook from Dr Wang's class and puts together a cheat sheet with definitions for future reference.

Definitions Cheat Sheet:

1. Null hypothesis (H_0): By convention, H_0 is almost always a statement of no change or difference. It is assumed true until sufficient evidence is presented to reject it.
2. Alternative hypothesis (H_a): Statement of change or difference. This statement is considered true if H_0 is rejected.
3. Type I error: This is the error of concluding that there is a difference when, in fact, there is no difference.
4. Type II error: This is the error of concluding that there is no difference when there really is a difference.
5. Alpha (α) risk: This is the maximum risk, or probability, of making a type I error, sometimes also called a significance level. This probability is always greater than zero, and is often set at 0.05 (5%).
6. Beta (β) risk: This is the maximum risk or probability of making a type II error. This is often set at 0.20 (20%).
7. Confidence level ($1 - \alpha$): If the α-risk is 5%, then the confidence level for making the right decision is 95%.
8. Power ($1 - \beta$): This is the probability that a statistical test will detect a difference when there really is one.
9. Significant difference: This is the term used to describe the results of a statistical hypothesis test, where a difference is too large to be attributed to chance.
10. *p*-value: This is the probability that an observed difference is due to chance alone. Observed *p*-values equal to α or less are considered evidence to reject the null hypothesis.

A hypothesis test is a statistical means of answering a yes/no question. For many hypothesis tests, the yes/no question will be: *Is there a difference?* Maria is interested in ensuring that she is crystal clear on three important hypothesis-testing concepts.

The first key concept about hypothesis testing is that of comparison. Is the researcher comparing observed data to a standard (or benchmark) or directly comparing observed data from two populations (such as lines, processes, batches, etc.)?

1. A hypothesis test comparing observed data to a standard would look like this: Is the average viscosity for line 6 different than 4000?

   H_0: Line 6 average viscosity = 4000

   H_a: Line 6 average viscosity ≠ 4000

2. A hypothesis test comparing observed data from two populations would look like this: Is the average viscosity for line 4 different than the average viscosity for line 6?

   H_0: Line 4 average viscosity = Line 6 average viscosity

   H_a: Line 4 average viscosity ≠ Line 6 average viscosity

   Another way to specify these null and alternative hypotheses is as follows:

   H_0: Line 4 average viscosity − Line 6 average viscosity = 0

   H_a: Line 4 average viscosity − Line 6 average viscosity ≠ 0

The second key concept about hypothesis testing is that of direction. Are we only interested in one side of the comparison or both?

1. A hypothesis test with an one-sided comparison would look like this: Are line 4 average viscosities greater than line 6 viscosities?

   H_0: Line 4 average viscosity ≤ Line 6 average viscosity

   H_a: Line 4 average viscosity > Line 6 average viscosity

2. A hypothesis test with a two-sided comparison would look like this: Is the average viscosity for line 4 different than the average viscosity of line 6?

   H_0: Line 4 average viscosity = Line 6 average viscosity

   H_a: Line 4 average viscosity ≠ Line 6 average viscosity

In other words, the alternative hypothesis (H_a) could be specified as less than, not equal, or greater than, while the null hypothesis (H_0) is by convention equality. Many tests are set to check for any difference, either greater than or less than equality. These are called two-sided or two-tailed tests because the tests check for differences on both sides of the equality. When the alternative hypothesis is focused on only one side, the test is called one-sided or one-tailed test. In the two-sided test, the α-risk is split in half to account for both sides of the distribution. In the one-sided test, the entire α-risk is concentrated on a single side, which gives more statistical power than in the two-sided test to distinguish a difference on that side. However, any difference on the opposite side is indistinguishable from the null hypothesis. The graphs below illustrate how α-risk is divided to accommodate both sides of the equal sign.

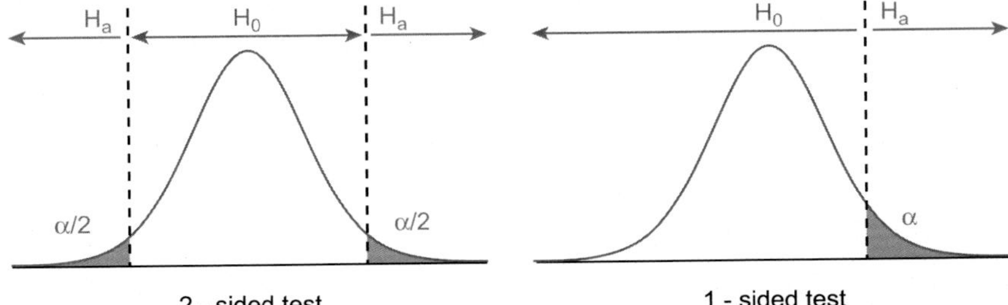

The third key concept about hypothesis testing is whether the alternative hypothesis is specific or general. The examples so far have all been general—they do not specify the size of the difference in the alternative hypothesis. An example of a specific alternative hypothesis is as follows:

1. Is line 4 viscosity greater than line 6 viscosity by at least 400?

   H_0: Line 4 viscosity − Line 6 viscosity = 0

   H_a: Line 4 viscosity − Line 6 viscosity > 400

Failing to reject the null hypothesis does not necessarily mean that the alternative hypothesis is accepted. If an alternative hypothesis is not specified, then the conclusion drawn for a given sample size, is that there is not sufficient evidence to reject the null hypothesis and accept the alternative hypothesis.

Once appropriate null and alternative hypotheses and a one- or two-sided test, along with a specified significance level (α) are established, then a *p*-value is calculated from the collected data. Two possibilities can occur:

1. p-value $< \alpha =>$ reject the null hypothesis.
2. p-value $\geq \alpha =>$ fail to reject the null hypothesis. If an alternative is specified and sample size based on specified alpha and beta risks is determined, then we accept the alternative hypothesis.

Maria recalls doing the math to calculate a test statistic and estimating *p*-value from a statistical table while in school. Calculating an exact *p*-value is a lot simpler using Excel. Nevertheless, Maria expands her definitions sheet with some formulas for calculating the test statistics:

1. One sample *t*-test:

$$t = \frac{\overline{X} - \mu}{\frac{s}{\sqrt{n}}}$$

2. Two sample *t*-test:

$$t = \frac{\overline{X}_1 - \overline{X}_2}{S_{X_1X_2} \cdot \sqrt{\frac{1}{n_1} + \frac{1}{n_2}}} \qquad S_{X_1X_2} = \sqrt{\frac{(n_1 - 1)S_{X_1}^2 + (n_2 - 1)S_{X_2}^2}{n_1 + n_2 - 2}}$$

3. Test for equal variances:

$$F = \frac{S_1{}^2}{S_2{}^2}$$

Depending on the question of interest, there is a variety of test statistics. Based on what is being compared and the data type (continuous or discrete), a number of different tests can be deployed.

In the case of continuous measures, for the two sample *t*-test and two variances *F*-test, there is an assumption that the data for both samples follows a normal distribution, that is, the histogram follows a bell shape. All the data that Maria has collected to date have followed a bell shape. When executing her tests, Maria had assumed that the data follow a normal distribution, and the histograms confirm that.

Armed with her findings, definitions sheet, and refreshed concepts, Maria feels confident reviewing the results with Lisa. She picks up the phone and calls Lisa.

Maria: "I'm very confident in stating that line 6 produces more consistent product. However, it is consistently lower in viscosity."

Lisa, who has also done some investigative work on the floor, has a hard time believing the results and thinks that the findings may be a fluke. She is convinced that there must have been a phenomenal day on line 4 or a really bad day on line 6 when the samples were collected.

Lisa: "What if we compare all plant lines to obtain a complete picture of the production?"

CHAPTER 4

Analysis of Variance (ANOVA)

Eager to continue with her work on BBQ sauce viscosity, Maria arrives at her office early the next day thinking she would look into statistical analysis techniques for comparing more than two samples before the additional data arrives. Her plan is to start with the textbook from Dr Wang's class. It had occurred to her that she could certainly perform *t*-tests as she had done before, one for each pair of lines. With five lines there are quite a few pairs to compare: line 2 versus line 3, line 2 versus line 4, and so on. She lists them all on a sheet of paper and counts 10 pairs. She has already performed the *t*-test for line 4 versus line 6, so nine more *t*-tests would be required. And then, she would have to compare the results of all 10 tests. It looks to be a lot of work even when using Excel. Maybe there is a more efficient way to tackle this problem.

She is surprised to find an email from Lisa with the data from three additional lines. The text of the email reads:

I have included viscosity data from lines 2, 3, and 5 to compare with the data you have for lines 4 and 6. I can't wait to see what you find!

"That woman must never sleep! With all of the things she is responsible for at the plant, I can't believe that she was able to get this information to me so quickly," Maria exclaims.

The data files look similar to the files she had received earlier. The first few lines in the data file look like this:

line	Date and Time	viscosity
2	4/28/2013 8:09	5109
2	4/28/2013 8:18	4978
2	4/28/2013 8:26	4929
2	4/28/2013 8:34	4916
2	4/28/2013 8:43	4430
2	4/28/2013 8:51	4822
2	4/28/2013 8:59	5006
2	4/28/2013 9:08	3987
2	4/28/2013 9:16	4487
2	4/28/2013 9:24	4219
2	4/28/2013 9:33	4494
2	4/28/2013 9:41	4633

Statistics for Food Scientists
http://dx.doi.org/10.1016/B978-0-12-417179-4.00004-4

There is now a column indicating the line where the product was produced. Maria checks and sees that there is the same number of data points from the same date as the previous files.

After taking care of a few minor tasks, Maria opens her textbook and scans the table of contents. She finds a chapter that contains the *t*-test that she performed. Two chapters after this, she finds another chapter titled "Comparing *k* Means—One-way Analysis of Variance (ANOVA)." Guessing that this may be what she is looking for, she turns to that chapter and begins reading.

Maria quickly learns that analysis of variance (often abbreviated as ANOVA) could be used for her comparison of the production lines. Its name relates to the concept that the variability in a set of data can be broken into different components. In the simplest form, there are two components:

1. Variability between the factor level means
2. Variability of the individual values within each factor level

Factors designate the group of things being compared; in her case, these are the production lines. Levels are the names of the elements in the factors; in her case, these are the line numbers 2 through 6. An illustration in the book demonstrates these concepts:

	Factor		
sample	**A**	**B**	**C**
1	90	94	100
2	92	96	115
3	88	98	97
mean score	90	96	104

Within Factor Level

Between Factor Levels

The statistical test for differences between the factor level means uses a ratio of these two variability components, with the variability between factor levels in the upper part of the ratio and the pooled within factor level variability in the lower part of the ratio. A larger ratio suggests treatment differences.

For her application, the factor levels would be the different production lines. Maria learns that if there are only two factor levels, ANOVA is identical to the *t*-test that she has performed. ANOVA is just an extension of the analysis she already completed. The technique will compare the variability between the lines and the variability within the lines.

Looking ahead in the book, she sees that the next chapter is titled "Multi-way Analysis of Variance." The ANOVA technique is quite flexible and can accommodate more complex problems, in particular additional factors. The example in the book describes a problem investigating different drugs (factor 1) for both male and female patients (factor 2).

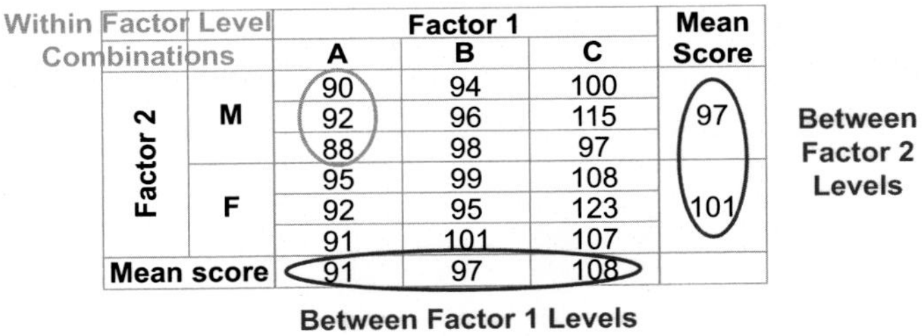

Within Factor Level Combinations

Factor 2		Factor 1 A	Factor 1 B	Factor 1 C	Mean Score
Factor 2	M	90	94	100	97
		92	96	115	
		88	98	97	
	F	95	99	108	101
		92	95	123	
		91	101	107	
Mean score		91	97	108	

Between Factor 2 Levels

Between Factor 1 Levels

This gets Maria thinking about the data Lisa provided her. It is from one production day, but from both the day and evening production shifts. There are different operators for the lines on each shift, and they may run the machinery differently. Perhaps she should consider the shift as an additional treatment in her analyses.

"Maybe I should not bite off more than I can chew," Maria thinks aloud. "Let me start with the simpler one-way ANOVA to compare the lines and see what I can learn from that analysis."

Maria sees that in the Analysis ToolPak add-in, there are three choices for ANOVA: Single Factor; Two-Factor with Replication; and Two-Factor without Replication.

Maria arranges the data in Excel so that the viscosities from the five lines are in separate columns. The first few lines of the rearranged data sheet look like this:

Line 2	Line 3	Line 4	Line 5	Line 6
5109	5172	5076	5289	4046
4978	5042	4618	5059	4724
4929	4796	4663	5125	4584
4916	4784	4747	4353	4541
4430	4840	4132	4807	4432
4822	5423	4726	4258	4489
5006	4918	4816	5113	4950
3987	4731	5385	4789	4603
4487	5070	4885	4801	4552
4219	4621	4793	4355	4643
4494	4423	5423	4260	4353
4633	4635	5001	5249	4500
4388	4473	4873	4784	4954
4425	4981	4490	4648	4849
4914	5066	4521	4565	4317
4607	4789	5374	4233	4719
4838	4586	4609	4688	4607
4179	5005	4689	4579	4876
4839	4948	4807	4028	4737
4269	4464	4617	4815	4911
4777	4666	5263	5421	4736
4221	4574	4108	4608	4372
5014	5048	4957	4776	4519

As she did when comparing only lines 4 and 6, Maria starts by making histograms of viscosity for each of the lines. Her graphs look like this:

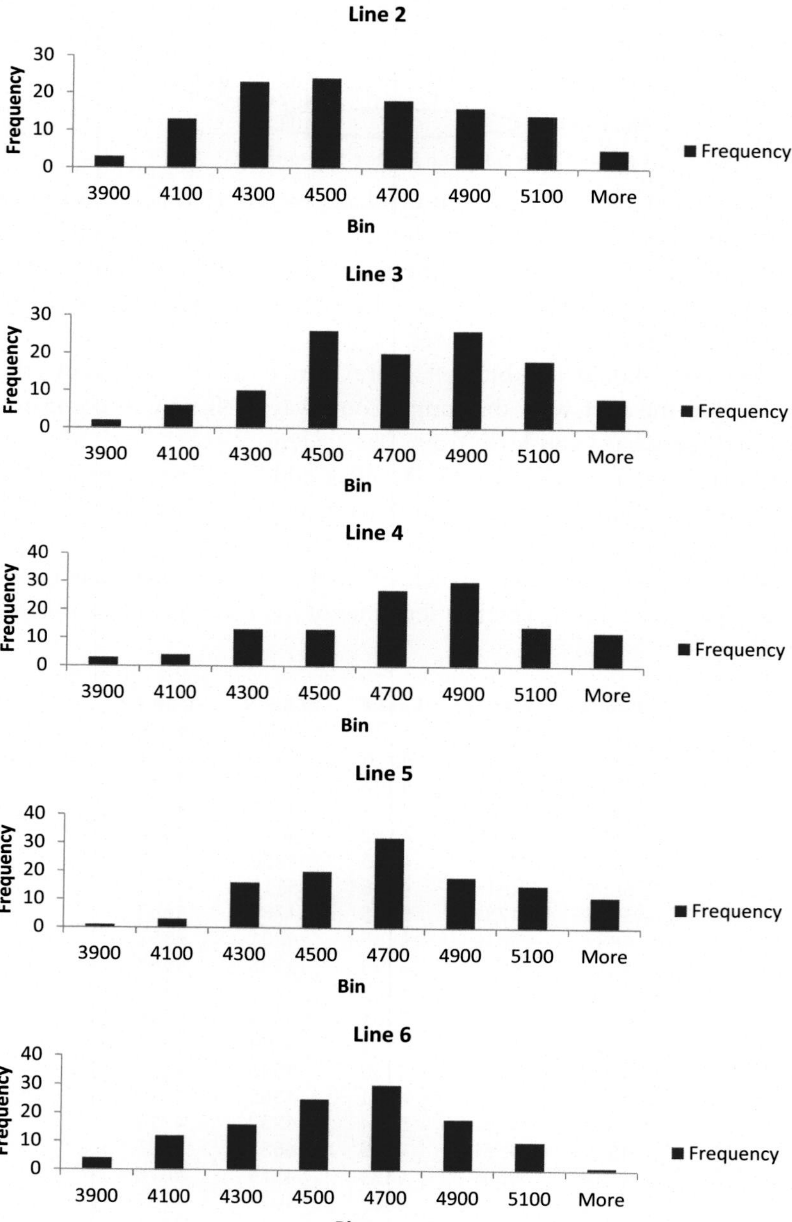

Maria notes that line 2 in addition to line 6 may have a lower average than the other lines. She also observes that the variability of line 6 still looks to be smaller than that of the other lines.

As she did when comparing only lines 4 and 6, Maria also makes plots of the viscosity data in time sequence for all five production lines. Her plots look like this:

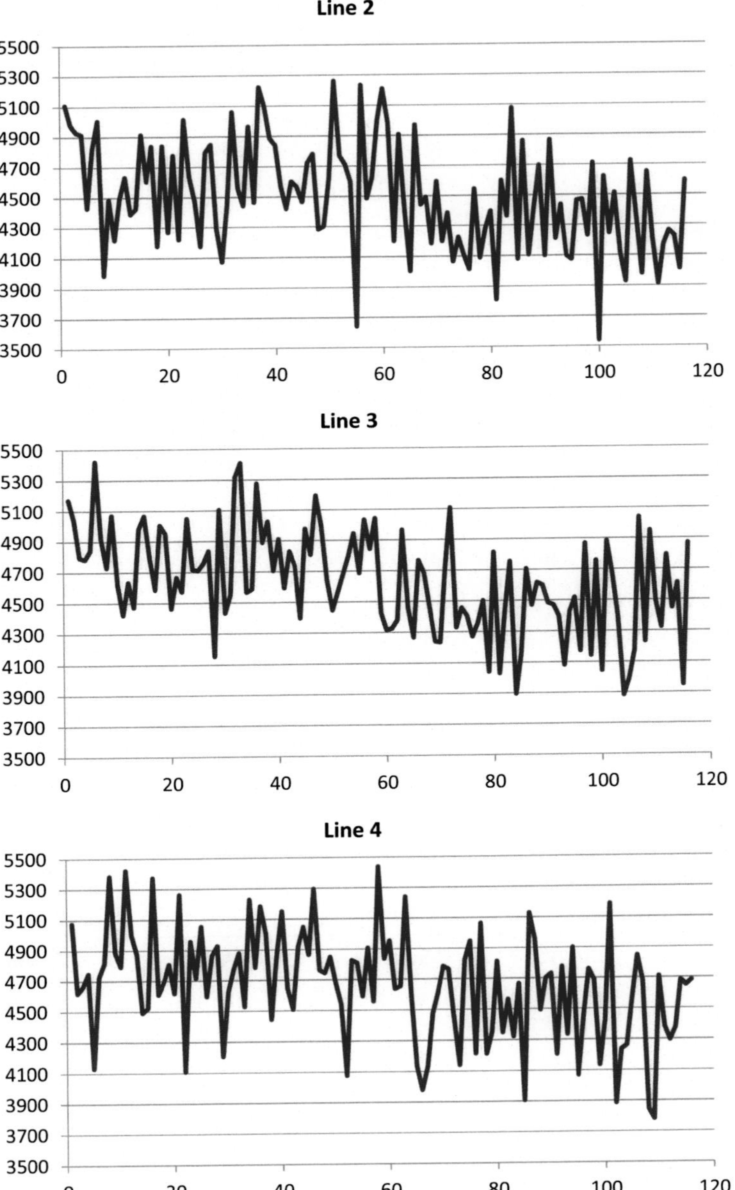

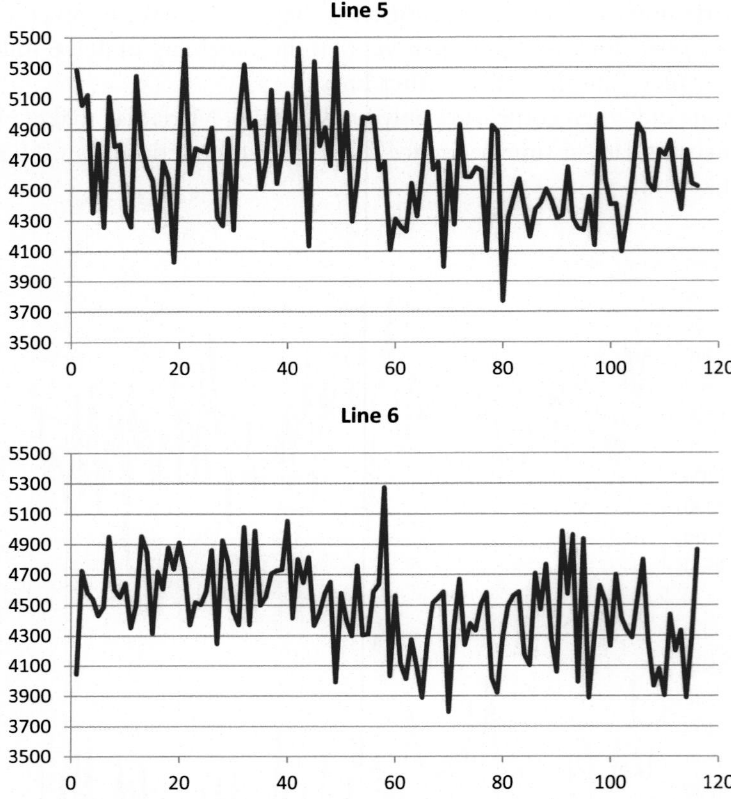

She does not see an increasing or decreasing trend for the lines except for line 3, which does seem to display diminishing viscosities over time. She makes a note to mention that to Lisa.

In the menu of her Analysis ToolPak add-in, she selects ANOVA: Single Factor. She selects the full input range and checks the box indicating that the column labels are in the first row. The analysis output that appears in a new work sheet looks like this:

ANOVA: Single Factor

Summary

Groups	Count	Sum	Average	Variance
Line 2	116	521,847	4499	129,823
Line 3	116	537,496	4634	111,617
Line 4	116	540,892	4663	125,955
Line 5	116	536,978	4629	113,420
Line 6	116	519,734	4480	91,052

ANOVA

Source of Variation	SS	df	MS	F	p-value	F crit
Between groups	3,325,273	4	831,318	7.27	0.00001	2.39
Within groups	65,764,690	575	114,373			
Total	69,089,962	579				

Most of the analysis details are easily understood. There is a table with summary statistics for each of the lines with the sum, average, and variance. The ANOVA table below the summary table contains a *p*-value for the between groups source of variation that seems to be the statistical test for the differences between the lines, exactly what she is looking for. She recalls that smaller *p*-values indicate significant differences. As the *p*-value here rounds down to zero, it would indicate the average viscosities are different between the lines.

"OK, now I'm getting somewhere!" Maria looks at the means for each of the five lines and sees that for line 3 and line 5, the means are pretty similar at 4634 and 4629. And line 2 and line 6 are also similar with means of 4499 and 4480. Line 4 has a mean of 4663, which is pretty close to the means for line 3 and line 5.

"Do I decide on my own what differences are significant? Lisa would want something more definitive I'm sure," Maria thinks out loud as she builds her analysis summary. "Maybe there is a way to determine this more objectively."

Maria goes back to her textbook to see if this question can be addressed. As she reads on, she comes across the concept of multiple paired comparison (MPC) procedures, which seems to be a way to determine individual treatment differences after an ANOVA analysis is performed. A number is calculated to determine the minimum significant difference between treatments. Any two treatments with a mean difference smaller than this number would not be considered different; any two treatments with a mean difference larger than this number would be considered different. There are several ways of calculating the number, and she is not sure which one to choose. The textbook mentions Fisher's least significant difference (LSD), Duncan's, Student-Newman-Keuls, Tukey's Honest Significant Difference (HSD), and Scheffe's as possible choices in certain situations. For some, the minimum significant difference is smaller and the significance test is more liberal. For others, the minimum significant difference is larger and the test is more conservative. But, she does see one statement that has relevance: If you are interested in all pairwise comparisons, you should use Tukey's HSD. She is interested in comparing all of the lines to each other, so she follows this advice.

Maria looks up MPC procedures in Excel and finds that none are available in the Analysis ToolPak add-in. "This is a problem; why not stop with the ANOVA analysis and try to determine the individual paired differences myself?" Maria grumbles in frustration. Fortunately the book describes the calculation for the minimum significant difference.

It uses a new tabled distribution that she is not familiar with: the studentized range. A value from this distribution is the q in this formula:

$$\frac{q_{k,v,\alpha/2}\, s\sqrt{\frac{1}{n_i}+\frac{1}{n_j}}}{\sqrt{2}}$$

In this formula s is the square root of within groups variance from the ANOVA; $q_{k,v,\alpha/2}$ is the alpha upper significance level of the studentized range for k means; v is the number of degrees of freedom for the within groups variance (labeled df in her ANOVA output); and n_i and n_j are the sample sizes for each of the treatments being compared.

Maria looks at the studentized range table in her textbook and sees that it stops at $v = 120$. The degrees of freedom for her analysis are 575, not even close to this! But, she also sees that there is a table entry for infinity, and that number is very close to the one for 120; so, for her degrees of freedom of 575, the entry for infinity should be OK to use. This value is 3.87 for an alpha = 0.05. Using this, she calculates the minimum significant difference to be 121.5.

Maria now examines the table of averages for each line. The two smallest averages are for line 2 and line 6 at 4499 and 4480, respectively. The difference between the two is smaller than 121.5, so the average viscosity of these two lines is not significantly different. Line 3, line 4, and line 5 have averages of 4634, 4663, and 4629, respectively. These three averages are all within 121.5 of each other and are then not significantly different. But, each of these three averages is greater than 121.5 from the averages of both line 2 and line 6. So, the lines form two groups: line 2 and line 6 have significantly smaller average viscosities than line 3, line 4, and line 5.

Maria adds these details to her analysis summary by creating a table ordering the lines by decreasing average viscosity. She feels confident now that she is prepared to meet with Lisa to deliver the results of her analyses. She is about to arrange a meeting for the following day when she remembers that she had thought about analyzing the data as a two-way

ANOVA with shift as the second treatment. The two-way ANOVA would add an additional dimension to the analysis that may be interesting.

Line	Average Viscosity	Group
4	4663	A
3	4634	A
5	4629	A
2	4499	B
6	4480	B
minimum significant difference	121.5	

Maria now starts working with the data to create a new column that indicates the shift that produced and collected the data based on the time stamp. All data from the first shift is from 8:00 AM to 4:00 PM. The second shift data is from 4:00 PM until midnight. Her data file now looks like this:

shift	Line 2	Line 3	Line 4	Line 5	Line 6
1	5109	5172	5076	5289	4046
1	4978	5042	4618	5059	4724
1	4929	4796	4663	5125	4584
1	4916	4784	4747	4353	4541
1	4430	4840	4132	4807	4432
1	4822	5423	4726	4258	4489
1	5006	4918	4816	5113	4950
1	3987	4731	5385	4789	4603
1	4487	5070	4885	4801	4552
1	4219	4621	4793	4355	4643
1	4494	4423	5423	4260	4353
...	...	...	...	...	...
2	5002	4426	4836	4110	4032
2	5209	4316	4953	4310	4561
2	4981	4325	4638	4260	4116
2	4203	4379	4654	4230	4013

In the menu of her Data Analysis add-in, she selects ANOVA: Two-Factor with Replication since she has repeated viscosity measurements for each line and shift combination. She chooses the full input range and indicates that there are 58 rows for each sample (Shift). The analysis output that appears in a new window looks like this:

ANOVA: Two-Factor with Replication

Summary	**Line 2**	**Line 3**	**Line 4**	**Line 5**	**Line 6**	**Total**
Shift 1						
Count	58	58	58	58	58	290
Sum	268,403	279,231	278,452	276,659	267,248	1,369,993
Average	4628	4814	4801	4770	4608	4724
Variance	109,425	70,285	95,764	119,372	61,583	97,848

Continued

ANOVA: Two-Factor with Replication—cont'd

Summary	Line 2	Line 3	Line 4	Line 5	Line 6	Total
Shift 2						
Count	58	58	58	58	58	290
Sum	253,444	258,265	262,439	260,319	252,486	1,286,953
Average	4370	4453	4525	4488	4353	4438
Variance	118,653	88,423	119,576	69,080	89,163	100,079
Total						
Count	116	116	116	116	116	
Sum	521,847	537,496	540,892	536,978	519,734	
Average	4499	4634	4663	4629	4480	
Variance	129,823	111,617	125,955	113,420	91,052	

ANOVA

Source of Variation	SS	*df*	MS	*F*	*p*-value	*F* crit
Sample	11,889,085	1	11,889,085	126.30	0.000000	3.86
Columns	3,325,273	4	831,318	8.83	0.000001	2.39
Interaction	220,237	4	55,059	0.58	0.673685	2.39
Within	53,655,367	570	94,132			
Total	69,089,962	579				

In her analysis, the "Sample" source of variation is from the shifts and the "Column" source of variation is from the lines. This appears to be the standard naming convention in Excel. The analysis validates her earlier determination about the line differences since the *p*-value listed for "Columns" is smaller than 0.05. To her surprise, the analysis also indicates that there is a significant difference between the two shifts since the *p*-value for "Sample" is also smaller than 0.05. Shift 1 has a higher average viscosity than shift 2. She looks at the means for the two shifts for each of the production lines and sees that the differences between shift 1 and shift 2 for each of the lines seem to be much larger than the differences between the lines!

There is also a source of variation listed as "Interaction." She looks back at her textbook and finds out that a significant interaction is when the effect of one factor is not the same for different levels of another factor. In her analysis, the interaction source of variation is not significant. So this difference between shifts is consistent across the lines. Why would the viscosities be so different between the shifts?

Maria is now even more puzzled about the line performances at the plant. She is happy that she performed the additional analysis to account for shifts since that seems to be an important finding. She is certain she will need to defend this finding to Lisa when she shares the information.

Maria sets up a meeting with Lisa for the next morning. She carefully describes the steps in her analysis and how she has come to the conclusion that while lines do perform differently, there is an even bigger difference in viscosity for products produced between shift 1 and shift 2.

"I think that you have made an important discovery Maria," Lisa says after Maria presents her case. "I'm convinced that your analysis is thorough and correct from the details you have shown. But let's think this through some more. The data show clearly that the viscosities are consistently higher on shift 1 regardless of the production line. Although there are different operators on each line, the viscosity measurements for all lines are made in the quality lab by the same technician. The technician is different between the shifts, can we be sure that this is a true shift difference and not some sort of measurement problem?"

Maria realizes that Lisa is correct and that what she had assumed was a production difference could be a problem with the measurements. "I'll see what I can do to determine what is really going on here and get back to you with what I find," she promises Lisa.

CHAPTER 5

Measurement Systems Analysis (MSA)

Maria discovered large differences in performance of all production lines across the two production shifts, with generally consistent production within a shift. First shift production is consistently higher in viscosity. But, as Lisa pointed out, it is not clear whether the differences in viscosities between the shifts are real differences or due to the measurements being taken by different lab technicians.

Maria has never given any thought to the viscosity measurement process. She doesn't even know how the measurements are made. She just assumed that the measurements were accurate. She decides that an understanding of the measurement process is critical at this time.

She sets up a meeting with Chad, the lab supervisor, to have him explain the viscosity measurement process. He is happy to explain the measurement system, and he mentions that she is the first to demonstrate an interest. It turns out that the measurement process is somewhat involved.

Chad starts a lengthy description, "The measurement device is a Brookfield viscometer. This device is used for texture measurements in a wide variety of products. The device rotates a spindle that is submerged in a sample of the product. The force required to rotate the spindle at a specific speed, measured in centipoise (CP), is reported. Higher values indicate a thicker product. A large number of spindles can be used to perform measurements, since the device can measure equally well the texture of low viscosity products, such as salad dressing, and high viscosity products, such as mayonnaise. It is also used outside of the food industry for products like paints and oils."

Chad discusses with Maria the development of the measurement system using the Brookfield viscometer. A key step in developing a measurement process is determining the specific spindle to use and at which speed the measurements will be made. For BBQ sauce, Chad had selected a number of products that represent the range of viscosities (from thin to thick) for which the measurement system would need to apply. The company's BBQ sauce

Statistics for Food Scientists
http://dx.doi.org/10.1016/B978-0-12-417179-4.00005-6

portfolio contains several products with viscosities varying from thin to thick: Classic, the original product developed; Thick and Spicy, which contains higher levels of tomato paste; and Honey BBQ, the thickest in the set due to the addition of honey in the formulation. He had augmented the set of products with two competitor products that expanded this range, as future development efforts may involve the creation of products with viscosities similar to these competitors' products. One competitor's product is a Carolina-style sauce, which is quite thin. Another is a very thick sauce with a claim on the bottle that it "clings better to meat than any other sauce."

Chad's development of the measurement system involved subjecting these five products to the Brookfield device at a range of speeds, using several different spindles. The goal was to have a clear and consistent differentiation between the products. Consistency meant that repeated measurements of the same product were producing similar results. And a clear difference meant that the typical measurement of each product was not near that of another product.

Chad had started by using the Brookfield with different spindles on all of the sauces, at four different speeds. He had quickly determined that one spindle, the RV-3, worked well across the set of products since it could spin freely even in the thickest sauce. He then made repeated measurements of all of the products at four speeds: 6, 12, 30, and 60 rpm. The data he collected look like this:

Product	RPM	Test 1	Test 2	Test 3	Average	Stdev	CV
Competitor 2	6	36000	36267	32000	34756	2390	6.9%
Competitor 2	12	21333	21600	19600	20844	1086	5.2%
Competitor 2	30	10987	11093	10560	10880	282	2.6%
Competitor 2	60	6720	6720	6640	6693	46	0.7%
Original	6	23600	21733	23867	23067	1163	5.0%
Original	12	14067	13533	14267	13956	379	2.7%
Original	30	7387	7333	7467	7396	67	0.9%
Original	60	4600	4693	4667	4653	48	1.0%
Competitor 1	6	9600	9653	9120	9458	294	3.1%
Competitor 1	12	5600	5627	5333	5520	163	2.9%
Competitor 1	30	2830	2773	2688	2764	71	2.6%
Competitor 1	60	1720	1610	1685	1672	56	3.4%
Thick and Spicy	6	21867	24667	21733	22756	1657	7.3%
Thick and Spicy	12	13200	13800	13067	13356	391	2.9%
Thick and Spicy	30	6933	7120	6880	6978	126	1.8%
Thick and Spicy	60	4360	4387	4347	4365	20	0.5%
Honey	6	32933	32667	32667	32756	154	0.5%
Honey	12	19667	19867	19467	19667	200	1.0%
Honey	30	10160	10347	10133	10213	117	1.1%
Honey	60	6307	6333	6307	6316	15	0.2%

Chad had looked to determine a speed that provides the most stable and reliable measurement system. From the data he collected, he calculated the coefficient of variation, or CV, for each sample at each speed. CV is the standard deviation divided by the mean (and multiplied by 100 to show in %). Since measurements with higher values often display more variation, the CV gives a basis for comparing the measurement variability "corrected" for the part of the scale where the measurements reside.

Smaller CVs are most desirable, since they indicate a more repeatable measurement. The graph below displays the CVs for each sample by RPM. For nearly every sample, the lowest CVs are for measurements made with the highest speed, 60 rpm. Chad decided that the most reliable measurements would be made using this speed.

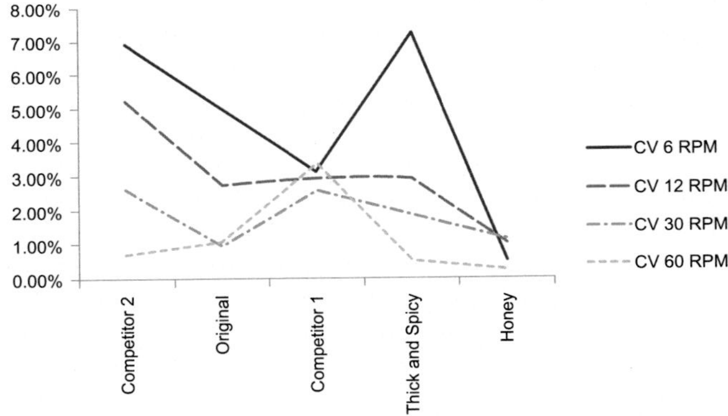

Having determined the best speed for making viscosity measurement using the Brookfield device, Chad then documented the procedure for making the measurements. The documentation included the volume of the sample being measured and the depth at which the spindle should be submerged. The analysts making the viscosity measurements were then given a brief training session on using the Brookfield viscometer to make viscosity measurements for BBQ sauce.

Maria shares with Chad her recent data analyses showing the consistently higher viscosities on the first shift. She mentions her discussion with Lisa about the measurements being made by different analysts between the two shifts. It has been a number of years since Chad developed the viscosity measurement system and trained the analysts. "Could something have changed since then?" Maria asks Chad. While he does not directly answer the question, he agrees that they should look into the situation.

Maria starts looking for a way to assess the current state of the viscosity measurement process. She decides that a conversation with Dr Wang, her graduate advisor, would be beneficial as she seemed to be familiar with a wide variety of topics related to the food industry. She must have run into this type of problem before. She sends Dr Wang an email to schedule a call and is pleased when Dr Wang responds quickly with a suggested time for the next morning.

"Maria, how very nice to hear from you!" Dr Wang exclaims when she picks up the phone the next morning. "How is the BBQ sauce business?"

"I'm learning so much every day," Maria explains. "It seems like I will never know enough, and there is always a new challenge." Maria describes her recent work with the viscosity and her suspicion about the measurement process. "How can I determine if the measurements are valid?"

Dr Wang explains that engineers and scientists across all industries face similar questions and that the study of measurement processes has a specific name: measurement systems analysis. Methodologies were initially developed in the 1950s from the National Bureau of Standards, a US Government Agency (now called the National Institute of Standards and Technology). Methodologies were published by the Automotive Industry Action Group in the early 1990s. These have been refined over time, and a language protocol has been formalized.

Dr Wang goes on to explain that there are two components to measurement system variability: accuracy and precision. Accuracy is the difference between a sample's average measurement value and the sample's true value. Precision is the variation in measurements of the same sample measured repeatedly with the same measurement system. Dr Wang shares a diagram that illustrates these concepts. The measurements in set one are both accurate and precise. The measurements in set two are precise but not accurate. The measurements in set three are accurate but not precise. The measurements in set four are neither precise nor accurate.

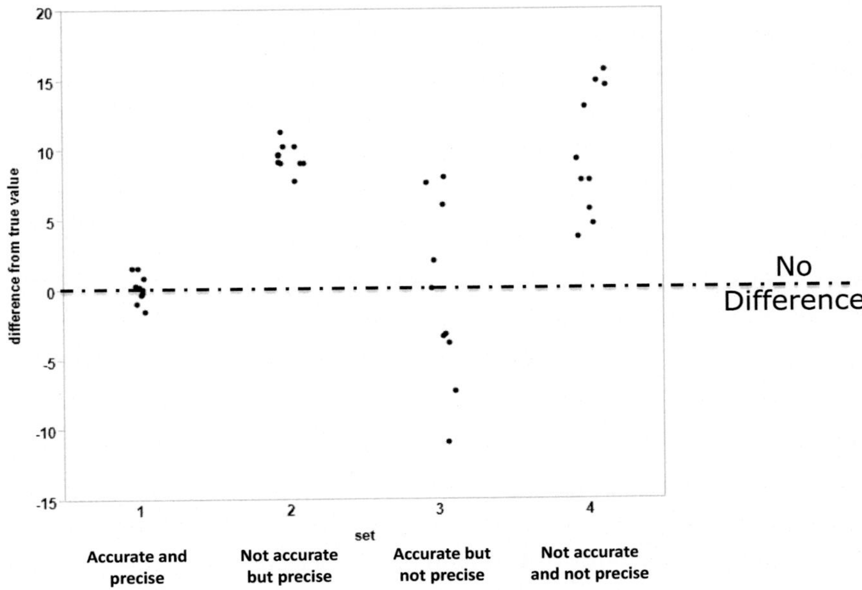

Dr Wang further explains that there are two components to precision: repeatability and reproducibility. Repeatability is the variation in measurements obtained by one analyst using the same measurement device several times on the same sample. Reproducibility in its simplest form is the variation in the average of measurements made by different analysts using the same measurement device several times on the same sample. When laboratories have several of the same measurement devices, reproducibility may also include differences in the average measurements made by analysts across the set of devices.

"There is a very specific way in which measurement system precision is determined that I can share with you," Dr Wang concludes. I'll send to you a link that explains gage R&R studies."

"Have you been using Excel to analyze your data?" Dr Wang asks. "I know that we used Excel in class, but you are now getting into more sophisticated statistical procedures, so I think that you should invest in data analysis software that will enable you to do more. There are several good options. Any one of them will be very useful with your immediate measurement system task, and they will also have many more analysis tools that you will likely find helpful down the road."

"The gage R&R examples show output from one of the data analysis software packages, though the others will have a similar output format," adds Dr Wang. After a few more minutes of discussion the call concludes.

Not long after, Maria receives an email from Dr Wang with links to several sites for software packages that Dr Wang had mentioned and another site that describes gage R&R studies. She is excited after reading about the software options and the many data analysis tools she will have available and emails Steve requesting permission to purchase. Before long, his email response with permission arrives. A few clicks later, she has purchased and downloaded the data analysis software.

Maria clicks on the link describing gage R&R studies and starts reading:

Gage R&R (repeatability and reproducibility) studies are the method for determining how much of the observed total variation is due to measurement system precision. A gage is any device or measurement method. The basic procedure for conducting a gage R&R study is very straightforward. A number of samples (N) are selected to represent the full range of long-term variation for a process. Each sample will be measured M times by K analysts. The samples are given generic codes to hide their identity, and typically, the first replicate measurements are made on the N samples, then the second replicate measurements are made on the N samples, with this procedure repeated until all M replicates are completed. An alternative to this is to completely randomize the M replicates of the N samples.

The data analysis plan for a gage R&R study is as follows:

1. Plot the data.
2. Examine the components of variation: sample-to-sample, R&R.
3. Calculate analysis of variance (ANOVA).
4. Evaluate gage R&R error.

Maria is familiar with some of these steps, as she has been actively involved in analyzing data for her process improvement work. However, there are some specific outputs to a gage R&R study that she familiarizes herself with through an example. In this example, three operators perform two replicate measurements on four samples for water activity. There are several data plots that are typically used in these studies.

The "by Operator" graph shows differences in operators' averages (represented by the "x" on the graphs), while the individual sample measurements (represented by the dots) give insight into operator reproducibility.

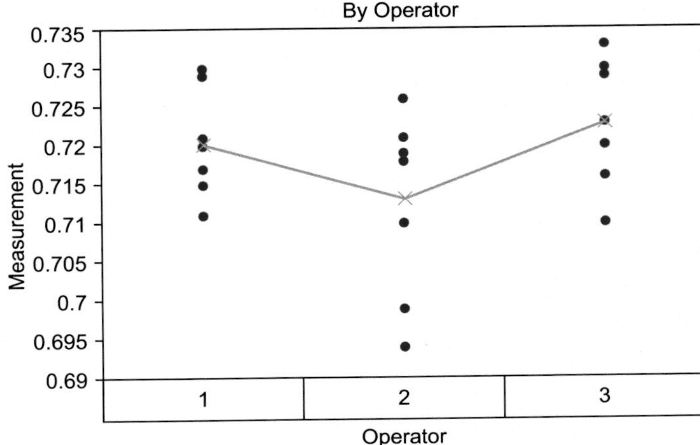

In this example, it appears that operator 2 gives lower results than the other two operators, due to its lower average value.

The "by Sample" graph shows differences in average between samples.

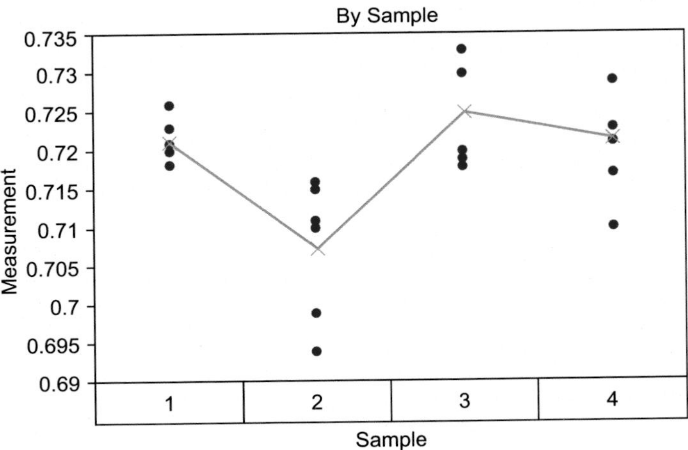

In this example, it appears that there are measured differences between the samples. This is expected since the samples were chosen to represent a wide range.

The "Operator by Sample" graph shows whether the operators measure the same sample consistently. In this example, operator 2 has lower measurements than the other operators for three of the four samples, and operators 1 and 3 differ on two of the samples (sample 3 and sample 4).

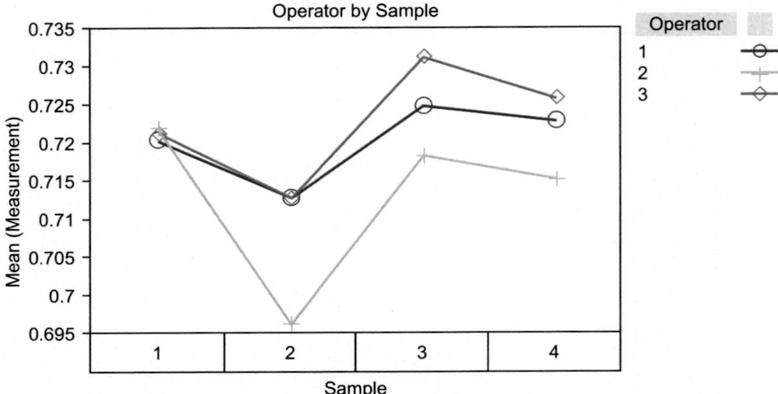

The "Sample within Operator" graphs gives insight into the consistency of the operators in measuring the samples. In the example below, each of the operators has a problem in repeating measurements for one or more samples. For example operator 1 has produced two different results, when measuring sample 3. The replicates are joined by the vertical line on the graph.

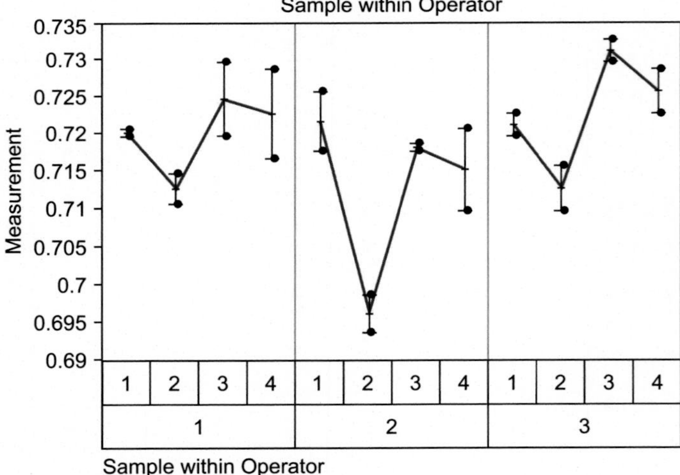

Maria is familiar with ANOVA and how the output from this analysis can help her determine the statistical significance. ANOVA can help quantify what has been perceived visually from the graphs.

Analysis of Variance

Source	DF	SS	Mean Square	F Ratio	Prob > F
Operator	2	0.000419	0.00021	5.39527	0.0456*
Sample	3	0.001079	0.00036	9.27175	0.0114*
Operator*Sample	6	0.000233	3.88e-5	1.65658	0.2150
Within	12	0.000281	2.34e-5		
Total	23	0.002011	8.74e-5		

Maria observes small p-values for both the sample and operator effects. These confirm the expected sample differences and also the operator differences she observed in the "by Operator" graph.

The components of variation are related to the ANOVA model and represent a partitioning of the total response variability into repeatability, reproducibility, and sample-to-sample.

Variance Components for Gauge R&R

Component	Var Component	% of Total	20 40 60 80
Gauge R&R	0.00005242	49.50	
Repeatability	0.00002342	22.11	
Reproducibility	0.00002900	27.39	
Part-to-Part	0.00005348	50.50	

In the example, Maria sees that 50% of the variability is due to sample differences (denoted by part-to-part), while nearly 50% of the total variability is due to R&R error. She notes that guidelines for gage R&R error suggest that less than 10% would be considered acceptable, and that between 10% and 30% might be considered acceptable in some instances. In the example, the measurement system is obviously unacceptable.

A second example Maria has displays output for an acceptable measurement system. Below are "by Operator," "by Sample," "Operator by Sample," and "Sample within Operator" graphs for the acceptable measurement system.

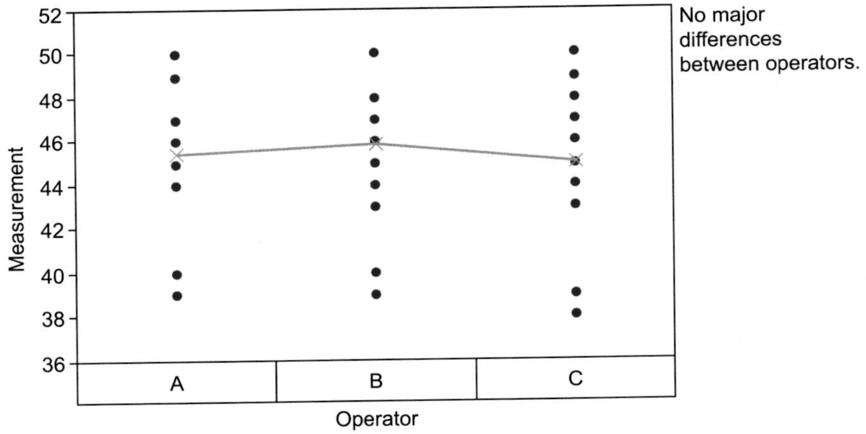

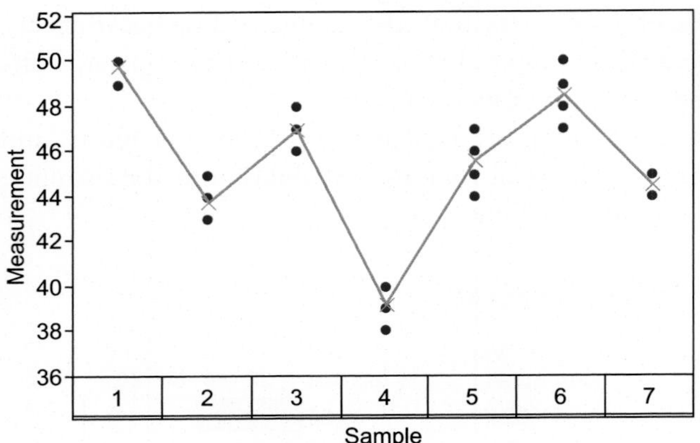

Consistent Repeatability when the same sample is measured multiple times.

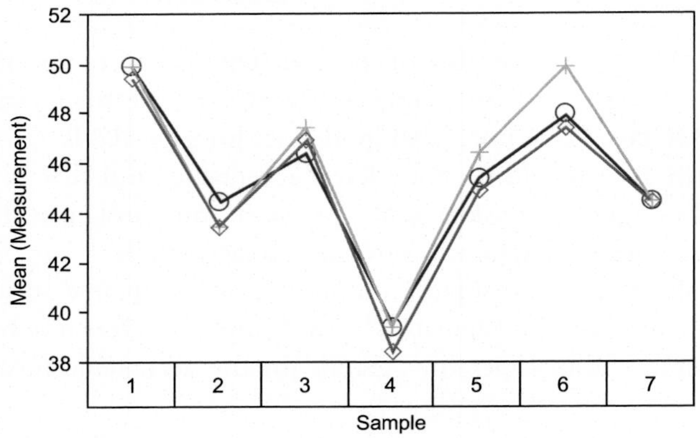

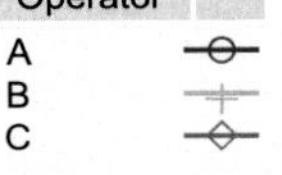

Consistent sample differences identified by each operator.

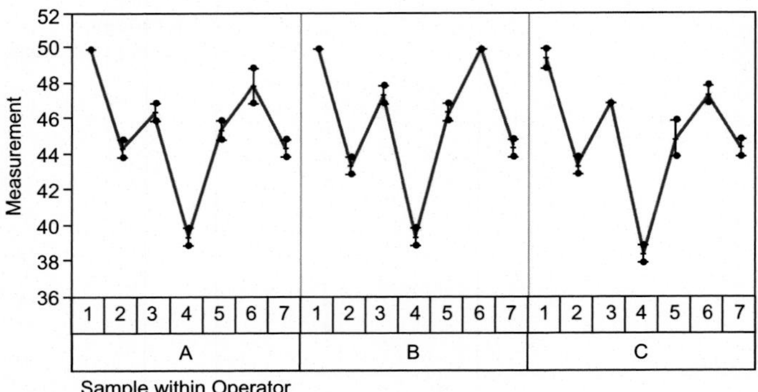

General agreement on sample differences, good repeatability and good reproducibility.

Variance Components for Gauge R&R

Component	Var Component	% of Total	20 40 60 80
Gauge R&R	0.750000	5.79	
Repeatability	0.547619	4.23	
Reproducibility	0.202381	1.56	
Part-to-Part	12.210317	94.21	

She notes the obvious consistency of the data in each graph. The variance components summary shows that less than 6% of the total variability is due to R&R error, well below the 10% threshold of acceptability.

Maria discusses what she has learned about measurement systems analysis with Chad and shares the examples. They decide to conduct a gage R&R study with the analysts for each of the two production shifts, another lab technician who occasionally performs viscosity measurements when needed, and Chad himself. Chad points out that they will need to be clear that the "operators" in the study are really the different lab analysts and not the actual line operators.

Since the quality lab makes viscosity measurements for all of the BBQ sauce varieties produced by the company, Maria selects five products that span the range of viscosities from thinnest to thickest. The ideal gage R&R study has analysts making repeated measurements on the same sample. The Brookfield viscometer shears the sample as the spindle rotates during the measurement process. This means that the sample composition may change as a consequence of the measurement process. A true replicate measurement is not really possible since the additional measurement will be made on a sample that may now be different due to the process of making a measurement on it.

Maria and Chad consider the meaning of a replicate in this specific instance. Samples from the same bottle should be very similar. Bottles produced at the same time should also be similar: they have the same composition, were produced at the same time, and cooled in the bottle at the same rate. They decide to define a replicate as the same product formulation produced on the same line at the same time.

Maria selects a number of bottles of each variety produced at nearly the same time. She creates three samples for each of the five products for each of the four analysts and assigns them random three digit codes to blind them from the analysts. Her design worksheet looks like this:

sample	replicate 1	replicate 2	replicate 3
1	339	132	675
2	671	493	872
3	102	239	925
4	815	326	524
5	489	563	310

She also creates a separate data collection spreadsheet for 15 observations for each of the analysts, one of which looks like this:

sample	viscosity
102	
489	
815	
339	
671	
239	
132	
563	
326	
493	
524	
872	
675	
925	
310	

For each analyst, she randomly orders the samples within the replicates so that there will be no bias due to the order in which the samples are measured within the replicates.

The data Maria receives from the operators look like this:

Sample	Analyst	Viscosity 1	Viscosity 2	Viscosity 3
1	1	2586	4074	2712
1	2	8152	6989	4150
1	3	5443	5475	5612
1	4	3041	3399	2407
2	1	4825	6266	5667
2	2	7829	7173	10064
2	3	7339	8151	7444
2	4	6097	6167	6213
3	1	4396	3715	3836
3	2	5150	7764	7230
3	3	4495	4868	5406
3	4	4438	3068	3993
4	1	31	1251	1187
4	2	1166	2768	1853
4	3	2127	2328	1404
4	4	1144	1017	4148
5	1	5501	5833	5739
5	2	9087	9646	8027
5	3	7195	6734	7552
5	4	6265	3862	5883

She repeats the data analysis steps described earlier. It very quickly becomes obvious that there are problems with the current measurement system. The "by Operator" plot shows noticeable operator differences.

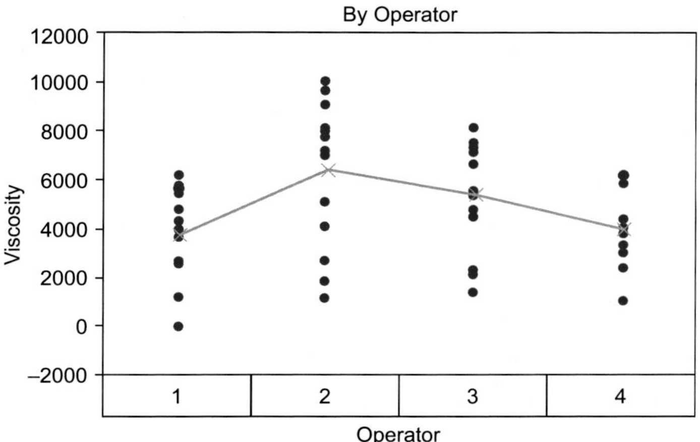

The "Operator by Sample" graph shows clearly that the operators' measurements (of the same sample) are not consistent for any set of samples. It is desirable to observe sample-to-sample differences (indicating a selection of a good range of values); however, "within" sample (operator-to-operator) differences ought to be negligible.

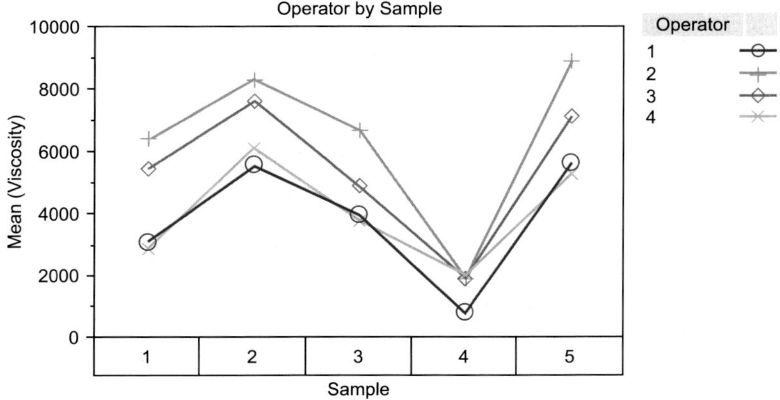

Lack of repeatability seems to be a problem for all of the operators as demonstrated in the "Sample within Operator" graph. Operator 3 seems to have the best repeatability.

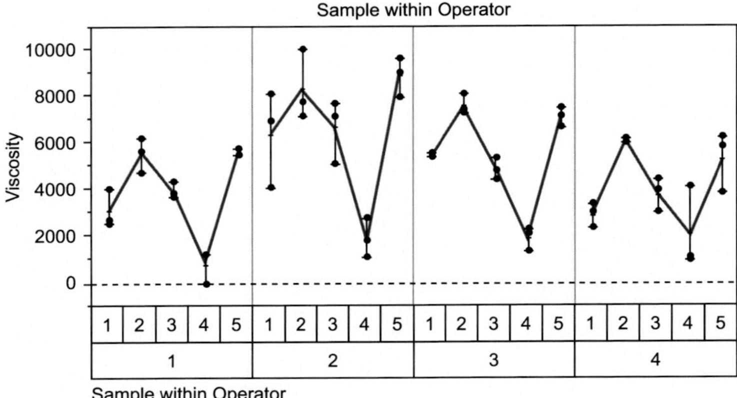

The variance components analysis results show a total gage R&R variance component of 36%, well above the suggested threshold of 10%. The component breakdown indicates both repeatability and reproducibility concerns.

Variance Components for Gauge R&R

Component	Var Component	% of Total	20 40 60 80
Gauge R&R	2485664.3	36.13	
Repeatability	898607.1	13.06	
Reproducibility	1587057.2	23.07	
Part-to-Part	4393299.6	63.87	

Maria and Chad discuss potential ways to fix the R&R problems with the current measurement systems. It has been a while since the operators had received any training in the measurement system. Chad conducts training in the proper execution of the measurement process with each of the operators, highlighting the need for consistency in all aspects of the process.

Maria and Chad repeat the gage R&R study after the operator training. They use new samples of each of the products in the original study, and create new three-digit codes and randomizations for the study design. The data from the second study is captured in the following table:

Sample	Analyst	Viscosity 1	Viscosity 2	Viscosity 3
1	1	4649	4373	3947
1	2	4748	4984	4410
1	3	4404	4546	4606
1	4	4893	3830	4269
2	1	6347	6643	6593
2	2	6810	7355	7186
2	3	7053	6828	7055
2	4	6628	7086	6410
3	1	4369	4442	4668
3	2	4850	4378	5335
3	3	4570	4825	4958
3	4	4315	4647	4350
4	1	1718	1284	1670
4	2	1391	2062	2063
4	3	1803	1957	2238
4	4	2115	1334	1879
5	1	6358	6420	6288
5	2	6619	6669	6473
5	3	6413	6328	6331
5	4	6340	6020	5946

Maria repeats the gage R&R study analysis process with the new data. The "Sample within Operator" graph for the new data and the variance components are shown below.

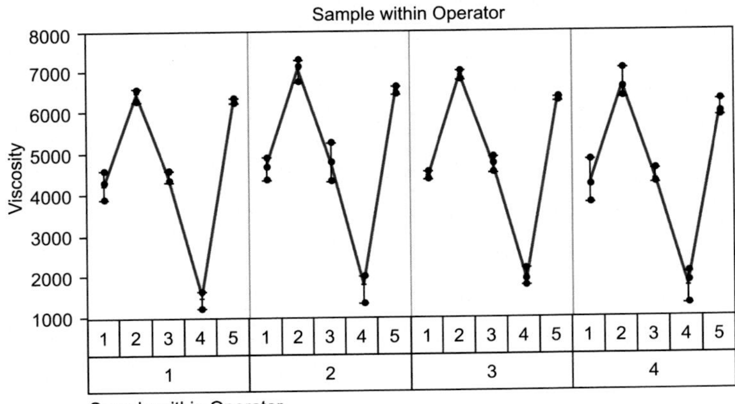

Variance Components for Gauge R&R			
Component	Var Component	% of Total	20 40 60 80
Gauge R&R	83532.0	3.11	
Repeatability	63548.0	2.36	
Reproducibility	19984.0	0.74	
Part-to-Part	2606597.3	96.89	

Now, only slightly more than 3% of the variability is due to the gage R&R, and the graphs display a high degree of repeatability and better consistency across the operators.

Maria and Chad are now satisfied with the measurement system and put plans into place for auditing and occasional retraining if necessary. Maria now feels confident that the measurements that she receives in the future will be useful in guiding her improvement efforts.

Now, Maria turns her attention back to the assessment of the viscosities for each of the lines. Has the lack of a good measurement system potentially masked the line-to-line differences? She decides to repeat her earlier work and collects, as earlier two shifts of data from each of the lines. Before using her software to perform the ANOVA modeling, she sees that she can create graphs to visualize the data much more easily that she could in Excel. She first creates histograms for each line and shift combination. She can create these at once and have them put together in one presentation with the same scale as the default. Her histograms look like this:

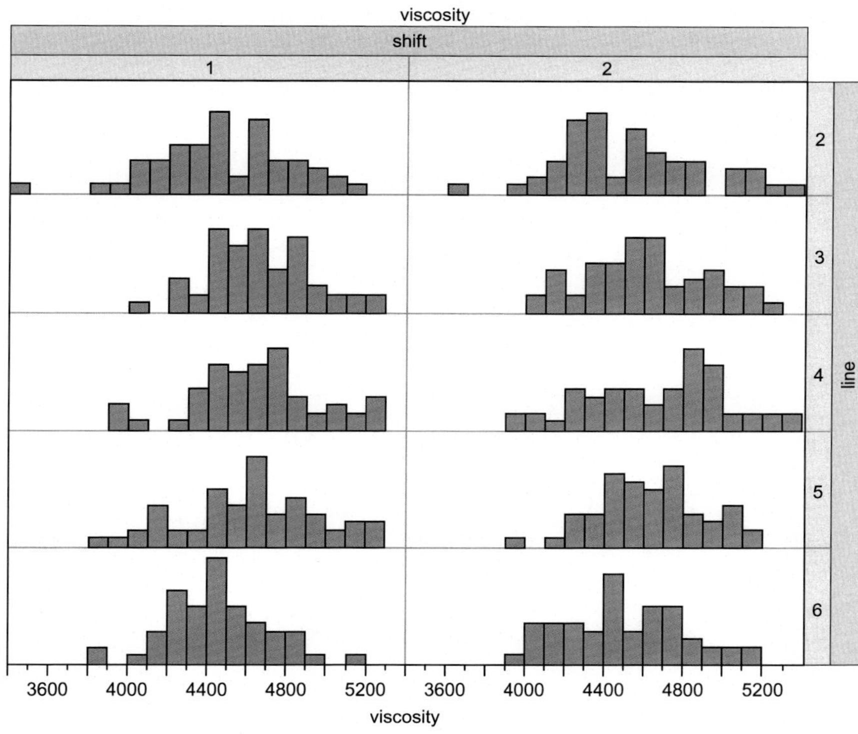

The histograms are laid out so that each row is a shift and each column is a production line. They are presented on their sides to provide a better comparison across the production lines. Maria notes the similarity in shape of the distributions and that the center of these distributions seem smaller for lines 2 and 6.

Maria also creates the time series plots for the new data using her software. She creates a separate set of graphs for shifts 1 and 2 since they cover different time frames and stacks them by line for easier comparison. The graphs look like this:

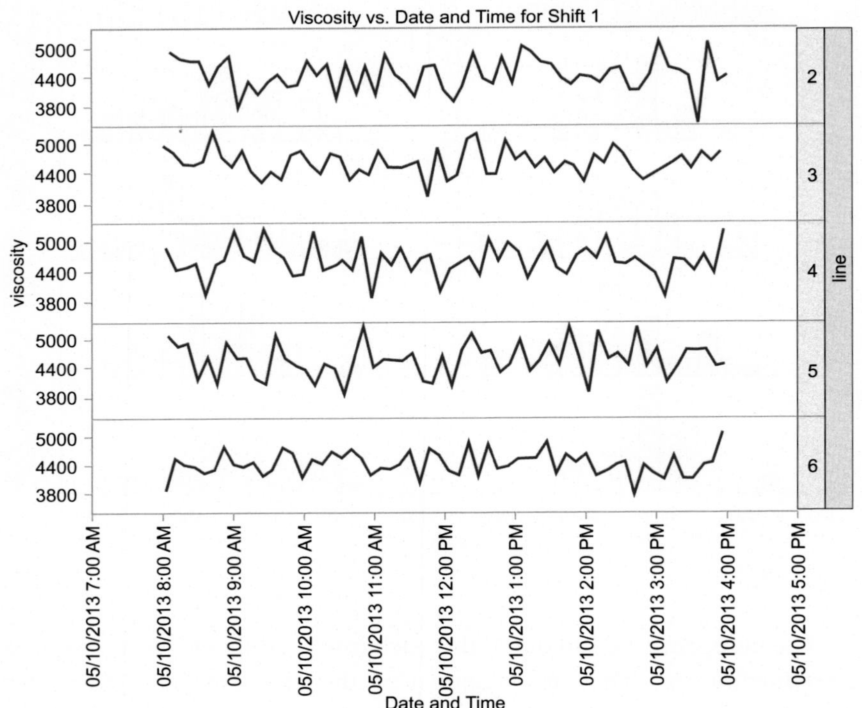

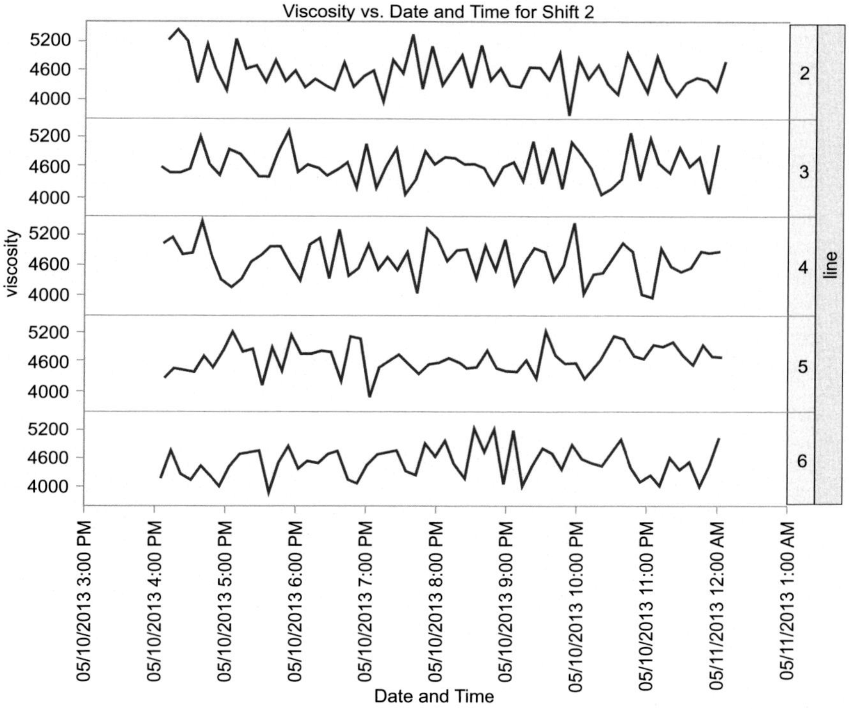

She sees that there are no increasing or decreasing trends in viscosity for any of the lines for both shifts.

Maria finds another graph that she finds useful for visualizing the viscosity data. A box plot shows the distribution of the data somewhat like the histogram but can sometimes make it easier to compare the center and spread. She arranges box plots of the viscosity data by line and shift.

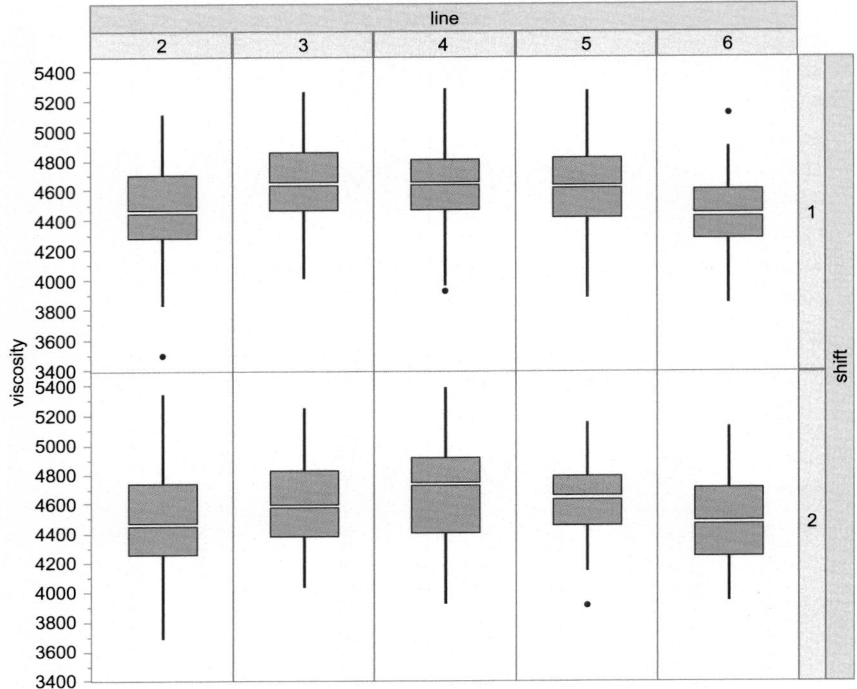

The box in a box plot is the middle 50% of the data, and the line in the middle of the box is the median. The lines extending from the boxes either represent the range of the data or, in some instances, individual data points are left in the box plot to indicate that they stand out from the rest of the data distribution. From Maria's box plots, she can more readily see that lines 2 and 6 are centered lower than the rest for both shifts, and that perhaps line 6, the recently installed line, does have less variation in viscosity.

Maria also sees that the ANOVA modeling, to compare lines and shifts, is easy to create and that the Tukey's method for comparing means is an easy option to include in the analysis. Her results look like this:

Summary of Fit

RSquare	0.062346
RSquare Adj	0.047541
Root Mean Square Error	306.8063
Mean of Response	4580.975
Observations (or Sum Wgts)	580

Analysis of Variance

Source	DF	Sum of Squares	Mean Square	F Ratio
Model	9	3567544	396394	4.2111
Error	570	53654151	94130	**Prob > F**
C. Total	579	57221696		<.0001*

Effect Tests

Source	Nparm	DF	Sum of Squares	F Ratio	Prob > F
Shift	1	1	27265.9	0.2897	0.5906
Line	4	4	3320601.6	8.8192	<.0001*
Shift*line	4	4	219676.8	0.5834	0.6748

LS Means Plot

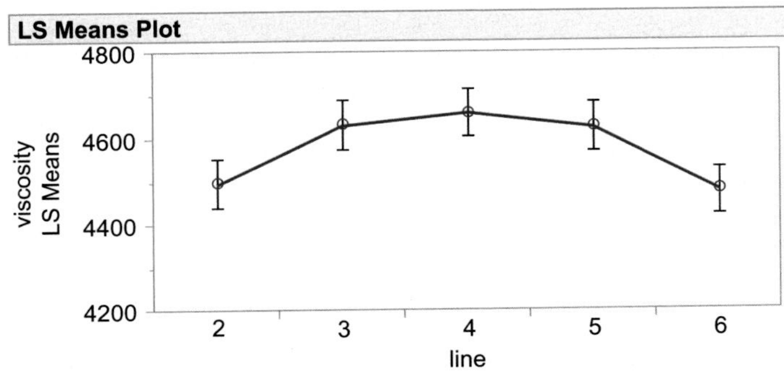

LSMeans Differences Tukey HSD

α= 0.050 Q= 2.73655

Level			Least Sq Mean
4	A		4662.7433
3	A		4633.6256
5	A		4629.1960
2		B	4498.7070
6		B	4480.6035

Levels not connected by same letter are significantly different.

With no significant shift differences, it is clear now how the lines differ in their performance. The Tukey's means comparison lists the mean viscosities for the lines in descending order. In the last output table, it states that "levels not connected by the same letter are significantly different." It is clear that lines 2 and 6 have significantly lower viscosity levels than the other three lines. Though these findings are no different from what she observed before, Maria now has confidence in her performance assessment and can develop a strategy for improving performance across the lines. "Lesson learned," Maria notes to herself. One needs to be confident that the measurement system is reliable before any analyses are performed. A bad measurement system will lead to bad conclusions!

CHAPTER 6

Regression and Correlation

Maria now better understands the BBQ sauce process and how the individual lines perform. However, she still does not have an understanding of why viscosity varies. There are many things that may account for the variability in viscosity levels, including processing and ingredient variation.

Maria gathers a small team of experts to help determine the potential factors that may affect BBQ sauce viscosity. The team includes her manager, Steve; Louise, the quality manager; and several of the line operators. The team produces a long list! The factors fall into two groups: processing factors such as pump pressure, steam pressure, and line speed; and formulation factors such as the amounts of sugar, tomato paste, and fruit paste. Maria suggests continuing with the viscosity sampling throughout the upcoming shift, but also to have one of the line operators record the levels of these processing and formulation factors at the same time as taking the viscosity measurements.

BBQ sauce is made in batches, with the ingredients mixed in large vats that are then pumped to the bottling lines. Consequently, the ingredients only vary batch to batch, whereas the pressures and line speed can vary as the batch is pumped and bottled. Once the ingredients are mixed, it takes around half an hour to process the batch into bottles. Maria suggests taking measurements five times as each batch is bottled and proceeds to create a data collection template for the operator.

The next morning, the collected data arrive in Maria's inbox. The sample data for the first two batches look like this:

date and time	batch	steam pressure	pump pressure	line speed	sugar level	tomato paste	fruit paste	viscosity
5/23/2013 8:14	1	246.95	395.38	26.90	62.81	1032.17	10.14	4125
5/23/2013 8:23	1	248.82	391.80	32.16	62.81	1032.17	10.14	4126
5/23/2013 8:31	1	244.60	397.82	31.31	62.81	1032.17	10.14	4130
5/23/2013 8:39	1	243.89	401.87	28.44	62.81	1032.17	10.14	4135
5/23/2013 8:48	1	245.07	387.82	32.00	62.81	1032.17	10.14	4142
5/23/2013 8:56	2	237.32	399.06	22.60	63.12	1042.06	10.96	4216
5/23/2013 9:04	2	240.51	404.06	32.63	63.12	1042.06	10.96	4221
5/23/2013 9:13	2	241.20	399.29	31.59	63.12	1042.06	10.96	4221
5/23/2013 9:21	2	238.68	396.43	29.81	63.12	1042.06	10.96	4225
5/23/2013 9:29	2	241.07	398.42	34.39	63.12	1042.06	10.96	4225

Maria first creates histograms for viscosity and each of the six factors.

Statistics for Food Scientists
http://dx.doi.org/10.1016/B978-0-12-417179-4.00006-8

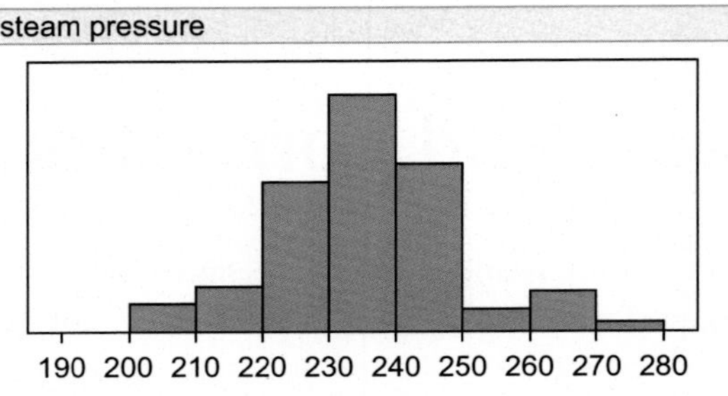
steam pressure
190 200 210 220 230 240 250 260 270 280

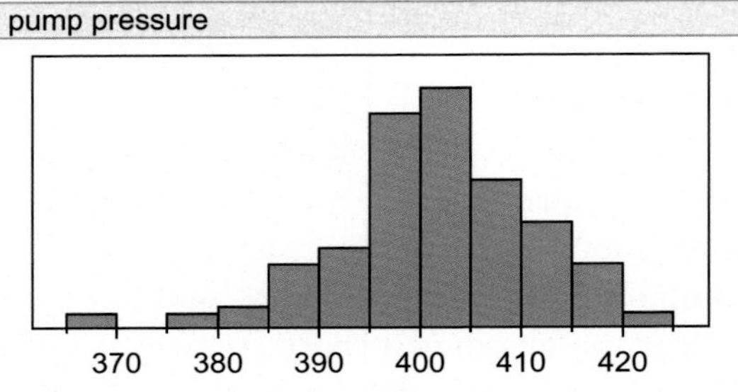
pump pressure
370 380 390 400 410 420

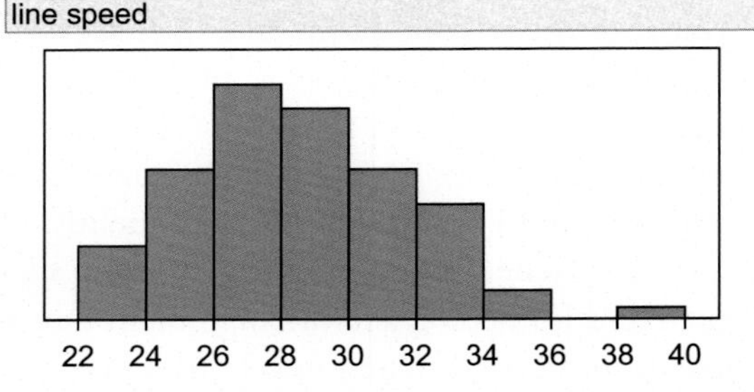
line speed
22 24 26 28 30 32 34 36 38 40

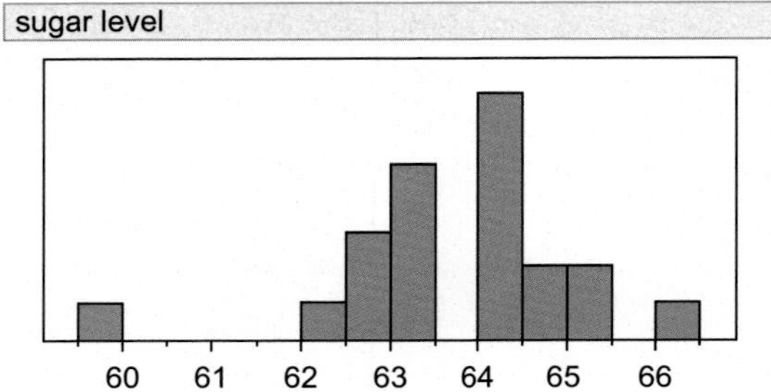
sugar level
60 61 62 63 64 65 66

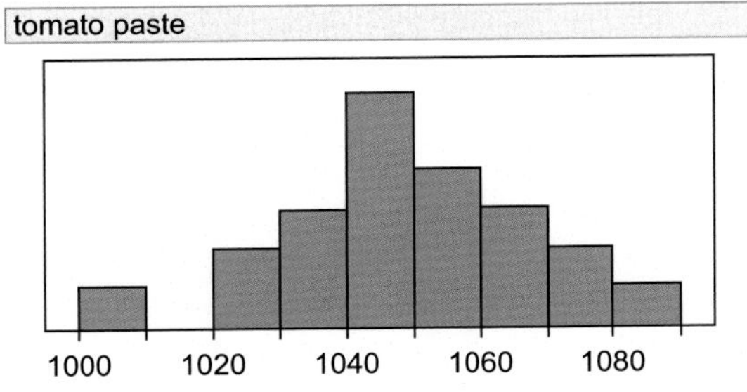

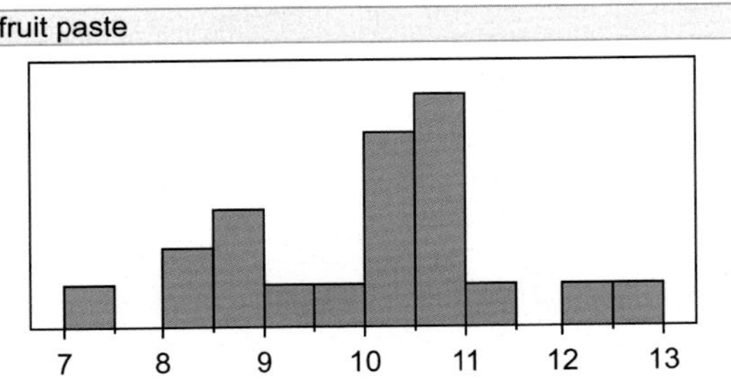

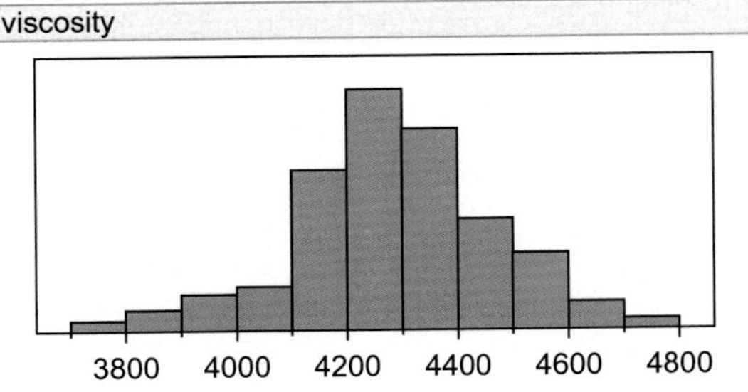

She observes no major anomalies and sees that each of the data sets seem to follow approximately a bell shape.

To understand the potential relationships between the individual factors and viscosity, Maria creates scatter plots of each factor with viscosity, with the factors on the horizontal axis and viscosity on the vertical axis. Her scatter plots look like this:

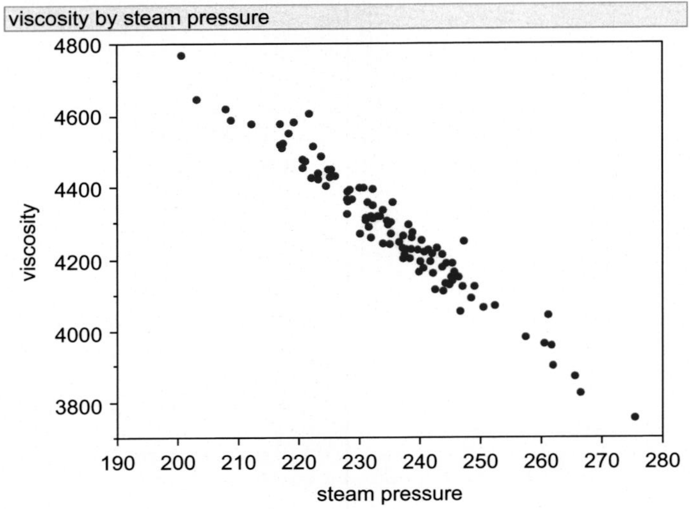

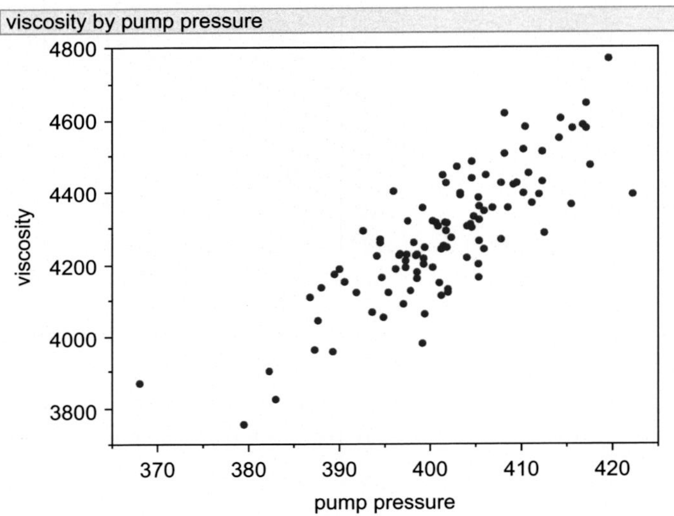

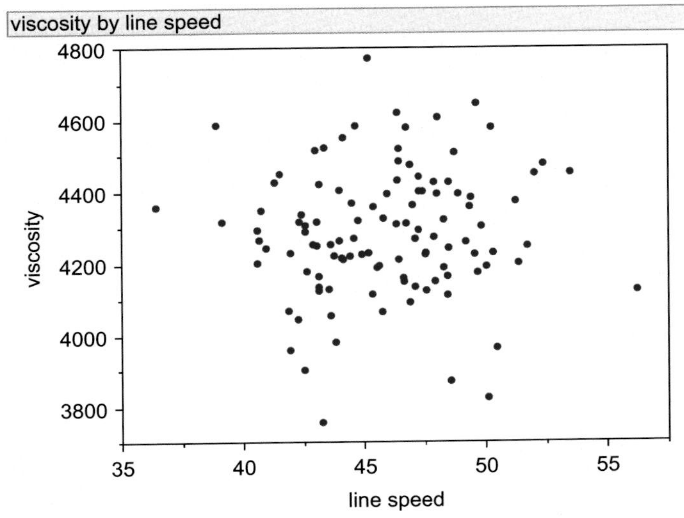
viscosity by line speed
viscosity
line speed
4800
4600
4400
4200
4000
3800
35
40
45
50
55

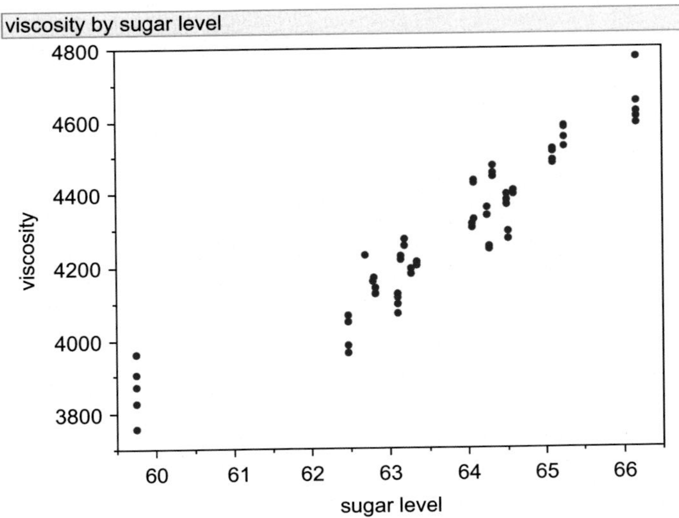
viscosity by sugar level
viscosity
sugar level
4800
4600
4400
4200
4000
3800
60
61
62
63
64
65
66

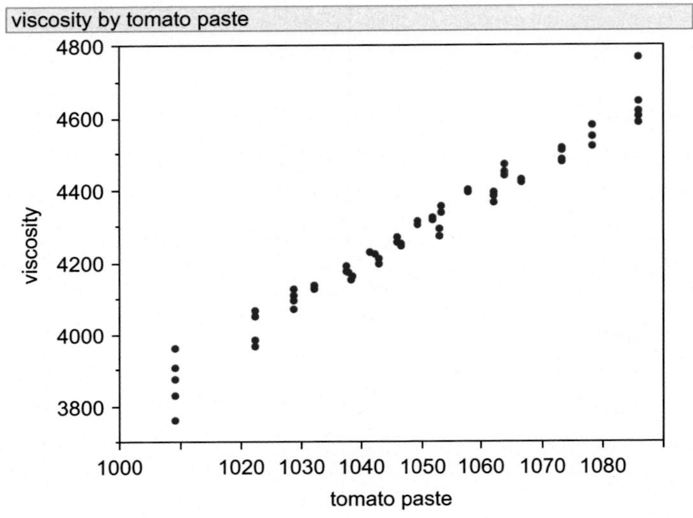

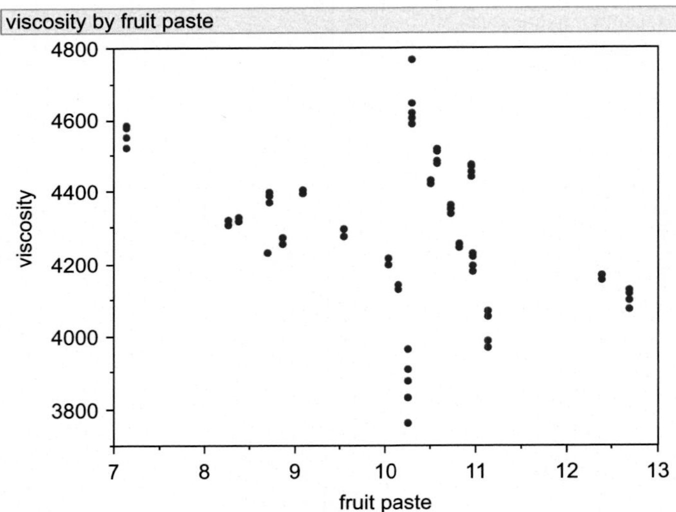

Some of the scatterplots seem to show stronger relationships than others, though it is difficult to objectively determine the strength of some of the relationships based on the plots alone.

Maria recalls from her notes of data analysis sessions with Dr Wang that the correlation coefficient is a measure of the strength of the relationship between two measures. This measure is represented by the letter r, with a value between -1 and 1. The sign of the correlation coefficient indicates the direction of the relationship. An $r = 1$ indicates a perfect positive linear relationship, while $r = -1$ indicates a perfect negative linear relationship. Her data analysis software lets her easily create a matrix of correlation coefficients, with all possible pairs of her input variables. She has correlations not only between the factors and viscosity, but also of the factors with each other. Her correlation matrix looks like this:

	steam pressure	pump pressure	line speed	sugar level	tomato paste	fruit paste	viscosity
steam pressure	1.00						
pump pressure	-0.82	1.00					
line speed	0.53	-0.44	1.00				
sugar level	-0.90	0.83	-0.50	1.00			
tomato paste	-0.96	0.83	-0.49	0.92	1.00		
fruit paste	0.36	-0.28	0.22	-0.28	-0.38	1.00	
viscosity	-0.97	0.83	-0.50	0.92	0.98	-0.37	1.00

Maria notes both positive and negative correlation coefficients, and some with values close to ± 1. The correlation between tomato paste and viscosity is 0.98, indicating that as tomato paste level increases, viscosity increases. This makes sense to her as more paste would seem to make a thicker product. The correlation between steam pressure and viscosity is -0.97, indicating that as steam pressure increases, viscosity decreases. This too makes sense since the higher steam pressure would introduce more water into the product, making it thinner.

The other correlations with viscosity are smaller, but they may still be meaningful. Also she sees that processing and formulation factors are correlated with each other in varying degrees. Maria recalls that Dr Wang used the following guidelines for interpreting correlations:

$|r| \geq 0.7 \rightarrow$ Strong correlations

$0.4 < |r| < 0.7 \rightarrow$ Moderate correlation

$0 < |r| \leq 0.4 \rightarrow$ Weak correlation

Her data analysis software also provides her with a matrix of scatter plots for each of the pairs of variables. That matrix of plots looks like this:

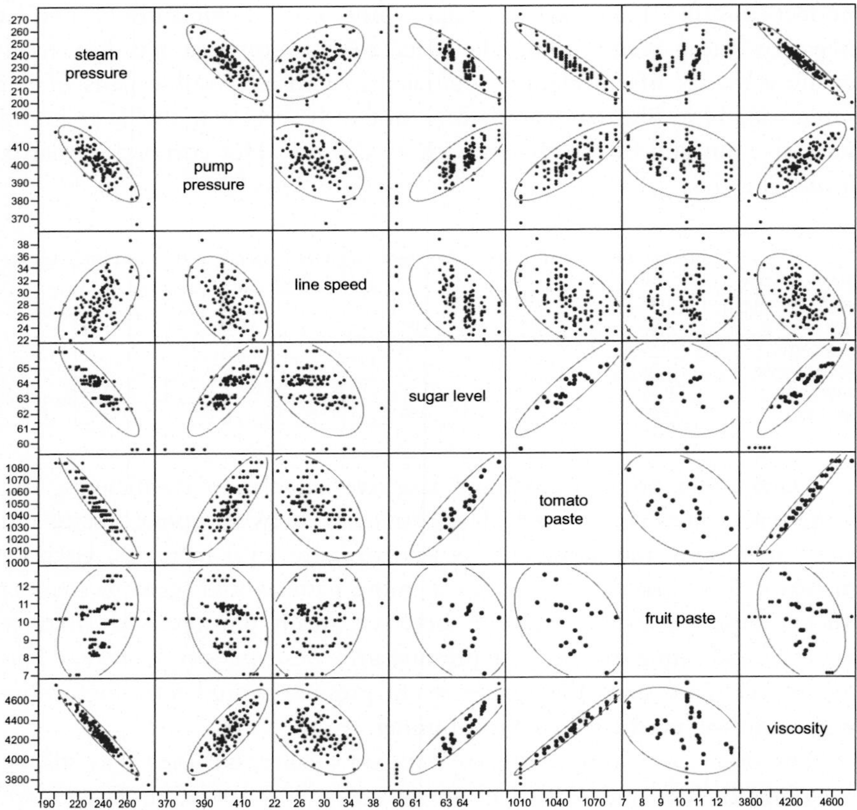

Each plot has an ellipse added that contains most of the data points. Ellipses provide a guide in understanding the relationships between the measures. The more circular ellipse shapes, the weaker the relationships are between the measures.

Maria recalls that Dr Wang had stressed that correlation is not the same as causation. For example, she cannot state that the slower line speed is causing higher viscosity. It may be that a more viscous product causes the line to run slower or that there is something else that is causing both viscosity and line speed to increase or decrease at the same time. Maria also recalls that correlation is a measure of a linear relationship, and if the relationship is not linear, the correlation coefficient would likely show a weak relationship. This is another reason it is useful to visualize the relationships.

The correlation coefficient can be calculated using the following formula:

$$r = \frac{\sum_i (X_i - \overline{X}) \sum_i (Y_i - \overline{Y})}{\sqrt{\sum_i (X_i - \overline{X})^2 \sum_i (Y_i - \overline{Y})^2}}$$

Maria by now knows that graphing the data each and every time is the first step in analyzing any data set. This time she made no exception and is highly rewarded for it. She sees that line speed and fruit paste seem to have weak relationships with viscosity, as also indicated by their corresponding correlation coefficients. At the same time, she notices that there is a strong linear relationship between viscosity and steam pressure as well as between viscosity and tomato paste level. Maybe these are the keys to controlling viscosity.

A related statistical analysis technique is linear regression. It expands on correlation by fitting a line relating a predictor variable to a response. Maria considers viscosity a response and all of these factors as potential predictors. Estimating the regression line could serve as an additional control to check if the fit of the relationship is strong.

In its simplest form, linear regression relates one factor X to a continuous response variable Y. The fitted line has the following equation:

$$Y = \beta_0 + \beta_1 X + \varepsilon$$

where

$Y =$ Output or response (dependent variable)

$\beta_0 =$ Intercept (the point at which the line crosses the Y-axis)

$\beta_1 =$ Slope

$X =$ Input factor or predictor (independent variable)

$\varepsilon =$ Error or noise

Multiple regression expands this simple form to include more than one input factor. The fitted line has the following equation:

$$Y = \beta_0 + \beta_1 X_1 + \beta_2 X_2 + \beta_3 X_3 + \ldots + \beta_n X_n + \varepsilon$$

where

Y = Output or response (dependent variable)

β_i = Model parameters to be estimated

X_i = Inputs or predictors (independent variables)

ε = Error or noise

As per Maria's notes, the main purpose of regression analysis is to find the line that best fits the data. The equation for the line relates the quantitative predictor variable to a response variable that can be used to make predictions. With a multiple regression equation, the relationships between Y and the Xs is still linear. The purpose of multiple regression is also to make predictions about a response variable.

Regression can be used to determine if a factor is a true predictor or if the relationship is observed by chance. With one factor, there is a p-value associated with a hypothesis test of whether the slope of the line is significantly different from zero. While a $p < 0.05$ does indicate a significant relationship between the factor and response, it does not alone determine if the relationship is strong enough to accurately predict the response from the factor. Another measure from the regression, R^2, describes the strength of the relationship between the response and the factor. It measures how close the individual data points are to the fitted regression line. Deviations between the points and the line can be considered error. The smaller the errors, the closer R^2 is to 1. If the line does not fit well, then R^2, approaches zero.

R^2 is also called the coefficient of determination. It represents the percentage of variability in the response that can be explained by the factor(s). R^2 is the square of the correlation coefficient. So, $R^2 = r^2$.

After going through her notes, Maria feels confident about fitting regression lines to the data collected from the production line. She fits a line for each of the processing factors with viscosity.

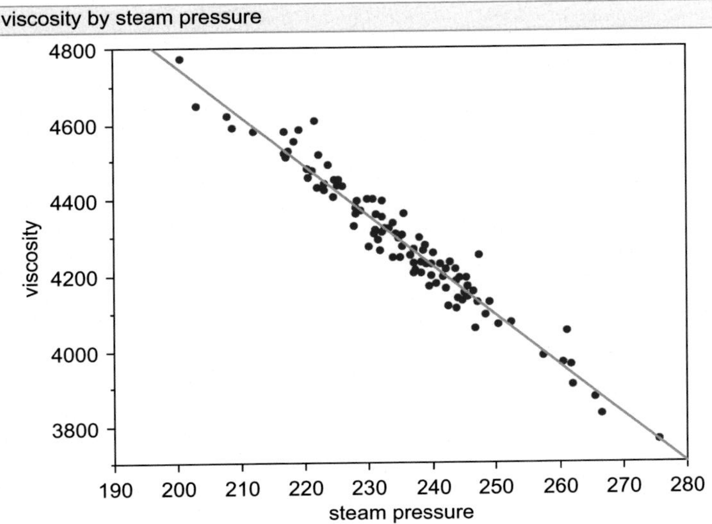

Linear Fit

Linear Fit

viscosity = 7388.1034 – 13.194431*steam pressure

Summary of Fit

R Square	0.950291
R Square Adj	0.949831
Root Mean Square Error	39.79612
Mean of Response	4286.281
Observations (or Sum Wgts	110

Analysis of Variance

Source	DF	Sum of Squares	Mean Square	F Ratio
Model	1	3269859.3	3269859	2064.656
Error	108	171043.0	1584	**Prob > F**
C. Total	109	3440902.3		<.0001*

Parameter Estimates

Term	Estimate	Std Error	t Ratio	Prob>|t|
Intercept	7388.1034	68.3696	108.06	<.0001*
steam pressure	–13.19443	0.29038	–45.44	<.0001*

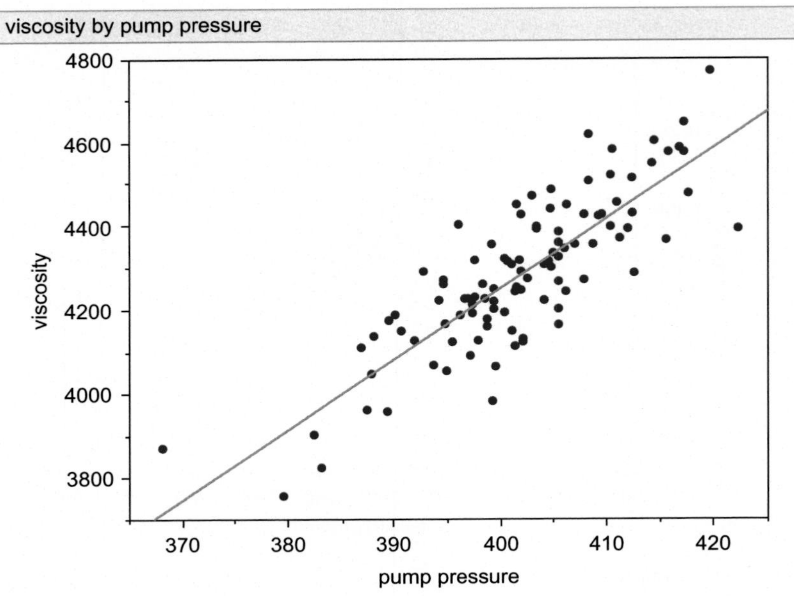

Linear Fit

Linear Fit

viscosity = –2449.113 + 16.769609*pump pressure

Summary of Fit

R Square	0.692845
R Square Adj	0.690001
Root Mean Square Error	98.9244
Mean of Response	4286.281
Observations (or Sum Wgts	110

Analysis of Variance

Source	DF	Sum of Squares	Mean Square	F Ratio
Model	1	2384010.3	2384010	243.6135
Error	108	1056891.9	9786	**Prob > F**
C. Total	109	3440902.3		<.0001*

Parameter Estimates

Term	Estimate	Std Error	t Ratio	Prob>|t|
Intercept	–2449.113	431.6344	–5.67	<.0001*
pump pressure	16.769609	1.074416	15.61	<.0001*

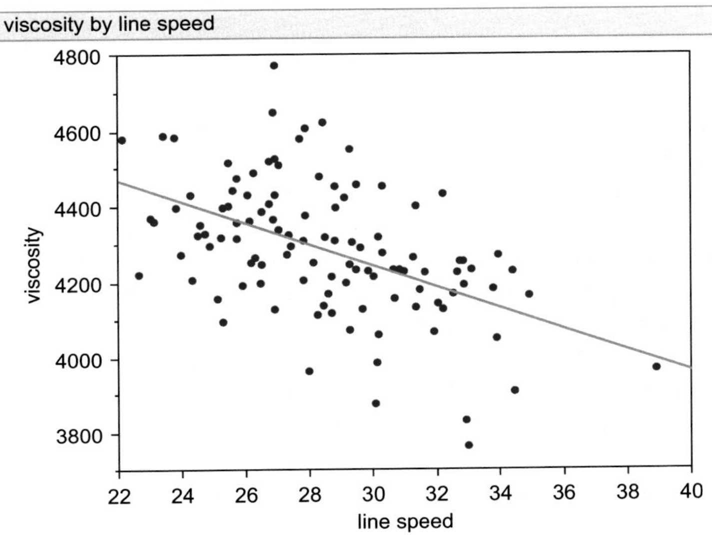

—— Linear Fit

Linear Fit

viscosity = 5091.8942 – 28.291634*line speed

Summary of Fit

R Square	0.253509
R Square Adj	0.246597
Root Mean Square Error	154.2185
Mean of Response	4286.281
Observations (or Sum Wgts	110

Analysis of Variance

Source	DF	Sum of Squares	Mean Square	F Ratio
Model	1	872300.6	872301	36.6769
Error	108	2568601.7	23783	**Prob > F**
C. Total	109	3440902.3		<.0001*

Parameter Estimates

Term	Estimate	Std Error	t Ratio	Prob>|t|
Intercept	5091.8942	133.8343	38.05	<.0001*
line speed	–28.29163	4.671555	–6.06	<.0001*

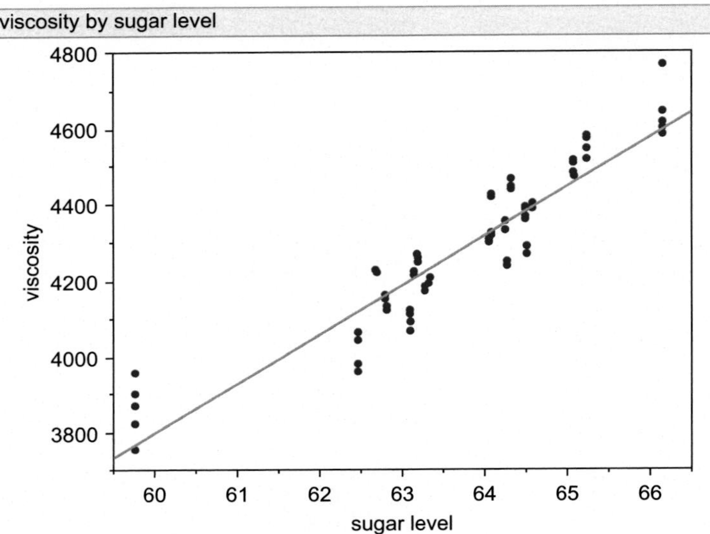

—— Linear Fit

Linear Fit

viscosity = –3952.097 + 129.32095*sugar level

Summary of Fit

R Square	0.848672
R Square Adj	0.84727
Root Mean Square Error	69.43596
Mean of Response	4286.281
Observations (or Sum Wgts	110

Analysis of Variance

Source	DF	Sum of Squares	Mean Square	F Ratio
Model	1	2920196.2	2920196	605.6798
Error	108	520706.1	4821	Prob > F
C. Total	109	3440902.3		<.0001*

Parameter Estimates

Term	Estimate	Std Error	t Ratio	Prob>|t|
Intercept	–3952.097	334.8151	–11.80	<.0001*
sugar level	129.32095	5.254693	24.61	<.0001*

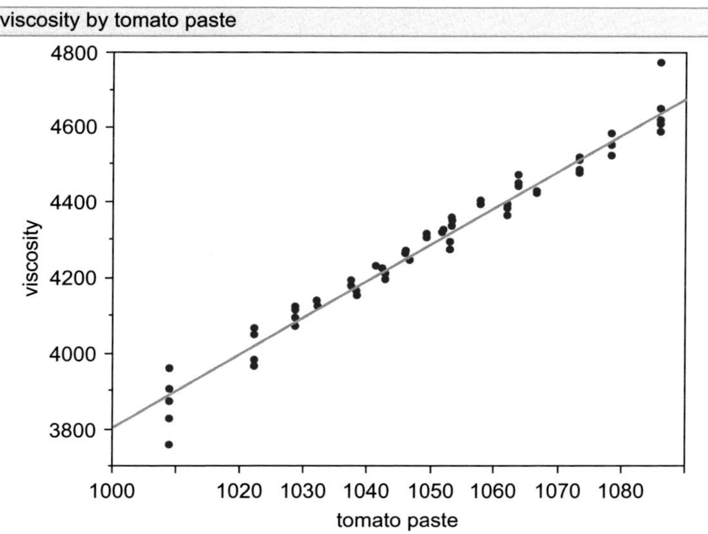

—— Linear Fit

Linear Fit

viscosity = –5933.165 + 9.7413042*tomato paste

Summary of Fit

R Square	0.968854
R Square Adj	0.968566
Root Mean Square Error	31.50089
Mean of Response	4286.281
Observations (or Sum Wgts	110

Analysis of Variance

Source	DF	Sum of Squares	Mean Square	F Ratio
Model	1	3333733.2	3333733	3359.581
Error	108	107169.1	992	**Prob > F**
C. Total	109	3440902.3		<.0001*

Parameter Estimates

Term	Estimate	Std Error	t Ratio	Prob>|t|
Intercept	–5933.165	176.3388	–33.65	<.0001*
tomato paste	9.7413042	0.168064	57.96	<.0001*

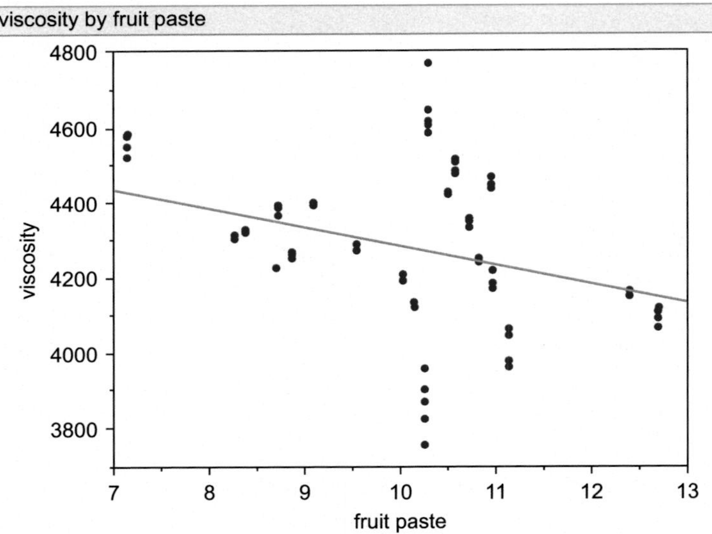

—— Linear Fit

Linear Fit

viscosity = 4787.9859 – 49.963258*fruit paste

Summary of Fit

R Square	0.13958
R Square Adj	0.131613
Root Mean Square Error	165.5692
Mean of Response	4286.281
Observations (or Sum Wgts	110

Analysis of Variance

Source	DF	Sum of Squares	Mean Square	F Ratio
Model	1	480281.4	480281	17.5201
Error	108	2960620.9	27413	**Prob > F**
C. Total	109	3440902.3		<.0001*

Parameter Estimates

Term	Estimate	Std Error	t Ratio	Prob>|t|
Intercept	4787.9859	120.8967	39.60	<.0001*
fruit paste	–49.96326	11.93665	–4.19	<.0001*

There is a lot in the output from her statistics software! She is familiar with the analysis of variance (ANOVA) table since that was in the output from her ANOVA models. She sees in the parameter estimates table the intercept and slope for each of the lines fitted with the individual factors. The Prob > $|t|$ column lists the *p*-values for the intercept and slope. For each of her processing and ingredient factors, the *p*-values for the slope are very small, so all of the factors are significant. But she can see from the graphs that some relationships are stronger than others and the R^2 values differ greatly. Maria makes the following observations:

1. Strong relationships exist between tomato paste and viscosity ($R^2 = 0.97$) and steam pressure and viscosity ($R^2 = 0.95$).
2. Moderate relationships are seen between sugar level and viscosity ($R^2 = 0.85$) and pump pressure and viscosity ($R^2 = 0.69$).
3. Weak relationships are present between line speed and viscosity ($R^2 = 0.25$) and fruit paste and viscosity ($R^2 = 0.14$).

Maria brings her expert team back together to discuss the results. Since both tomato paste level and steam pressure have strong relationships with viscosity, she suggests that these two need to be better controlled to keep viscosity on target and to reduce variation.

Connie, one of the line operators, mentions that the ingredient factors are adjusted for each batch to meet food safety control points. "Acid and sugar levels can vary with the input tomato paste. We add the ingredients to the batch tank based on the formula. We measure the acid level of each batch and often need to adjust sugar or tomato paste levels to meet the acid target. We cannot use them to control viscosity too. But, we can adjust steam pressure if we need to" adds Connie.

The regression line between steam pressure and viscosity will be the basis of the control plan Maria puts in place. The slope of the regression line is -13.2. The units for steam pressure are millibars (mb). So, an increase in steam pressure of 1 mb will decrease viscosity about 13 cP. Maria uses this finding to document the control strategy as a standard operating procedure for the line operators to use going forward.

Steve is pleased with Maria's work on developing the control strategy. "This is the first time we have used actual data to develop controls for the plant. Your data analysis skills are proving to be incredibly valuable!"

Maria blushes when she hears this praise from her boss.

CHAPTER 7

Statistical Process Control (SPC)

Now that the team has determined a way to adjust the process to change viscosity by adjusting the steam pressure, Maria has been asked to determine a strategy for monitoring and controlling viscosity during production. She has seen that there is not much consistency in what is currently being done and that each line operator has their own practice for how often they monitor viscosity and how they adjust process settings based on what they see.

She remembers that Dr Wang had discussed statistical process control (often abbreviated SPC) in one of her classes as one way to monitor and control processes, and Maria decides to set up a call with her. Before the call Maria does a bit of homework on this topic, so that she can make best use of their time. A search on Google gives enough background to make her feel smart!

Process capability is a methodology to quantify process performance with respect to specifications. SPC is a methodology for monitoring and adjusting processes to keep them on target and within specification limits. They are both quality tools that use statistical methods.

SPC emphasizes early detection and prevention of quality defects rather than correcting defects after they occur. It can provide real-time monitoring and control of process variability and operator adjustments to keep a process on target. A key tool in SPC is a control chart.

For quantitative measures like viscosity, the definition of a target is required. The target is the most desirable value for the process measure. Often when a target is determined, a tolerance around that target is also determined. The tolerance, or specification range, is the range of the measure on the product that is still considered acceptable. This concept acknowledges that some level of variation in the product is considered acceptable. This target is used in the SPC charts, and the tolerance is used in determining process capability.

As Maria starts thinking about BBQ sauce viscosity, she realizes that though she has been told the ideal value, she has no idea how this was determined and if an appropriate specification range exists! Is what she now believes to be the target level well defined? What level of viscosity below

Statistics for Food Scientists
http://dx.doi.org/10.1016/B978-0-12-417179-4.00007-X

or above the target would consumers even notice? Furthermore, even if they noticed it, would they care? Maybe consumers have a wide tolerance to viscosity variation. Would higher or lower viscosity product affect production efficiencies?

Maria contacts the person responsible for understanding consumer perceptions of the company's products, a marketing professional named James. She asks: "Do we know how consumer preferences for our BBQ sauces relate to our analytical viscosity measurements?"

James's answer is both surprising and puzzling: "No, not really. We have never tried to relate viscosity measurements to consumer acceptance of the product. We don't really know the ideal viscosity level and how much tolerance consumers have to variation in viscosity."

"I need to understand these to create a viscosity control strategy," Maria explains. "Is there research that we can conduct to better understand this?"

"Can you create or provide products of the same variety with different viscosities?" James asks. "I could then design a consumer test to evaluate these products and determine the acceptable range and also verify your target."

"I think I can do this. I'll get back to you soon with some ideas," Maria quickly replies, though she really does not know how she will approach James's request.

Maria starts thinking about the product and how to determine a viscosity specification. Ideally, she would need to have products that span a wide range of viscosity to provide to James for his consumer test. Maybe she can start evaluating product lots to identify those with different viscosities and select a range. That would be time-consuming and could potentially add other sources of variability like ingredient variation and product age. Maybe there is a way to force viscosity differences?

"Wait, I know how to do this!" Maria exclaims. "Since steam pressure affects viscosity, I can make systematic changes so that viscosity will vary too!"

Maria starts thinking about how she would design a trial to produce products with different viscosities by changing levels of steam pressure. She checks back in with James to see how many products he can handle in the consumer test. The consumers will rate their liking of each product on a 1–9 scale that is commonly used in product testing and also answer a number of other questions on the products. Since this process takes some time, he is not comfortable giving a consumer more than four samples in the same testing day, but he can recruit the consumers to return for an additional day if needed.

Maria envisions how she and James can relate the product viscosities to the consumer liking measurement. She will have measured viscosity on each of the products being tested. Since she now knows there is variability across bottles, she will measure viscosity on more than one bottle for each of the products. James will have individual consumer ratings for each of the products. By relating the average consumer ratings with the average viscosity measurements for each product, they should be able to understand consumer tolerance to viscosity differences and determine a viscosity range where the product is well liked by consumers. A plot would visualize this relationship, and a fitted curve could be used to determine the specific lower and upper viscosity limits for acceptance. She expects the relationship will look something like this:

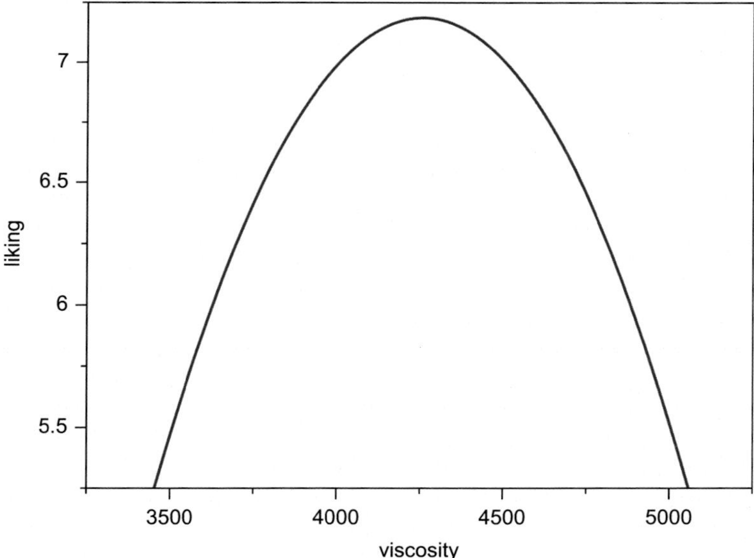

The real relationship may be more or less peaked; that is what she is really after. She expects that to get a good idea of the relationship between consumer liking and viscosity, she will need more than four products. So, Maria decides that she will create and provide eight products to James for the consumer test. She sets aside time in the pilot plant to create these products.

Maria produces the products for the consumer across the anticipated viscosity range of interest by varying the steam pressure in seven equally spaced levels. As a test, she runs a duplicate of the middle steam pressure setting to see how similar these samples will be. She expects their measured viscosities to be similar. If they are very different, then something unexpected happened when she produced the products, and she will need to determine what happened, correct it, and make the samples again. She recalls that Dr Wang had repeatedly mentioned how inadvertent bias can influence research results, so she does not produce the products by varying the steam pressure in an increasing or decreasing order, but shuffles these levels in a random way to minimize any effect on viscosity over the production order.

She measures the viscosity on these eight samples, and they do indeed have quite a large range. The samples from the two runs with the same target steam pressure have similar viscosities, so she feels good about her execution of the trial. The extremes of this range are well beyond what she thinks consumers would tolerate, but she is guessing somewhat, since she really does not know how the consumers will react. If she confidently knew the consumer tolerance ranges, she would not need this consumer study!

James is very specific on how he has consumers evaluate the eight product samples. He conducts the consumer evaluation at a time when the products Maria has produced are of an age (shelf life) typical to that of when consumers buy and use the product. Each consumer evaluates the samples in a different order based on a pattern James has created to minimize tasting order bias. The average acceptance scores for each of the eight products are calculated and presented to Maria. She has already had viscosity measurements made on five bottles of each of the eight samples and sets up a table with the average viscosity for each of the samples in one column and the average acceptance score in a second column.

viscosity	liking
3528	5.44
3812	6.41
3971	6.96
4178	7.09
4240	7.22
4516	6.98
4728	6.47
5048	5.42

Maria creates a graph of these two measures and sees the anticipated inverted U relationship. She knows that to fit a curvilinear relationship between viscosity and acceptance, she will need to perform a regression analysis as she had done before, but now include the squared term for viscosity as well in order to describe the curved nature of the relationship.

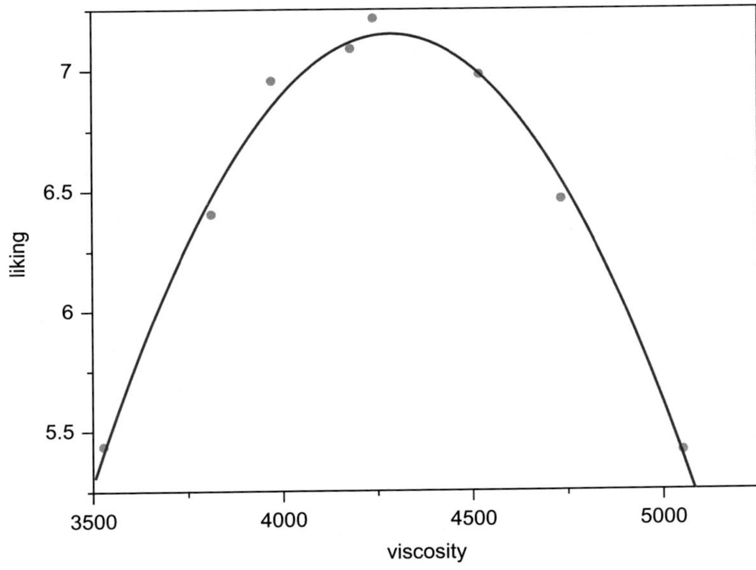

Polynomial Fit Degree=2

liking = 6.6279473 + 0.0001238*viscosity - 3.0274e-6*(viscosity-4261)^2

Summary of Fit

RSquare	0.991681
RSquare Adj	0.988354
Root Mean Square Error	0.077472
Mean of Response	6.49875
Observations (or Sum Wgts)	8

Analysis of Variance

Source	DF	Sum of Squares	Mean Square	F Ratio
Model	2	3.5774777	1.78874	298.0256
Error	5	0.0300098	0.00600	**Prob > F**
C. Total	7	3.6074875		<.0001*

Parameter Estimates

Term	Estimate	Std Error	t Ratio	Prob>|t|
Intercept	6.6279473	0.251913	26.31	<.0001*
viscosity	0.0001238	5.94e-5	2.08	0.0915
(viscosity-4261)^2	–3.027e-6	1.241e-7	–24.38	<.0001*

Maria is familiar with the analysis output, since she has had recent experience with linear regression models. She observes in the graph that her fitted line is close to the data points. The R-Square for the regression model is very close to 1, another indication that the model represents the individual data points well. And, her model has a very small p-value for both the overall model and the parameter estimate for the squared term, which indicates the curvilinear relationship between viscosity and liking.

Maria shares her analysis with James, and they both agree that the results show a strong relationship between viscosity and liking and that the model can be used to develop a target and specification limits. The peak of the curve is for a liking of around 7.15 and corresponds to a viscosity close to 4300; this will serve as the target. He indicates that liking mean differences less than 0.50 are not likely to be important. Using the model to predict liking, the viscosity range of 3900 to 4700 ($\pm$ 400 from the target) should have a predicted liking not less than 6.65 (0.50 less than the ideal value of 7.15 target).

Now that she feels confident with the specification she has developed, Maria turns her focus on developing a monitoring and control strategy. She arranges a call with Dr Wang to discuss her basic understanding of SPC and get some direction.

Dr Wang is her usual cheery self. "How is my favorite BBQ sauce developer?" is her greeting to Maria when the call connects.

Maria catches her up on her recent work developing the viscosity specification and her need to now create a monitoring and control strategy. Dr Wang recommends a specific SPC textbook that Maria can use as guidance, but advises that Maria alone will be able to gather the necessary information on how to best choose and implement the tools for her project. Maria purchases the textbook that Dr Wang recommends and dives in.

The primary SPC tool is a control chart. The chart displays a measure over time. For Maria, the measure of interest is viscosity. There are control charts to monitor individual viscosity measurements or averages of small groups to understand if a process is centered on the target. Variability should also be monitored over time in a separate control chart. Tracking variability may be just as important as tracking viscosity, since if a process becomes more variable, even if it is still centered at the target, then a defective product (a product with viscosity outside the specification range) is more likely to be produced. There are also control charts for discrete characteristics, such as number of defective samples or number of defects per unit.

In the ideal world, every bottle produced would be measured for viscosity. With this perfect information, an operator would be able to adjust the process if it was shifting to lower or higher viscosity levels. Measuring viscosity for every bottle produced would be prohibitively expensive, so Maria will have to choose both how often samples will be taken for measurement and how many samples to take each sampling period. Both sampling frequency and sampling amount affect the resources that will be needed for this monitoring program, as well as the effectiveness of the program, so Maria wants to get it right!

She reads that it is best to collect a small number of samples at each sampling time. A control chart of the sample averages (called an X-bar chart) is used to evaluate process centering. A control chart of either the sample standard deviations (called an s-chart) or the sample range (called an R-chart) is used to evaluate process variation. There are two reasons that this is best:

1. Short-term process variation is best understood when the samples are taken at the same time, rather than accumulated over a number of sampling times.
2. The control charts operate best when the data follows a normal distribution, and even if the individual data points do not follow a normal distribution, sample averages do based on the central limit theorem in statistics.

Sometimes, the measurement process is so involved or costly that only one sample can be measured at each sampling time. In this case, a control chart of the individual measurements (I-chart) is used to monitor how the process is performing with respect to the target. To monitor variation, a moving range chart (MR chart) is used, with the range calculated over a number of sampling time points. If the individual data points do not follow a normal distribution, then other types of charts may be needed.

Maria thinks about her BBQ production. It is easy enough to measure a number of samples taken at the same time since the measurement process is relatively easy (and reliable now!) and takes no more than a minute.

She would like to have samples taken relatively frequently so that if the process is shifting up or down, an adjustment can be made before too much product is produced. Since the line operator will be collecting the samples and then those samples will be taken to the Quality Lab to make the measurements, she cannot have the line operators performing this task more frequently than every half hour, considering the number of lines and lab resources. There are five filler heads in each bottling line, and it occurs to Maria that collecting samples from each filler head would be prudent—as there maybe some differences between them. So, her plan is to propose that five samples, one from each filler head, be taken and measured every half hour.

Maria arranges to discuss her plan with the line operators at their next weekly meeting. Also present at this meeting are the Quality Lab staff members. She talks through her proposed monitoring plan and asks for questions. The operators see no problems with pulling the samples at specified times, and the lab staff can handle the measurements if the lines stagger their sampling times. The measurement values will be transmitted back to each line operator through the data management system, and each line operator will see the results within 10 min of the sampling time.

Maria is feeling good about it all until Connie, one of the line operators, says "So, it is still up to us to decide when and how to adjust the process, right?"

Maria was so focused on the data collection that she had forgotten about establishing clear guidelines for process adjustments! She asks the group how and when they currently make adjustments. She recalls that she had been told that each operator has his/her own strategy, and those adjustment strategies differ widely. She tells the group that she wants them to operate consistently and will need to do some additional work to determine what that practice will be.

Maria remembers that she had seen a chapter in her SPC text about run rules and thinks that it may give her some idea on how to react to the control chart output. She looks into this chapter and finds out that there are indeed ways to determine when to make an adjustment to the process based on what is viewed in the control chart. There are several different sets of rules (Western Electric, Nelson, Montgomery, and so on), and as she reads through them, she sees that some of them are the same or similar. They all seem to work in a consistent way. Control zones are created on the charts based on the process standard deviation. Control limits are defined as ±3 standard deviations from the target or center line. A blank control chart with zones looks like this:

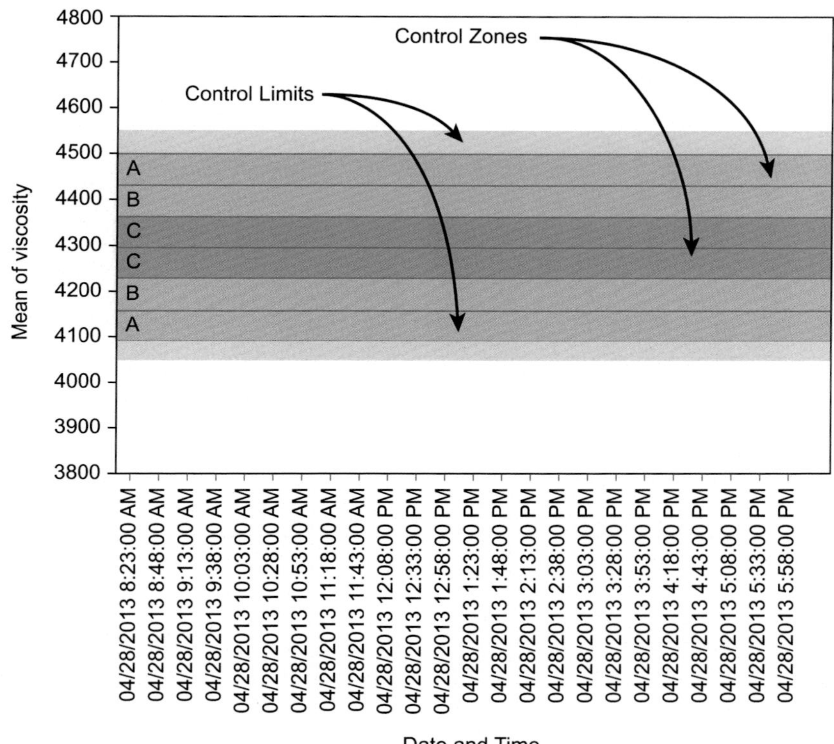

Further in the chapter, she sees a summary: the top five run rules for SPC charts to detect process shifts are as follow:

1. One point outside of the control limits
2. Two out of three points beyond two-sigma on the same side of the target
3. Four out of five points beyond one-sigma on the same side of the target
4. Eight consecutive points on either side of the center line
5. A steadily increasing or decreasing pattern of six points

These are all situations that are unlikely to occur if the process is still in control and not deviating from the target and has not changed in some way. A control chart with a run rule violation on rule 2 would look like this:

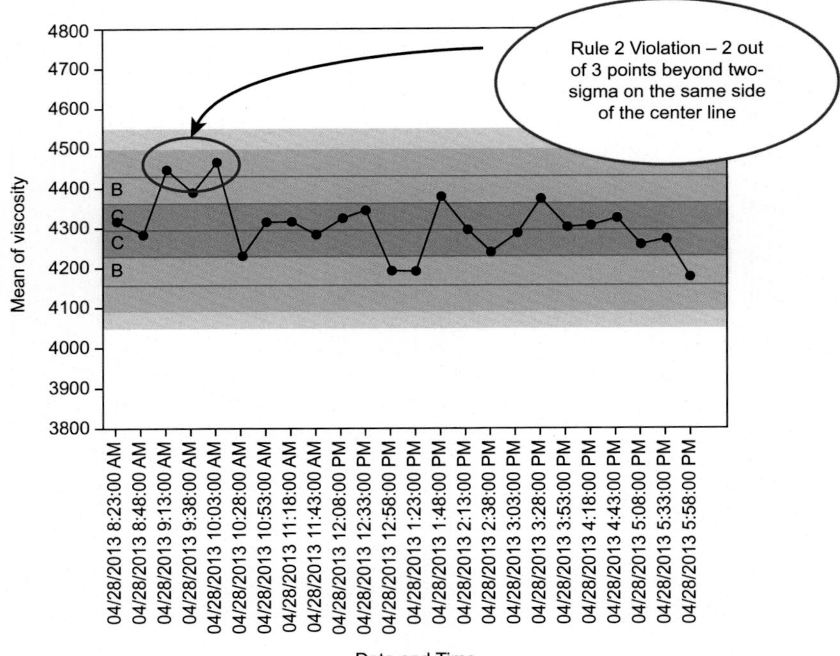

Now that she has some rules for determining when a process is shifting away from the target, she needs to determine a strategy for process adjustment. It seems best to her to use the information in the data that triggers the rule—that is telling her something about how the process may be shifting. She is also aware that bold adjustments to the process will likely be an overreaction. Whatever she decides, it needs to be easily understood and implemented by the operators.

To use the information from the data, she decides that she will simply average the data points that triggered the rule. For rule 2, this method means averaging the three relevant points in "the two out of three." For rule 3, this means averaging the five data points in "the four out of five." For rules 4 and 5 this means the eight or six points, respectively, that create the pattern. And for rule 1, there is no averaging; the information is contained in the single data point.

Maria feels confident about the regression relationship that she had determined between steam pressure and viscosity in her earlier work, so she decides to use that regression to help her determine how to adjust the process. The equation for the regression line is as follows:

$$\text{Viscosity} = 7388.1 - 13.19 \times \text{Steam Pressure}$$

The average viscosity change above or below the target will determine the increase or decrease in steam pressure. For example, an average viscosity of 4500 determined from the run rule violation is 200 above the target of 4300. Using the equation, Maria calculates that an increase in steam pressure of 15 would bring the viscosity back to the target. It may well be that different actions may be needed for different run rule violations, so as this strategy is implemented, Maria will evaluate process performance and make changes to the correction strategy if the performance is not adequate.

Maria writes up, as a work procedure, the data collection strategy, run rules, and adjustment method using the process data. With the work procedure in place, the current and any future line operators can consistently monitor and make process adjustments when needed.

CHAPTER 8

Sampling

Maria arrives at the plant one morning, and she becomes painfully aware that her statistical process control (SPC) program and the processes that she implemented just a few weeks ago are having an impact. During the previous night's production, many run rule violations were triggered. Although process adjustments were made according to the SPC strategy, the viscosity could not be brought back to target. The production manager placed the entire shift's production on hold. Maria is excited to see that the SPC process is well embraced by the plant management but also sad as she realizes that a large amount of product is put on hold as a direct result of the SPC program. She is wondering what could have caused the run rule violations and the product hold. She goes to the Quality Lab and bumps into the analytical lab manager, Chad, and the plant quality manager, Louise.

Maria: "Chad and Louise, I heard about the product hold!"

Louise: "I'm so glad to see you here today. We really need your help with this situation."

Maria: "Do you know what happened last night? Why is the product on hold?"

Chad: "The Brookfield measurement device in the Quality Lab failed verification at the end of the shift. That means it stopped measuring viscosity correctly at some point after the verification at the end of the previous shift. The operator made process adjustments based on these inaccurate measures and may have inadvertently caused the viscosity to go off target. I just fixed the Brookfield measurement device and verified it with the device in the R&D analytical lab. So this will no longer be a problem, and hopefully today's production will be fine."

Louise: "I'm getting a lot of pressure from Corporate to release the product on hold. I received four phone calls, asking me about our last night's BBQ production, including threats for not loading the product on the trucks this morning. So, I need your help to develop a sampling plan to determine if last night's production is of acceptable quality and can be shipped!"

Maria: "I understand. Our sales team must be feeling pressure from our customers, especially now that the BBQ grilling season has begun.

Statistics for Food Scientists
http://dx.doi.org/10.1016/B978-0-12-417179-4.00008-1

The plant has been running at full capacity for a month to prepare for the peak-season, and now that the demand has arrived, we cannot afford not to fulfill orders. But we cannot send bad quality product out the door either. Do you have any more specifics on the product on hold?"

Louise: "Could we sample 10 bottles and release the product to get Corporate off my back?"

Maria: "We need to do what is right for the customer. That includes not releasing the product if it does not pass our quality standards. I would not think that this could be considered a physical hazard to consumers?"

Louise: "No. The product viscosity during our night shift was not in line with our specifications. The device was reading higher than it should have so we are worried that the product produced is not thick enough. I already pulled around 10 bottles to measure the viscosity. Do you think that those will be enough to make a statistically sound decision whether or not to release the product?"

Maria: "Are you able to isolate a time frame when the instrument was working properly?"

Chad: "We had not thought about that, but you make a good point. Potentially we could release the first half of the production lot, since the line operator started making adjustments to the process around the middle of the shift based on the bad measurements from the Brookfield viscosity meter. Product viscosity was likely under control before the operator started making adjustments."

Maria: "If it is true that during the second half of the shift we are uncertain about the results and during the first half our process was still in control, we may consider evaluating the product on hold in two separate phases. We will do our due diligence with the first half of the shift to confirm our hypothesis and sample more frequently from the second half of the shift where we suspect the quality issue may have occurred due to potentially incorrect adjustments. For starters, could you please find out how much product was made during the first and the second halves of the shifts? What is the overall product dollar value? What kind of sampling resources do we have? How many samples can be evaluated per hour?"

Louise: "I will collect that information right away."

Maria: "Thanks Louise, if you find out anything new during your snooping around, please call or email me."

Maria rushes out of the lab to the cafeteria. In order for her to function properly in this crisis, she needs a double espresso. As she waits for the coffee brewer to fill her cup, she thinks about how she may be able to determine the number of samples needed and how to collect the samples. But the answers to both of these questions are based on many considerations, including scope, risk, measurement capability, and the specific production line.

Maria is interested in evaluating a certain number of samples from last night's production (the population) and drawing conclusions about what last night's production defect rate likely is.

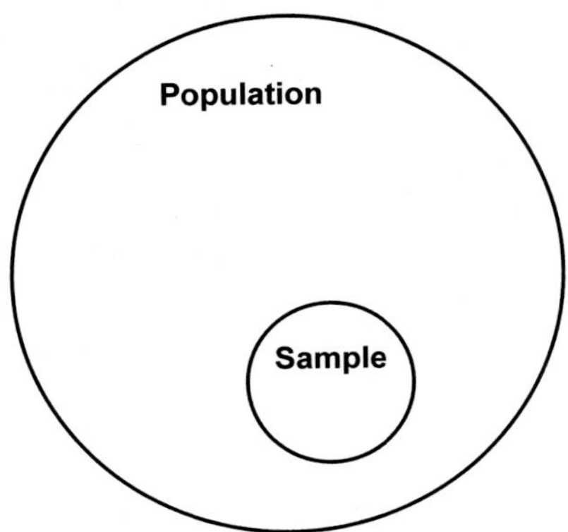

She recalls what she learned about sampling at school. Dr. Wang had always said that it is not just how many samples, but also how they are collected that is important. She digs out her notes from the class where sampling had been discussed.

HOW MANY SAMPLES?

There are two common approaches to determining the number of samples evaluated:

1. **Acceptance sampling plan**: As the name suggests, this approach is typically used to accept or reject the lot of raw materials or finished product based on predetermined acceptable and unacceptable defect levels.
2. **Minimum sample size sampling plan**: The approach is used to quantify the maximum level of a defect with a specified level of confidence.

Acceptance sampling plans are based on a statistical hypothesis test set up to answer the question: Should a lot be accepted and shipped based on the observed number of defects. There are two types of incorrect decisions that can be made, and risks associated with each:

On one hand, an unacceptable lot can be falsely accepted; while on the other hand, an acceptable lot can be falsely rejected. The hypothesis test has certain components. First, the null and alternative hypotheses are defined based on the proportion of defectives. The null hypothesis is shown next:

$$H_0 : p_{\text{defective}} = p_1$$

p_1 is called the acceptable quality level (AQL). For example, a 1% defect level could be considered acceptable.

The alternative hypothesis is this:

$$H_1 : p_{\text{defective}} = p_2$$

p_2 is called the rejectable quality level (RQL). For example, a 5% defect level could be considered unacceptable.

The choices for p_1 and p_2 are specific to each sampling situation and should be dictated by business considerations.

The acceptance sampling plan can be designed to hold the probability of accepting a "bad" lot and rejecting a "good" lot based on specific levels. These probabilities are as follows:

Alpha: This is the probability of rejecting a lot with a defect level of p_1. For example, this probability could be set at 10%.

Beta: This is the probability of accepting a lot with a defect level of p_2. For example, this probability could be set at 10%.

A sampling plan with AQL = 1%, RQL = 5%, alpha = 10%, and beta = 10% would have a sample size of 100 and a decision rule to reject the lot if three or more defectives were found. This example has a balanced risk profile where alpha = beta. There is willingness to tolerate the same amount of risk of not releasing a good lot (a cost to the business), as of releasing a bad lot (a risk to the business). As with the choices for p_1 and p_2, the choices of alpha and beta are specific to each sampling situation and should be dictated by business considerations.

A visualization of the hypothesis test looks like this:

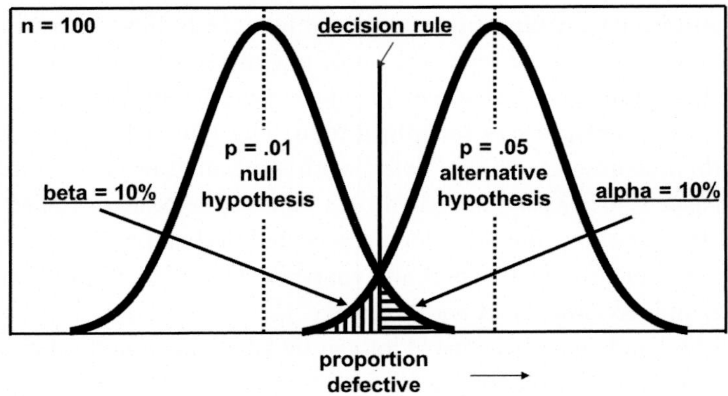

If the decision rule is moved to the right, alpha will decrease and beta will increase. If the decision rule is moved to the left, beta will decrease and alpha will increase. If the sample size is increased, both curves will become narrower and more peaked, reducing both risks equally.

Minimum sample size plans allow for the smallest possible sample to be taken to answer the question: Is the percent defective products no greater than a predetermined maximum allowable defect level (MADL), and at what level of confidence? The table here is used to determine the number of samples:

	Number of Defective Units in Sample (**c**)					
Confidence Level	0	1	2	3	4	5
80	160/n	299/n	428/n	552/n	673/n	791/n
90	230/n	389/n	532/n	668/n	799/n	927/n
95	300/n	474/n	630/n	775/n	915/n	1051/n
99	461/n	664/n	841/n	1005/n	1160/n	1311/n

To estimate the number of samples needed, a maximum allowable defect level and the desired level of confidence is chosen. Then, a minimum number of samples can be determined using the table above.

For example, in a sample of $n = 300$:

- If zero defects are found, we can be 95% confident that the true defect rate is no greater than 1% (300/300 = 1%).
- If one defect is found (out of 300 samples), we can be 80% confident that the true defect rate is no greater than approximately 1% (299/300 = 0.997%).

For example, if a 95% level of confidence is desired and no more than 1% defect rate is acceptable, then a minimum of 300 samples would be evaluated, and no defects would be allowed to accept the production lot. Alternatively, 474 samples could be evaluated, and one defect would be acceptable to yield the same conclusion of 95% level of confidence and no more than 1% defect rate estimate.

HOW TO COLLECT SAMPLES?

How samples are collected is just as important (and perhaps even more important) than the number of samples. It is important to collect samples without any form of bias. A well-chosen smaller number of samples is better than a poorly chosen large number of samples. Randomly choosing units for inspection from the lot will ensure that there is no bias, but this can be difficult to implement with a production lot of units in multiple cases and cases stacked into many pallets. If a defect can appear anywhere in a lot, then a stratified sampling approach with strategically chosen subgroups may ensure that units selected represent the production lot without bias.

Maria's incoming email signal chimes. The email from Louise with information about the product on hold has finally arrived. The lot on hold consists of 120 pallets, 12,000 cases, or 120,000 bottles (10 bottles per case and 100 cases per pallet, with 10 cases on each of 10 layers) of BBQ produced over an 8-hours shift. The pallets are numbered and arranged so that the order in which they were produced is easy to follow. The last 4 hours are more highly suspect, while the first 4 hours carry a lower risk. There is no consumer safety hazard, but some consumers may express dissatisfaction from a less viscous, or thinner, product.

The hold product value is substantial, but if a large product inspection is required, then the sampling costs will outweigh the benefits due to the large labor costs.

Maria thinks about how to approach the situation at hand. The defect levels in the acceptance sampling plan example from her class notes seem to be much larger than would be tolerable in this situation. And she is not sure that Louise or anyone else would easily be able to determine an AQL and RQL for this situation. The minimum sample size plan approach seems like it would be easier to use in this situation since the concepts of maximum allowable defect level and confidence level are more easily understood.

Maria emails Louise and explains MADL and confidence level and suggests that since the first half of the shift is less suspect separate plans, with different MADL levels, should be created.

Louise states in the email that for the more suspect being in the second half of the shift, she is willing to accept at most a 0.5% defect rate at 95% level of certainty. Since the first half of the shift is less suspect, the business is willing to have reduced sampling and looks to a recommendation from Maria.

Based on the information Louise provides, Maria recommends starting by sampling from the first 4 hours of production so some customer orders can be fulfilled quickly, since this part of the lot has a higher chance of being released. Maria proposes the following sampling strategy:

1. For the first 4 hours of production, sample 60 bottles. If no defects are found, then the team will be 95% confident that the defect rate is no larger than 5%, and this portion of production can be shipped. This sample plan may not seem to be very rigorous, but the sample size is higher than what is collected in the current SPC program. Since the

team feels that the production during this timeframe is likely of acceptable quality, the proposed plan is essentially repeating the measurements now that the Brookfield has been fixed.

2. For the second 4 hours of production, sample 600 bottles. If no defects are found, then the conclusion will be the defect rate is no larger than .5% with 95% confidence. Then, the second 4 hours of the production lot can be shipped as the defect rate would fall in line with the plant historical tolerance levels for viscosity. However, if one, even one, defect is identified, then this portion of production will not be shipped.
3. How will the 60 or 600 samples be chosen? Equal numbers of bottles should be from each hour of production. Since the production is continuous, a sample chosen from every n^{th} bottle to get to the target sample size would be desired. However, to minimize sampling effort, Maria recommends sampling cases and then taking a sample of bottles from each case.
   - **a.** For the first 4 hours of production, 60 bottles are required. Since there are 120 pallets, a case can be taken from every other pallet; this will result in 60. Select one bottle from each case to provide the 60 bottles. Since the potential for a defect is not related to the layer or position on the pallet, any case from the top layer in the pallet can be chosen, as this will be easiest to pull without taking apart the pallet.
   - **b.** For the second 4 hours of production, 600 bottles are required. Again there are 120 pallets, so sampling one case from each pallet and selecting five bottles will provide the 600 bottles. Again selecting any case from the top layer is easiest.

Maria's sampling plan is quickly put into effect. She joins the team to help select the bottles and helps Chad record the viscosity levels for each of the selected BBQ bottles. Some of the viscosity levels are high and some are very low, but none of the samples seem to be outside of the identified specification limits. The first 4 hours of production is quickly approved for release. A few hours later, the second 4 hours of production is also approved for release.

Maria and the team are pleased with the results. Since this is the first time she has constructed a sampling plan, she decides to set up a call with Dr Wang for a quick debrief on the situation. It is a few days before they are able to connect.

After listening to Maria's update, Dr Wang says, "Maria, you have done a great job and I'm sure your company appreciates your efforts. But I think that your sampling plan could have been more efficient. When you have a continuous response like viscosity, you often need far fewer samples than with a binary (or pass/fail) response. You may be able to use as few as 30 representative samples to be able to estimate the defect rate if the data is normally distributed. There is a great benefit in using continuous response data if at all possible and exploiting that continuity."

"Wow, I didn't even think of that," responds Maria. "There was such a push to have an answer that I was happy to find the table I used from your class notes!"

Dr Wang continues, "If you create a histogram of the viscosity data, you should be able to estimate this by adding the target and specification limits as if you were calculating process capability measures. Then, it will be a part of the standard output."

Maria is excited, "I think I know how to do this in my software. I wish I had called you before we started, we could have finished much faster. But I'll be better prepared next time thanks to you!"

After hanging up, Maria performs the analysis Dr Wang had suggested. She creates a histogram using the 60 viscosity measurements from the first 4 hours of production. She adds the current viscosity target and upper and lower specification limits.

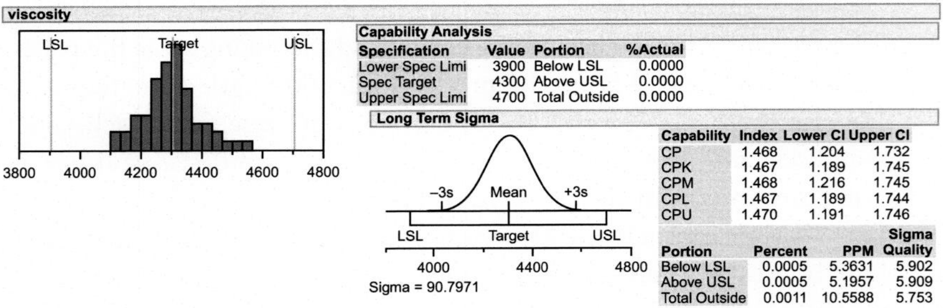

Capability Analysis

Specification	Value	Portion	%Actual
Lower Spec Limi	3900	Below LSL	0.0000
Spec Target	4300	Above USL	0.0000
Upper Spec Limi	4700	Total Outside	0.0000

Capability	Index	Lower CI	Upper CI
CP	1.468	1.204	1.732
CPK	1.467	1.189	1.745
CPM	1.468	1.216	1.745
CPL	1.467	1.189	1.744
CPU	1.470	1.191	1.746

Portion	Percent	PPM	Sigma Quality
Below LSL	0.0005	5.3631	5.902
Above USL	0.0005	5.1957	5.909
Total Outside	0.0011	10.5588	5.753

From examining the histogram, Maria notes that the data seem to follow a normal distribution and confirms that there are no actual values both below and above the specification limits. The long-term projection based on the normal distribution suggests only a tiny percentage (0.0005%) of product produced would be below the specification limit. She repeats this analysis with the 600 viscosity measurements from the second 4 hours of production with similar results. She certainly would not have needed this many measurements to understand the viscosity distribution for the second half of the held product. This is a valuable lesson she will not forget!

Maria now feels much more confident in her sampling skills and hopes that she will not need to use them again any time soon!

CHAPTER 9

Process Capability

With the issue of product dispensation past, the team focus on delivering the best product on consistent basis has resumed. Lisa's recent feedback had been that there was a positive impact on the BBQ lines since the statistical process control (SPC) system was implemented. She asks Maria to present the dynamic SPC monitoring system implemented earlier on the BBQ lines to the operations management team.

Lisa: "This will be a great opportunity to demonstrate to the operations management team all of the improvements we have made. The operations Vice President, my boss, along with the managers of the other production facilities will be present along with the research vice president, Steve's boss. You should make sure that Steve can attend. Their next meeting is next Tuesday; I can make sure you are added to the agenda. Let's plan on a half hour, with a 20 min presentation and time afterward for questions."

The rest of the week went by quickly and the weekend past even faster, as Maria spent most of it setting up tables, decorating them with flowers, and arranging seating for her Grandma's 90th birthday celebration. Most importantly, everyone at the party had a good time, and Grandma Maria (who Maria is named after) had the time of her life on the dance floor. When Maria arrived at work Monday morning, her immediate task was to create her presentation for the senior management team. She decided that a live demonstration of the SPC monitoring system would be most impactful and made arrangements to have that as the focus of her presentation. She spent most of the day preparing for the presentation, and by the time she left the office, she felt confident.

Maria was a bit nervous as she did not personally know most of the meeting participants. Armed with her SPC knowledge and a smile, she entered the room and set up the projector, so that everyone could see her screen. Then she projected an example of the "live" BBQ viscosity SPC monitoring system for one of the production lines. In the demonstration she showed how a run rule violation would trigger an alarm and how the system would then advise on the corrective action.

Statistics for Food Scientists
http://dx.doi.org/10.1016/B978-0-12-417179-4.00009-3

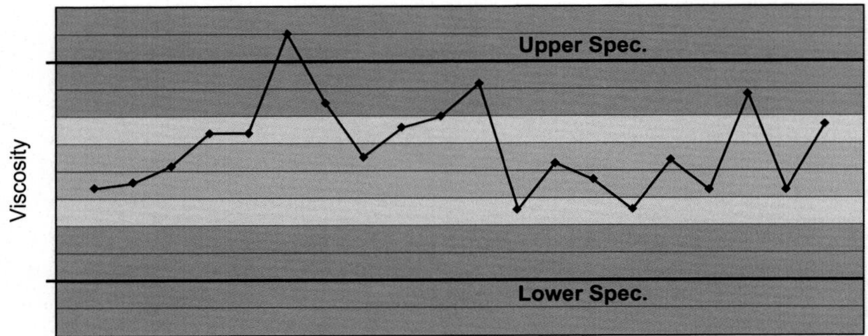

Maria began with defining SPC as a process that helps operators monitor the production performance as well as a system that helps operators determine when to adjust a process and to what degree. Most plant managers were eager to be early SPC adopters and to implement SPC on their respective production lines. At that moment, something unexpected happened: Joe, the operations Vice President, asked Maria, a question in a soft voice.

Joe: "I'm puzzled that your system only looks at immediate control. I assume that the data you are capturing is saved on the plant server. Can't you summarize this data to produce a report that Lisa can use to demonstrate overall performance versus specifications on each line? We can use this report to determine which lines perform better or worse and focus improvement efforts on those that are performing poorly. Also, we monitor other quality characteristics like pH, salt, and sugar levels. Shouldn't we use this SPC system to monitor and control these too? The performance reporting would include these measurements."

Maria is stunned that she had not considered performance reporting! But before she can respond, Lisa comes to her rescue.

Lisa: "We had considered this of course. We are so excited about what we have started that we did not want to wait to present it to you and the team. And since this will take considerable resources to put in place, we wanted to make sure everyone is in agreement."

Steve jumped in: "I have pretty good knowledge of process capability measures and had planned to suggest these to Maria. I wanted her to keep focused on the dynamic monitoring to make sure that was correct before moving to the reporting stage."

Lisa: "Maria, Steve, and I will partner to implement a performance reporting system for the monthly plant performance meetings. We will focus on viscosity for now, and once we have your and the team's approval, we can replicate the approach for the other quality measures. We'll present this at next months' meeting."

The very next day, Maria meets with Steve. He shares some of the positive feedback he had received on Maria since she started working with the team at the plant. Maria acknowledges the feedback and relates the plant team's desire to implement SPC and adopt it across the floor.

Steve: "Maria, I'm really proud with what you have been able to achieve in such a short time."

Maria: "Actually I was blindsided by Joe's comments yesterday. I feel that I should have anticipated that there is so much more to do. You and Lisa saved my butt for sure! If I remember correctly, last month you led a seminar on process capability."

Steve nods his head in an agreement, as Maria continues.

Maria: "However, at the time of the presentation, process capability did not mean much to me, and honestly, some of the information went way over my head. Now, my absent-mindedness during your presentation is haunting me. Can you give me a quick summary of what it is about? I want to be able to put it to practice soon. I know I can do this using my statistics software. Dr Wang had suggested to look at the distribution of the viscosity measurements of the held product to project the proportion of a lot that would be out of specification. Lisa said we will partner to implement this, but she is so busy that I think that it will be mostly up to you and me."

Steve responded, "I'll be glad to help you out with this. Actually, I have a lot of passion for process capability measures. They are great measurement tools that are underutilized in our plants. Since you have established a great rapport with Lisa and her teams, it looks like now is the time for a successful implementation. Let me introduce the concepts to you, and you can bring forward process capability measures to your friends at the plant."

Luckily, although the name may be perplexing, the process capability approach is not. Process capability analysis is a standard output from many statistical packages, and the concepts are very simple.

Process capability analysis allows researchers to find the linkage between the voice of the customer (what the customer wants, based on defined specification limits) and the voice of the process (what the process can deliver). To quantify the relationship, a capability index is calculated:

$$\text{Capbility Index} = \frac{\text{Voice of the Customer}}{\text{Voice of the Process}}$$

Capability indices help us focus on the nature of the problem: Can we provide what the customer is asking us to deliver? In the example below, the customer tolerance (upper specification limit—lower specification limit) is smaller than what the process is capable to deliver, and some proportion of the finished product is expected to meet customer expectations.

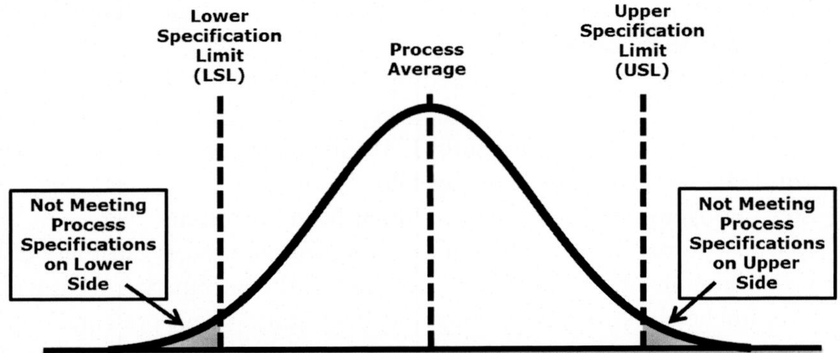

For process capability evaluation to be meaningful, the specifications used must be customer based. There are four groups of potential customers:

1. Consumers
2. Customers that sell a company's products (i.e., a grocery store chain)
3. The government (through regulations)
4. The company producing the product (i.e., manufacturing efficiencies, food safety requirements)

If a specification is not based on a customer need, then two adverse conditions may occur:

- Under-specification: We think we are producing good products, but in fact, we are failing to meet the customer needs.
- Over-specification: We are operating to overly tight requirements that were not generated by the customer and have no impact on the customer experience.

The approach for estimating if a process is capable is straightforward. Three things are needed to measure process capability:

- The customer specifications
- The process typical values—what the process is centered on (measured by the process average)
- The process variation—how much variation there is around the center (measured by the process standard variation)

A process researcher compares the data representative of the process to the predefined customer tolerances (specifications). Four possible outcomes from comparing the process performance with the customer specifications are shown below:

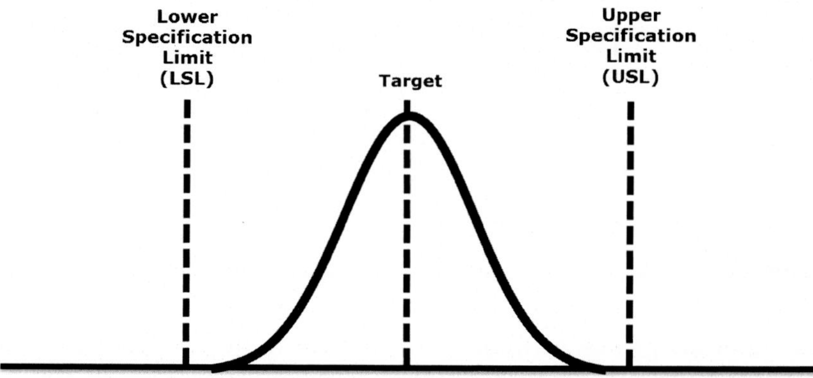

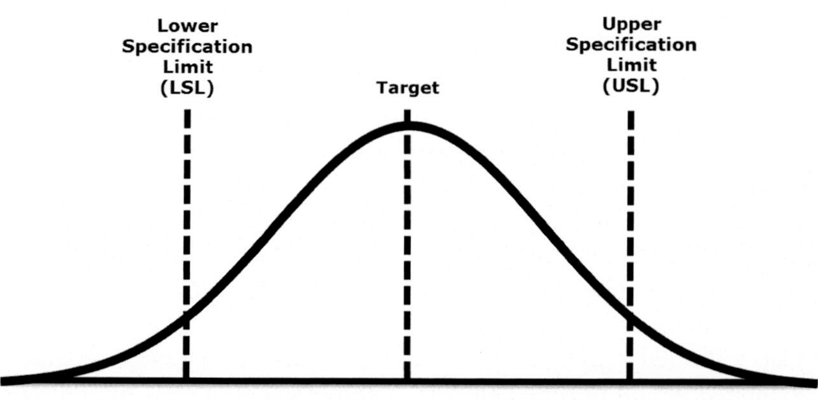

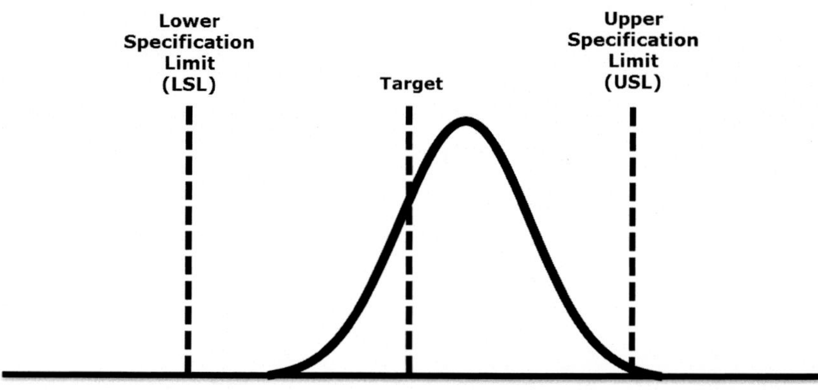

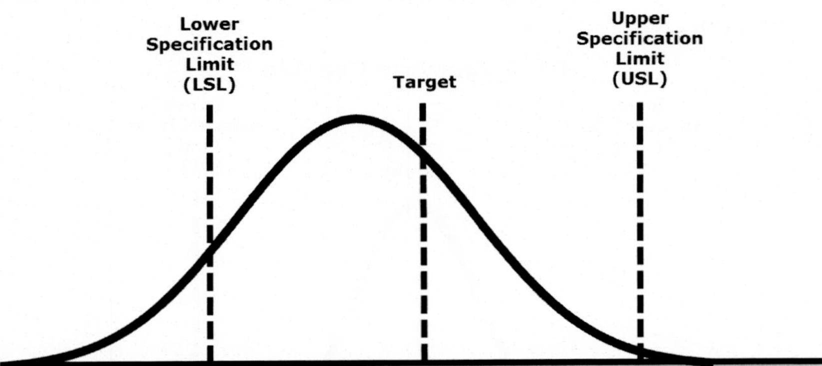

Process capability gives us visibility to the breadth of the process—a measure of the process performance in relation to the customer specifications, while also providing a perspective on the process center—a measure of the process performance in relation to a target (often the middle point between the lower and upper specification limits).

If more than a visual interpretation is needed, process capability indices are used to further quantify the process ability to meet the specifications. These indices allow the researcher to predict defect levels.

To properly set up the analysis and to measure capability, one needs the following:

1. A clearly defined customer specification. Upper Specification Limit (USL) - the maximum acceptable level and Lower Specification Limit (LSL) - the minimum acceptable level. Both of them ought to be meaningfully defined by the customer.
2. Data collected as part of a SPC plan—collected at time intervals over an extended period of time that comes from a stable process and follows a normal distribution

Two capability indices frequently used are C_p and C_{pk} (and similarly P_p and P_{pk}).

C_p and P_p relate customer tolerance (USL minus LSL) to the process variability. However, they do not take into consideration process centering.

$$C_p = \frac{\text{USL} - \text{LSL}}{6 \times \sigma_{\text{Within}}}$$

$$P_p = \frac{\text{USL} - \text{LSL}}{6 \times \sigma_{\text{Overall}}}$$

In these formulas the σ's represent process standard deviation. C_{pk} and P_{pk} take into account the location of the mean (process center), and a performance index is calculated relative to the specification limit nearest to the mean.

$$C_{PK} = \min(C_{PL}, C_{PU})$$

$$C_{PU} = \frac{USL - \overline{X}}{3 \times \sigma_{Within}} \qquad C_{PL} = \frac{\overline{X} - LSL}{3 \times \sigma_{Within}}$$

$$P_{PK} = \min(P_{PL}, P_{PU})$$

$$P_{PU} = \frac{USL - \overline{X}}{3 \times \sigma_{Overall}} \qquad P_{PL} = \frac{\overline{X} - LSL}{3 \times \sigma_{Overall}}$$

The difference between C_p and P_p and C_{pk} and P_{pk} is in the way the standard deviation used in the measure is calculated. C_p and C_{pk} represent short-term performance; P_p and P_{pk} represent long-term performance.

C_{pk} and P_{pk} quantify how many defects are being produced (product out of specification). C_p and P_p quantify the spread of the process in relation to the consumer-defined specification range, a measure of process *potential*, but they do not quantify how many defects are actually produced. This is evident in the illustration here:

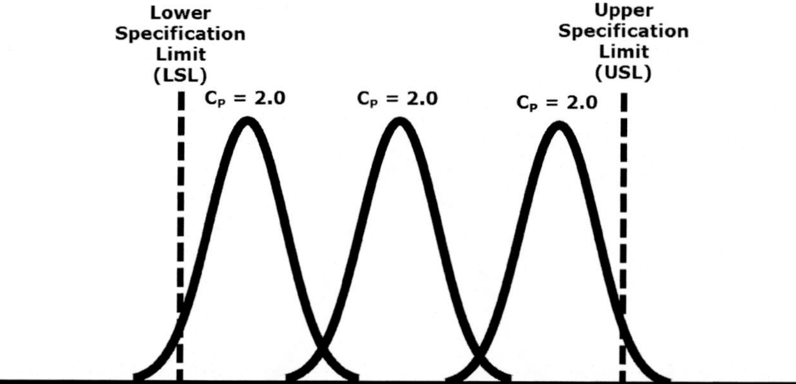

Short-term variation is sometimes thought of in very small time intervals. Process control data are often collected in subgroups in frequent time intervals, for example, five samples every 15 min. In these instances, C_{pk} uses a pooled standard deviation based on these subgroups; P_{pk} uses one overall standard deviation across the time intervals without regard for subgroupings. Subgroups can be determined by a number of factors:

- Time period: Five samples taken every 15 min.
- Lot numbers: Six samples taken from each ingredient blend.
- Tooling: Eight samples each taken from a different filler head.

Processes can vary even between short periods of time, due to ingredient changes or variation or physical sources such as heat build-up. Each process can have its own unique structure that can be a source of variation, and the data used to evaluate process capability should be collected to adequately reflect this structure. Examples of elements of structure that should be considered include filler heads, locations in ovens and driers, mixing tanks, and physical tooling such as dies.

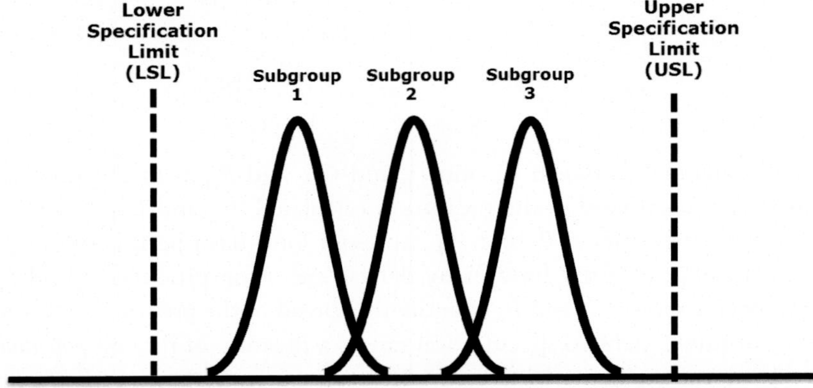

C_{PK} **is calculated using the pooled standard deviation of the individual subgroups**

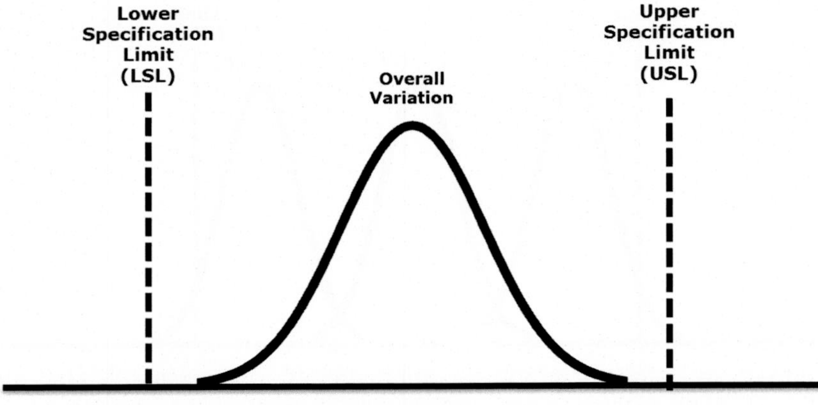

P_{PK} **is calculated using the overall standard deviation across the subgroups**

P_{pk} is also sometimes used to estimate performance over much longer periods of time to account for seasonal variation in ingredient streams and atmospheric conditions like temperature and humidity in production facilities that have little control of these elements.

C_p can be considered as a measure of process entitlement—how well we can potentially perform if our process is centered and all sources of long-term and structural variation are eliminated.

General guidelines for all of these capability measures are the following:

- Excellent process capability: measures > 1.67 ($P_{pk} = 1.67$ corresponds to an expectation of observing one defect per million units manufactured)
- Good process capability: measures between 1.33 and 1.67 ($P_{pk} = 1.33$ corresponds to an expectation of observing 63 defects per million units manufactured)
- Fair process capability: measures between 1.0 and 1.33 ($P_{pk} = 1.00$ corresponds to an expectation of observing 2700 defects per million units manufactured)
- Poor process capability: measures < 1.0

The larger the C_p index, the "tighter" the process is running compared to the specification limits. For example, a manufacturer may have a good control of the production process, or customers have a wide tolerance in their definition of acceptable product. In such situations, there may be an opportunity for a manufacturer to operate in the part of the specification range that requires a minimal cost or effort.

After Steve finishes his lecture to Maria, they consider how they could use capability measures in the performance reporting system.

Maria: "Our current SPC data collection plan is already well designed for understanding process capability. I had not thought of the idea of structure when we came up with it. It just felt right to include data from all filler heads each time, and to collect samples at roughly equal time intervals. We can easily look at long- and short-term variation."

Steve: "I think that for each line we can use data from within a shift. We can use C_{pk} to understand the performance of that specific shift and P_{pk} to report performance for the month, which will include variation between the shifts as they have different operators and also long-term variation sources like ingredient batches. C_p calculated within each shift can tell us how far we are from perfection; we can look at the individual measures and may be average them in some way to give us an overall picture."

Maria: "I think my data analysis software will let me calculate these measures. I'll design a report and show it to Lisa first to see what she thinks before we respond to Joe."

Maria pulls the most recent shift of SPC data for viscosity from one of the lines to use as a sample data set. This data has five samples at roughly equal intervals of around 25 min. Her software does indeed calculate the capability measures. She sees that she can look at measures for both short-term and long-term variations.

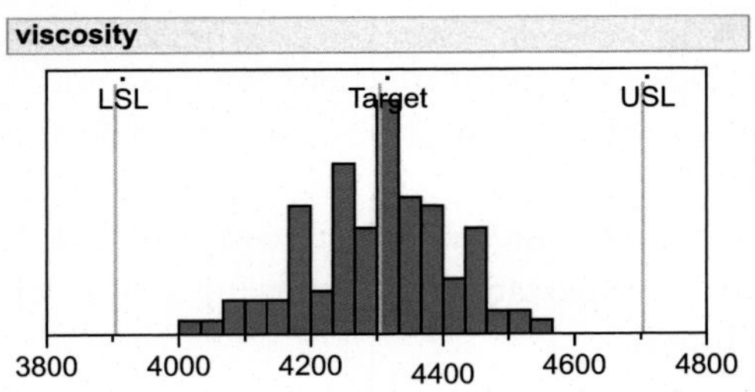

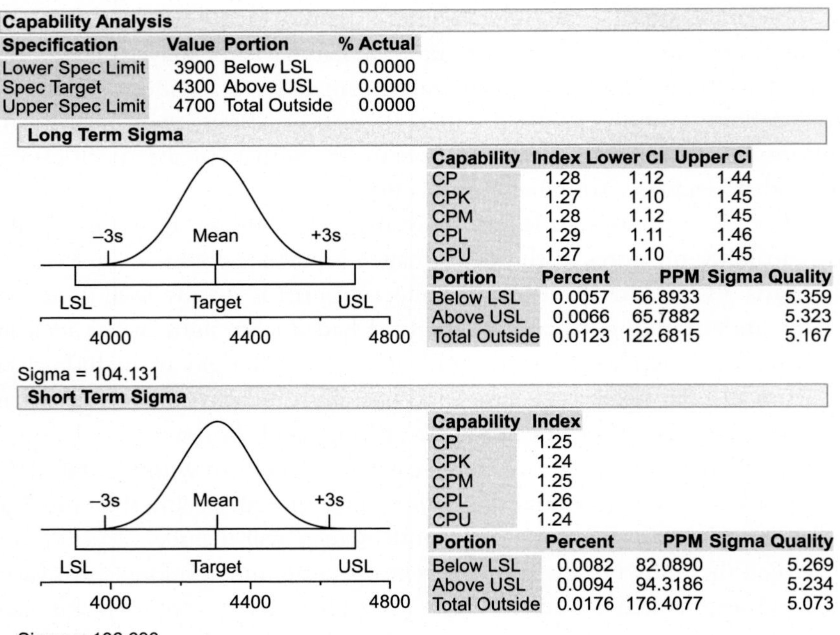

Capability Analysis

Specification	Value	Portion	% Actual
Lower Spec Limit	3900	Below LSL	0.0000
Spec Target	4300	Above USL	0.0000
Upper Spec Limit	4700	Total Outside	0.0000

Long Term Sigma

Capability	Index	Lower CI	Upper CI
CP	1.28	1.12	1.44
CPK	1.27	1.10	1.45
CPM	1.28	1.12	1.45
CPL	1.29	1.11	1.46
CPU	1.27	1.10	1.45

Portion	Percent	PPM	Sigma Quality
Below LSL	0.0057	56.8933	5.359
Above USL	0.0066	65.7882	5.323
Total Outside	0.0123	122.6815	5.167

Sigma = 104.131

Short Term Sigma

Capability	Index
CP	1.25
CPK	1.24
CPM	1.25
CPL	1.26
CPU	1.24

Portion	Percent	PPM	Sigma Quality
Below LSL	0.0082	82.0890	5.269
Above USL	0.0094	94.3186	5.234
Total Outside	0.0176	176.4077	5.073

Sigma = 106.633

Maria likes the visual display in the output. She sees that her individual samples all fall within the viscosity specification; both in the histogram with the specification limits added and in the "% Actual" for both below and above, the specification limit is zero. She also sees that the percent projected in both the long-term sigma and short-term sigma portions of the output is not zero but very small. Steve had explained that since the capability indices are calculated using the sample average and standard deviation, this could occur—in this case, the indices suggest that even though none of the individual bottles measured were outside of the specification limits, there is a small proportion of the production that likely is. Maria sees that the output uses C_p and C_{pk} in the long-term sigma and short-term sigma parts of the output, a difference from Steve's terminology. This is not a problem since the calculations are correct, but she should make sure this terminology difference does not confuse everyone moving forward.

The biggest surprise for Maria is that there is little difference between short-term and long-term performance, at least within this shift. Sigma, the estimate of the standard deviation, is slightly larger short-term than long-term. That would suggest that the variation in viscosity is due to the filler heads rather than over time. She creates a graph of the data over time to see if this is truly the case. Her graph looks like this:

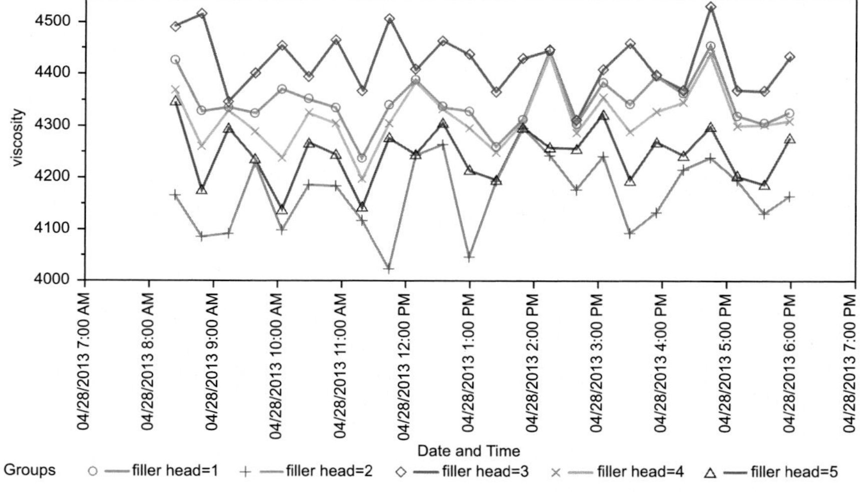

From her graph, it is obvious that the filler heads are delivering different viscosities. Filler head 3 with the diamond symbol is consistently highest, and filler head 2 with the plus symbol is consistently the lowest. There is some overlap between the other filler heads, but it seems obvious that filler head is a large source of variation, at least for the shift of data that Maria had selected for her process capability analyses.

Maria pulls data from a longer time period for that line and repeats the capability analyses and graphs the data with similar results. She then uses ANOVA to confirm her findings.

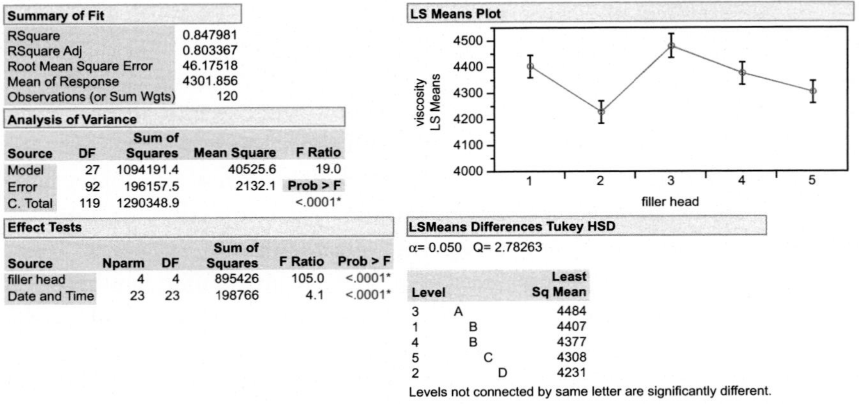

Summary of Fit

RSquare	0.847981
RSquare Adj	0.803367
Root Mean Square Error	46.17518
Mean of Response	4301.856
Observations (or Sum Wgts)	120

Analysis of Variance

Source	DF	Sum of Squares	Mean Square	F Ratio
Model	27	1094191.4	40525.6	19.0
Error	92	196157.5	2132.1	Prob > F
C. Total	119	1290348.9		<.0001*

Effect Tests

Source	Nparm	DF	Sum of Squares	F Ratio	Prob > F
filler head	4	4	895426	105.0	<.0001*
Date and Time	23	23	198766	4.1	<.0001*

LSMeans Differences Tukey HSD

α= 0.050 Q= 2.78263

Level					Least Sq Mean
3	A				4484
1		B			4407
4		B			4377
5			C		4308
2				D	4231

Levels not connected by same letter are significantly different.

The ANOVA analysis clearly shows that though viscosity does vary significantly over time, much of the viscosity variation is due to the filler heads (the sum of squares due to filler head is much greater than that for data and time). In the Tukey's analysis, she can clearly see how the filler heads viscosity averages differ; confirming that filler head 3 delivers the thickest product and filler head 2 the thinnest.

Maria is excited both with her findings on the filler head differences and the potential performance assessment report that she can provide to the operations management team. She sets up a meeting with Lisa to review.

Next morning, as Lisa walks through the plant doors with a hot cup of coffee, the receptionist tells her that she has a visitor waiting in her office.

Maria: "Good morning Lisa, I hope that you do not mind me coming in so early and waiting in your office. I have been putting together the monthly performance summary using the capability measures with Steve's guidance. And, I found something important that may help us to reduce viscosity variation. I was so eager to share the results with you that I could barely sleep last night."

"Slow down, Maria," said Lisa. "Catch your breath, or you will get both of us in trouble. Then, tell me how was Granny Maria's birthday last weekend. Was she dancing like no one is watching?"

Maria: "So sweet of you to ask about my Grandma. I think that Granny had one too many martinis with her friends, and later that night she had the time of her life on the dance floor. Rumor has it that her party this year was the best yet!"

Lisa smiles and continues the conversation: "So what have you discovered, Maria? Will it help our performance for viscosity?"

Maria shows Lisa her graph of viscosity by filler head over time and her ANOVA analysis and describes how it demonstrates that the filler heads deliver different viscosities. Lisa agrees with the interpretation.

Lisa: "I can have the maintenance crew look into the filler heads on this line to determine why this is happening. However, before I do so, it would be a good idea to see if this is the case for all of the lines. You are becoming quite a process detective!"

Maria: "I'll perform similar analyses with data from the other lines. But now, let me show you what I have been working on for the performance summary. Steve schooled me in process capability measures, and they are a perfect complement to our SPC system. SPC is typically used to establish control through increasing visibility to short-term process variability and establishing directives for corrective actions. Process capability measures allow us to monitor long-term process compliance to specifications. I think this is what Joe had asked about at the meeting the other day."

Lisa: "Yes, and it sounds like you have something for the management team. Have you had a chance to look at our viscosity data using these measures?"

Maria: "Take a look at my laptop's screen, and let me explain what you are looking at."

Maria describes the process capability output she had generated the day before and how the individual capability measures are interpreted. Lisa too likes the visual aspect of the analysis.

Maria: "It was this analysis that led me to look into the filler heads. I would have thought that short-term variation would have been much smaller than the longer term variation. When it was not, I recalled that we collected data from each of the filler heads at each time point for the SPC data collection. Fortunately, the filler head is part of that data record."

Maria continues after taking a sip from her water bottle: "First take a look at the histogram, this shows the distribution of last month's viscosity measurements for two production days exported from our SPC daily data. The LSL of 3900 and the USL of 4700 represent the specification limits that we determined from our consumer research. The target is simply the midpoint between the LSL and the USL. What do you think of our performance?"

Lisa: "I know that last month did not have any quality concerns with viscosity, and your chart suggests that we did well; nothing is outside of the specification limits."

Maria: "That is correct; we had no individual bottles with measured viscosity outside of the tolerance. But, the chart under "Long-Term Sigma" shows what we would expect from the process in a long run. Based on our current process variation, we would expect that around 120 bottles out of each million would fall outside of the specification range. Slightly more above the upper limit than the lower limit. The larger the process capability measures C_p and C_{pk} are, the less likely it is that the process will produce defects."

Lisa: "Are you saying that we shipped out of spec product?"

Maria: "The data analysis suggests that based on the measured data collected, a small proportion of production could have been out of specification. Maybe this is acceptable? If not, then we may have to do some work to reduce the viscosity variation and ultimately the proportion of product produced out of specification."

Lisa: "Maria, these measures are exactly what Joe and the management team is looking for. I love your analytical and practical approach to problem solving. I think that for this line the risk of producing product out of specification is low. I'm not sure about the other lines. Can you produce similar analyses for the rest of the lines so that we can identify where our biggest opportunities are?"

Maria: "Yes! With these analyses we will be able to make some intelligent decisions as to where to put our resources for maximum impact."

Lisa: "If we can implement these in a report that we can call up for any line for any time period, then we will have fulfilled Joe's request. I cannot describe to you how happy I am with the approach that you shared with me. It seems that process capability is a quality professional's 'best friend.' I wish that I knew about it years ago."

Maria: "Thank you for your time, Lisa, and for the opportunity to implement these measures at the plant. I'll work with Steve and our plant IT applications developer to create the reporting structure."

CHAPTER 10

Design of Experiments Foundation

Maria's process control strategy has been in place for a number of months now with mixed results. In many instances, when viscosity starts to drift, the process changes in steam pressure based on her regression equation are effective. But in other instances, this strategy does not seem to work, and the operators are making adjustments to the steam pressure without improving performance.

Maria recalls that when she had process and ingredient data collected to determine her process control strategy, several of the measures were related. Maybe controlling viscosity is not as simple as monitoring and adjusting steam pressure. She repeats the data collection exercise she performed when she first developed regression relationships and reaches similar conclusions. Steam pressure has the strongest relationship with viscosity, and the ingredient and process measures are correlated with each other.

Maria sets up a call with Dr Wang to discuss her dilemma. Dr Wang is as usual accommodating to her "favorite BBQ sauce student."

Dr Wang listens patiently to Maria's update, and then says, "I'm impressed with how much you have been using your data analysis skills!"

Maria thanks her and continues, "While I have made some great advances in our understanding of viscosity, I think I may have run into a wall. The ingredient and processing variable relationships are probably more complicated than I thought."

"I agree that the variable relationships are probably quite complicated," said Dr Wang. "The best way to understand how they work is to study them independently and see how they work together. How familiar are you with the approach of statistical design of experiments?"

"You may have mentioned this in one of our classes," Maria tried to search her memory, "but otherwise, I know nothing."

"I'm really not an expert in this area," said Dr Wang. "You would be better served taking a class so that you will have a proper understanding. I'll send you links to some that I know to be good."

Maria receives information on classes in statistical design of experiments from Dr Wang and sees that one is being offered nearby in the next few

Statistics for Food Scientists
http://dx.doi.org/10.1016/B978-0-12-417179-4.00010-X

weeks. In the class description, she sees that they will use the same statistical software that she has been using for her recent data analysis work. She checks with her manager Steve, who readily agrees that this will be time and money well spent.

Maria returns from her three-day class full of excitement. She creates another cheat sheet that summarizes what she learned from the course to help prepare her for a discussion with the team.

Maria summarizes her key takeaways from class as follows:

- In a statistically designed experiment, factors (usually ingredient levels and/or processing conditions) are systematically varied so that their effects can be quantified in the most efficient way.
- Statistical analysis of designed studies focuses on the development of models that relate how formulation and/or process variable changes affect the responses.
- These models can provide a wide variety of information, such as the following:
  - An ordering of important formulation/processing variables
  - Optimal product formulations (subject to cost and production constraints, if applicable)
  - Consumer-based specifications
- There are a number of experimental types each with a different objective:

  Screening designs are used to collect data to quantify the effect of factors, with a goal of either reducing the number of factors under investigation or identifying factor ranges for further study.

  Response surface designs are used to collect data to develop a detailed model of the space defined by the factor ranges. These models can be used to determine optimal factor settings.

  Mixture designs are special forms of screening or response surface designs where the factor levels are constrained to a sum total.

  Robust (Taguchi) designs are used to determine the factor settings that reduce end product variability due to uncontrollable noise.

In the class, she saw specific examples of each of these design types. The instructor led the class through the process of creating the designs for each of the examples using the statistical software and then provided data to follow through the data analysis for each example. Maria is eager to use the design of experiments approach to better understand how elements of the BBQ production process affect viscosity.

She recalls the list of formulation and processing variables that her team had come up with in their first attempt to better control viscosity. There were three ingredient variables: sugar level, tomato paste level, and fruit paste level. There were also three processing variables: pump pressure, steam pressure, and line speed. When she had analyzed the process data collected over a period of time, there were correlations between many of these variables. But they can be independently changed for her experiment, which is an important consideration in the statistical design of experiments approach.

Maria ultimately wants to find the combination of these factors that will produce the target viscosity and also develop a control strategy that will work consistently. In other words, she realizes that she wants to create a response surface design and use the resulting model to determine the optimal factor settings. At this point, she does not want to rule out any of the six ingredients and process variables as potential factors in any experiments she conducts on the BBQ process.

Using her software, she looks at her options for a six-factor response surface design. The smallest design she sees is very large with 46 runs—this means 46 different combinations of the six ingredient and processing variables. It seems like a lot of work, and she is not even sure if it can be done. In particular, she doubts that production can be stopped for experimentation for very long. She decides to discuss the matter with the team at the morning production meeting the next day.

The next morning Maria shares with the team, "I have an approach for determining how to better keep viscosity under control. It will require us to stop production and perform an experiment where we systematically change six factors in a very specific pattern. The factors are: sugar level, tomato paste level, fruit paste level, line speed, pump pressure, and steam pressure. There will be 46 different combinations of these six factors. Is this something we can do?"

Alex is the production supervisor for the daytime shift. His eyes widen considerably when Maria mentions 46 combinations.

He says, "I guess anything is possible, but what you are suggesting would mean stopping production 46 times to make changes. Then we will need to run the line at the new levels long enough for the line to stabilize before we collect the samples to measure for viscosity. I would guess that each run would take around 30 min. That is a lot of production interruption! Will the product you produce with all of these changes be something that we can sell? Will viscosity be in specification?"

"I don't know for sure," answers Maria, "So we should probably assume that much of it will not be within specification. So there is a cost to this work for sure. But what we learn from this experiment will mean that future production will be better, so there is a cost advantage in that."

"I will check with Lisa, the plant manager," says Alex. "I'm sure she will see the benefit to some experimentation, but I think she will ask if something less costly will get us the information we need."

Maria hears back from Alex later in the day. He confirms what he had thought the reaction from Lisa would be. She is willing to devote some resources to experimentation, but would like some additional options from Maria.

Maria had already been thinking that this was a possibility. She had recalled from her class that screening designs were often used as a step before a response surface design to eliminate factors that had little impact. She uses her software to look into screening design options for six factors.

Screening designs typically have two levels for each factor. The designs are factorial combinations of these levels. A full factorial design will have all possible combinations for each factor. For Maria's six factors this means $2^6 = 64$ combinations or design runs. But many screening designs are fractions of the set of all possible combinations. A specially determined fraction may provide sufficient information about the individual factors. These fractions are most always in powers of two: $^1/_2$, $^1/_4$, $^1/_8$, and so on. For Maria's six factors, a half fraction would be 32 combinations or runs, a quarter fraction would be 16 runs, and an eighth fraction would be eight runs.

Maria recalls from her class that the concept of design resolution was important in determining design fractions. This concept considered the idea of factor interactions. A factor interaction means that the effect of one factor is different for different levels of another factor. For Maria, an example would be that the effect of steam pressure would be different for different levels of pump pressure. She tries to imagine what this would look like.

From her earlier work, Maria had determined that increasing steam pressure caused a decrease in viscosity. An interaction between steam pressure and pump pressure would mean that the rate of decrease would be different for different pump pressures. A graph of this might look like this:

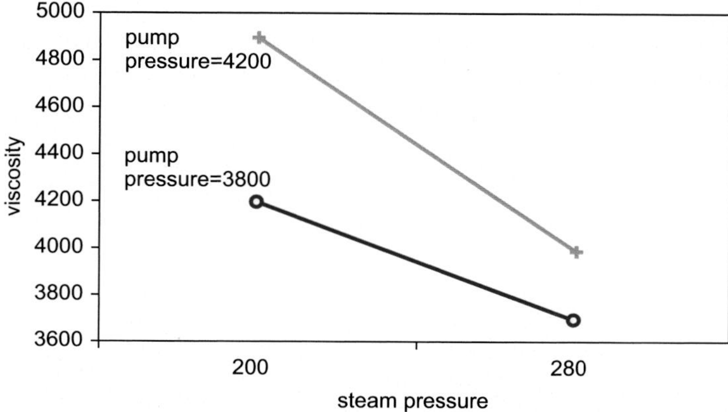

Maria considers that each factor could interact with any other factor. With six factors, this means that there are 15 potential pairwise interactions! Maria thinks that though there is the potential for this to happen for some of the factor pairs, the goal of her screening stage is to identify the most important factors. If there are interactions between these important factors, then they would be identified in the subsequent response surface design.

The screening design options are identified by their resolution in her software. Resolution is listed as a number: 3, 4, 5, or greater. She looks back at her notes from her class and finds these definitions:

- Resolution 3 designs: Main effects are not confounded with other main effects but are confounded with pairwise interactions.
- Resolution 4 designs: Main effects are not confounded with other main effects or pairwise interactions; pairwise interactions are confounded with each other.
- Resolution 5 designs: There is no confounding between main effects, between pairwise interactions, or between main effects and pairwise interactions.

Confounding is when effects cannot be separated from each other. To her, it seems that a resolution 4 design would be best, since she is most interested in the main effects to reduce the number of factors for her second design. At least the main effects would not be confounded with any pairwise interactions.

Her software output shows that there is a 16-run resolution 4 screening design for her six factors. This seems like a good option for her first design. She also looks into her options for response surface designs with fewer than six factors. A design with just two factors could be as small as 10 runs. A design with three factors could be as small as 16 runs. If with screening design she can eliminate half of the factors for the next phase, then she would have reduced the total number of runs for experimentation from 46 runs to 32 or less. This two-step approach seems like a better strategy.

At the next morning's production team meeting, she sees that Lisa, the plant manager, has joined the meeting. Maria presents her new experimentation strategy to the team.

Maria explains, "If we conduct our experimentation in two stages, we will likely need fewer resources and have less interruption in production. Stage one would consist of a screening design with all six factors, each at a low and high level. This design would have 16 runs and the goal would be to identify the factors that are most important in affecting viscosity. Stage two would consist of a response surface design with only the important factors identified in stage one. I'm hoping that we can focus on three or fewer at this stage. If so, the second design could be as few as 16 runs. So our total effort will be around 2/3 of the 46 runs I had originally proposed."

"I like this approach," Lisa says. "We do need to fix this viscosity problem, and this is a much more efficient plan. According to Alex, these two experiments would each take 8 h to execute. I'm willing to devote one of the production lines for a full shift for each experiment. Alex, would you please work with Maria to set this in motion."

Maria is pleased that she can proceed. She and Alex agree to meet later in the day to work out the details of the design and make a schedule for the execution.

CHAPTER 11

Screening Experimental Designs

Maria and Alex meet the next morning to plan the screening design. Maria had already decided on a 16-run resolution 4 design for her six factors. She and Alex will also need to decide on the ranges for each factor.

Maria: "A while back, I had one of the operators collect both ingredient and process variable measurements during production. The variation in each of these might help us set a range for these factors in the design."

Alex: "Over how much time was that data collected? Do you think it represents production overall?"

Maria: "It was just one production shift. It would be good to know how representative that is. For the processing variation, it may be representative, but I don't have a good idea about ingredient variation."

Alex: "I think the operators would have some insight. Let's ask a few of them."

Maria and Alex walk onto the plant floor and visit with a few of the operators for the production lines. There is consistency to what they hear:

1. For ingredients: Tomato paste, fruit paste, and sugar levels can all vary quite a bit, even within a single shift of production. Actually, the operators change these levels based on how they think a batch looks in the mix tank.
2. For processing factors: The operators change the pump pressure as they see fit to make a good product. Lately they were instructed to adjust steam pressure according to the SPC plan. But, prior to that they had used their intuition to adjust the steam pressure. Line speed varies based on line performance, but higher line speeds are desirable to meet production targets.

Armed with this information Maria and Alex again discuss the factor ranges. They both agree that the data that Maria had already collected probably represents typical variation for these factors. If anything, the

http://dx.doi.org/10.1016/B978-0-12-417179-4.00011-1

variation may be more due to the operator adjustments than anything else!

Maria then looks back at the process data she worked with earlier. She calculates the ranges of each of the measures to come up with these ranges for the design:

- Steam pressure: 200–280
- Pump pressure: 370–420
- Line speed: 22–40
- Sugar level: 60–66
- Tomato paste: 1000–1100
- Fruit paste: 7–13

Maria creates her design based on these levels. It looks like this:

steam pressure	pump pressure	line speed	sugar level	tomato paste	fruit paste
200	370	22	60	1000	7
200	370	22	66	1100	13
200	370	40	60	1100	13
200	370	40	66	1000	7
200	420	22	60	1100	7
200	420	22	66	1000	13
200	420	40	60	1000	13
200	420	40	66	1100	7
280	370	22	60	1000	13
280	370	22	66	1100	7
280	370	40	60	1100	7
280	370	40	66	1000	13
280	420	22	60	1100	13
280	420	22	66	1000	7
280	420	40	60	1000	7
280	420	40	66	1100	13

There are two levels for each of the factors. For steam pressure, the first eight runs are at the low level of 200, and the next eight runs are at the high level of 280. For pump pressure, the first four runs are at the low level of 370, and the next four are at the high level of 420; then this pattern repeats for the remaining eight runs. There are also repeating patterns for line speed and sugar level, and somewhat odd patterns for tomato paste and fruit paste.

Maria recalls from her class the importance of randomizing the experimental design runs. This ensures that any unknown factors that may influence the viscosity will not be mixed up with the factors in the design. If for some unknown reason the viscosity increases or decreases over the time the design runs are executed, then with the design above the increase or decrease would look like an effect of steam pressure. Maria is fairly sure that the design runs could be executed in any order but thinks it is best to confirm this with Alex.

Another concept that Maria recalls from her class is replication, or repeating certain design runs. This is an important check—the resulting data from these runs should be similar, and if they are not, this would indicate a problem in the execution of the design runs or possibly with the measurement system. Often the repeated runs are at the center of the design space so that the design is still balanced. This factor level combination is called the center point, the middle level of each factor. This will certainly add to the size of the design, but Maria feels that this is important enough to devote additional resources.

She discusses both randomization and replication with Alex. He agrees that replication is important here since this is the first time that any experimentation has been done at the production facility.

Alex: "I'll need to confirm with Lisa that we can make commit the additional resources, but I think that she will understand the importance too. It will add an hour to the shift when we run the experiment, so I'll ask the operators to stay on to complete the experiment. Executing the experiment in a randomized order should not be a problem for these factors. Fortunately, we are not changing the cook temperature in this design, since we would want to make as few changes as possible to that. Any time we change cook temperature, it takes nearly 30 min to reach the new set point. That would virtually double the execution time for each design run."

Maria: "I think in class the instructor mentioned that randomization can sometimes be restricted for these types of situations. I'm glad we don't have that complication for our very first designed experiment!"

A short time later, Alex emails Maria confirming that Lisa has agreed to the added center point replicates. He adds that after consulting the production schedule, the experiment can be executed the following week. Maria creates a modified version of her earlier design with the added runs and randomizes the run order.

steam pressure	pump pressure	line speed	sugar level	tomato paste	fruit paste
200	370	22	66	1100	13
280	370	40	66	1000	13
280	420	22	60	1100	13
240	395	31	63	1050	10
200	420	40	66	1100	7
280	370	22	66	1100	7
200	420	40	60	1000	13
200	370	40	60	1100	13
200	370	40	66	1000	7
280	420	40	66	1100	13
280	420	22	66	1000	7
240	395	31	63	1050	10
280	370	22	60	1000	13
200	370	22	60	1000	7
280	420	40	60	1000	7
200	420	22	60	1100	7
280	370	40	60	1100	7
200	420	22	66	1000	13

Maria wants to leave nothing to chance for her first experiment. She is confident that Alex will ensure that the execution on the line will be conducted properly. She wants the same for the viscosity measurements that will be the response in the experiment, so she sets up time to discuss this with Chad, the analytical scientist she had worked with a few months back on the viscosity measurement system. She describes the upcoming experiment and the upcoming execution process.

Chad: "I can see why you will want to ensure that you have reliable data since a considerable amount of resources will be spent on the execution. If Lisa is willing to devote these resources, she is expecting that the resulting information will be useful. I can make the needed viscosity measurements for your study rather than interrupting the daily operations of the plant analytical lab. We should discuss the experimental execution and how many samples you will need to have measured."

Maria: "I had not thought much about the number of samples we would be collecting. Since Alex is coordinating the execution of the design runs we should include him in this discussion. I'll set up a meeting where we can go through the details."

Alex responds to Maria's email request suggesting a meeting the following day, mentioning that he needs more time to work out his execution plan. When Maria, Chad, and Alex meet late the following morning, Alex has a notepad with several sheets filled.

Alex: "It took some work to figure out how best to execute the design runs, but I like the plan I've worked out. Connie is probably our best operator, so I want to use her and her production line. Her line is scheduled for sanitation next Tuesday night, so we can execute the experiment on Wednesday during her shift. I want us to start with a clean line since we are changing ingredient levels along with the processing conditions."

Alex continues: "We will create a separate batch for each design run with the ingredient levels specified in the design. The mixing and cook cycle should take about 15 min for each batch. We will pump each vat to be bottled with the steam pressure, pump pressure, and line speed specified in the design. That too will take about 15 min. Since each vat is emptied to a holding tank before bottling, we can rinse the batch tank and fill it with the ingredients for the next design run while the previous design run is being bottled. While the next batch is mixed and cooked, we will rinse the bottling line and set it at the steam pressure, pump pressure, and line speed for the next design run. We will repeat this sequence until we have completed all of the design runs."

Maria: "Wow you have really thought this through! It sounds like a great plan, but how does Connie feel? She may have a very long day if this does not go as planned."

Alex: "Connie is committed to seeing the execution through. She will likely get some overtime pay and also knows that if we can better understand how to control the production lines it will make her life easier in the future."

Chad: "I too am committed to seeing this design through regardless of how long I have to stay to make the measurements. We need to determine how many samples and how to collect them during the bottling process. I'm sure you are expecting more than one bottle measured for each design run."

Maria: "I'm sure there is some level of variation in viscosity between the bottles no matter how well we control the mixing, cooking, and bottling. For our process monitoring, we collect samples from each of the five filler heads each time. Since it takes around 15 min to bottle a batch, maybe we can collect 10 bottles, one sample from each filler head at two times during the bottling and have Chad measure them. Is that a reasonable ask?"

Alex feels that this will not be a burden for Connie to pull the samples, and Chad has no objections to the number of measurements he will perform, so all are agreed to the data collection plan. Next Wednesday looks to be a very busy day!

Maria: "I'll plan to spend the entire day with you all. If something unexpected comes up, I can lend a hand."

The design execution day turns out to be very intense, with a lot of work. And while the execution did not go perfectly and a few times they had to stop the line to make the factor adjustments, there were no major issues. Connie, Alex, Chad, and Maria felt good about the day overall and hoped that they will have useful information.

Maria transfers Chad's viscosity measurements for each of the design runs to her spreadsheet. She has 10 viscosity measurements for each of her 16 design runs. The raw data looks like this:

Sample	V 01	V 02	V 03	V 04	V 05	V 06	V 07	V 08	V 09	V 10
1	4588	4585	4599	4607	4595	4581	4601	4596	4577	4595
2	4639	4641	4648	4639	4654	4637	4647	4644	4644	4648
3	4711	4694	4710	4715	4694	4700	4703	4704	4687	4700
4	4500	4483	4487	4483	4493	4478	4468	4481	4477	4504
5	5091	5093	5101	5099	5086	5083	5084	5088	5090	5071
6	5063	5074	5047	5071	5058	5072	5061	5082	5062	5065
7	4939	4934	4947	4959	4958	4967	4960	4938	4958	4948
8	5046	5052	5042	5047	5049	5034	5040	5029	5055	5042
9	4164	4163	4174	4160	4167	4165	4161	4155	4166	4179
10	4389	4379	4374	4387	4393	4378	4369	4375	4363	4346
11	3936	3931	3927	3944	3937	3947	3924	3918	3936	3934
12	4098	4132	4116	4123	4109	4100	4115	4106	4110	4122
13	4606	4598	4590	4615	4608	4604	4598	4615	4593	4591
14	4340	4356	4364	4368	4350	4367	4360	4356	4379	4367
15	4285	4280	4296	4280	4279	4282	4281	4286	4292	4267
16	4283	4282	4291	4282	4286	4279	4301	4275	4296	4297

She again starts by visualizing the individual viscosity measurements for each of her 16 design runs. She creates histograms of the 10 viscosity measurements by run.

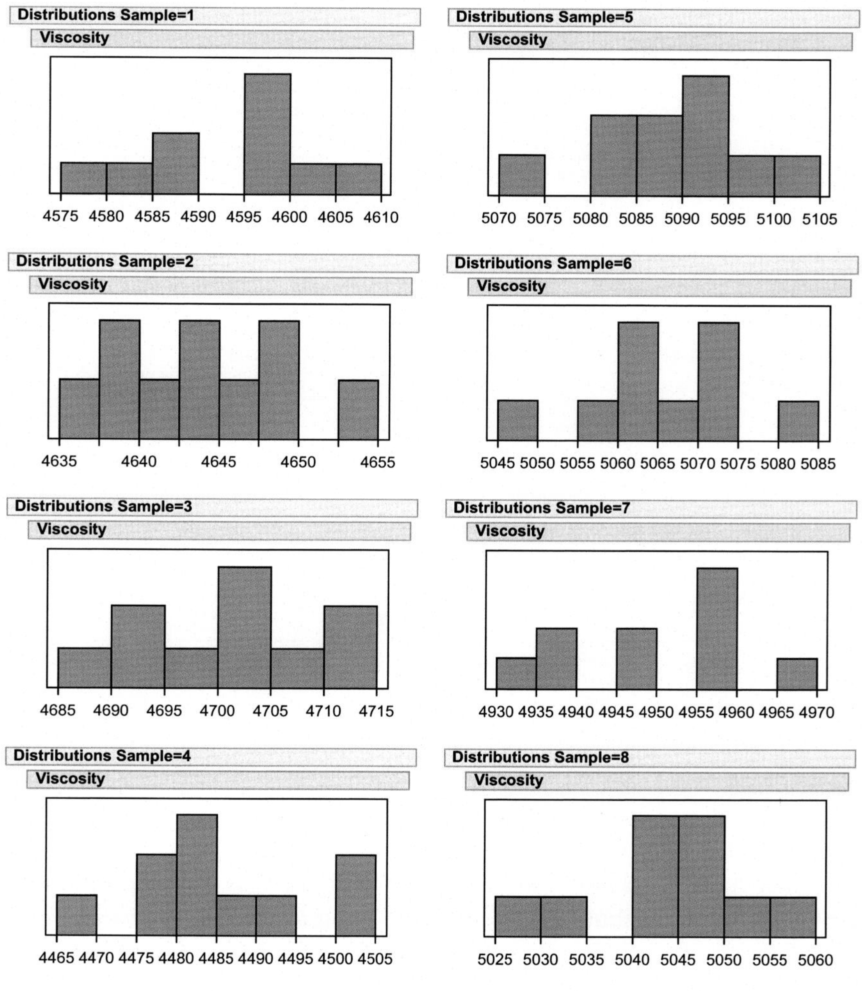

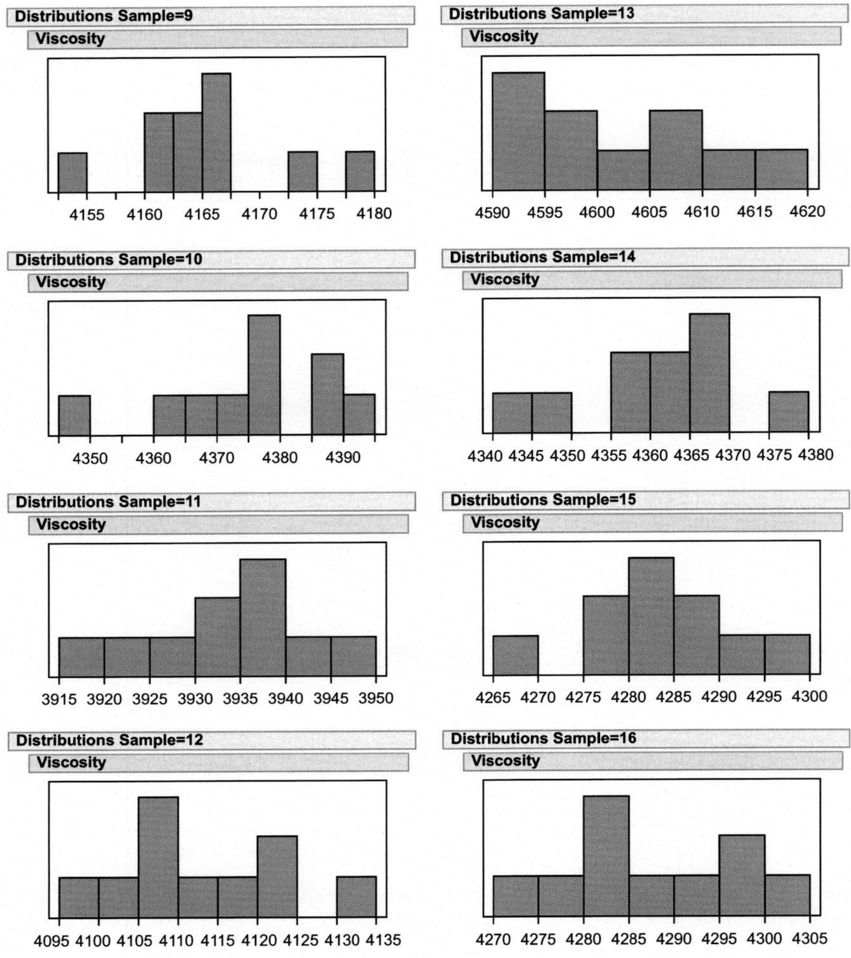

Maria looks at these histograms to see if there is any indication of skewness in the distributions. She is also hoping to see that the variation in measurements for the samples is similar. The graphs have very different ranges since they also have very different centers (due to the factors that were varied in her experiment she hopes!), so it is difficult to see if this is true. She quickly calculates the average, range, and standard deviation in viscosity measurements for the 16 samples and arranges them in a table. Her table looks like this:

Sample	Average Viscosity	Range	Standard Deviaton
1	4592	30	9
2	4644	16	5
3	4702	28	9
4	4485	36	11
5	5089	30	9
6	5065	35	10
7	4951	33	11
8	5044	26	8
9	4165	24	7
10	4375	47	14
11	3933	29	9
12	4113	34	11
13	4602	25	9
14	4361	38	11
15	4283	29	8
16	4287	26	9

The ranges and standard deviations for each of the 16 design runs are similar enough that she feels comfortable in thinking that there are no substantial differences in variability between the runs. The averages are considerably different; these averages are what she intends to use as a response for modeling the factor effects. Maria is happy to see these differences; otherwise, the factors in her design do not likely affect viscosity!

For her resolution 4 design, the intended model for Maria's design has main effects for each of the six factors but no pairwise interactions. Maria fits this model to her experimental data.

Summary of Fit

RSquare	0.93
RSquare Adj	0.89
Root Mean Square Error	120.28
Mean of Response	4543.25
Observations (or Sum Wgts)	16.00

Analysis of Variance

Source	DF	Sum of Squares	Mean Square	F Ratio
Model	6	1795714	299286	20.7
Error	9	130215	14468	**Prob > F**
C. Total	15	1925929		<.0001*

Sorted Parameter Estimates

Term	Estimate	Std Error	t Ratio	Prob>|t|
steam pressure(200,280)	–278.3	30.1	–9.25	<.0001*
pump pressure(370,420)	166.9	30.1	5.55	0.0004*
line speed(22,40)	–68.5	30.1	–2.28	0.0488*
tomato paste(1000, 1100)	41.2	30.1	1.37	0.2036
fruit paste(7,13)	22.9	30.1	0.76	0.4651
sugar level(60,66)	3.6	30.1	0.12	0.9067

Maria's software orders the factor effects in decreasing levels of significance. These factor effects have been standardized to the range of each of the factors so that the size of their effects can be directly compared. She sees that steam pressure is the most significant effect, followed by pump pressure, and then line speed. The ingredient factors of tomato paste, fruit paste, and sugar level all have effects that are not significant as their *p*-values are all much greater than 0.05, the level commonly used as a cutoff to determine significance. The direction of the factor effects can be seen in the signs of the estimates, but Maria produces a plot to better visualize not just the direction of the effect but also the relative impact of each factor.

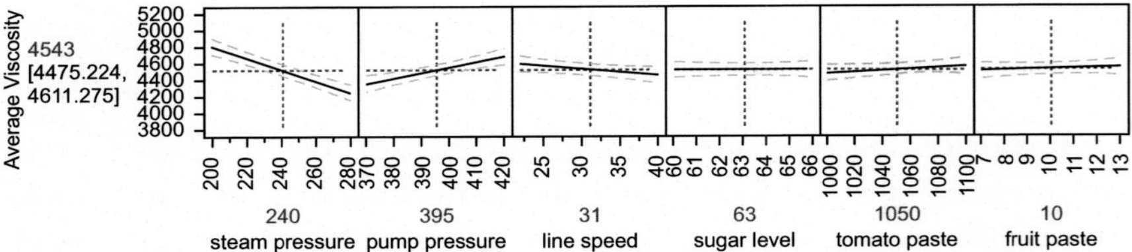

She can see from the slopes of the lines for each factor the different impacts. Steam pressure and pump pressure have pronounced effects. The effect of line speed is somewhat modest. The ingredient level effects are not significant. This is a very important finding for Maria since the ingredient levels only change for each batch, but the processing factor levels can be adjusted as each batch is filled. This suggests more opportunities to control the process to deliver the desired viscosity.

There is one more step in the analysis that Maria recalls from her class—verifying that her model is not missing anything critical by analyzing the model's residuals. Residuals are the difference between the predicted viscosity from her model and the actual viscosity measurements of her design runs. Though these model predictions may not be very accurate for a screening design like hers, there should be no pattern in them if she plots them against the order of the design runs and the predicted values from the model. They should follow a normal distribution centered at zero. Maria's software allows her to easily create these plots since they are an important step in the analysis process.

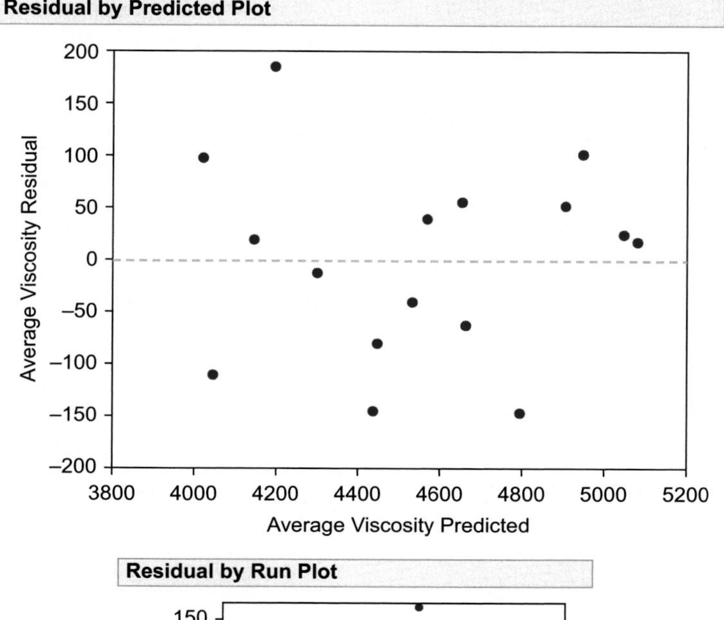
Residual by Predicted Plot
Average Viscosity Residual
200
150
100
50
0
−50
−100
−150
−200
3800
4000
4200
4400
4600
4800
5000
5200
Average Viscosity Predicted

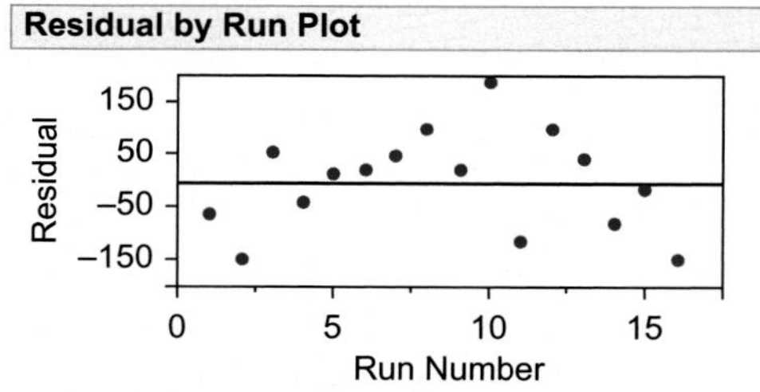
Residual by Run Plot
Residual
150
50
−50
−150
0
5
10
15
Run Number

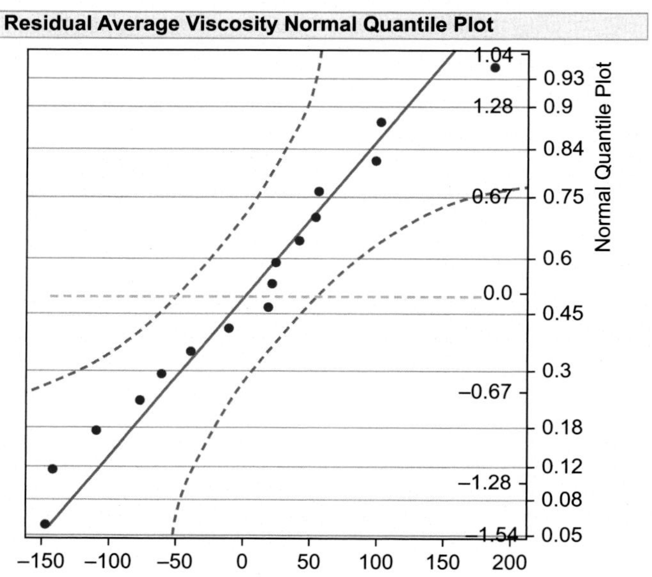
Residual Average Viscosity Normal Quantile Plot
Normal Quantile Plot
1.04
1.28
0.67
0.0
−0.67
−1.28
−1.54
0.93
0.9
0.84
0.75
0.6
0.45
0.3
0.18
0.12
0.08
0.05
−150
−100
−50
0
50
100
150
200

If in these plots Maria was to see any kind of pattern, it would indicate that something in addition to the factor effects might have influenced the viscosity data from her design. A pattern could be a steady increasing or decreasing of the residuals over the ranges, or more or less variation in them across the ranges. She sees no pattern in either the residual by predicted or residual by run number plot. Maria also creates a normal quantile plot of the residuals. This is a special type of plot to assess if a set of data follows a normal distribution, with the data values on the horizontal axis and normal quantiles on the vertical axis. The points should fall on a straight line if they do follow a normal distribution. Maria sees this is the case in her plot, and that the residuals are more or less centered at zero. Her residual analyses leave her feeling confident that the viscosity differences in her experiment are due to her experiment's factors and nothing else.

Maria collects her data analysis output to create a report for the team. She also starts thinking about her subsequent design to optimize her process and determine the factor levels that deliver the target viscosity. She is not sure whether this new design she should vary only steam pressure and pump pressure or should also include line speed. It would certainly be simpler to vary only the first two. She is thinking that this will be her suggestion for the team. She sets up a meeting with Alex, Chad, and Connie for the next morning to discuss the results.

Maria shares her analysis results with the team and concludes by stating:

> "Our design has given us some clear direction on what strongly influences viscosity. Steam pressure, pump pressure, and line speed all have significant effects on viscosity. All the ingredient factors have minimal effects. The effect of line speed is not nearly as large as the other two, so maybe we do not need to vary that in the next experiment. I can design a follow-up experiment using the three or just the two. What do you all think?"

Alex and Chad are quick to point out that eliminating line speed will make the next experiment much smaller. "Lisa will be much happier with less interruption," Alex points out.

> Connie speaks up: "I have noticed that line speed sometimes seems to be important on certain days when I'm operating the line. Also, we sometimes want to run at higher speeds to meet product demand, especially in the warmer weather months when people are barbecuing more. I think we should continue with this factor in the next design so that we will better understand if it interacts with pump pressure and steam pressure. And I'm willing to do the extra work for a larger design."

Alex: "Connie makes a good point. I think if I explain this reasoning to Lisa, she will commit the resources for the larger design. How many runs would we need for three factors versus two?"

Maria: "16 runs would be the minimum for three factors versus 10 runs for two factors. When I first proposed the approach of screening and then the response surface design, I had suggested that if we could get down to three factors for the second design we would still be better off than a 46-run response surface design with all six factors. Please remind her that what we are asking her for now is no different than that!"

Alex emails the team later that day stating that Lisa has agreed to move forward with the three-factor, 16-run design and that he will set up time with Maria to plan this next step.

CHAPTER 12

Response Surface (Optimization) Experimental Designs

Maria and Alex meet to discuss the response surface design. Again, the first step is to determine the factor ranges.

Alex: "I think we can keep the same ranges from the screening design, since the viscosity data was more or less centered around our target of 4300."

Maria: "I think we will have some choices to make in the design. Let me work with my software and see what I can come up with and we can meet to discuss the options."

Maria starts by looking at the design possibilities for three factors. She sees that the two choices with the smallest number of runs have 15 or 16 runs. She recalls that when she had first looked at this in designing the overall experimental approach, the designs had different names. At that point, she had not looked into the difference between the two, but she had suggested 16 as the minimum number of runs needed since one more run would not make much of a difference in the overall effort. Now, she needs to better understand her options to make the best choice of a design.

Two typically used response surface designs are the central composite design (CCD) and the Box-Behnken design. Both can be used to fit the quadratic response surface model, which has linear and curvilinear (quadratic) effects for each of the factors along with all pairwise interaction effects. But the designs look different, so she wants to understand the differences. And with the CCD, there are some choices that she also needs to understand.

Maria creates a three-factor Box-Behnken design and a three-factor CCD in their generic form with −1, 0, and 1 coding for the low, medium, and high levels. She has the designs created in the standard order so that she can more easily compare them.

Statistics for Food Scientists
http://dx.doi.org/10.1016/B978-0-12-417179-4.00012-3

The Box-Behnken design looks like this:

Run	X1	X2	X3
1	-1	-1	0
2	-1	1	0
3	1	-1	0
4	1	1	0
5	0	-1	-1
6	0	-1	1
7	0	1	-1
8	0	1	1
9	-1	0	-1
10	1	0	-1
11	-1	0	1
12	1	0	1
13	0	0	0
14	0	0	0
15	0	0	0

The CCD looks like this:

Run	X1	X2	X3
1	-1	-1	-1
2	-1	-1	1
3	-1	1	-1
4	-1	1	1
5	1	-1	-1
6	1	-1	1
7	1	1	-1
8	1	1	1
9	-1	0	0
10	1	0	0
11	0	-1	0
12	0	1	0
13	0	0	-1
14	0	0	1
15	0	0	0
16	0	0	0

She immediately notices some differences in the two designs. First is that the number of center points is different, three for the Box-Behnken versus two for the CCD. But the biggest difference is how the other design runs are structured. In the CCD, there are runs that have all low or high levels for all three factors. For example, in row 1 the levels for all three factors are set to -1. There are eight runs like this—they are the factorial combinations of the low and high levels for each of the factors. For the Box-Behnken design, every run contains the middle level for at least one of the factors. Maria thinks about the space that the design covers are represented as a cube. She draws a cube and then draws circles where the design points fall in the cube for both design types. She compares the two drawings and sees that for the CCD, the design reaches to the corners of the cube, but the Box-Behnken design does not.

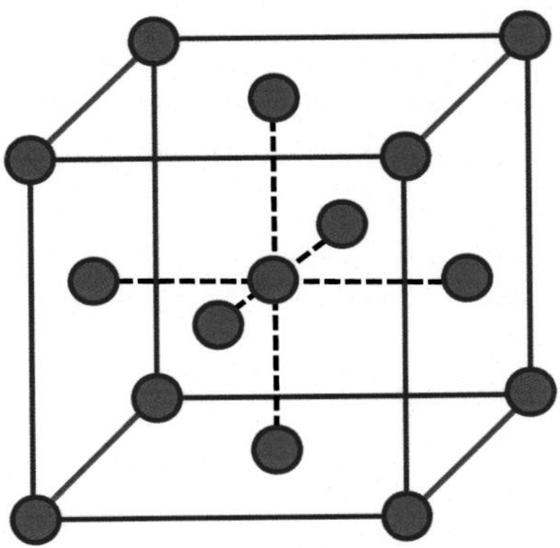

Central Composite Design

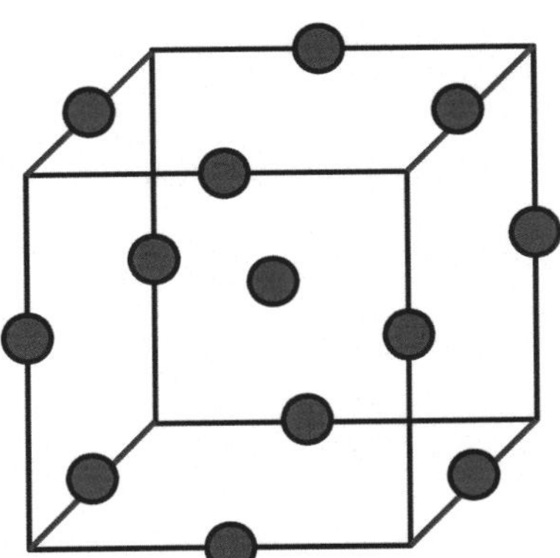

Box-Behnken Design

"I think I would want my design to cover as much area as possible," Maria thinks aloud. She decides to create a CCD for her factors.

Maria sees in her software that she has a choice to make with the construction of her CCD. There are three types of points in the design:

Factorial points that are high and low-level combinations for each of the factors, which are like in her earlier screening design. These are the first eight runs in her design above.

Axial points, where the levels of two of the three factors are at the middle, and the third factor is set to either a low or high level. These are runs 9–14 in her design above.

Center points that are the middle level of all factors at the same time, which are also like in her earlier screening design. These are runs 15 and 16 in her design above.

There is a choice to be made as to the level of the axial points, how low or high that the level is set. Her choices are on face, rotatable, and orthogonal. The design she has already created has the axial points on face. This means they are in the middle of the face of the six sides of the cube that defines the design region. She creates designs with the rotatable and orthogonal axial values to compare.

Run	X1	X2	X3
1	-1	-1	-1
2	-1	-1	1
3	-1	1	-1
4	-1	1	1
5	1	-1	-1
6	1	-1	1
7	1	1	-1
8	1	1	1
9	-1.68	0	0
10	1.68	0	0
11	0	-1.68	0
12	0	1.68	0
13	0	0	-1.68
14	0	0	1.68
15	0	0	0
16	0	0	0

The CCD with the orthogonal axial values looks like this:

Run	X1	X2	X3
1	-1	-1	-1
2	-1	-1	1
3	-1	1	-1
4	-1	1	1
5	1	-1	-1
6	1	-1	1
7	1	1	-1
8	1	1	1
9	-1.29	0	0
10	1.29	0	0
11	0	-1.29	0
12	0	1.29	0
13	0	0	-1.29
14	0	0	1.29
15	0	0	0
16	0	0	0

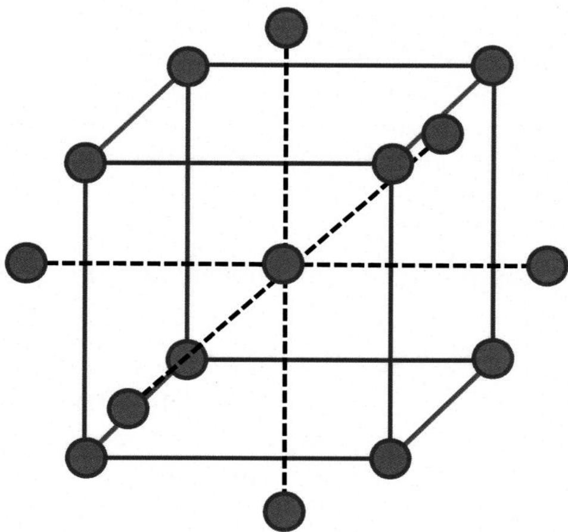

Central Composite Design with axial points off the face of the cube

Maria again draws a cube and places the design points in the space and sees how the axial points would now be pulled away from the six faces of the cube. Maria sees that for the rotatable design, the axial levels are set at ±1.68, and for the orthogonal design, the axial levels are set at ±1.29. The help feature in her software describes what is special about these level choices. The rotatable axial levels are set so that each point in the design is equidistant from the center point and that the increase in the precision of any model predictions from the center of the design is the same in all directions. The orthogonal axial levels are set so that each of the response surface model effects is orthogonal, or independent from the others.

Maria: "I get that these are both desirable properties. I like the idea of increasing the size of the design space without increasing the number of design runs. But I'm not sure how far we can go with the ranges. I'll create each of the designs and talk them over with Alex."

The design with the rotatable axial levels looks like this:

run	steam pressure	pump pressure	line speed
1	200	370	22
2	200	370	40
3	200	420	22
4	200	420	40
5	280	370	22
6	280	370	40
7	280	420	22
8	280	420	40
9	173	395	31
10	307	395	31
11	240	353	31
12	240	437	31
13	240	395	16
14	240	395	46
15	240	395	31
16	240	395	31

To illustrate the differences in the axial level choices, she arranges them in a table that looks like this:

	steam pressure		pump pressure		line speed	
axial type	**low**	**high**	**low**	**high**	**low**	**high**
face	200	280	370	420	22	40
orthogonal	189	291	363	427	19	43
rotatable	173	307	353	437	16	46

Maria and Alex discuss the design and the potential axial-level choices. They both agree that expanding the design space by using the widest axial levels would be best. But Alex has a concern.

Alex: "I'm sure that like in the screening design we will be making some product that we can't sell. Lisa had agreed that would be OK for this stage of experimentation too. But I don't know if we can operate at the extremes. I think we should see what Connie thinks since she has been a line operator for many years. She may know if this will be a problem. I'll check with her at the end of her shift today. I'll email you with her response."

Maria: "You can point out that these extreme levels for each factor are run at the middle level of all of the other factors. That may influence her thinking."

Maria receives the email from Alex just before shutting down her computer for the evening. In it, Alex explains that Connie's only concern with the most extreme axial levels is the line speed level of 46. She can't recall ever running the line at such a high speed. She will make a small test run at the start of her shift tomorrow at that speed to confirm that it would work. Alex will join Connie for the test and get back to Maria with a decision.

By the middle of the next morning, Maria again receives an email from Alex confirming that the line can be run at a speed of 46. Maria can now finalize the design creation. As with her screening design, she randomizes run order. Her final design looks like this:

run	steam pressure	pump pressure	line speed
1	280	420	22
2	240	395	31
3	200	420	40
4	200	370	40
5	280	370	22
6	240	395	16
7	280	370	40
8	240	353	31
9	307	395	31
10	200	420	22
11	240	395	46
12	280	420	40
13	200	370	22
14	173	395	31
15	240	395	31
16	240	437	31

All of the same team members are on board for this second design. Alex will work with Connie executing the runs. They will again collect 10 bottles while running the line at each of the factor settings. Chad will perform the viscosity measurements himself so that their experiment does not interfere in the Quality Lab's daily operations. And, Maria will be at the line to assist in any way needed.

The design execution day is again very intense but runs more smoothly overall than their first design execution. Their experience from the first execution helped greatly, and they did not need to stop the line to make the factor level adjustments. Maria and the team are pleased with themselves.

Maria's spreadsheet with the viscosity measurements for the design runs looks very similar to the sheet for her first design.

Sample	V 01	V 02	V 03	V 04	V 05	V 06	V 07	V 08	V 09	V 10
1	4597	4606	4593	4593	4604	4613	4589	4607	4586	4579
2	4253	4225	4224	4229	4242	4238	4232	4235	4223	4233
3	4699	4693	4688	4689	4673	4678	4674	4673	4701	4668
4	4540	4528	4528	4510	4529	4541	4525	4539	4536	4537
5	4006	4028	4011	4005	4024	4012	4002	4004	4011	4004
6	4360	4349	4351	4336	4344	4346	4338	4339	4346	4359
7	3905	3897	3897	3879	3911	3904	3896	3892	3894	3908
8	4084	4085	4081	4090	4092	4091	4090	4101	4079	4099
9	4032	4023	4039	4032	4029	4025	4044	4026	4033	4037
10	4824	4830	4828	4821	4829	4798	4819	4834	4828	4811
11	4129	4124	4143	4142	4137	4132	4113	4133	4132	4145
12	4254	4257	4256	4245	4256	4245	4259	4253	4257	4262
13	4867	4856	4854	4866	4859	4853	4841	4862	4858	4857
14	4908	4916	4913	4899	4899	4899	4919	4913	4889	4913
15	4297	4322	4309	4317	4301	4331	4308	4319	4289	4315
16	4811	4808	4824	4813	4823	4810	4803	4812	4813	4819

Maria again starts her analysis process by visualizing the distributions of the individual viscosities for each of the design runs. She creates histograms of the 10 viscosity measurements by run.

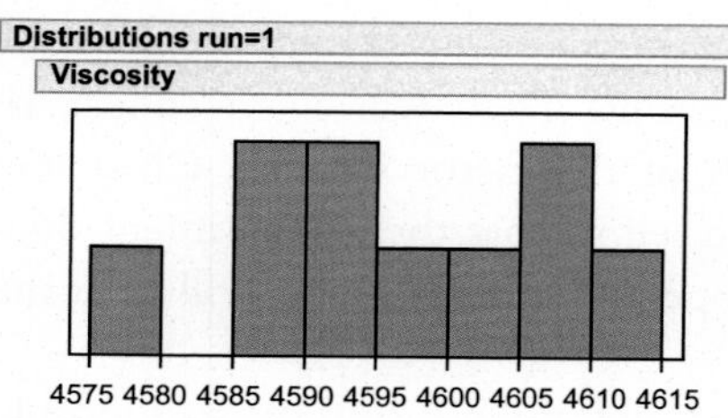
Distributions run=1
Viscosity
4575 4580 4585 4590 4595 4600 4605 4610 4615

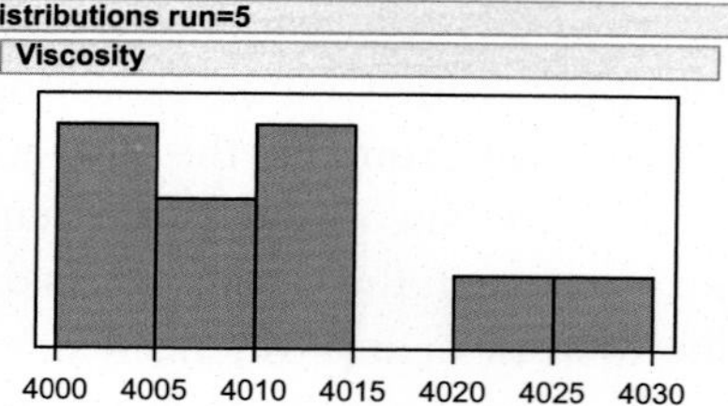
Distributions run=5
Viscosity
4000 4005 4010 4015 4020 4025 4030

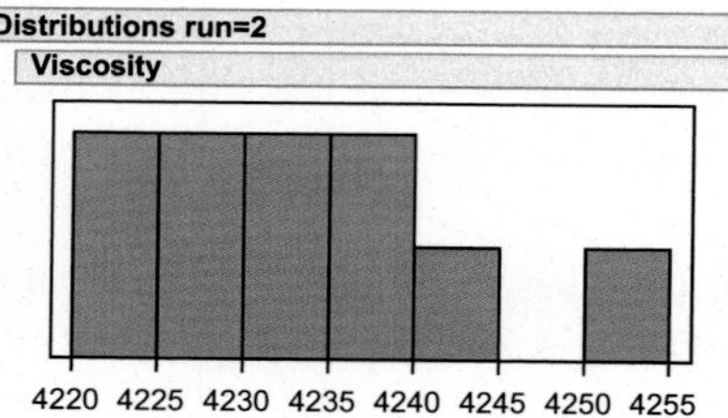
Distributions run=2
Viscosity
4220 4225 4230 4235 4240 4245 4250 4255

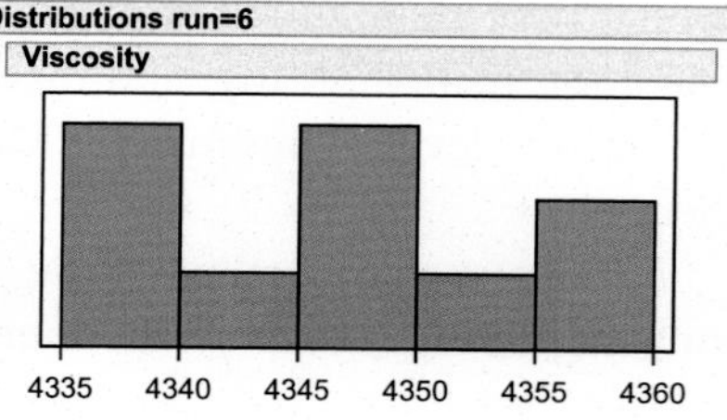
Distributions run=6
Viscosity
4335 4340 4345 4350 4355 4360

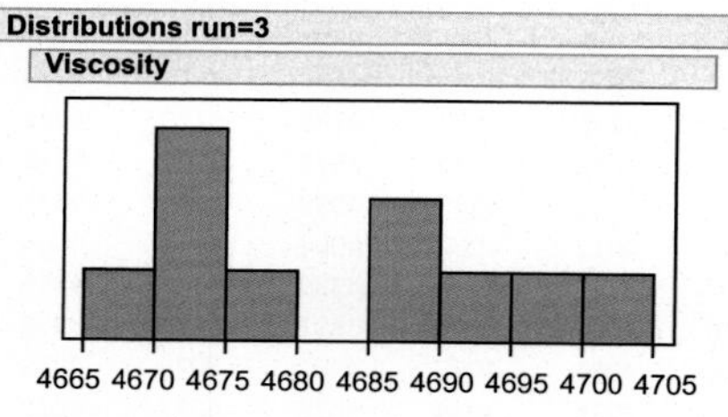
Distributions run=3
Viscosity
4665 4670 4675 4680 4685 4690 4695 4700 4705

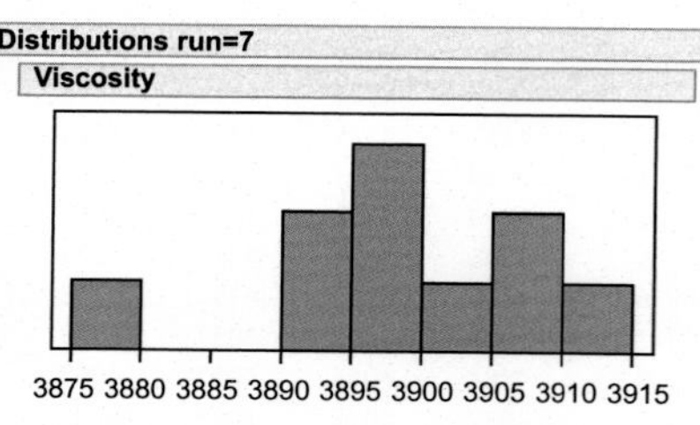
Distributions run=7
Viscosity
3875 3880 3885 3890 3895 3900 3905 3910 3915

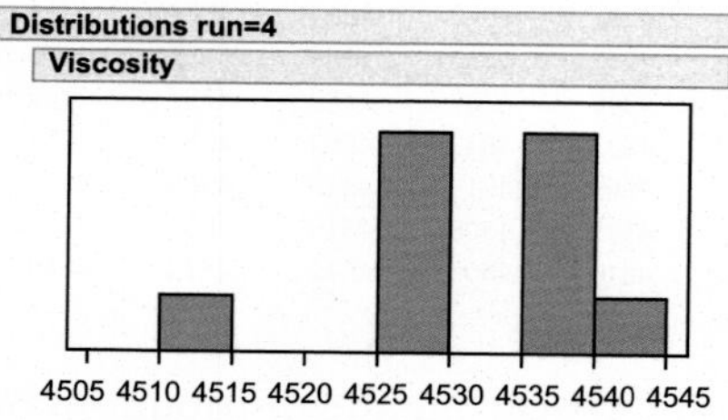
Distributions run=4
Viscosity
4505 4510 4515 4520 4525 4530 4535 4540 4545

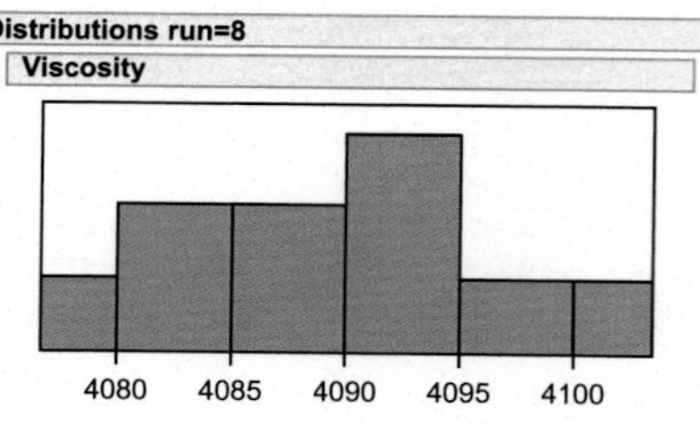
Distributions run=8
Viscosity
4080 4085 4090 4095 4100

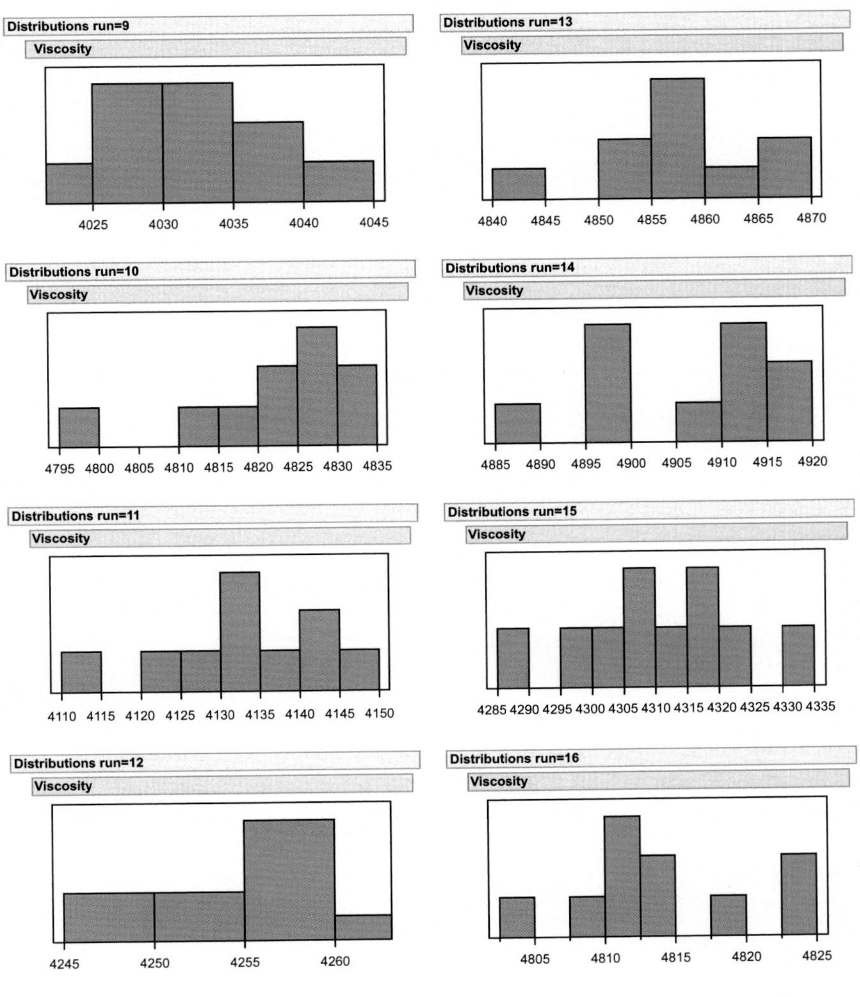

Maria looks at these histograms to see if there is any indication of skewness in the distributions. She is also hoping to see that the variation in measurements for the samples is similar. As with her screening design, the graphs have very different ranges since they also have very different centers, so it is difficult to see if this is true. So again, she calculates the average, range, and standard deviation in viscosity measurements for the 16 samples and arranges them in a table.

Sample	Average Viscosity	Range	Standard Deviaton
1	4597	34	11
2	4233	30	9
3	4684	33	12
4	4531	31	9
5	4011	26	9
6	4347	24	8
7	3898	32	9
8	4089	22	7
9	4032	21	7
10	4822	36	11
11	4133	32	10
12	4254	17	6
13	4857	26	7
14	4907	30	10
15	4311	42	13
16	4814	21	7

The ranges and standard deviations for each of the 16 design runs are similar enough that she feels comfortable in thinking that there are no differences in variability between the runs. As with her screening design, the averages are substantially different, and these are what she will use as a response for modeling the factor effects.

The model that typically fits to data from a response surface design has main effects, pairwise interactions, and squared terms. The squared terms are to accommodate any curvature in the factor effects. For three factors, this means three main effects, three pairwise interactions, and three squared terms. Maria fits this model to her experimental data.

Summary of Fit

RSquare	0.96
RSquare Adj	0.90
Root Mean Square Error	107.30
Mean of Response	4407.53
Observations (or Sum Wgts)	16.00

Analysis of Variance

Source	DF	Sum of Squares	Mean Sqaure	F Ratio
Model	9	1647863	183096	15.90
Error	6	69074	11512	Prob > F
C. Total	15	1716937		0.0016*

Sorted Parameter Estimates

Term	Estimate	Std Error	t Ratio	Prob>|t|
steam pressure(200,280)	−264.0	29.0	−9.09	<.0001*
pump pressure(370,420)	166.8	29.0	5.74	0.0012*
line speed(22,40)	−93.6	29.0	−3.22	0.0180*
steam pressure*pump pressure	103.1	37.9	2.72	0.0347*
steam pressure*steam pressure	84.5	35.3	2.40	0.0536
pump pressure*pump pressure	78.1	35.3	2.22	0.0686
pump pressure*line speed	−5.3	37.9	−0.14	0.8927
line speed*line speed	3.3	35.3	0.09	0.9283
steam pressure*line speed	1.3	37.9	0.03	0.9739

The output from Maria's software is similar to what she saw for the screening design. The factor effects are again ordered in decreasing levels of significance and have been standardized to the range of each of the factors so that the size of their effects can be directly compared. She sees that steam pressure is the most significant effect, followed by pump pressure, and then line speed. All of these have *p*-values smaller than 0.05, the level commonly used as a cutoff for significance. In addition, the interaction of steam pressure and pump pressure also has a *p*-value smaller than 0.05, and the squared terms for both steam pressure and pump pressure have *p*-values only slightly larger than 0.05. With the more complicated model with interactions and squared terms, it is not obvious how all of these factors affect viscosity; so Maria now creates interaction plots to visualize these effects.

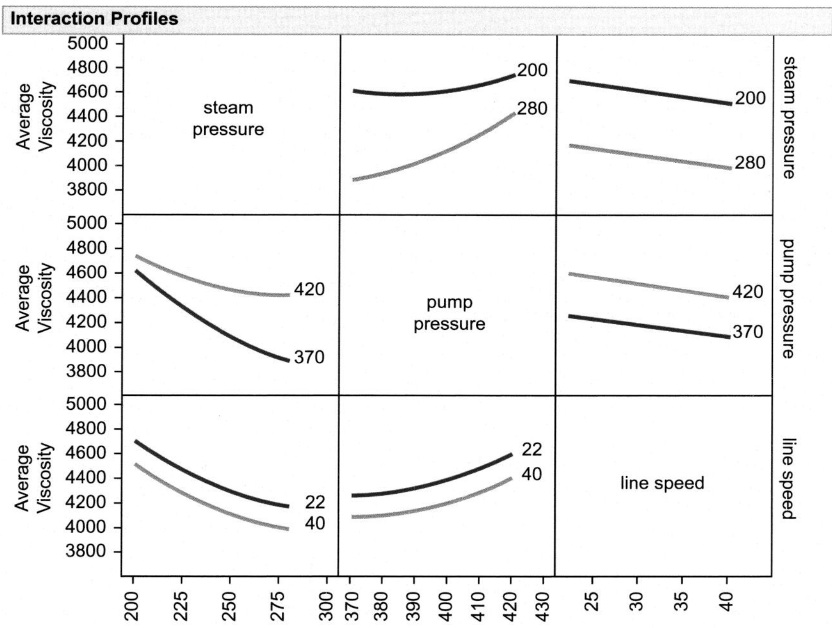

These graphs show all of the potential factor effects and their interactions. The plots in the first column all have steam pressure on the horizontal axis; those in the second column have pump pressure on the horizontal axis; and those in the third column have line speed on the horizontal axis. There are separate lines in each plot for the levels of the factor listed for each row. Maria examines the plot in the first column, second row.

This plot shows the effect of steam pressure at two different levels of pump pressure. There is a dramatic decrease in viscosity as steam pressure increases at the lower pump pressure level of 370, but only a modest decrease at the higher pump pressure level of 420. This clearly shows the significant interaction of steam pressure and pump pressure. The decreases have a curve to them rather than being strictly linear. The graph in the second column, first row shows the same interaction effect, with the pump pressure on the horizontal axis and curves for two different steam pressures. The graphs in column three show that the effect of line speed appears strictly linear and is consistent regardless of the levels of both steam pressure and pump pressure. Though the lines for the different levels of steam pressure and pump pressure are far apart, the slopes of the lines are nearly identical. This confirms the lack of significance of the interaction effects between line speed and steam pressure and line speed and pump pressure.

As with the screening design, Maria verifies that her model is not missing anything critical by analyzing the residuals. As with the screening design, there should be no pattern in them if she plots them against the order of the design runs and the predicted values from the model, and they should follow a normal distribution centered at zero. She again creates plots to perform this assessment.

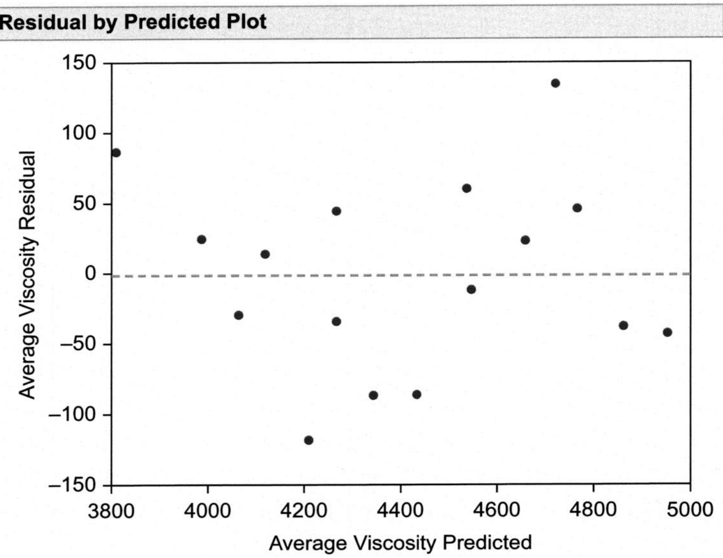

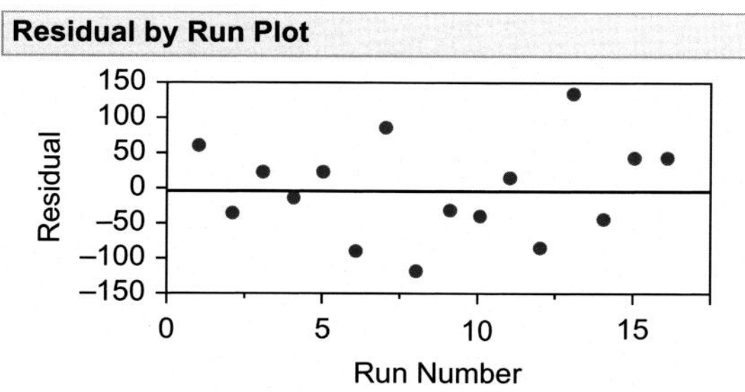

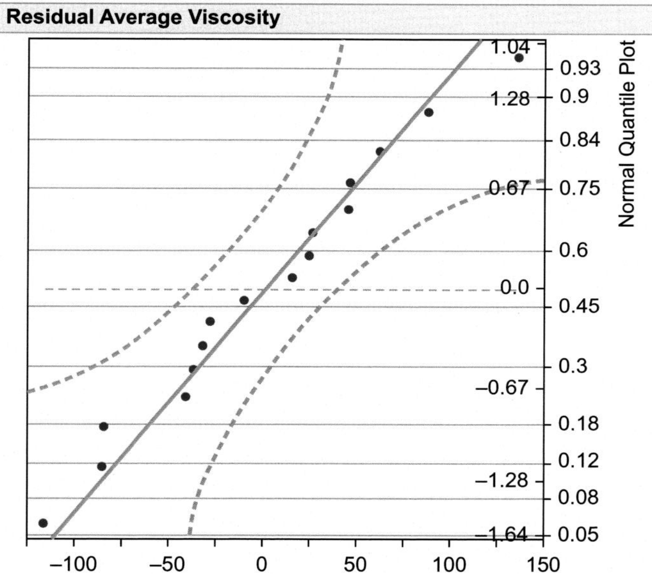

Maria sees nothing in these graphs to suggest that her model is not adequately describing the factor effects on viscosity.

Maria recalls from her class that often response surface models are refined to remove the effects that are not significant. This simplifies the models somewhat prior to using them to make predictions. In her analysis, she had seen that three of the effects had very large *p*-values: the interaction of steam pressure by line speed; the interaction of pump pressure by line speed; and the squared term for line speed. She will remove these terms from her model. The squared terms for steam pressure and pump pressure were close to the 0.05 level of significance, so she will keep these in the model, at least for this initial refinement.

Summary of Fit	
RSquare	0.96
RSquare Adj	0.93
Root Mean Square Error	87.82
Mean of Response	4407.53
Observations (or Sum Wgts)	16.00

Analysis of Variance

Source	DF	Sum of Squares	Mean Sqaure	F Ratio
Model	6	1647520	274587	35.60
Error	9	69417	7713	Prob > F
C. Total	15	1716937		<.0001*

Sorted Parameter Estimates

Term	Estimate	Std Error	t Ratio	Prob>|t|
steam pressure(200,280)	−264.0	23.8	−11.11	<.0001*
pump pressure(370,420)	166.8	23.8	7.02	<.0001*
line speed(22,40)	−93.6	23.8	−3.94	0.0034*
steam pressure*pump pressure	103.1	31.1	3.32	0.0089*
steam pressure*steam pressure	83.1	26.2	3.17	0.0113*
pump pressure*pump pressure	76.7	26.2	2.93	0.0167*

Maria notices that much in this refined model output is similar or identical to her initial model output. The estimated effects for steam pressure, pump pressure, line speed, and the interaction of steam pressure and pump pressure are the same. The *p*-values are mostly smaller, and those for the squared terms for steam pressure and pump pressure are well below the 0.05 level of significance, suggesting that they do belong in the model. She notices that the R Square for the model is the same at 0.96, meaning that her model explains 96% of the total variability in the response. The root mean square error is also somewhat smaller. This is essentially the standard deviation of her model's predictions, so smaller is better since she will use the model to make predictions. She again creates interaction plots.

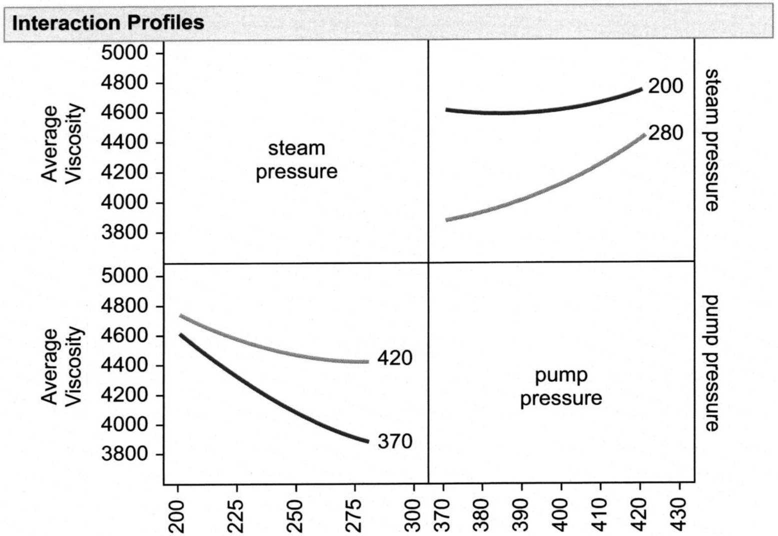

This is a simpler visualization now since the interaction effects of steam pressure and pump pressure with line speed have been removed from the model. The plots look nearly identical to those in her initial visualization, so her interpretation of how steam pressure and pump pressure affect viscosity is still the same. And since the estimate of the effect of line speed is the same as in her initial model, her interpretation of that factor effect is also the same. The refined model has not changed her understanding of the factor effects but made the model easier to understand and better able to predict.

Maria creates additional visualizations to better understand the factor effects. Her software enables her to create a three-dimensional representation of the model for any two factors and the viscosity response for set levels of any additional factors. Since line speed does not interact with the other factors, she chooses to set this at the middle level of 31. Her visualization looks like this:

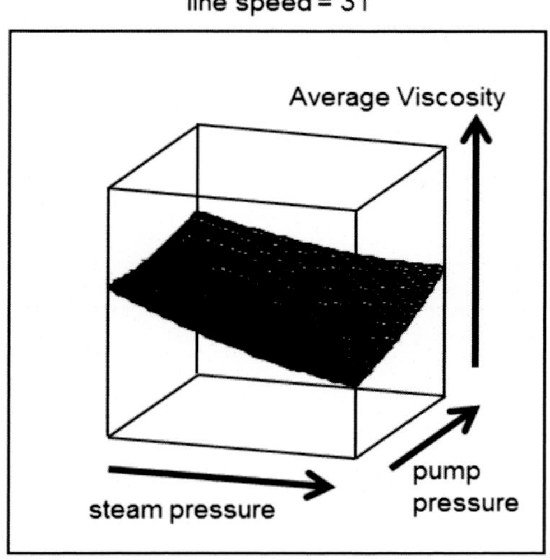

In this graph, steam pressure increases from left to right in the cube, and pump pressure increases from front to back in the cube. Viscosity increases from top to bottom. She can easily see that the viscosity is highest at low steam pressure and high pump pressure, and that viscosity is lowest at high steam pressure and low pump pressure. Since the effect of line speed is linear and negative, this shape would shift up for a lower line speed and down for a higher line speed. There are no axis labels, but even with them, it would be difficult to determine factor levels that would produce a specific viscosity.

A second visualization is better for that determination. It again presents the model for any two factors, but now the viscosity response is represented as contours. She also creates this visualization for a line speed of 31.

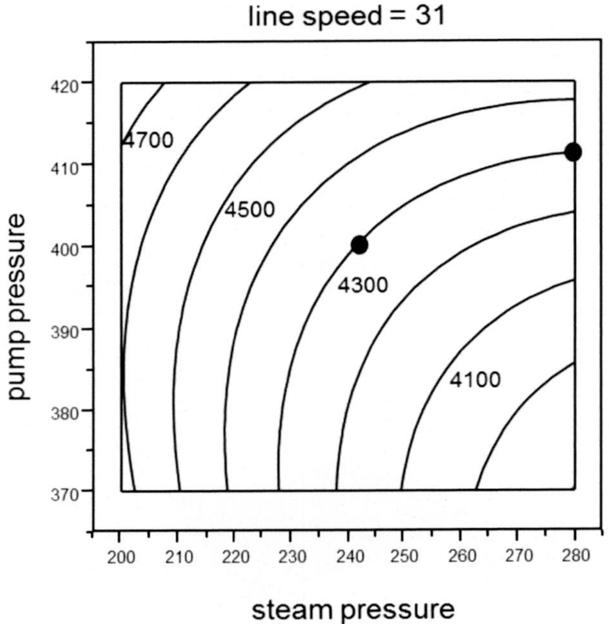

Maria can now easily see what combinations of steam pressure and pump pressure are predicted to deliver a viscosity of 4300, her target. There are many, and they fall along the line labeled 4300. A steam pressure of 240 and pump pressure of 400 near the middle of the plot looks to come very close to this target. Also, at the far right, a steam pressure of 280 and a pump pressure of 410 also predicts close to the target. Maria notes these potential factor level combinations on her plot with a heavy dot.

Maria creates similar plots for line speeds of 22 and 40.

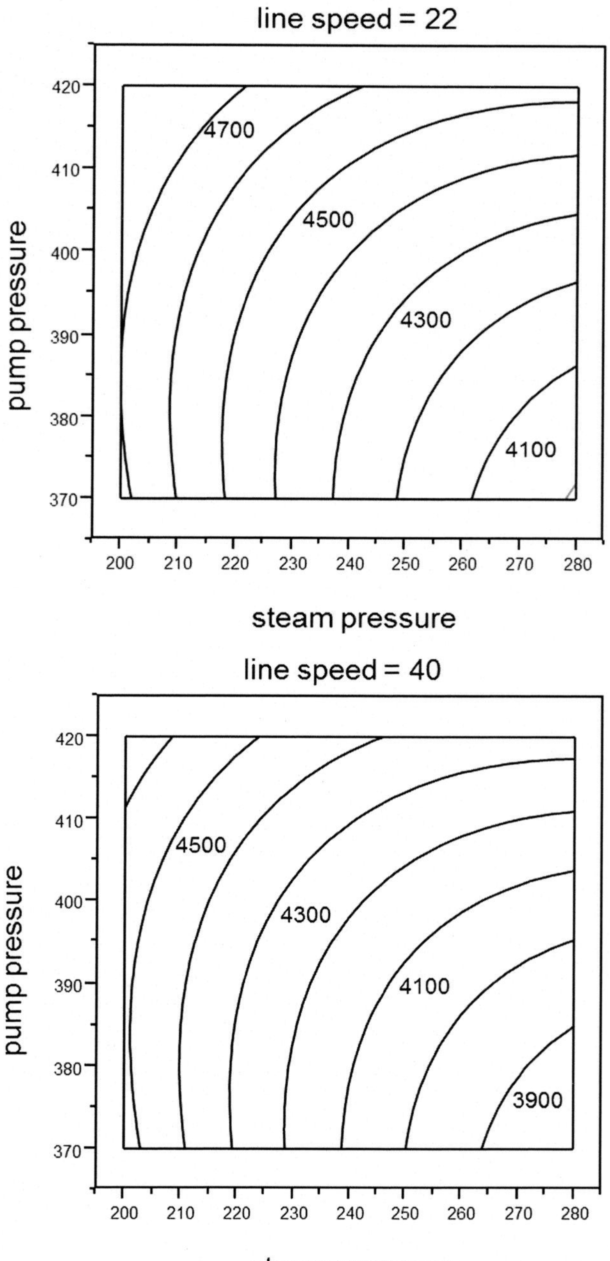
line speed = 22
4700
4500
4300
4100
pump pressure
420
410
400
390
380
370
200 210 220 230 240 250 260 270 280
steam pressure
line speed = 40
4500
4300
4100
3900
pump pressure
420
410
400
390
380
370
200 210 220 230 240 250 260 270 280
steam pressure

From these graphs, Maria can see now the curve at her target of 4300 moves from the lower left to the upper right as line speed increases. Her software will allow her to save the model formula so that she can use it to predict the viscosity anywhere within her design space.

Maria has many things to share with the team. Since her refined model is so similar to the initial model, she does not expect to see anything in her residual plots. She again creates them just to be sure. They look like this:

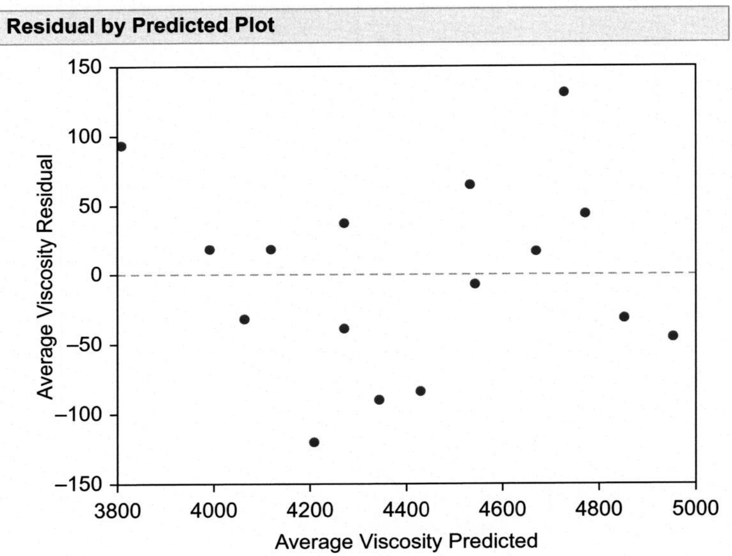

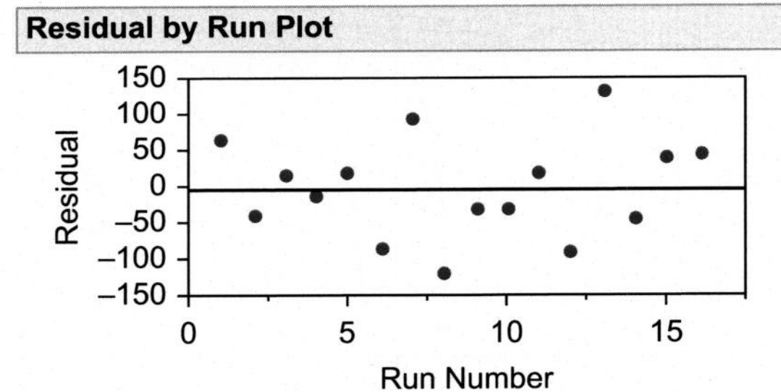

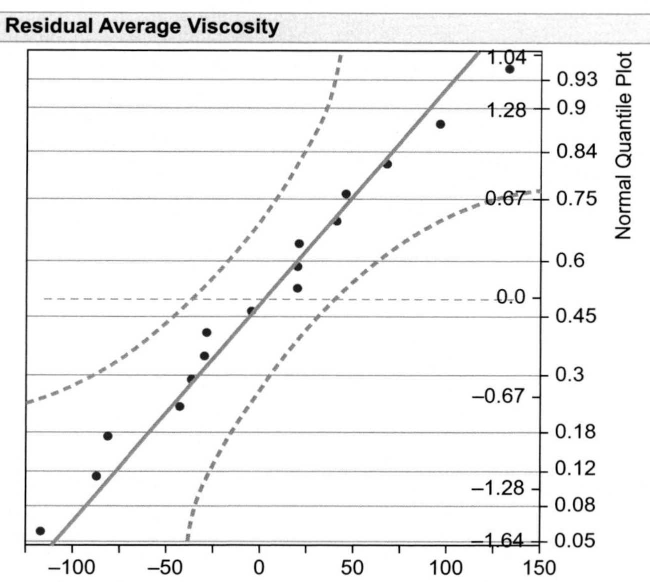

As expected, Maria sees no patterns in the residuals. She is confident with her model and prepares to share the results. She arranges a meeting for the next day with Alex, Connie, Chad, and most importantly, Lisa.

Maria shares the analysis results of the second phase of the designed experiment to the team the following day. All are excited with the results and are eager to implement them. Lisa questions the team.

Lisa: "We now have a model that relates how steam pressure, pump pressure, and line speed affect viscosity. Is there anything more we need to do before we implement it in our operations? And how do we do that?"

Maria: "While I'm confident with our model, it is important that we do not neglect the important step of validating the model. At this point, we only have paper predictions."

Lisa: "What do we need to do to validate the model? Would we need to make some additional runs?"

Maria: "Yes, that is the only way to truly validate the model. It would not need to be very many."

Lisa: "How do we determine what to do here?"

Alex: "I would like to see if the model predictions are valid over a range of line speeds. We often run faster or slower depending on demand for the product. If the model predicts that certain steam pressure and pump pressure combinations will deliver the target viscosity, I would like to verify that for slow, moderate, and fast line speeds. This way, we know that our process adjustments will deliver the desired response."

Maria: "I think that this is a good approach, we can validate the model where it matters the most to us."

Chad: "I would like for us to consider something else in the validation. The process settings that we are most interested in are not likely to be identical to design runs from our experiment. We should also include some validation runs that are identical to design runs. That way, we will be sure that nothing else has changed substantially with this second set of runs."

Lisa: "This all makes sense to me. I'm willing to commit resources for the validation runs. I like Alex's suggestion that we validate at process settings that should deliver the target at low, medium, and high line speeds. And I agree with Chad that we would want to include some repeated runs from the design. What should they be?"

Maria: "Since the other runs are all at our target of 4300 with different line speeds, I think these other runs should be at predictions below and above our target. This way, we will know that our model will still be accurate at a different target."

Lisa: "I see a plan forming here. If we have six validation runs, we can use three to validate the process settings at low, medium, and high line speeds. And then have three repeats of design points that delivered low, medium, and high viscosity levels. Maria, can you put that together for us? Alex, please help Maria with this."

Maria: "I'll send out a plan to the team before the end of the day. Alex, let's find a time to sit down and work out the details of the validation."

Maria and Alex meet later in the day to work out the specifics of the validation plan. They look at the observed data from their design and notice that there are three design points that have both different process settings and substantially different observed viscosities. They use the prediction equation from the model to determine the predicted viscosity for these runs. It is easy enough to determine runs that will meet the target by examining the model's contour plots for each of the line speeds and using the model's prediction equation.

Maria and Alex decide on the following runs for validation:

run	steam pressure	pump pressure	line speed	average viscosity from design	predicted viscosity
1	200	420.0	22	4822	4852
2	240	395.0	31	4311	4271
3	280	370.0	40	3899	3803
4	260	400.0	22		4300
5	228	380.0	31		4304
6	248	412.5	40		4297

The first three runs in the validation set are repeats from the experimental design. The second three are model predictions that are near the target of 4300 at the low, middle, and high line speeds from the design.

Maria shares the recommendations for validation that she and Alex have determined with the team. All are in agreement, so Maria and Alex schedule the validation runs for the following week.

The execution plan for the validation runs is the same as what has been used for both designs. Chad again makes all of the measurements for runs. There are again 10 measurements for each of the six runs. The validation data looks like this:

run	V 01	V 02	V 03	V 04	V 05	V 06	V 07	V 08	V 09	V 10
1	4846	4842	4831	4837	4853	4828	4845	4841	4833	4850
2	4309	4289	4280	4289	4268	4276	4283	4273	4280	4288
3	3849	3848	3843	3824	3842	3835	3837	3839	3818	3831
4	4289	4298	4293	4288	4280	4293	4280	4306	4291	4282
5	4317	4330	4311	4294	4320	4303	4317	4308	4318	4316
6	4295	4295	4283	4269	4301	4291	4302	4303	4296	4277

Maria again starts by creating distributions of the data by run. Her histograms look like this:

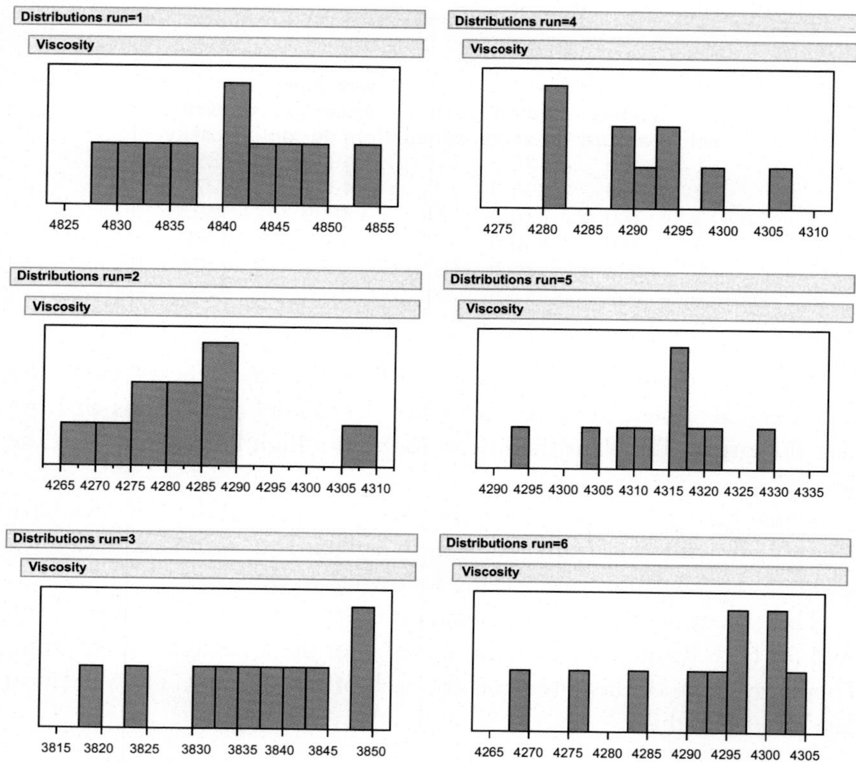

She sees no substantial differences in the distributions for the runs. Like before, she also calculates the range and standard deviations for the runs. These summary statistics look like this:

Run	Average Viscosity	Range	Standard Deviation
1	4841	25	8
2	4284	41	11
3	3837	31	10
4	4290	26	8
5	4313	36	10
6	4291	34	11

Maria sees no substantial differences in these summary measures of variation and now feels confident that she can compare the run averages to the model predictions and for the first three the observed values from the design. She adds these means to her earlier table; it now looks like this:

run	steam pressure	pump pressure	line speed	average viscosity from design	predicted viscosity	average validation viscosity
1	200	420.0	22	4822	4852	4841
2	240	395.0	31	4311	4271	4284
3	280	370.0	40	3899	3803	3837
4	260	400.0	22		4300	4290
5	228	380.0	31		4304	4313
6	248	412.5	40		4297	4291

For the first three validation runs, Maria has three values to compare: the average viscosity from the design; the predicted viscosity from the model; and the average viscosity from the validation. For all three runs, the differences between any of these is not more than 40, which is similar to the range of the 10 individual bottle measurements for the design runs in both her screening and response surface designs. The last three validation runs have predictions near her target of 4300, and the average viscosity of the validation runs is within 10 units, even closer than in the first three validation runs. Based on these observations, Maria feels very confident that her model predictions are accurate enough for her purpose of determining steam pressure and pump pressure levels that will deliver the target viscosity across a range of line speeds. Maria shares the validation results with the team at the next morning's production meeting.

Maria: "The average viscosities from the validation runs were very close to the model predictions. I'm confident that we can use our model as a basis for a control strategy in production going forward."

Lisa: "This is great work! Besides improving our control strategy, I feel we now have a much better understanding of our production process. I will share this work with our management so that they will not only understand the progress that we have made but also to demonstrate the effectiveness of the systematic experimental approach. I can already think of a few other places where we can apply this approach."

CHAPTER 13

Mixture Experimental Design

Steve opens up the research and development leadership team meeting bragging about the value of the DOE work that Maria has just finished. Design of experiments has presented a powerful approach that had not been used before. Maria's work has helped the plant not only to deliver more consistent BBQ viscosity, but also the systematic approach has led to a better understanding of how factors such as steam pressure and line speed affect viscosity. Steve has several improvement opportunities in mind, and he hopes that by applying the design of experiments methodology to them, the results will be similarly positive.

A few days later, Maria receives a call from Adam, one of the procurement managers. The cost of one of the three starches used in the base BBQ sauce has increased dramatically because of a material shortage. The price is now nearly three times its historical average.

Adam: "We are hoping to use a different starch in the product, so that we don't have a cost increase. Joe, our starch expert, has identified a potential alternative. Can you replace the current high-cost starch with this alternative?"

Maria: "We will need to perform some experiments to answer this question. I'll engage the development team to create a research plan and will get back to you."

Maria starts by discussing the problem with Steve. He mentions that there had been some research done with the alternative starch in the past which indicated it was not a direct replacement in the current starch blend. Prior development work showed that finished product would not be thick enough if the starches are simply swapped on a one-to-one basis. There may be a blend of this new starch with the others that could deliver the desired product viscosity.

Maria: "So, we need to figure out what the best three-starch blend is."

Steve: "Can we use the design of experiments approach that we used to optimize viscosity in the past?"

Maria: "That is exactly what I am thinking. But, there is a big difference here, the three starches sum to a fixed percentage in the formula. I remember in my design of experiments class that mixture

http://dx.doi.org/10.1016/B978-0-12-417179-4.00013-5

designs are used in these situations. I will have to look at my class notes to recall more details about this approach."

Maria has time to look at her notes later in the day and then finds Steve at his desk.

Maria: "I think I have some ideas on how to approach this. We are really interested in the best proportional mixture of the three starches. This is similar to mixing a cocktail. I'll use a Margarita as an example."

Steve: "That is one of my favorites!"

Maria: "The classic recipe for a Margarita is two parts tequila, one part lime juice, and one part triple sec. This is the same as 50% tequila, 25% lime juice, and 25% triple sec. But maybe, there is a different mixture of these three that will taste even better—maybe a little less tequila and a little more triple sec. But, the proportions will always add to 100%. Each mixture will be a different design run."

Steve: "I'm following so far."

Maria: "In a mixture design, we could have a design run with 100% one component and 0% of the others. This would make a terrible tasting Margarita. So, with a Margarita, we would have some constraints, like tequila needs to be between 40% and 60%, lime juice between 20% and 30%, and the same for triple sec. Maybe there are no constraints for our starch blend. Would we ever use only one of the starches?"

Steve: "That is a great question. I am not sure if our BBQ formula needs three starches; this is how we have been making it for as long as I can remember. Since we are changing to a new starch, we don't likely know. If we can make a good product with only two or even one starch, that would simplify the process and the ingredient line and also make Adam in Procurement happy. Maybe there are some constraints with the starches, like one of them cannot exceed 50% or another can be no less than 30%, but no more than 50%. Joe would be better able to tell us this."

Maria moves closer to the white board in Steve's office and starts drawing.

Maria: "Let me help you visualize how the three starch design space would look, and how it would change if we add constraints. The mixture of three components can be represented by a ternary plot. A ternary plot looks like this:"

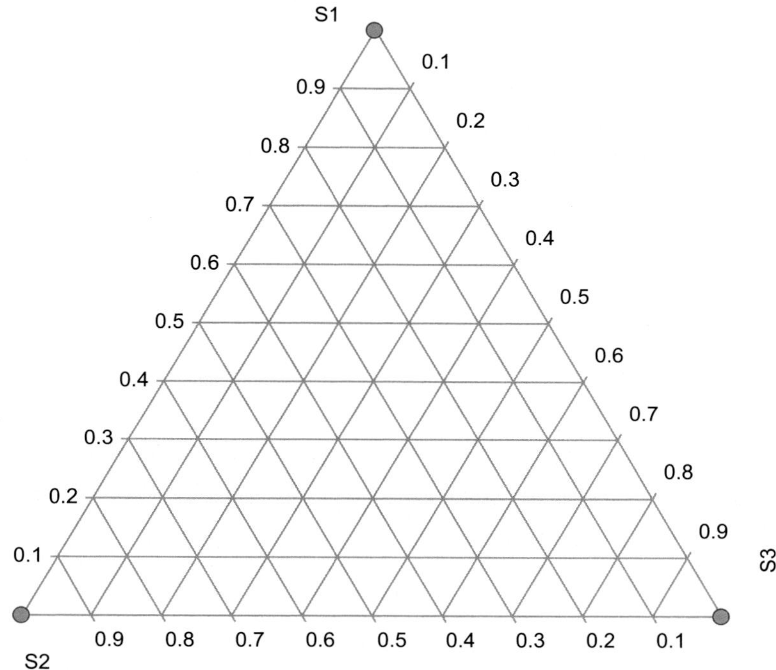

In the plot, any blend of the three starches can be represented, and the sum of the three starches will be equal to 100%. The corners of the triangle represent a sample with only one of the starches. The one at the top is 100% of starch 1. The one in the left corner is 100% of starch 2. The one in the right corner is 100% of starch 3. The tick marks on the plot face in specific directions indicate the level of each component in the starch blend.

The middle of the edge of the triangle represents the 50/50 blends of the two components on that edge of the triangle. In the next graph, points in the middle of the edges in the triangle have been added. The point in the middle of this right edge of the triangle is the 50/50 blend of starch 1 and starch 3. The point in the middle of the bottom edge is the 50/50 blend of starch 2 and starch 3. The point in the middle of the left edge is the 50/50 blend of starch 1 and starch 2.

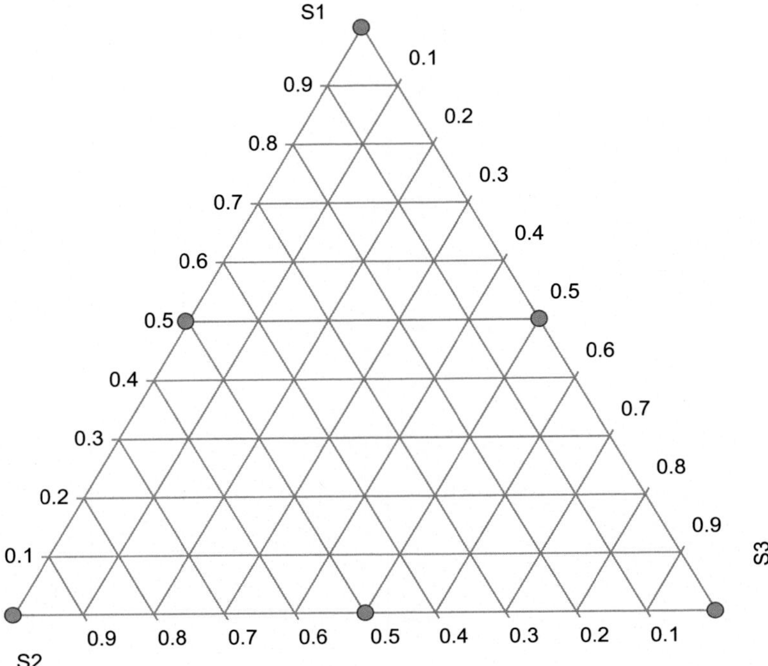

Finally in the next graph, the point in the middle of the triangle is the 33/33/33 blend of the three starches. Like in an optimization design, it is the center point in the design space.

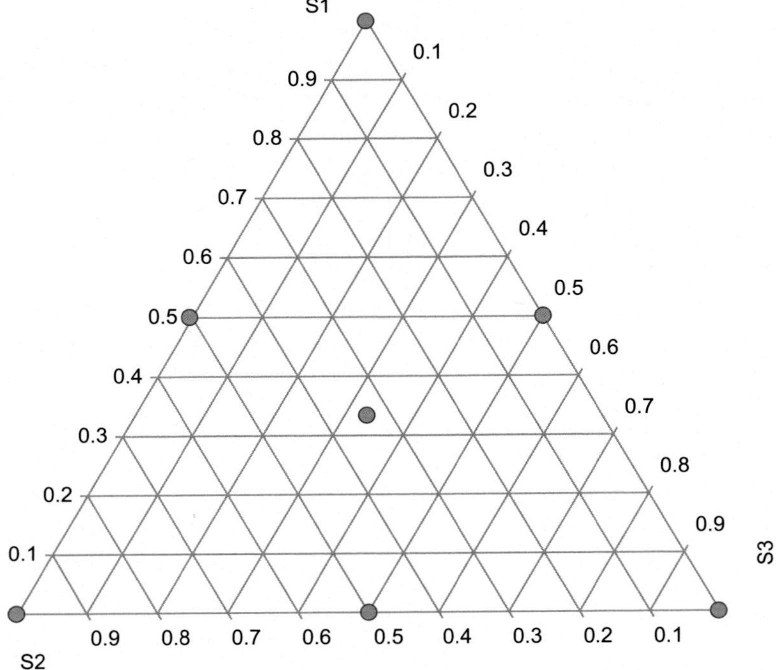

If we add a constraint on any one of the starches, the design space changes. The next graph illustrates the situation where starch 1 cannot be less than 40%. This means that the other starches cannot be greater than 60%. The area outside of the design space is a darkened.

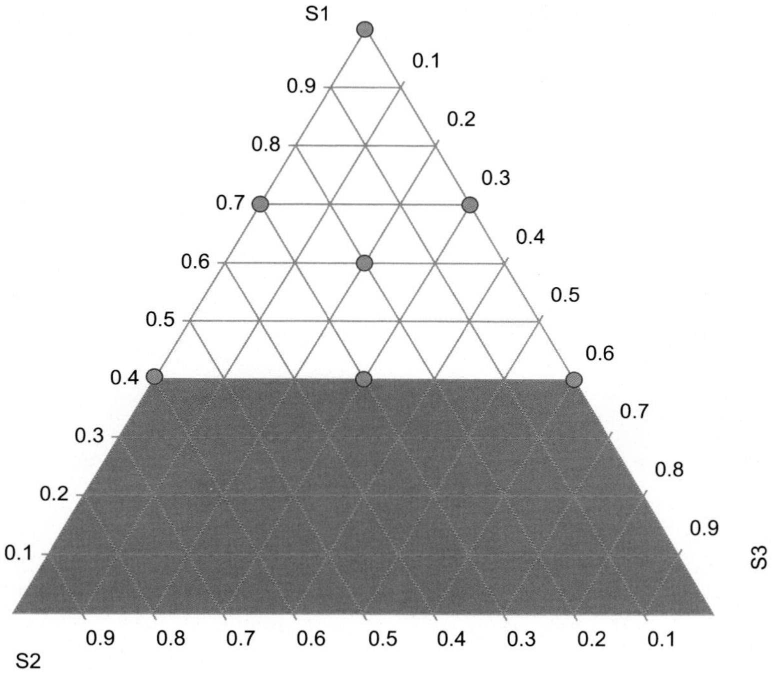

If all three starches have constraints, so that each cannot take values above or below certain parameters, the design space will be constrained. In the next graph, starch 1 must be less than 80%, starch 2 must be less than 70%, and starch 3 must be less than 50%.

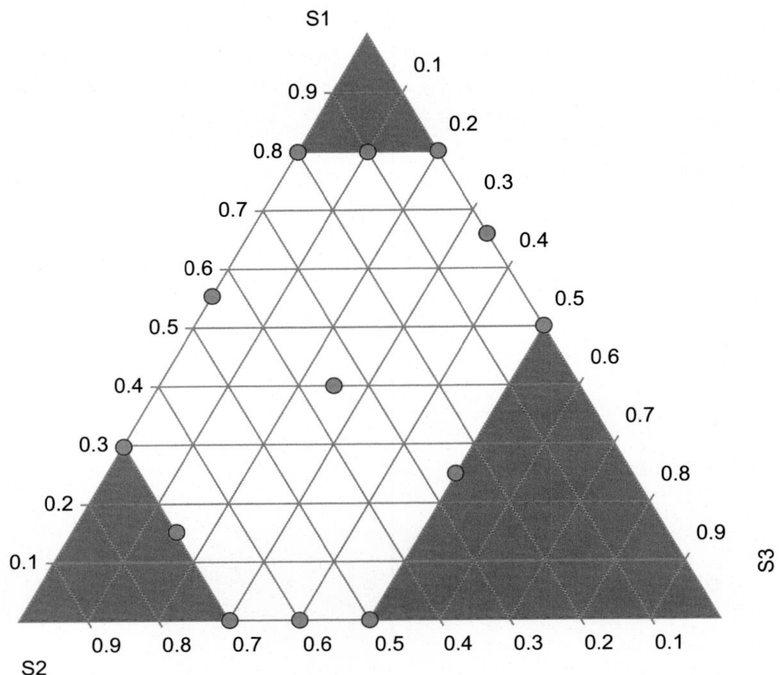

Steve: "Wow, I never thought that a simple triangle could help me visualize the mixture space and any potentially complex constraints. We need to find out if we need all three starches and if there are constraints on any one of them."

Maria: "Let's call Joe and see what he says."

Joe picks up before the third ring.

Joe: "Hi Steve, how are you doing?"

Steve: "Greetings Joe. I am doing great, thank you for asking. I have Maria with me in my office, and we have been talking about our BBQ starch mix."

Maria: "Joe, you suggested a new starch for BBQ sauce to replace the starch with the increased cost in our formula, and we need your help with answering a few questions. Do you think that one or two starches may be sufficient to match the current viscosity instead of using all three starches? And do you have any concerns in terms of starch ranges? Are there any constraints that we should be aware of, specific ranges for any of the starches that we will need to operate within?"

Joe: "To tell you the truth, anything is possible. A two-starch mix may be sufficient, though the work we did a few years ago indicated that all three types of starches may be needed. However, based on that research and since we have not worked with this new starch, I would not eliminate the opportunity to simplify our ingredient line. As far as starch constraints, I would look at all possibilities, as there is no reason we should have more or less of any one starch. Please let me know what you learn, and feel free to reach out again if you have any other questions."

Steve: "Thanks Joe. Have a great day."

Joe: "You too. Have a productive rest of the week."

Maria: "It sounds like we should look at all possibilities. I will set up the design across the entire 0–1 range."

Steve: "Do you mean 0–100% range?"

Maria: "Yes. The range could be referred to both ways: 0–1 if a proportion or 0–100% if a percentage. I'll use proportions. The design we can use will look like this:"

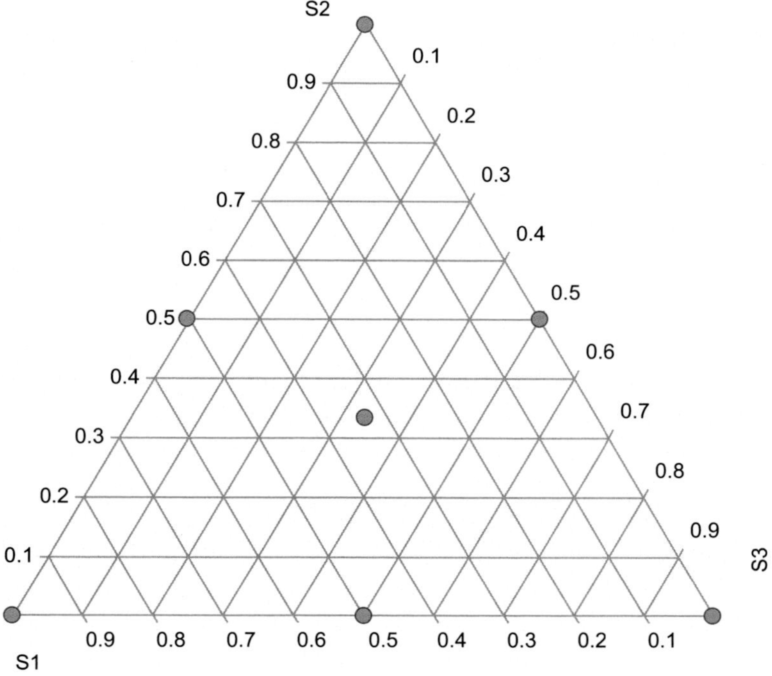

Steve: "I recall that you replicated the design run at the center in your earlier experiment. If you do that here, we would need to create just eight samples to evaluate all three starches? That seems like a very small number! Is it possible that we are forgetting something? Your earlier design had 16 runs to fully understand the linear and curvilinear effects of the ingredients and their interactions."

Maria: "You are correct; your memory is quite good. With a mixture design where two of the three factor levels are known, the last is known too, since they sum to 100%. The model we use is a different form than what we used when the factors were completely independent, and this smaller number of runs is sufficient."

Steve: "The mixture design is similar to the response surface design, yet so different. On one side, the thought process is the same, but on the other side, the structure of the design and the visualization of the design space is completely different. Fortunately, instead of 16 samples to evaluate the three starches, we will only need to produce and evaluate eight since we are interested in the optimal blend of the three starches."

Maria sets up the design runs. As with her earlier design, she decides to produce the runs in a random order. Her design looks like this:

Sample	S1	S2	S3	Cost
1	0.33	0.33	0.33	0.32
2	1.00	0.00	0.00	0.25
3	0.00	0.00	1.00	0.30
4	0.50	0.50	0.00	0.33
5	0.00	1.00	0.00	0.40
6	0.33	0.33	0.33	0.32
7	0.00	0.50	0.50	0.35
8	0.50	0.00	0.50	0.28

Joe had provided Maria with the cost of the individual starches, and she easily determines the cost of any starch blend. This information will be helpful in determining the best blend if she has more than one that meets the viscosity target.

All of the same team members are on board for this new design. Connie will run the line, but since there are no changes in the operating conditions, she will need no additional help. Maria will make sure that the different starch blends will be incorporated during the ingredient mixing stage of the process. Alex will again collect 10 bottles while running the line at each of the factor settings just as in the earlier experiments, two pulls from each of the five filler heads. Chad will again perform the viscosity measurements himself so that their experiment does not interfere with the Quality Lab's daily operations.

The design execution day is again very busy but runs smoothly as the team is now even more experienced in execution. When Chad has completed the measurement of all of the samples, he provides the data to Maria.

Sample	V 01	V 02	V 03	V 04	V 05	V 06	V 07	V 08	V 09	V 10
1	4311	4311	4295	4299	4298	4302	4306	4285	4296	4281
2	3654	3666	3639	3662	3645	3648	3655	3659	3656	3645
3	5026	5036	5026	5041	5007	5043	5024	5038	5031	5031
4	4177	4167	4146	4153	4167	4154	4147	4158	4145	4138
5	4576	4605	4593	4584	4581	4562	4572	4559	4567	4579
6	4278	4315	4292	4293	4286	4262	4298	4278	4291	4287
7	4371	4390	4383	4396	4395	4391	4399	4396	4395	4405
8	4369	4365	4383	4374	4381	4375	4371	4364	4379	4395

Maria again starts the data analysis process by creating histograms of the 10 individual viscosities for each of the design runs.

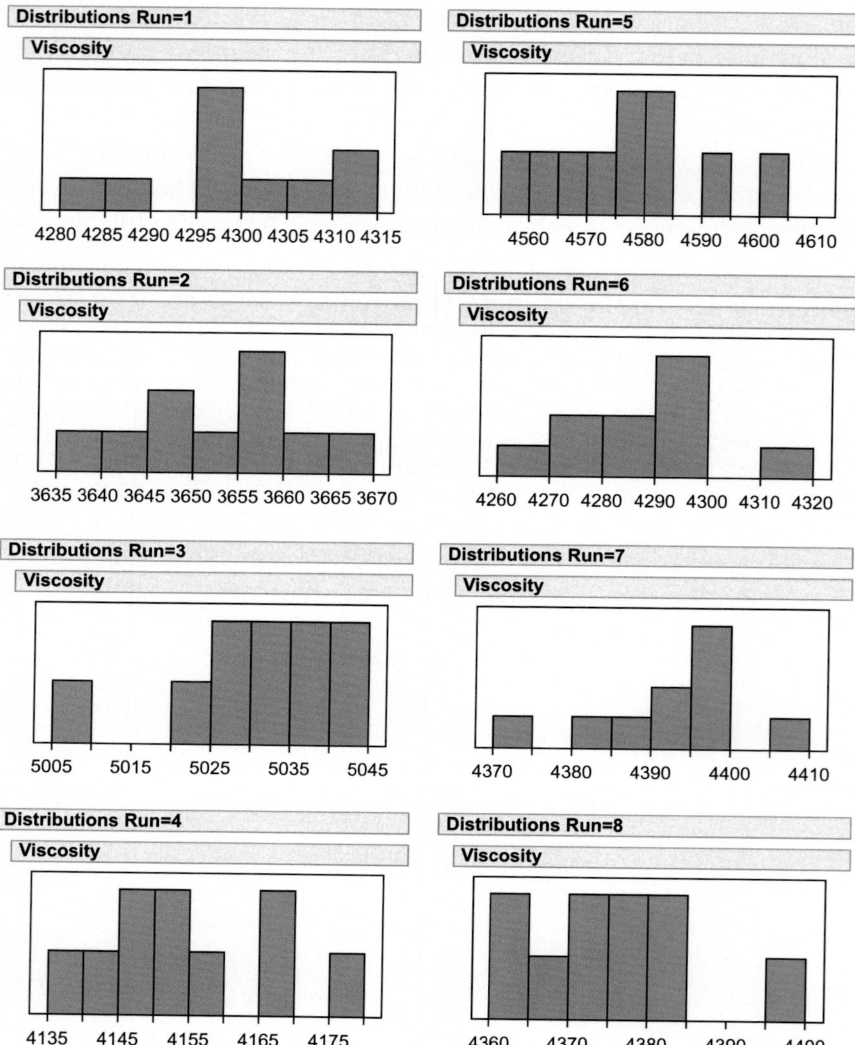

Maria sees that the variation in measurements for the samples is similar, and there is no indication of skewness or unusually high or low measurements. As before, she calculates the average, range, and standard deviation in viscosity measurements for the eight design runs and arranges them in a table.

Sample	Average Viscosity	Range	Standard Deviation
1	4298	30	10
2	3653	27	8
3	5030	36	10
4	4155	39	12
5	4578	46	14
6	4288	53	14
7	4392	34	9
8	4376	31	9

Being satisfied that the ranges and standard deviations are not substantially different for the design runs and seeing the average viscosities, she is eager to start the modeling process. She first creates a ternary plot with the average viscosities for the design runs at the design points.

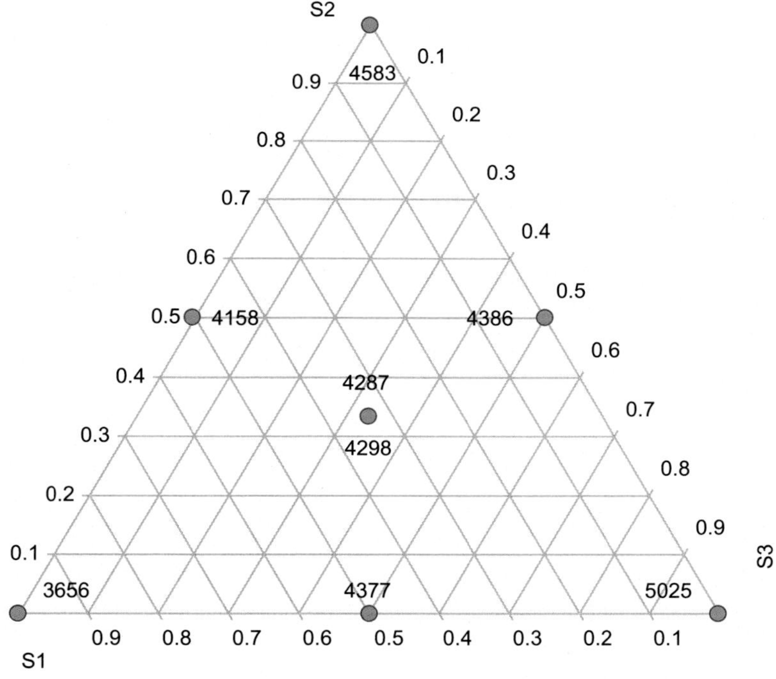

In the plot, it is clear that starch 3 alone delivers the thickest product since the highest viscosity average is highest for that design point. Similarly, starch 2 alone delivers the thinnest product since the viscosity average is lowest for that product. None of the average viscosities are at her target of 4300 exactly, but the two at the middle come very close, and several others are not too far off of the target. Maria is hoping that the modeling will reveal a number of starch blends predicted to meet the target.

Maria recalls from her design of experiments class notes that the model for a mixture design like hers is a bit different than the typical regression models used with the screening and response surface designs. For her mixture, the model includes effects for the three individual starches and also interaction terms for the three pairs of starches. The intercept is dropped from the model and squared terms are not included, this is also called a Scheffe model. Maria fits this model to her experimental data.

Summary of Fit

RSquare	1.00
RSquare Adj	1.00
Root Mean Square Error	16.03
Mean of Response	4346.25
Observations (or Sum Wgts)	8.00

Analysis of Variance

Source	DF	Sum of Squares	Mean Square	F Ratio
Model	5	1046580	209316	814.8
Error	2	514	257	Prob > F
C. Total	7	1047094		0.0012*

Tested against reduced model: Y=mean

Sorted Parameter Estimates

Term	Estimate	Std Error	t Ratio	Prob>|t|
S3	5027.75	15.94	315.42	<.0001*
S2	4575.75	15.94	287.06	<.0001*
S1	3650.75	15.94	229.03	<.0001*
S2*S3	−1602.97	71.01	−22.57	0.0020*
S1*S2	203.03	71.01	2.86	0.1037
S1*S3	183.03	71.01	2.58	0.1233

The output from her software is again similar to the output from her earlier designs. Maria is surprised to see that the R-square for her model is 1.00—a perfect fit! Looking closer, she sees that this is really 0.9996, and her software has rounded it up for the output. It seems that the execution of this design was very good and likely due to the teams experience with the process and that changing ingredient blends is easier to control than changing the process settings.

The individual starch effects in the model correspond to the viscosity levels where each is the only starch used. The interaction of starch 2 and starch 3 is highly significant with a *p*-value much smaller than 0.05. Maria is not sure how to interpret that without some type of visualization. The other two interactions have *p*-values higher than 0.05. Since the model already fits so well, it does not seem that refining the model by removing these terms will affect her interpretation and predictions, so she does not make that effort for this design.

Her software has visualizations of her model that help her to understand the effects of the three starches. One is a profiler that shows the effects of increasing or decreasing any one of the starches while keeping the other two in the same proportions. The profiler looks like this:

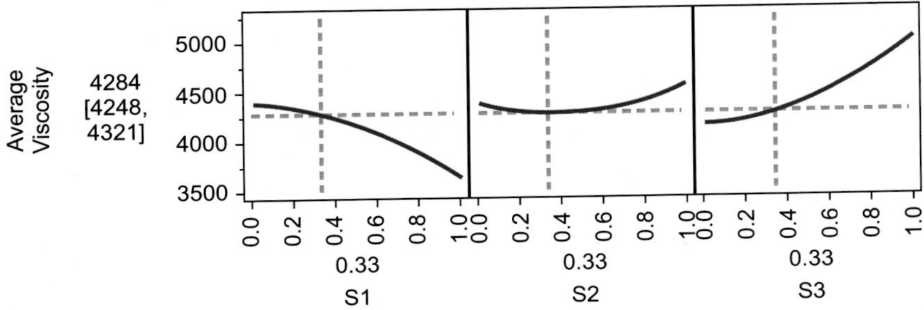

In this plot, each of the starches is in the blend in equal proportions—0.33 for each. The predicted viscosity for this blend is 4284, and a 95% confidence interval for this prediction is also displayed. Maria can see that by increasing starch 1, while keeping the other two in equal proportions, the viscosity will decrease. By increasing starch 3, while keeping the other two in equal proportions, the viscosity will increase. The effect is similar for starch 2, though the increase in viscosity is much smaller.

The second visualization is a contour plot of the ternary space. Maria creates the plot so that it displays the contour for her viscosity target of 4300 with other contours in increments of 200. The plot looks like this:

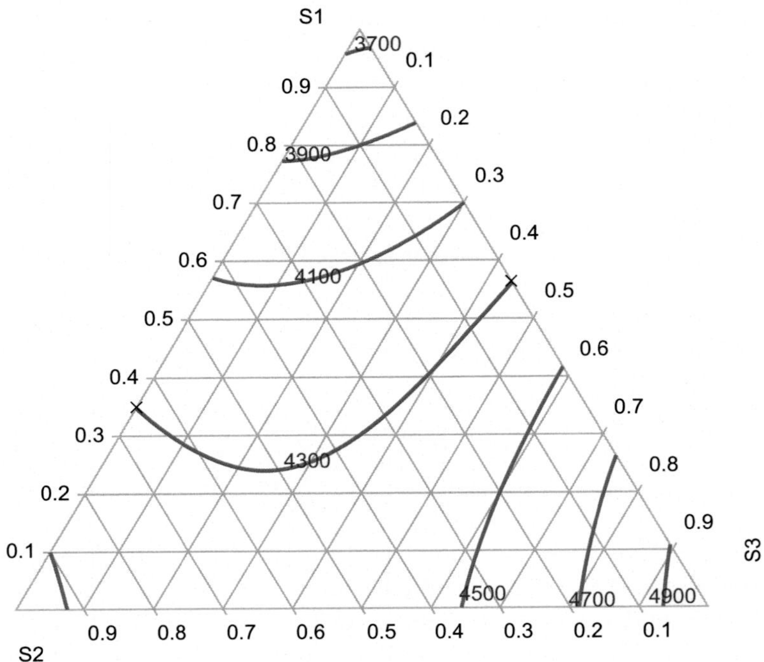

Maria sees that there are many blends that will result in a predicted viscosity of 4300, including two at the edges of the triangle that would be blends of only two starches. On the left side of the triangle, the blend of 35% starch 1 and 65% of starch 2 is predicted to meet this target. On the right side of the triangle, the blend of 56% starch 1 and 44% of starch 3 is also predicted to meet this target.

Maria now creates a visualization of the cost of the blends in the ternary space. She has contours displayed at intervals of 0.02. Since these contours are straight lines, she removes the reference lines on the graph so that the cost contours can be more easily seen. The plot with formula cost contours looks like this:

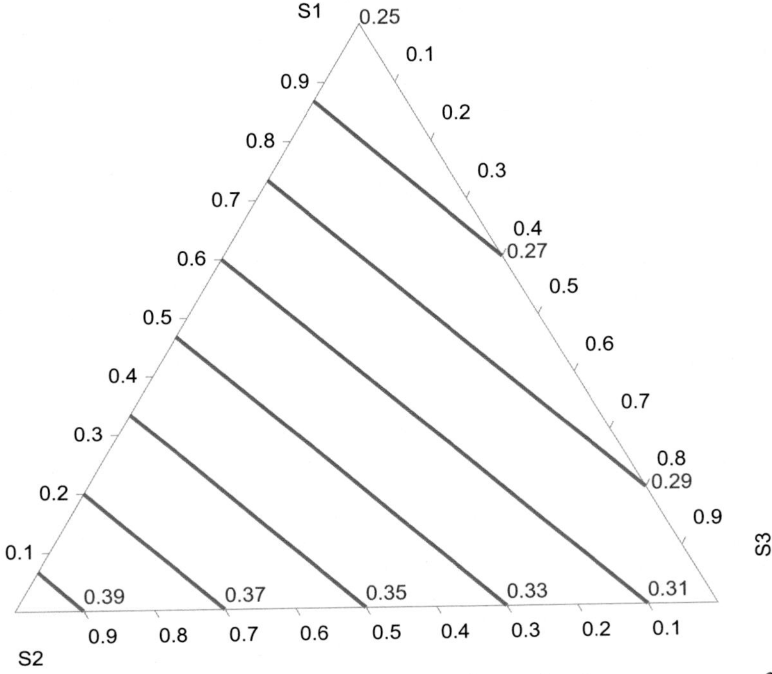

Maria sees that cost decreases moving from the lower corner of the triangle to the top. This makes sense since starch 2 is most expensive, and moving in this direction decreases the amount of that starch in the blend. Putting this information together with the viscosity prediction, she determines that the blend of 56% starch 1 and 44% of starch 3 is the lowest cost blend to meet her target.

As with her other designs, Maria verifies that her model is not missing anything critical by analyzing the residuals. There should be no pattern in the residuals if she plots them against the order of the design runs and the predicted values from the model, and they should follow a normal distribution centered at zero. She again creates plots to perform this assessment.

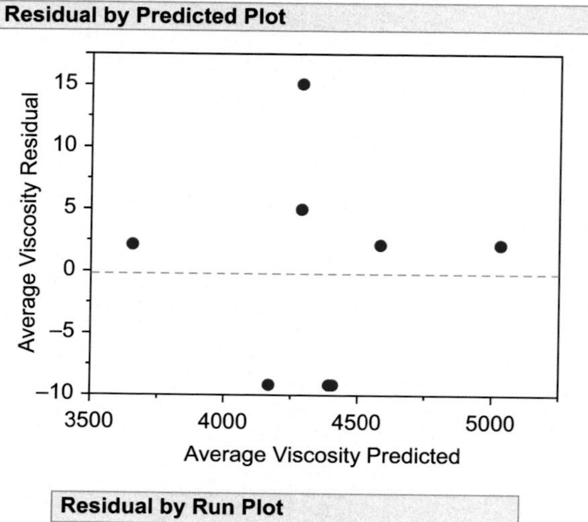
Residual by Predicted Plot
Average Viscosity Residual
15
10
5
0
–5
–10
3500
4000
4500
5000
Average Viscosity Predicted

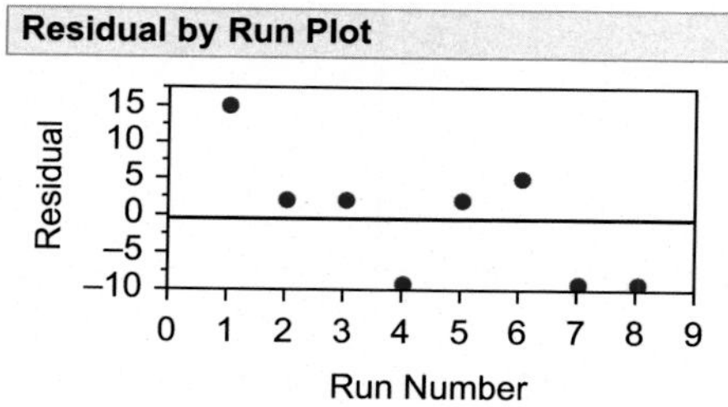
Residual by Run Plot
Residual
15
10
5
0
–5
–10
0 1 2 3 4 5 6 7 8 9
Run Number

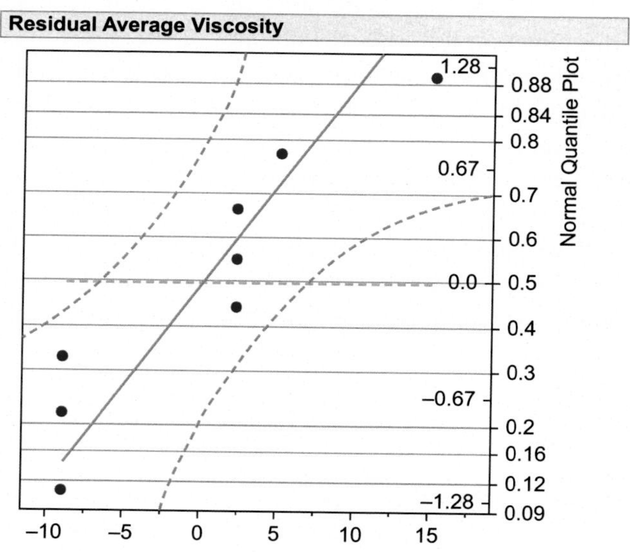
Residual Average Viscosity
Normal Quantile Plot
1.28
0.67
0.0
–0.67
–1.28
0.88
0.84
0.8
0.7
0.6
0.5
0.4
0.3
0.2
0.16
0.12
0.09
–10
–5
0
5
10
15

Maria sees nothing in these graphs to suggest that her model is not adequately describing the factor effects on viscosity. She conveniently finds Steve at his desk and shares the modeling results with him.

Steve: "This is a great result. Not only have we reduced the cost from the current formula, but also using only two starches in the blend simplifies production."

Maria: "While I'm confident with our model, like with our response surface model, it is important that we validate the model predictions. I think that we should make four validation runs, two blends at the target of 4300 and repeat two design runs from the experiment. This is the approach we took before."

Steve: "I agree. Let's use the low-cost two-starch blend and a three-starch blend with a cost similar to the current for the two predicted to meet the target. Pick two from the design that span a wider range of viscosity."

Using this guidance Maria comes up with the following runs for the validation study:

run	S1	S2	S3	Cost	observed viscosity from design	predicted viscosity
1	1.00	0.00	0.00	0.25	3656	3651
2	0.00	0.00	1.00	0.30	5025	5028
3	0.56	0.00	0.44	0.27		4302
4	0.40	0.20	0.40	0.30		4303

Steve agrees with the plan and arranges with Connie, Alex, and Chad the time to execute these runs the following week.

Maria receives the viscosity data from Chad shortly after the execution.

run	V 01	V 02	V 03	V 04	V 05	V 06	V 07	V 08	V 09	V 10
1	3658	3634	3652	3634	3653	3656	3642	3663	3652	3648
2	5026	5007	5012	5022	5022	5034	5025	5048	5024	5030
3	4292	4326	4300	4282	4287	4300	4295	4309	4284	4298
4	4306	4297	4312	4317	4300	4317	4311	4323	4292	4317

Maria again starts by creating distributions of the data by run.

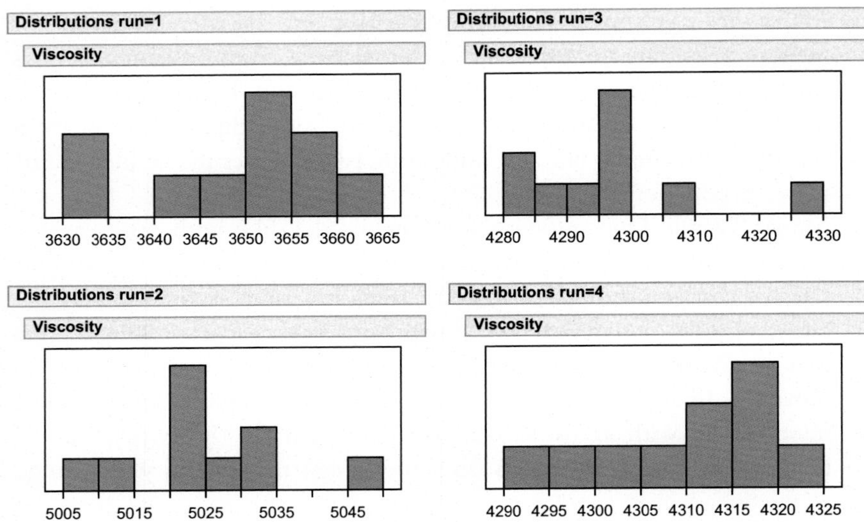

She sees no substantial differences in the distributions for the runs. Like before, she also calculates the range and standard deviations for the runs. Thus the summary statistics look like this:

run	Average Viscosity	Range	Standard Deviation
1	3649	29	10
2	5025	41	11
3	4297	44	13
4	4309	31	10

Maria sees no substantial differences in these summary measures of variation and now feels confident that she can compare the run averages to the model predictions and for the first two observed values from the design. She adds these means to her earlier table, and it now looks like this:

run	S1	S2	S3	Cost	observed viscosity from design	predicted viscosity	observed validation viscosity
1	1.00	0.00	0.00	0.25	3656	3651	3649
2	0.00	0.00	1.00	0.30	5025	5028	5025
3	0.56	0.00	0.44	0.27		4302	4297
4	0.40	0.20	0.40	0.30		4303	4309

It is clear that the results validate the predictions and that the blend of 56% starch 1 and 44% starch 3 can be used. Maria shares the validation results with Steve.

Steve: "Maria, this is great news. If we had to continue to use the existing starch blend our formula would have more than doubled to $0.63.

We would have to pass this larger ingredient cost onto the consumers, or severely reduce our margins. As a result of your team's great work, we can now share that we will save a few cents on a bottle instead of facing the unpleasant moment of increasing BBQ sauce retail shelf price. Your mixture design is a lovely thing!"

CHAPTER 14

Wrapping It All Up!

As Maria approaches her end of year performance review at the end of her first year at Ultimate BBQ, Maria tries to list all of the things she has learned. Of course, she had never been in a food production facility before, and all of the sanitation and safety rules that have become second nature were new at first. Her day-to-day work has been different than what she had expected—she did not realize how much time she would be spending in the production facility. And, she is most proud to have developed her statistics and data analysis skills and applied them to solve problems and make a real impact.

Maria is planning to visit her university in a few weeks for homecoming weekend. She emails Dr Wang, mentioning her visit and inviting her to lunch to celebrate her first successful year as a food scientist. She is pleased with Dr Wang's enthusiastic response, and after exchanging a few more brief emails, they settle on a time and place. A few weeks pass, and on the agreed upon date, Maria arrives only a moment before Dr Wang at Little Smokey's BBQ Shack.

Dr Wang greets Maria "I laughed out loud when you suggested this place in your email! I would think by now that you are tired of BBQ!"

Maria chuckles "I guess I've become obsessed by BBQ!"

The host leads them to a table, and soon after, they place their orders with a waiter. Maria insists that they order two dishes with very different sauces so that she can check them both out.

While they are waiting for their order to arrive, Maria says "I'm amazed how much I've learned in just one year. You prepared me well here at the university and away from the classroom throughout my first year. Thank you so much!"

"I've been happy to do so. "Dr Wang responds." It is wonderful to see you grow into a solid food scientist. What was your biggest surprise?"

Maria had already been thinking about this and is quick to respond. "I would not have thought that so much of the work would involve production. However, an even bigger surprise is how much I use statistical tools."

Statistics for Food Scientists
http://dx.doi.org/10.1016/B978-0-12-417179-4.00014-7

Dr Wang smiles and says "From your phone calls and emails, I know some of the tools you have been using. And I think you have picked up some more from colleagues. Let's see if we can make a list of these skills and how you have used them."

Maria starts her list by saying "You taught me about summary statistics and the importance of visualizing data rather than just crunching numbers. And we had performed some hypothesis tests in your class, but I did not really see how they would be used to answer real-life questions. Early on, when I used these tools to compare the production lines, hypothesis tests were important in helping us make decisions about how the lines were performing. This led us to focus on where the real problems were."

Dr Wang adds "I remember too that you had to determine if the viscosity measurements were reliable."

"Oh yes, that was so important!" Maria agrees "I just assumed that the numbers coming from an expensive instrument are correct. But it is people who are operating the instruments, so we had to look at the entire system, including operators and sample preparation. We now repeat that assessment of the measurement system every quarter since if the measurements are not reliable, everything else we are doing is a mess."

Dr Wang adds "So, once you ensured that the measurement process was good then you made a control plan for viscosity, right?"

Maria quickly agrees "I thought I had determined that it was steam pressure that most affects viscosity and thought that a control plan based on adjusting that alone would be sufficient. But it was not as simple as that since other things in addition to steam pressure affect viscosity."

Dr Wang nods in agreement "I think that was when I suggested that you learn more about experimental design."

"Yes" Maria agrees "but even before that I had the situation where we had a whole shift of production on hold because of the problem with the measurement system. I had thought of the product as being defective or not defective, a binary result, and came up with a sampling plan based on this consideration. You gave me some great advice on how to think about exploiting that the viscosity measurements are continuous in order to come up with a more efficient sampling scheme in the future."

"I remember that! "says Dr Wang." We all learn how to do things better based on hindsight. But, you were still able to give your colleagues good advice on how to determine what to do with the product you had on hold."

Maria agrees and continues "Even though I had created the control plan and also a monthly reporting system, our process still was not producing at the needed levels. The design of experiments approach was instrumental in understanding the factors influencing this and ultimately understanding how to better control viscosity."

Dr Wang smiles "I think you now know more about designing experiments than I do! Not only did you go through the typical screening and optimization steps, but you also used a mixture design at some point, right?"

"Yes "Maris says enthusiastically" we needed to change the starch blend when one of the ingredients became unavailable because of a shortage. I just finished up that work a few weeks ago."

Dr Wang concludes "Your statistical skills have certainly developed greatly in your first year working as a food scientist. Your questions made me realize the need for a resource list for my students. I consulted several professors at the University's Statistics department. I'll email you the list. And I'll be sure to mention your experiences in my classes in hopes of stressing to my students the importance of these skills in becoming effective food scientists."

"That is so flattering to me! "Maria gratefully responds" I was planning to pick up the check anyway; now, I'm certainly going to do so!"

When Maria returns from her trip, she finds the email from Dr Wang with the resource list. She will discuss purchasing these texts with Steve at her next development meeting.

Dr Wang's Resource List:

David S. Moors, George P. McCabe: Introduction to the Practice of Statistics.

Douglas C. Montgomery: Introduction to Statistical Quality Control, 7th Edition.

George E. P. Box, J. Stuart Hunter, William G. Hunter: Statistics for Experimenters: Design, Innovation, and Discovery, 2nd Edition.

Douglas C. Montgomery: Design and Analysis of Experiments, 8th Edition.

INDEX

Note: Page numbers followed by "f" and "t" indicate figures and tables, respectively.

Q

R

S

T

Printed in the United States
By Bookmasters

Medicinal Plants for Holistic Health and Well-Being

Medicinal Plants for Holistic Health and Well-Being

Edited by

Namrita Lall

Academic Press is an imprint of Elsevier
125 London Wall, London EC2Y 5AS, United Kingdom
525 B Street, Suite 1800, San Diego, CA 92101-4495, United States
50 Hampshire Street, 5th Floor, Cambridge, MA 02139, United States
The Boulevard, Langford Lane, Kidlington, Oxford OX5 1GB, United Kingdom

Library of Congress Cataloging-in-Publication Data
A catalog record for this book is available from the Library of Congress

British Library Cataloguing-in-Publication Data
A catalogue record for this book is available from the British Library

ISBN: 978-0-12-812475-8

For information on all Academic Press publications visit our website at https://www.elsevier.com/books-and-journals

Publisher: Mica Haley
Acquisition Editor: Erin Hill-Parks
Editorial Project Manager: Timothy Bennett
Production Project Manager: Priya Kumaraguruparan
Designer: Victoria Pearson

Typeset by TNQ Books and Journals

Contents

List of Contributors

Isa Anina Lambrechts, University of Pretoria, Pretoria, South Africa

Analike Blom van Staden, University of Pretoria, Pretoria, South Africa

Dikonketso Bodiba, University of Pretoria, Pretoria, South Africa

Bianca D. Fibrich, University of Pretoria, Pretoria, South Africa

Quenton Kritzinger, University of Pretoria, Pretoria, South Africa

Namrita Lall, University of Pretoria, Pretoria, South Africa

Murunwa Madzinga, University of Pretoria, Pretoria, South Africa

Marco Nuno de Canha, University of Pretoria, Pretoria, South Africa

Carel B. Oosthuizen, University of Pretoria, Pretoria, South Africa

Sunelle Rademan, University of Pretoria, Pretoria, South Africa

Anna-Mari Reid, University of Pretoria, Pretoria, South Africa

Karina Mariam Szuman, University of Pretoria, Pretoria, South Africa

Danielle Twilley, University of Pretoria, Pretoria, South Africa

Foreword

Yet another book about plants as medicine? Surely an overabundance of books on the subject already exists? However, I am confident that this new title, *Medicinal plants for holistic health and well-being*, is in a league of its own. The promotion of plant-derived products to treat all kinds of maladies under the banner of being "natural," and therefore by implication safe and desirable to use, is the order of the day. Unfortunately, claiming such health benefits without proper scientific support has undeniably caused some reputational damage to the noble idea of promoting plants as medicine. This new book by Prof. Lall et al., however, is sure to not only bring much credibility to the scientific study of plant medicines but would also contribute toward its promotion as an exciting field of research, especially among students. The book is also special in that it is a true team effort, conceived and written by a research leader of note and several of her postgraduate students and colleagues. It is a celebration of the trials and tribulations of scientific investigation, ultimately culminating in the excitement and reward of discovering a new pharmacologically active compound or of a new commercially released natural product that can be recommended with confidence.

This publication also bears testimony to what can be achieved by a supervisor and mentor of postgraduate students through sound and inspiring leadership and much hard work. I met Namrita Lall when she was still a botany student and was struck by her enthusiasm for her studies. Through hard work and several years later, Namrita Lall is a distinguished professor in Medicinal Plant Science at the University of Pretoria. She is truly passionate about evaluating the wonders of medicinal plants and not only values and respects available information on biocultural usage in her research but also takes it beyond by proving the efficacy of plant-derived products and eventually developing valuable pharmaceutical and cosmeceutical products. This is further substantiated by the various recognitions she has already received, including the Distinguished Young Women in Science Award (2011) from the Department of Science and Technology, South Africa, and more recently, the Order of Mapungubwe (2014), South Africa's highest civilian honor granted by the President of the country, for achievements in the international arena that have served the country's interests.

Not only are we as humans dependent on green plants for our existence but also, like all organisms, we share with them what Charles Darwin has aptly referred to in a broad and metaphorical sense as the "Struggle for Existence."

It always amazes that green plants, being essentially converters of radiant energy from the sun into a chemical form that serves as potential food for innumerable consumers, still manage to exist out there. Their survival, in the face of the threat of being consumed, is due in no small part to their remarkable ability to create a defense arsenal of chemicals that far exceeds our own. Fortunately some health threats faced by both plants and humans have broadly similar causes, making plants excellent candidates to search for chemicals to serve our own needs. Moreover, the chemical diversity in plants is so vast that they are the ideal source for chemical prospecting to discover novel compounds for human application. Through trail and error over millennia, much useful information about the health benefits of plants has been discovered by humans of all cultures. Such indigenous knowledge, though sadly fast disappearing if not already extinct in many societies, provides a rich source of ideas for scientific studies aimed at the development of safe and effective products for the pharmaceutical and cosmetic industries. Using the ethnomedicinal approach as a point of departure not only saves a great deal of time and resources in modern drug discovery but also encourages the forging of relationships between scientists and the communities holding the key to invaluable traditional knowledge.

This accessible and well-illustrated book gives a clear, comprehensive, and up-to-date introduction to the various usages and scientific validation of medicinal plants against a selection of common health conditions from across the globe. It brings together scientific studies conducted worldwide by scientists and students, including Prof. Lall's research team, and reports on several leads derived from plants used for specific purposes in traditional medicine. It provides a hands-on approach for evaluating the efficacy of medicinal plants for various conditions. The book eloquently captures the excitement of understanding natural and social phenomena of plants as a valuable resource with exceptional promise in the field of medicine. Current trends show increased popularity of natural products over their heavily commercialized synthetic counterparts. Many factors have spurred this popularity, including increased perceived safety. However, the misconception that plants pose no hazardous toxicity or other negative side effects needs to be corrected. Despite this, plants hold many advantages. For instance, they often offer multiple benefits in the treatment of a health condition and not only target specific symptoms.

True to its objectives and especially by endorsing the value of traditional knowledge and the power of the scientific method, this book indeed succeeds in highlighting the wonder of plants in holding solutions for many human health problems. I highly recommend it to prospective students, teachers, and scientists at all levels, to entrepreneurs, practitioners, and people working in the pharmaceutical and cosmeceutical industries, and to anyone else interested in medicinal plant science. It should prove, especially, useful to students embarking on a research project and also to established researchers as information on the usage of medicinal plants and the scientific validation thereof has been culled from various sources. Although the work is in the first instance aimed at an academic audience, informed members of the public keen to learn more about the discovery and testing of new medicines should also find it a highly informative read. I would like to congratulate Prof. Lall and her team on a book I believe makes a significant contribution to the field of modern drug discovery.

Abraham E. van Wyk
University of Pretoria
May 2017

Chapter 1

Traditional Medicine: The Ancient Roots of Modern Practice

Anna-Mari Reid, Carel B. Oosthuizen, Bianca D. Fibrich, Danielle Twilley, Isa Anina Lambrechts, Marco Nuno de Canha, Sunelle Rademan, Namrita Lall
University of Pretoria, Pretoria, South Africa

Chapter Outline

1.1 THE USE OF PLANTS IN MEDICINE: A HISTORIC TALE

Plants have been around longer than mankind; naturally, they would be one of the first resources exploited for their medicinal value. In ancient times, the etiological agents of disease were unknown to man, and hence the use of plants as medicinal sources became the product of trial and error, ultimately unveiling valuable medicinal plants. The resulting knowledge of this trial and error formed the root system from which modern medicine would eventually stem. The earliest records of plants used in medicine date back 5000 years to a Sumerian clay slab describing 12 recipes for drug preparation, referring to over 250 plants! Even more interestingly, it describes plants still commonly used in modern times such as the poppy and mandrake (Petrovska, 2012). The Mandrake, *Mandragora officinarum*, contains the alkaloid hyoscine, responsible for the use of Mandrake in Greek medicine as a surgical anesthetic. The presence of this powerful hallucinogenic alkaloid is what led to the association of the Mandrake with the supernatural (Carter, 2003). The effects of the mandrake, supernatural or not, are what led to the incorporation of mandrake as a

Medicinal Plants for Holistic Health and Well-Being. http://dx.doi.org/10.1016/B978-0-12-812475-8.00001-9

medicinal plant into many additional rituals and traditional medicine practices. The popularity of mandrake and the legends surrounding its use extend beyond medicine into the world of film, where it was interestingly incorporated into the popular film, *Harry Potter and the chamber of secrets*.

Although difficult, the World Health Organization (WHO) has harnessed the characteristics and elements of traditional medicine, drawing from descriptions as far back as the evidence allows even to modern traditional medicinal practices, to formulate a definition, which describes traditional medicine as "a holistic term encompassing diverse health practices, approaches, knowledge and beliefs incorporating plant, animal and/or mineral based medicines, spiritual therapies, manual techniques and exercises applied singularly or in combination to maintain well-being, as well as to treat, diagnose or prevent illness." Ultimately and more simplistically, traditional medicine refers to any form of indigenous healthcare system with ancient roots, cultural bonds, trained healers, and a theoretical construct. Examples include Ayurveda, Unani, Kampo, Shamanism, and traditional Chinese medicine (Fabricant & Farnsworth, 2001). The use of plants in medicine by specific ethnic groups is termed ethnobotanical medicine (Farnsworth, 1994a,b). A medicinal plant is thus any plant with alleged medicinal value that is used by Western standards or which contains components that are used as drugs (Laird & Kate, 2002). In industrialized countries, adaptations of traditional or ethnobotanical medicine are termed "Complementary" or "Alternative" medicines where complementary medicine refers to the use of medicinal plants in conjunction with a Western medicinal treatment, and alternative medicine refers to the use of a medicinal plant or plant-based medicine in place of Western medicine (WHO, 2003). Western medicine is distinguished from traditional medicine by the fact that it considers only ailments of the physical body, based on the principles of technology, science, knowledge, and clinical analysis developed in Northern America and Western Europe (Richter, 2003).

Medicinal plants and plant-derived medicines are used globally to treat various ailments. In modern society, these medicines are not only used by traditional cultures but are also gaining popularity under Western civilization. Currently there is a trend in natural alternatives as a source of new commercial products to synthetic chemicals. Although the use of plants as medicine is often underestimated, awareness of the active components of plants, which enable them to perform healing functions, needs to be increased. These active components, called phytochemicals, exist as they are required by the plant itself for specific functions. Generally, these phytochemicals are classified as secondary metabolites as they are not crucial for the survival of the plant but rather render a secondary beneficial function that provides the plant with a selective advantage. More than 100,000 such secondary metabolites have been isolated and characterized, many of which have been implemented as active ingredients in common medicines or have provided a basis for the development of synthetic actives that are incorporated into medicines (Rates, 2001).

More than 50% of drugs that are clinically used in the world are derived from natural products and their derivatives, with 25% being from higher plants.

When considering ~250,000 higher plant species globally, only 10,000 have documented medicinal uses, of which only 150–200 are actually incorporated into Western medicine (McChesney, Venkataraman, & Henri, 2007), it becomes increasingly evident that plants possess a great potential avenue for exploration.

The WHO considers 252 drugs "basic" 11% of these (~28) originate exclusively from plants, with a significant amount of the remaining synthetic drugs being derived from natural precursors (WHO, 2003). This further exemplifies the importance of plants in healing, across both pharmaceutical and cosmeceutical industries. The vast amount of plants that have yet to be explored outlines the importance of research in medicinal plants, and when considering what has been accomplished with what knowledge we do possess, it becomes very clear that the possibilities are endless and much to be sought after (Verpoorte, 1998).

Natural and herbal medicines have shown their potential over synthetic drugs by mostly having fewer side effects and lower levels of toxicity. Pharmacologically natural medicines have also shown their importance in their use as starting material for drug synthesis or directly as therapeutics. Natural medicines can also function as models for pharmacologically active compounds that may possess higher activity and less toxicity than their synthetic counterparts (Verma & Singh, 2008).

Not only do plants serve as potentially great alternatives to synthetic drugs and therapies, but natural plant-based therapies have also become increasingly popular, setting the benchmark for a more natural and safe platform. Reasons for this increased popularity include inefficiencies or hazards posed by conventional synthetic therapies, adverse side effects, affordability, and accessibility. Folk medicine and increased ecological awareness recently have also encouraged the idea that natural products may be less harmless (Rates, 2001).

1.2 TRADITIONAL MEDICINE IS A CRUCIAL PART OF AFRICAN HERITAGE

Traditional medicine, on the African continent, dates back ~4000 years. It was once the sole medicinal system; however, even in recent times for some, it remains the dominant system with an estimated 80% of the African Member States population use traditional medicine as a primary source of health care. Africans have used traditional medicine for hundreds of years as a form of health care. In South Africa it is estimated that there are ~27 million individuals who use traditional medicine. It has been calculated that from these individuals who use traditional medicine, they use it ~4.8 times/year, which equates to an average of about 157 g of plant material for each treatment and 750 g/year. In South Africa, indigenous plants are mainly used and this equates to ~20,000 tons being used each year from at least 771 plant species that have been recorded. There are a range of plant parts used, which are collected from forests, grasslands, woodlands, and thickets (Fig. 1.1; HST, 2015).

The WHO estimates that of the plants harvested for medicinal use, about 86% of the plant parts harvested results in death of the plant, which has a huge impact on sustainability. The decrease in availability of plants increases the time taken to find these plants and increases cost of the material. Common plants such as *Scilla natalensis* (Fig. 1.2) cost about R53/kg and scarce plants such as *Salacia kraussii* can reach up to R4,800/kg. There is also a big trade in plants from South Africa with neighboring countries such as Zimbabwe and Mozambique (HST, 2015).

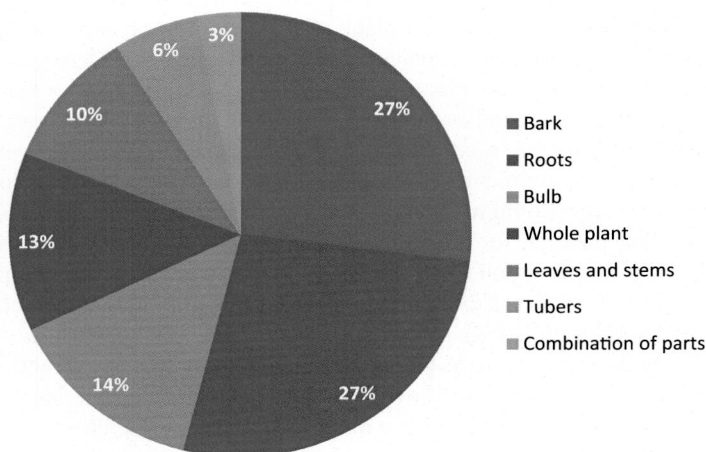

FIGURE 1.1 Percentage of plant parts used in traditional medicine in South Africa (HST, 2015).

FIGURE 1.2 *Scilla natalensis*, a commonly used medicinal plant traded at approximately R50/kg (Shebs, 2007a,b).

1.3 TRADITIONAL MEDICINE IN SOUTH AFRICA

South Africa has much to be explored as it is one of the richest temperate flora globally, comprising 24,000 specific and infraspecific taxa from 368 plant families. More than 10% of the Earth's vascular plant flora is offered by Southern Africa on less than 2.5% of its total surface area (Germishuizen & Meyer, 2003). South Africa, being one of the richest centers of plant biodiversity, has an excellent historic research base to begin with, gathered from a long history of traditional medicine implementing ~5700 different plant taxa (out of 24,000), of which 3000 are medicinally recognized (Mulholland, 2005).

Some popular South African plants often used include *Aloe ferox*, also commonly known as the bitter aloe, Kaapse aalwyn, or umhlaba (Fig. 1.3A). Traditionally it has been reported to be used as a laxative, for conjunctivitis, eczema, arthritis, and stress and hypertension and is now sold commercially as a laxative. An interesting medicinal plant reported to have many uses traditionally is *Boophone disticha*, commonly referred to as gifbol, bushman poison bulb or leshoma. Dried outer scales are used as a dressing for boils, septic wounds, and postcircumcision wounds. It is also thought to draw out pus. Mild decoctions are prepared for relief from headaches, eye conditions, weakness, and abdominal complaints, while stronger decoctions are used as hallucinogens and sedatives. Interestingly, examination of a 2000-year-old Khoisan mummy revealed scales

FIGURE 1.3 A few popular medicinal plants reported in folk medicine for their use in treating various conditions, from skin conditions to stomach complains and infections. (A) *Aloe ferox*, bitter aloe (Shebs, 2009), (B) *Cinnamomum camphora*, the camphor tree (Starr & Starr, 2009), (C) *Datura stramonium*, thornapple (Zell, 2009), (D) *Harpagophytum procumbens*, devil's claw (Pidoux, 2005), (E) *Hypericum perforatum*, St John's wort (Blanc, 2011), and (F) *Sutherlandia frutescens*, cancer bush (Humert, 2009).

of the bulb to be used for mummification purposes. *Catharanthus roseus*, the Madagascar periwinkle or isisushlungu has traditionally been used to treat diabetes; however, two alkaloids present within the plant are commercially available for the treatment of cancer in combination with chemotherapy. The camphor tree, *Cinnamomum camphora*, has been reported to be used traditionally for the treatment of heart conditions, colds and fevers, respiratory complaints such as pneumonia, inflammatory conditions, infections, diarrhea, and hysteria (Fig. 1.3B). Topical applications act as a counterirritant and antiseptic. The thornapple, *Datura stramonium* has traditionally been used for pain associated with boils, gout, abscesses, rheumatism, asthma, wounds, and as a hypnotic, and aphrodisiac (Fig. 1.3C). Commercially, alkaloids from this plant are used for the treatment of motion sickness, Parkinsonism as well as eye drops. *Harpagophytum procumbens*, the devil's claw or sengaparile, is traditionally used in the treatment of arthritis, rheumatism, digestive complaints (Fig. 1.3D). Pharmacologically it has been confirmed as an analgesic and antiinflammatory agent. St John's wort, *Hypericum perforatum* has been used for diarrhea, gout, rheumatism, and as an antidepressant (Fig. 1.3E). Oily extracts may be topically applied to wounds. Pharmacological evaluation has revealed *H. perforatum* to act as an antidepressant and antimicrobial agent. The cancer bush, *Sutherlandia frutescens*, has commonly been used in the Cape region of South Africa for the treatment of cancers and stomach complaints (Fig. 1.3F). Topically it has also been applied to ailments concerning the eye as well as wounds, and it has also been noted to be effective in the treatment of colds and influenza, varicose veins, chicken pox, piles, liver problems, inflammation, and backache. It has been attributed anticancer activity too.

For the provision of health care and medicine, many options exist within South Africa ranging from Western medicine, Western fringe practitioners, self-treatment, pharmaceuticals, faith healing, and traditional medicine (Dauskardt, 1990). Traditional medicine is a network of knowledge strongly supported by many individuals. It is therefore, very valuable for the information that is known to be passed on for many generations. Many modern Western medicines owe their existence due to the exploitation of all the components and structures of active substances that form part of medicinal plants.

South African traditional medicine is strongly associated with the use of herbs and multiremedies that are bestowed by the traditional healer known as a "sangoma" or "izinyanga" with strong spiritual components (Richter, 2003). Traditional healers may be divided into diviner-mediums (diviner-diagnostician) and herbalists (healers) (Jolles & Jolles, 2002). The differences between these divisions can be highlighted when considering the roles of each; for example, the diviner or *isangoma* diagnoses the patient generally through a spiritual means, therefore, serving as a communication medium between the patient and ancestral powers. This may be achieved through a trance state or through the throwing of bones (Dauskardt, 1990). The herbalist, or *inyanga*, is trained through an apprenticeship with an established herbalist and provides the correct herbs to be used for treatment. Although theoretically they appear easily distinguished, various sets of legislation, the history of South Africa and the encroachment of Western medicinal practices have blurred the lines separating the roles between these two (Richter, 2003).

The history of investigating the phytochemical potential of plants in South Africa can be attributed to the importance of stock farming, the risk that many diseases posed toward the stock farming industry, and the catastrophic effects this could have on the economy. Diseases such as "slangkop" poisoning, "stywe siekte," "vermeersiekte," and "gifblaar" poisoning resulted in the deaths of livestock due to their grazing on poisonous plants (Marais, 1944; Rindl, 1924; Theiler, 1911; Vahrmeyer, 1982). For that reason, halfway through the 20th century, extensive research was done to understand the nature of these deaths. However, it was only in the 1990's that a noticeable shift occurred in the field of research that wished to explore the use of traditional medicine and plants in the development of many products from pharmaceuticals to cosmeceuticals (Mulholland, 2005).

Several collaborative efforts between research organizations and traditional healers have found hundreds of plant species with the potential for drug development. Medicinal plants have played a major role in the traditions and lives of many people all over the world and from all walks of life. Plants that were used medicinally by various people and tribes had initiated the development of traditional healthcare systems such as Ayurveda and Unani, which have formed a key part of mankind for thousands of years (Gurib-Fakim, 2006). Today the medicinal importance of higher plants is still recognized with natural products and their derivatives having a 50% share of all drugs in clinical use (Gurib-Fakim, 2006).

1.4 DEVELOPMENT OF MODERN HERBAL PREPARATIONS

The development of modern herbal preparations involves many steps, namely identification of the plant material, authentication, collection, testing, standardization, processing, and finally, marketing. Each step has been developed in response to questions and concerns surrounding the use of herbal products.

The identification of medicinal plants may be through traditional knowledge systems (ethnopharmacology), on the basis of their plant family or through random selection. While random selection is somewhat inefficient in terms of time and resources, selecting a plant based on the traditional knowledge or plant family is more effective. Selection based on the plant family would be based on literature revealing other members of the plant family showing some medicinal value. The ethnomedicinal approach provides the greatest advantage above the aforementioned methods in that, it gives specific information about a specific species. Utilizing such methods, however, may lead to problems when moving forward into the second step of herbal production, authentication. This is because misidentification of the plant material may occur, especially in instances where closely related species exist. In such instances, a trained taxonomist is employed to correctly identify the plant species. The implications of misidentification may be severe where toxic plants are concerned and have implications in the treatment of the condition in question, showing no activity and ultimately being considered an ineffective treatment.

Once the correct plant material has been identified as authentic, material should be collected for testing. Collection should be sustainable, a lot of controversy has arisen surrounding the use of medicinal plant parts that cannot be harvested sustainably, such as roots. The material collected for testing may then be comprised of only plant parts, which have previously been reported on, or may be nonselective. For the plant material to be tested, a plant extract needs to be formulated. This is to concentrate the phytochemicals present within the plant material and ensure homogeneity in the sample to be tested. The plant may be extracted using water or organic solvents such as ethanol. Extraction may be through a variety of different mechanisms, including maceration, reflux, solid phase extraction, and supercritical fluid extraction. The extraction and concentration step may result in the concentration of phenolic or acidic compounds, and the effect of these on other active compounds present in the extract should be taken into consideration (Panichayupakaranant, 2011, Chap. 25). The mixture is then put through a filter such that the solvent containing the phytochemicals is separated from the plant pulp. The phytochemical solution is then allowed to dry such that all the solvents are evaporated to dryness, yielding a sticky syrup or powder of concentrated phytochemicals for testing. Authentication of the material as well as quality control, qualitative, and quantitative phytochemical determination should be carried out at all steps, that is to say on the fresh, dried, and powdered plant material as well as on the prepared extract (Panichayupakaranant, 2011, Chap. 25). The consequences involved with not adhering to these important parameters is firstly and most obviously the production of an unstandardized herbal formulation, with diminished or no efficacy, as well as increased potential for toxicity or adverse side effects when consumed (Panichayupakaranant, 2011, Chap. 25). Although an unstandardized herbal formulation may hold reduced efficacy as a disadvantage, standardization of all of the components that go into an herbal formulation should also confirm the absence of adulterants that may be added to enhance the efficacy without the consumer being aware of it. It is very important for the components to be specifically listed as the addition of adulterants do not give a clear indication of the medicinal value of the herbal formulation as an individual entity and may pose side effects to the consumer further down the line if, for example, the adulterant in question poses its own contraindications. This is different from a pure compound in that it is a cocktail of multiple molecules within the plant material.

Although whole extracts containing a phytochemical cocktail often offer the benefit of targeting more than one step in the pathogenesis or progression of any given condition, often also acting as a valued antioxidant, the application is somewhat limited due to the introduction of a vast number of variables that need to be considered. The specificity and target-orientated application of a single compound allow treatment to exist in a much more comfortable realm, with a drastically lowered number of variables when compared to a whole extract and far fewer biological implications, especially for consumption as a medicine.

Scientific testing is a crucial part of validating the traditional uses associated with the plants concerned. It normally involves a target-specific approach in certain conditions, closely examining the biological pathways involved and the effect of the plant sample on this pathway. Often, more than one test will be performed to evaluate the total activity of the plant sample. In this step toxicity and the adverse side effects of medicinal plants in plant-based formulations compared to synthetic counterparts are evaluated to determine the viability of the plant sample in any given application. Drew and Mayers (1997) classify adverse effects of an herbal formulation as being the result of the intrinsic characteristics (phytochemistry) of the sample or extrinsic factors (which may be the result of processing and manufacturing). For samples showing toxicity the therapeutic efficacy should be determined to determine if the benefit outweighs the cost and by how much. Samples showing a low therapeutic efficacy where the risk of increased adverse side effects exists should be discouraged. Depending on the method of administration, the active components may be incorporated into the medicine either as a pure compound or in the crude extract form as a cocktail containing multiple phyto-constituents. In the instance of compound isolation, the structure of the compound should be identified and checked for batch-to-batch consistency. In the instances where the crude extract is utilized, batch-to-batch consistency is attained through the metabolomic fingerprint of the species or via the identification of biomarkers, which are molecules that are associated with the compound of interest. The formulation of a standardized medicine thus involves both quantitative and qualitative phytochemical analysis.

Generally, two types of standardizations exist, either which aim to quantify the pure compound within an extract or the identification of a single or multiple actives within the extract. The efficacy of the pure compound needs to be evaluated versus the crude extract. The activity may differ greatly as the presence of other components in an extract may enhance or diminish the activity of the compound because the total value of a biological system is not always the sum of its components. Testing of the medicine also includes clinical trials to substantiate the scientific claims. Once it has gone through these steps and remains a good sample, the sample needs to be processed. Processing refers to the processing of the plant material into the form in which it is to be consumed (tablet, cream, ointment, syrup, tea, etc.) and includes cleaning the material and the removal of biological contaminants that may be present. The removal of heavy metals that may be toxic is also crucial in this step. Marketing, of course, is the final step and includes advertising the product to the relevant consumer. At this point, the medicine may be considered commercialized.

Although a great leap for nature, the commercialization step is not without hurdles. Upscaling of growing the plant requires the use of arable land for production. In Agro-countries, which focus a great majority of their resources on farming and the production of valuable crops aimed at diminishing world hunger, a balance needs to be struck to include medicinal plants. Obtaining the necessary permits and presenting the sustainability is also a great hurdle. The addition of medicinal plant farming does present great opportunity in terms of job creation, especially for unskilled labor abundant in third world countries. Not only are jobs created but employees providing this "unskilled labor" also gain valuable training surrounding the agricultural practices involved in growing medicinal crops, and also present opportunity for growth into a more skilled labor zone for more advanced steps in the production such as standardization. When implemented in the correct communities, commercialization of the medicinal plant crop allows for economic growth at the very basis of a community, instead of the traditional *trickle down* economics in which the rich get richer. Although modern medicine and traditional medicine both have a common goal, the general health of the population, traditional medicine, and the use of plants as a medicine should be done at a low-cost low-return level, where this form of medicine remains both easily accessible and affordable to the greater proportion of the population.

The benefits of natural products remain the fact that they offer complex phytochemical mixtures exhibiting great structural complexity and biological potency readily available for investigation (Panichayupakaranant, 2011, Chap. 25). The chapters that follow discuss a specific condition as well as the use of plants in its treatment, using traditional medicine as the basis for plant selection and validating these uses scientifically.

REFERENCES

Blanc, I. (2011). *Hypericum perforatum*. Online available: https://commons.wikimedia.org/wiki/Hypericum_perforatum#/media/File:HYPERICUM_PERFORATUM_-_MORROCURT_-_IB-477_(Peric%C3%B3).JPG.

Carter, A. J. (2003). Myths and mandrakes. *Journal of the Royal Society of Medicine*, *96*, 144–147.

Dauskardt, R. P. A. (1990). The changing geography of traditional medicine: Urban herbalism on the Witwatersrand, South Africa. *GeoJournal*, *22*(3), 275–283.

Drew, A. K., & Mayers, S. P. (1997). Safety issues in herbal medicine: Implications for health professionals. *Medical Journal of Australia*, *166*, 538–541.

Fabricant, D. S., & Farnsworth, N. R. (2001). The value of plants used in traditional medicine for drug discovery. *Environmental Health Perspectives*, *109*, 69–75.

Farnsworth, N. R. (1994a). Ethnopharmacology in drug development. *Ciba Foundation Symposium*, *185*, 42–51.

Farnsworth, N. R. (1994b). The role of ethnopharmacology in drug development. *Ciba Foundation Symposium*, *154*, 2–11.

Germishuizen, G., & Meyer, N. L. (2003). *Plants of Southern Africa: An annotated checklist. Strelitzia 14*. Pretoria: National Botanical Institute.

Gurib-Fakim, A. (2006). Medicinal plants: Traditions of yesterday and drugs of tomorrow. *Molecular Aspects of Medicine*, *27*, 1–93.

HST. (2015). *Chapter 13-economics of traditional medicine trade in South Africa*. Online available: http://www.hst.org.za/uploads/files/chap13_07.pdf.

Humert, C. (2009). *Sutherlandia frutescens*. Online available: https://commons.wikimedia.org/wiki/Lessertia_frutescens#/media/File:Sutherlandia_frutescens_1.jpg.

Jolles, F., & Jolles, S. (2002). Zulu ritual immunisation in perspective. *Africa*, *70*, 230.

Laird, S. A., & Kate, T. K. (2002). Linking biodiversity prospecting and forest conservation. In S. Pagiola, J. Bishop & N. Landell-Mills (Eds.), *Selling forest environmental services* (pp. 151–172). London, UK: Earthscan.

Marais, J. S. C. (1944). Monofluoro-acetic acid, the toxic principle of "gifblaar", *Dichapetalum cymosum*. *Onderstepoort Journal of Veterinary Science and Animal Industry*, *20*, 67–73.

McChesney, J. D., Venkataraman, S. K., & Henri, J. T. (2007). Plant natural products: Back to the future or into extinction? *Phytochemistry*, *68*, 2015–2022.

Mulholland, D. A. (2005). The future of ethnopharmacology: A Southern African perspective. *Journal of Ethnopharmacology*, *100*, 124–126.

Panichayupakaranant, P. (2011). Quality control, standardization, efficacy and safety of herbal medicines. In S. C. Mandal (Ed.), *Herbal drugs: A modern approach to understand them better* (pp. 387–435). Kolkata: New Central Book Agency.

Petrovska, B. B. (2012). Historical review of medicinal plants' usage. *Pharmacognosy Review*, *6*, 1–6.

Pidoux, H. (2005). *Harpagophytum procumbens*. Online available: https://commons.wikimedia.org/wiki/Harpagophytum_procumbens#/media/File:Harpagophytum_5.jpg.

Rates, S. M. K. (2001). Plants as source of drugs. *The International Society on Toxicology*, *5*, 603–613.

Richter, M. (2003). *Discussion paper prepared for the Treatment Action Campaign and AIDS Law Project: Traditional medicines and traditional healers in South Africa*.

Rindl, M. (1924). Preliminary note on a poisonous alkaloid from the above ground portions of the Transvaal yellow bulb *Homeria pallid*. *Transactions of the Royal Society of South Africa*, *11*, 251–256.

Shebs, S. (2007a). *Scilla natalensis*. Berkley, California: University of California Botanical Garden. Online available: https://commons.wikimedia.org/wiki/File:Scilla_natalensis_1.jpg.

Shebs, S. (2007b). *Scilla natalensis*. Berkley, California: University of California Botanical Garden. Online available: https://commons.wikimedia.org/wiki/File:Scilla_natalensis_1.jpg.

Shebs, S. (2009). *Aloe Ferox*. Online available: https://commons.wikimedia.org/wiki/Aloe_ferox#/media/File:Aloe_ferox_1.jpg.

Starr, F., & Starr, K. (2009). *Cinnamomum camphora*. Online available: https://commons.wikimedia.org/wiki/File:Starr_010419-0038_Cinnamomum_camphora.jpg.

Theiler, A. (1911). Stiff sickness in cattle. *Agricultural Journal of South Africa*, *1*, 10–21.

Vahrmeyer, J. (1982). *Gifplante van Suider Afrika*. Cape Town: Tafelberg Uitgewers.

Verma, S., & Singh, S. P. (2008). Current and future status of herbal medicines. *Veterinary World*, *1*(11), 347–350.

Verpoorte, R. (1998). Exploration of natures chemodiversity: The role of secondary metabolites as leads in drug development. *Drug Discovery Today*, *3*, 232–238.

World Health Organization. (May 2003). *Traditional medicine*. Online available: http://www.who.int/mediacentre/factsheets/2003/fs134/en/.

Zell, H. (2009). *Datura stramonium*. Online available: https://commons.wikimedia.org/wiki/Datura_stramonium#/media/File:Datura_stramonium_001.JPG.

Chapter 2

Are Medicinal Plants Effective for Skin Cancer?

Danielle Twilley, Sunelle Rademan, Namrita Lall
University of Pretoria, Pretoria, South Africa

Chapter Outline

2.1 CANCER

Cancer is a group of diseases occurring in all regions of the world and is the cause of millions of deaths each year. Different types of cancer have increased among certain regions of the world. The prevalence of cancer depends on a

Medicinal Plants for Holistic Health and Well-Being. http://dx.doi.org/10.1016/B978-0-12-812475-8.00002-0

variety of factors including the environment and the financial status of a country (WHO, 2014).

Cancer is the term used to describe a vast number of complex diseases, which can occur in almost all parts of the body. Different types of cancer differ in their behavioral aspects according to the types of cells from which they originate. They also differ from one another in terms of their growth rate, invasiveness, age of onset, prognosis, and their responsiveness to different treatments. These diseases are grouped together as cancer on the basis that all of these diseases display unregulated cell proliferation. Some cancers are also capable of metastasis, a process in which cancerous cells are able to spread to and invade other parts of the body. Angiogenesis, a process in which new blood vessels are formed, plays a pivotal role in the growth and metastatic potential of cancers (Angiogenic inhibitors, 2011; Klug, Cummings, Spencer, & Palladino, 2009).

2.1.1 Carcinogenesis

Henderson's dictionary of biology defines carcinogenesis as the "process by which a cancerous cell arises from a normal cell." Carcinogenesis is a complex process involving the division and differentiation of cells during the cell cycle and may also involve carcinogens. Carcinogens are defined as "any agent capable of causing cancer in humans or animals" by Henderson's dictionary of biology (Lawrence, 2008; Russell, 2010).

Differentiation and genetically controlled cell division is responsible for the development of specific tissues and organs. Cell proliferation takes place in a controlled manner to replace damaged and dead cells. When a cell deviates from its normal genetic program (mostly due to a mutation in one or more genes) during differentiation and division, tumors can start to develop. These cells are called transformed cells. Benign tumors occur when the transformed cells form a single mass, whereas malignant tumors can spread to other parts of the human body (Klug et al., 2009).

The most prevalent carcinogens are biological, chemical, and radiation carcinogens. Biological carcinogens involve the development of cancer in accordance with infection of a virus, bacteria, or parasite. Chronic infections with viruses such as the hepatitis B and C viruses and some types of the human papilloma virus (HPV) are the most significant causes of cancer via biological carcinogens. The HPVs are mainly associated with the development of cervical cancer. Chemical carcinogens include tobacco smoke, asbestos, arsenic, and aflatoxins found in food. Radiation exposure can come from various sources including the sun, radioactive radon gas, household appliances, etc. Ultraviolet (UV) light is the best known example of radiation carcinogens because of its role in the formation of skin cancers, which consist of basal cell carcinoma (BCC), squamous cell carcinoma (SCC), and malignant melanoma (MM). Both UVA and UVB rays from the sun can reach the Earth's surface and therefore, contribute toward the development of skin cancer; however, UVC rays are filtered by the ozone and cannot reach the Earth's surface and therefore, do not contribute toward skin cancer. UVB rays can directly induce skin cancer, whereas the UVA rays stimulate the carcinogenic effects of the UVB rays (Russell, 2010; WHO, 2015).

2.1.2 Cancer Prevalence Worldwide

Cancers are considered to be one of the major causes of death worldwide. In 2012 alone, cancer accounted for more than 8.2 million deaths. The five most significant cancers, relative to the death toll due to cancer, are lung cancer, liver cancer, stomach cancer, colorectal cancer, and breast cancer. The frequency with which these cancers occur among men and women may differ greatly. Cancer occurrence and cancer-related death numbers seem to be increasing in Africa, Asia, and Central and South America. Seventy percent of cancer deaths worldwide already occur within these regions and are expected to rise even more (WHO, 2014).

The rise in skin cancer cases across the world has been of major concern to health organizations. There has been a rise in the number of skin cancer cases over the past decade. New cases of skin cancer consist of 2–3 million nonmelanoma and 132,000 melanoma cases that occur globally each year. A major reason for the increase in skin cancer is the decreased levels of the ozone layer and therefore, is related to an increase in UV radiation and the casual attitude of the general public toward sun exposure. This occurrence is of even bigger concern in countries, such as Australia and South Africa, where UV radiation levels are already extremely high. Coincidently, Australia and South Africa are the top two countries in the world where skin cancer cases are prevalent (CANSA, 2010; WHO, 2002). It has been predicted that for a 10% depletion of the ozone layer there will be an increase of ~300,000 nonmelanoma and 4500 melanoma cases (WHO, 2015).

2.1.3 Cancer Prevalence in South Africa

The number of cancer cases reported has grown in frequency over the past few decades in South Africa. Currently, more than 100,000 cases are reported each year. In South Africa, cancer is accountable for more deaths than HIV/AIDS, tuberculosis, and malaria combined. The five most prevalent cancers found among South African men are prostate cancer, lung cancer, colorectal cancer, esophageal cancer, and Kaposi sarcoma. The five most prevalent cancers found among South African women are breast cancer, cervical cancer, colorectal cancer, Kaposi sarcoma, and melanoma. An increased number of these cancers are seen among the low-income population of South Africa (CANSA, 2013a). According to a recent study by Lancet, it has been predicted that the number of cancer cases in South Africa could increase by 78% by 2030 (Health.24, 2015).

Skin cancer is considered as one of the most common cancers in South Africa. About 20,000 skin cancer cases are reported each year and ~700 deaths each year are due to melanoma. With these numbers, relative to South Africa's population, South Africa has the second highest incidence rate of skin cancer in the world, after Australia. In South Africa, skin cancer is also considered to be the most common cancer present in the younger generation (CANSA, 2010; Wegner, 2011).

2.2 THE FUNDAMENTALS OF THE SKIN AND SKIN CANCER

2.2.1 The Skin

The skin is considered the largest organ in the human body. It is mostly adversely affected by UV radiation because it is directly exposed to sunlight. It is responsible for a variety of important functions of the human body and consists of three main layers, the subcutaneous; dermis; and epidermis. The subcutaneous layer of the skin is the base layer and is generally only exposed to UVA rays. This layer provides insulation and protection against knocks and falls. It also includes a layer of fat, which may act as an energy reservoir when the body is deprived of food (Balakrishnan & Narayanaswamy, 2011; National Geographic, 2014).

The dermis layer is accountable for the vigor and suppleness of the skin and can be found between the subcutaneous and epidermis layers. This layer contains fibers of collagen and elastin. Temperature regulation and signal transduction actions take place in this layer. The dermis is also responsible for various functions in the human body because it contains hair follicles and a range of glands such as sweat glands, apocrine glands, and sebaceous glands. Both UVA and UVB rays can affect the dermis layer of the skin. The epidermis layer is the outermost part of the skin. Keratinocytes are the most common cells found in the epidermis. These cells may grow outward continually. The stratum corneum is the layer of dead keratinocytes found on the surface of the skin. Melanin is a pigment, accountable for skin color, which is produced in melanocytes that are present in the epidermis. This pigment can absorb UV light and helps to protect the skin from UV damage. UV damage occurs when the amount of melanin produced is not sufficient to protect the skin and can be responsible for DNA damage; hyperplasia; immunosuppression; erythema; sunburn; photoaging; melanogenesis; edema; and the formation of reactive oxygen species (Balakrishnan & Narayanaswamy, 2011; National Geographic, 2014).

2.2.2 Skin Cancer

Skin cancer or skin neoplasia usually arises within the epidermis layer of the skin. The three main types of skin cancer are BCC, SCC, and MM (Wegner, 2011).

BCC occurs in the lower region of the epidermis in the basal cells and is the most common type of skin cancer (Fig. 2.1A). These carcinomas are most commonly found on sun-exposed areas of the body but may also be found in other areas. BCC is characterized by the superficial blood vessels that cover the

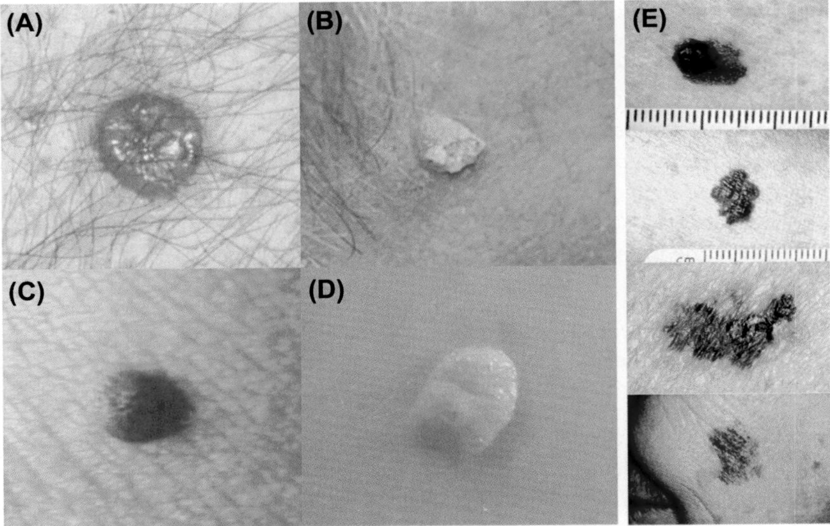

FIGURE 2.1 Basal cell carcinoma (A), squamous cell carcinoma (B), pigmented mole (C), nonpigmented mole (D), and the ABCDE model for detection of melanoma (E) (National Cancer Institute, 1985 & 1998; Nelson, 2012).

surface of the carcinoma. It usually forms tiny red bumps, which never heal and bleed occasionally. Even though BCC rarely spreads to other parts of the body, it could become dangerous if left untreated (CANSA, 2013b). The symptoms that are related to BCC include the following; however, a physician will need to give the final diagnosis (Skin Cancer Foundation, 2015):

- A persistent nonhealing sore
- An open sore that bleeds, oozes or crusts, and remains open and unhealed
- A patch that remains red and irritated on the face, chest, shoulders, arms, or legs
- A patch that may continuously itch or hurt
- A bump that is shiny or nodular that is clear or pearly and can be red, pink, or white
- A bump can be tan, black, or brown and can be confused with a mole
- A pink growth with an elevated border and an indentation, which is crusted in the center
- Formation of blood vessels on the surface of an enlarged growth
- White, yellow, or waxy appearances on the skin similar to a scar

SCC occurs in the epidermis layer, arising from the squamous epithelium (Fig. 2.1B). SCC has a higher probability of spreading to other parts of the body than BCCs. SCC is most commonly found on the regions of the body, which are exposed to the sun, such as the face, ears, lips, neck, and back of the hands.

It can also occur on the genitals, anus, and inside the mouth of both men and women. The appearance of SCC varies greatly, but the most noted appearances are red bumps, ulcers, or smooth lesions. If left untreated it may be fatal. This is especially relevant for people suffering from HIV/AIDS or chronic lymphocytic leukemia and people who have had organ transplants in the past because SCC is present in an aggressive form within these people. It can also spread to other organs, including the lymph nodes, and it may destroy healthy tissue surrounding the tumor (CANSA, 2013b). Symptoms of SCC are similar to that of BCC and include the following (American Cancer Society, 2015a):

- Wartlike growths
- Open sores that can ooze or become crusted and do not heal
- Raised bumps that have a lowered, indented area in the center
- Scaly patched or rough skin that can bleed or become crusted

MM is one of the top six cancers found among the women of South Africa, with around 627 cases reported in 2008. MM is a type of skin cancer occurring in the epidermis layer, arising as tumors in melanocytes. Melanomas are the most aggressive type of skin cancer due to the ability to spread to other parts of the body, such as the muscles, lungs, liver, brain, bone, and deeper into the skin. Melanomas occur when a gene(s) mutates in the melanocytes. This mutation will lead to a change in the protein produced, which in turn, cause these cells to multiply uncontrollably and form malignant tumors. These melanoma tumors may either bear a resemblance to moles or develop from a mole (Fig. 2.1C and D). Moles are usually dark brown to black but may occur in other colors as well (Fig. 2.1E). This type of skin cancer is primarily caused by frequent but extreme exposure to UV rays from both the sun and tanning beds. Investigations on the link between breast cancer and melanoma have indicated a 29%–46% higher risk of the development of melanoma in patients diagnosed with breast cancer (CANSA, 2013b, 2014). Symptoms of melanoma development are often based on the ABCDE rule, which describe the following features (Fig. 2.1E) (American Cancer Society, 2015b):

A. → **Asymmetry**: This is where the two halves of the mole are not the same
B. → **Border**: The border has an irregular or smooth appearance and can be notched or blurred
C. → **Color**: The mole consists of more than one color and can include colors such as black, tan, brown, white, pink, red, and sometimes blue
D. → **Diameter**: The diameter of the mole is more that 5 mm in size
E. → **Evolving**: The mole shows changes in color, size, or shape

Other symptoms are sometimes not associated with the above ABCDE rule and can include the following (American Cancer Society, 2015b):

- The mole starts to itch or becomes tender and painful
- An unhealing sore in the skin

- The pigmentation of the mole has spread outside of its borders
- Swelling and redness occurs next to the mole
- The mole starts to become scaly or nodular and starts to bleed or ooze

2.2.3 The Ultraviolet Index

The Global Solar UV Index (UVI) is a tool initiated by the World Health Organization (WHO) as a measurement of the UV radiation level at the Earth's surface. This endeavor is part of a larger program to raise public awareness toward the negative implication of overexposure to UV light. The UVI ranges from 0 to ≥11 (Table 2.1). The general thought is that the higher the UVI, the higher the risk will be for the general public to attain diseases associated with an overexposure to UV radiation, though other factors including the frequency of exposure to high levels of UV radiation also plays a huge role (WHO, 2002).

The South African Weather Service (SAWS) started measuring UVB radiation levels in 1994 as recommended by various institutions including the WHO. The sites where UVB radiation is measured in a few towns in South Africa includes Pretoria; Cape Town; Durban; Port Elizabeth; and De Aar. According to the measurements by the SAWS, it was found that all the sites except for Port Elizabeth show maximum UVB radiation levels. Mean radiation levels per month as measured by the SAWS from 1994 to 2009 have indicated that radiation levels fall in the high to extreme categories at all sites from August to April, with Pretoria and Durban having the highest levels of radiation. These readings indicate that South Africa is one of the countries with the highest UV radiation levels in the world. With such elevated levels of UV radiation, appropriate measures should be taken to educate the general public on how to minimize exposure and how to prevent diseases, such as skin cancer, associated with overexposure to UV radiation (SAWS, 2012).

TABLE 2.1 The Ultraviolet Index (UVI) Range and Its Adverse Effects (WHO, 2002)

Categories	UVI Range	Description
Low	0–2	No danger to the average person
Moderate	3–5	Little risk of harm from unprotected sun exposure
High	6–7	High risk of harm from unprotected sun exposure
Very high	8–10	Very high risk of harm from unprotected sun exposure
Extreme	11+	Extreme risk of harm from unprotected sun exposure

2.2.4 Conventional Treatments for Skin Cancer

There are various different types of treatments for patients that have been diagnosed with melanoma. Treatments that are currently being used include surgery, chemotherapy, radiation therapy, and immunotherapy. Table 2.2 summarizes the current drugs approved by the US Food and Drug Administration (FDA) for the treatment of various skin cancers as well as their target and the stage or type of cancer (AIM, 2014; Longley et al., 2001; Monroe, 2015; NIH, 2015).

2.3 PLANTS AS A SOURCE FOR CANCER PREVENTION AND TREATMENT

Plants have been used throughout history for their medicinal properties. The use of plants to treat various ailments can date as far back as to the era of the Egyptians (1500 before Christ), where onions were used for the treatment of inflamed wounds. A variety of ancient medicinal systems in India, Asia, and Africa also testifies to the immense and widely spread use of plants for their medicinal properties (Herr & Büchler, 2010).

2.3.1 Plant-Derived Compounds Currently Used for Cancer Treatment

Traditional herbal medicine has played a pivotal role in cancer treatment throughout history, though it was only from the 1950s that modern medicine turned its attention toward plants with anticancer activity. Anticancer compounds from plants such as vinca alkaloids and podophyllotoxins were discovered during this period. Because of the success of finding significant anticancer compounds in plants in the past, the United States National Cancer Institute (NCI) began a program in 1960 in which plants were collected from all over the world and screened for possible anticancer activities. Anticancer compounds, including the taxanes and camptothecins, were discovered by this program and are still used today for the treatment of a range of cancer types (Cragg & Newman, 2005).

2.3.2 Research on South African Plants for Novel Anticancer Treatments

In 1998, the NCI and the Council for Scientific and Industrial Research (CSIR) launched a research program, which involved the screening of plant extracts for their potential as anticancer agents in South Africa. These South African plants were randomly chosen, and their cytotoxicity was tested against three cell lines (breast, renal, and melanoma) by the CSIR. The CSIR has thus far tested the extracts of 700 plant species. It was found that the maximum number of plants tested for their research program belonged to the Asteraceae family. Other plant families such as Anacardiaceae, Apocynaceae, Araliaceae, Celastraceae,

TABLE 2.2 Food and Drug Administration Approved Drugs Used for the Treatment of Skin Cancer (Longley et al., 2001; Monroe, 2015; NIH, 2015; AIM, 2014)

Names	Common Names	Target	Types/Stages of Skin Cancer
Drugs for Basal and Squamous Cell Carcinoma			
Imiquimod	Aldara	Enhances the immune response by inducing the production of proinflammatory cytokines	Superficial basal cell carcinoma and squamous cell carcinoma
5-Fluorouracil	Efudex/Tolak/Carac/ Fluoroplex	Thymidylate synthase (TS) inhibitor which stops DNA replication	Superficial basal cell carcinoma that cannot be removed by surgery as well as superficial squamous cell carcinoma
Drugs for Melanoma			
Aldesleukin	Proleukin	Increases **IL-2**	Advanced metastatic melanoma
Dabrafenib	Tafinlar	Kinase inhibitor that blocks V600E (mutated forms of **BRAF**)	Metastatic melanoma/cannot surgically be removed or patients with BRAF mutation (stage III and IV melanoma)
Dacarbazine	DTIC-Dome	**Alkylating agent** (resting phase of cells)—damages DNA and RNA	Metastatic stage IV melanoma
Intron A	Recombinant interferon alfa-2b	Adjuvant therapy/possibly inhibits **angiogenesis**	Late-stage cancer/recurrent cancer/after surgery

Continued

TABLE 2.2 Food and Drug Administration Approved Drugs Used for the Treatment of Skin Cancer (Longley et al., 2001; Monroe, 2015; NIH, 2015; AIM, 2014)—cont'd

Names	Common Names	Target	Types/Stages of Skin Cancer
Ipilimumab	Yervoy	Blocks **CTLA-4** protein	Metastatic melanoma/cannot surgically be removed (stage III and IV melanoma)
Pembrolizumab	Keytruda	Blocks **PD-1** and ligands **PD-L1 and PD-L2**	Metastatic melanoma/cannot surgically be removed, patients treated with ipilimumab or BRAF inhibitor (stage III and stage IV melanoma)
Trametinib	Mekinist	Alone or in combination with dabrafenib- kinase inhibitor that blocks V600E and V600K (**mutated forms of BRAF**)	Metastatic melanoma/cannot surgically be removed or patients with BRAF mutation (stage III and stage IV melanoma)
Nivolumab	Opdivo	Blocks **PD-1** and ligands **PD-L1 and PD-L2**	Metastatic melanoma/cannot surgically be removed, patients treated with ipilimumab or BRAF inhibitor (stage III and stage IV melanoma or BRAF mutation)
Peginterferon alfa-2b	Sylatron/PEG-intron	Adjuvant therapy-possibly inhibits **angiogenesis**	Microscopic or macroscopic nodal melanoma/after surgical removals or lymph nodes dissection
Vemurafenib	Zelboraf	Kinase inhibitor that blocks V600E (**mutated form of BRAF**)	Metastatic melanoma/cannot surgically be removed or patients with BRAF mutation (stage III and IV melanoma)

Chrysobalanaceae, Crussulaceae, and Solanaceae were also represented. Of these families the plants belonging to the Asteraceae and Solanaceae families showed significant potential for cancer. The plant extracts that showed significant cytotoxicity on these cell lines were then screened by the NCI against 60 human cancer cell lines. Clinical trials are still required before any of these plant extracts can be incorporated into anticancer drugs for commercialization (Fouche et al., 2008).

2.3.3 Traditionally Used Plants From South Africa for Cancer and Their Associated Symptoms

In South Africa the number of cancer cases is increasing. There are many traditional healers who use medicinal plant for the treatment of various types of cancers as well as symptoms that might be associated with skin cancer, such as wounds, rashes, sores, blisters, boils, and abscesses. There are also traditional healers who use medicinal plants for the treatment of sunburns. As mentioned previously, exposure to the sun is one of the major risk factors that can lead to the contribution of skin cancer. Table 2.3 depicts a summary of plants traditionally used in South Africa for the treatment of cancer, skin cancer as well as symptoms, and skin disorders that could be associated with the various types of skin cancer (Fig. 2.2) (Coopoosamy & Naidoo, 2012; Koduru, Grierson, & Afolayan, 2007; Mahomoodally, 2013; Street & Prinsloo, 2013; Thring & Weitz, 2006).

2.3.4 Functional Foods and Nutraceuticals

Age-old statements such as "an apple a day keeps the doctor away" and "you are what you eat" have implied that the foods, including edible plants, play a huge role in the functioning of the body. Hippocrates (460–370 BC) himself stated "let your ailment be your medicament and your medicament be your food". With the increasing excitement regarding the search for medicines from plants, an increasing interest arose in the role that diet plays in disease, especially cancer. Today, foods that are consumed frequently and that demonstrate potential benefits toward human health or advocate a reduction in the risk of certain diseases are termed functional foods. Functional foods from plants include fruits; vegetables; herbs; spices; beans; oils; and beverages such as tea. Nutraceuticals refer to products that are sold over the counter, which contain phytochemicals (compounds found in plants) with potential benefits toward human health. Nutraceuticals are generally provided in the form of tablets or pills containing concentrated amounts of phytochemicals. Much research has been done on the ability of foods to be chemopreventive (Fig. 2.3; Table 2.4). Cancer chemoprevention generally refers to a pharmacologically active substance that has the ability to prevent, reverse, or inhibit the formation of cancerous cells or that can potentially inhibit the progression of cancerous cells to become invasive (Herr & Büchler, 2010; Nobili et al., 2009; Park & Pezzuto, 2002).

TABLE 2.3 A Few Plants Commonly Used in for the Treatment of Cancer, Skin Cancer and for Various Symptoms and Skin Disorders That Could Be Associated With Skin Cancer (Koduru et al., 2007; Mahomoodally, 2013; Street & Prinsloo, 2013; Coopoosamy & Naidoo, 2012; Thring & Weitz, 2006)

Species/ Families	Common Names	Tribe/Area	Plant Parts	Extract Preparation	Usage	References
Agapanthus africanus (L.) Hoffmanns/ Agapanthaceae	Agapanthus	Eastern Cape Province	Roots	Dried roots are powdered and infused in water and taken orally	Cancer	Koduru et al. (2007) and Braxmeier (2009) (Fig. 2.2A)
Agathosma betulina (Berg.)/ Rutaceae	Buchu	Khoi-San/ Western Cape Province	Whole plant/ leaves	Plant mixed with fat/lubrication/ tea of leaves taken orally	Keeps skin soft and moist/ prevention of cancer	Street and Prinsloo (2013) and Köhler (1987) (Fig. 2.2B)
Aloe ferox (Mill.)/Aloaceae	Cape Aloe/ bitter Aloe	Zulu's/Durban/ Western Cape Province	Leaves	Cut leaves and juice directly applied	Sunburns, burns, rashes, wounds, blisters, sores, and skin problems	Street and Prinsloo (2013), Coopoosamy and Naidoo (2012), Thring and Weitz (2006) and Pienaar (2015) (Fig. 2.2C)
Boophone disticha (L.f.) Herb./ Amaryllidaceae	Poison bulb/ sore-eye flower/gifbol	Zulu's/Durban/ Western Cape Province	Outer covering of bulb/fresh leaves	Applied directly to affected area	Boils and abscesses/stop bleeding of wounds and for sores	Coopoosamy and Naidoo (2012), Thring and Weitz (2006) and Hillewaert (1988) (Fig. 2.2D)
Bulbine frutescens L. (Willd.)/ Asphodelaceae	Snake flower	Zulu's/Durban	Leaves	Leaf juice is applied topically	Sunburn and rashes	Coopoosamy and Naidoo (2012) and image (Fig. 2.2E) taken by Fibrich (2016)

Bulbine lagopus (thumb.) N.E.Br./Asphodelaceae	Geel katstert	Western Cape Province	Leaves	Leaves broken and juice applied directly	Wounds, sores, and skin conditions	Thring and Weitz (2006) and Richfield (2010) (Fig. 2.2F)
Cannabis sativa L./Cannabaceae	Hemp	Eastern Cape Province	Leaves	Crushed leaves are taken orally	Cancer	Koduru et al. (2007) and Chmee2 (2010) (Fig. 2.2G)
Carpobrotus edulis L. L. Bolus./Mesembryanthemaceae	Suurvy/sour fig	Western Cape Province	Leaves	Leaf sap applied directly	Sunburn	Thring and Weitz (2006) and Tangopaso (2009) (Fig. 2.2H)
Celtis africana Burm. f./Ulmaceae	White stinkwood/witstinkhout	Eastern Cape Province	Bark and roots	Dried bark and roots are powdered and infuse and taken orally	Cancer	Koduru et al. (2007) and Shawka (2010a) (Fig. 2.2I)
Cotyledon orbiculata L./Crassulaceae	Kouterie/plakkie	Western Cape Province	Leaves	Leaf sap applied directly	Sores	Thring and Weitz (2006) and Shawka (2010b) (Fig. 2.2J)
Curtisia dentata (Burm. f.) C.A.Sm./Cornaceae	Assegai	Eastern Cape Province	Bark and roots	Bark and roots stamped and boiled in water and taken orally	Cancer	Koduru et al. (2007) and Shawka (2011a) (Fig. 2.2K)
Datura stramonium L./Solanaceae	Stinkblaar	Western Cape Province	Leaves	Place leaf directly on sores	Sores	Thring and Weitz (2006) and Blanc (2011) (Fig. 2.2L)

Continued

TABLE 2.3 A Few Plants Commonly Used in for the Treatment of Cancer, Skin Cancer and for Various Symptoms and Skin Disorders That Could Be Associated With Skin Cancer (Koduru et al., 2007; Mahomoodally, 2013; Street & Prinsloo, 2013; Coopoosamy & Naidoo, 2012; Thring & Weitz, 2006)—cont'd

Species/ Families	Common Names	Tribe/Area	Plant Parts	Extract Preparation	Usage	References
Elytropappus rhinocerotis (L.f.) Less./Asteraceae	Renosterbos	Western Cape Province	Leaves	Tea used to wash	Wounds and sores	Thring and Weitz (2006) and Shawka (2011b) (Fig. 2.2M)
Eucomis autumnalis (Mill.) Chitt/ Hyacinthaceae	Pineapple flower/ pineapple lily	Eastern Cape Province	Bulbs	Decoctions	Cancer	Koduru et al. (2007) and Massyn (2006) (Fig. 2.2N)
Euphorbia ingens E. Mey. ex Boiss/ Euphorbiaceae	Naboom	Eastern Cape Province	Latex	Latex applied topically for external cancers	External cancer	Koduru et al. (2007) and Hillewaert (2012) (Fig. 2.2O)
Gunnera perpensa L./Gunneraceae	River pumpkin	Eastern Cape Province	Rhizome	Aqueous infusions and decoctions taken orally	Cancer	Koduru et al. (2007) and Shawka (2011c) (Fig. 2.2P)
Harpagophytum procumbens (Burch.) DC/ Pedaliaceae	Devil's claw	San of the Kalahari	Water-storing secondary tuberous roots	Topical application from infusions, decoctions, tinctures, powders, and extracts	Boils, skin cancer, skin ulcers, sores, and skin injuries	Mahomoodally (2013), Street and Prinsloo (2013) and Culos (2015) (Fig. 2.2Q)

Harpephyllum caffrum Bernh./ Anacardiaceae	Wild plum	Zulu's/Durban	Bark	Extract	Eczema	Coopoosamy and Naidoo (2012) and image (Fig. 2.2R) taken by de Canha (2016)
Hypoxis argentea L./Hypoxidaceae	Inongwe	Eastern Cape Province	Corms	Corms are stamped, boiled in water, and taken orally	Cancer	Koduru et al. (2007)
Hypoxis colchicifolia Bak/ Hypoxidaceae	iLabatheka	Eastern Cape Province	Corms	Corms are crushed, boiled in water, and taken orally	Cancer	Koduru et al. (2007)
Hypoxis hemerocallidea Fisch., C.A. Mey. and Avé-Lall./ Hypoxidaceae	African potato	Variety of cultures including Zulu's/Durban/ Eastern Cape Province	Corms	Corms are pulverized, boiled in water, and taken orally	Cancer/internal tumors and antioxidant	Street and Prinsloo (2013), Coopoosamy and Naidoo (2012), Koduru et al. (2007) and Venter (2012) (Fig. 2.2S)
Knowltonia capensis (L.) Huth/ Ranunculaceae	Blistering leaves/ brandblare	Eastern Cape Province	Leaves and roots	Crushed leaves are directly applied/crushed roots are boiled and taken orally	Cancer/external cancer	Koduru et al. (2007)
Leonotis leonurus (L.) R. Br./ Lamiaceae	Klipdagga/ wild dagga	Western Cape Province	Leaves and flowers	Tea taken orally morning and evening	Cancer	Thring and Weitz (2006) and Otero (2012) (Fig. 2.2T)

Continued

TABLE 2.3 A Few Plants Commonly Used in for the Treatment of Cancer, Skin Cancer and for Various Symptoms and Skin Disorders That Could Be Associated With Skin Cancer (Koduru et al., 2007; Mahomoodally, 2013; Street & Prinsloo, 2013; Coopoosamy & Naidoo, 2012; Thring & Weitz, 2006)—cont'd

Species/Families	Common Names	Tribe/Area	Plant Parts	Extract Preparation	Usage	References
Lippia javanica (Burm. f.) Spreng/Verbenaceae	Fever tea/lemon bush	Zulu's/Durban	Leaves and stems	Tea is cooled and applied topically	Skin disorders, heat rashes, and other rashes	Coopoosamy and Naidoo (2012) and JMK (2013) (Fig. 2.2U)
Melianthus comosus Vahl./Melianthaceae	Kruidjie roer my nie	Western Cape Province	Leaves	Boiled in water and used to wash wounds and sores	Wounds and sores	Thring and Weitz (2006) and Shebs (2006) (Fig. 2.2V)
Merwilla plumbea (Lindl.) Speta/Hyacinthaceae also known as *Merwilla natalensis* (Planch.) Speta/Hyacinthaceae also known as *Scilla natalensis* Planch./Liliaceae	Wild squill	Eastern Cape Province	Bulbs	Decoctions	Cancer	Koduru et al. (2007) and BotBln (2010) (Fig. 2.2W)
	Inguduza	Zulu, Tswana, Swati, and Sotho	Bulb	Extract, powder, and cooked	Wounds, tumors, internal tumors	Street and Prinsloo (2013)
	Wild squill/blouslangkop	Zulu's/Durban/Eastern Cape Province	Plant/bulbs	Ash of burnt plant/decoctions	Open sores and skin conditions/cancer	Coopoosamy and Naidoo (2012), Koduru et al. (2007) and Shebs (2007) (Fig. 2.2X)
Pittosporum viridiflorum Sims/Pittosporaceae	Cheesewood/Kasuur	Eastern Cape Province	Bark and roots	Decoctions or infusions taken orally/dried barks and roots are pulverized and taken orally	Cancer	Koduru et al. (2007) and Starr and Starr (2007) (Fig. 2.2Y)

Psidium guajava L./Myrtaceae	Guava	Zulu's/Durban	Leaves	Crushed leaves applied topically	Wounds	Coopoosamy and Naidoo (2012) and Starr and Starr (2002) (Fig. 2.2Z)
Solanum aculeastrum Dunal/Solanaceae	Poison apple/ gifappel	Eastern Cape Province	Fruits, eaves, bark and roots	Fruits are boiled and filtered. Decoction of this taken orally	Cancer	Koduru et al. (2007)
Sutherlandia frutescens (L.) R. Br/Fabaceae	Cancer bush/ kankerbossie	Khoi-San/Cape Dutch/Western Cape Province	Leaves/ branches, flowers and seeds	Tea prepared and ingested morning and evening/ decoctions made from aerial parts and taken orally or administered topically	Internal cancer and wounds/ external cancer	Street and Prinsloo (2013), Koduru et al. (2007) and Hummert (2009) (Fig. 2.2AA)
Trema orientalis L. Blume./Ulmaceae	Pigeon wood	Zulu's/Durban	Leaves	Infusion	Sores and wounds	Coopoosamy and Naidoo, (2012) and Garg (2009) (Fig. 2.2AB)
Tulbaghia violacea Harv./Alliaceae	Wild garlic	Eastern Cape Province	Bulbs and leaves	Fresh bulbs are boiled and taken orally/leaves are crushed and infused in water and taken orally	Cancer	Koduru et al. (2007) and Ziarnek (2014) (Fig. 2.2AC)

FIGURE 2.2 A few plants traditionally used for the treatment of cancer, skin cancer, and for various symptoms and skin disorders that could be associated with skin cancer. (A) Agapanthus africanus, (B) Agathosma betulina, (C) Aloe ferox, (D) Boophone disticha, (E) Bulbine frutescens, (F) Bulbine lagopus, (G) Cannabis sativa, (H) Carpobrotus edulis, (I) Celtis africana, (J) Cotyledon orbiculata, (K) Curtisia dentata, (L) Datura stramonium, (M) Elytropappus rhinocerotis, (N) Eucomis autumnalis, (O) Euphorbia ingens, (P) Gunnera perpensa, (Q) Harpagophytum procumbens, (R) Harpephyllum caffrum, (S) Hypoxis hemerocallidea, (T) Leonotis leonurus, (U) Lippia javanica, (V) Melianthus comosus, (W&X) Merwilla plumbea, (Y) Pittosporum viridiflorum, (Z) Psidium guajava, (AA) Sutherlandia frutescens, (AB) Trema orientalis, and (AC) Tulbaghia violacea.

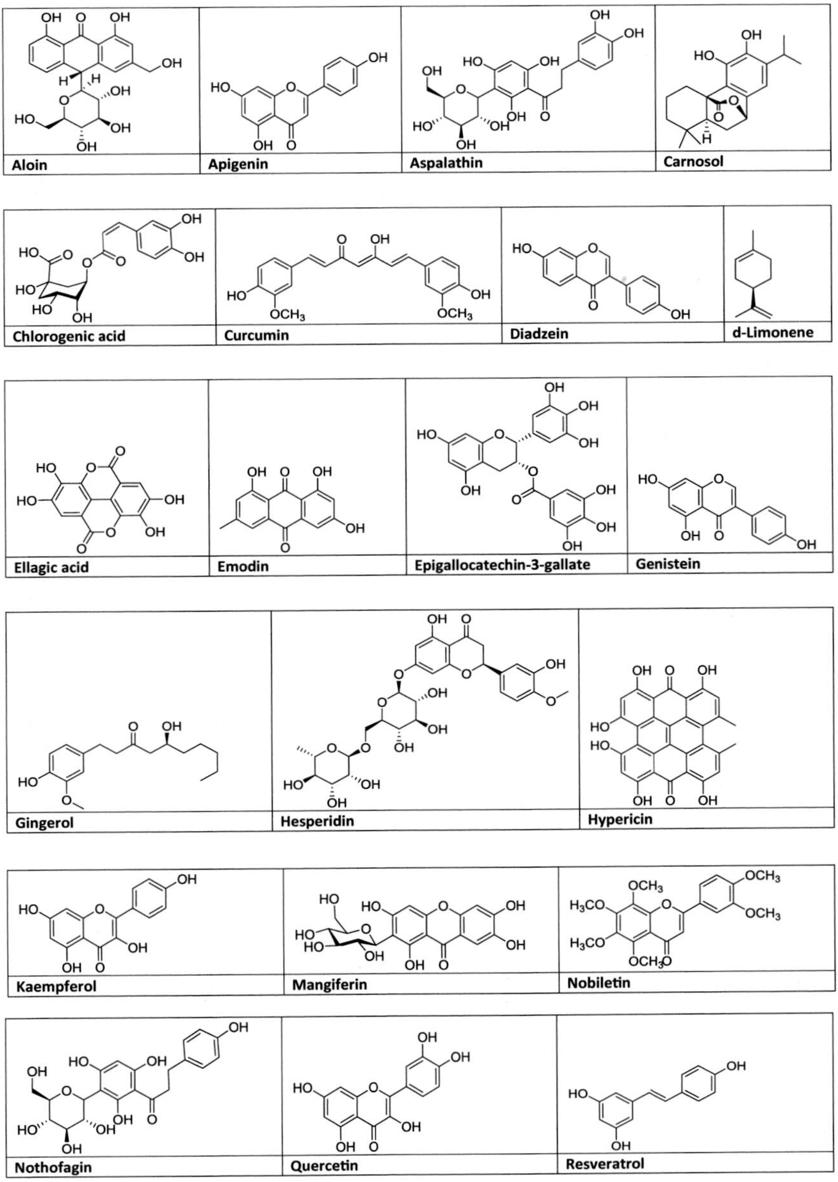

Aloin	Apigenin	Aspalathin	Carnosol

Chlorogenic acid	Curcumin	Diadzein	d-Limonene

Ellagic acid	Emodin	Epigallocatechin-3-gallate	Genistein

Gingerol	Hesperidin	Hypericin

Kaempferol	Mangiferin	Nobiletin

Nothofagin	Quercetin	Resveratrol

FIGURE 2.3 Chemical structures of known active phytochemical constituents found in functional foods and nutraceuticals.

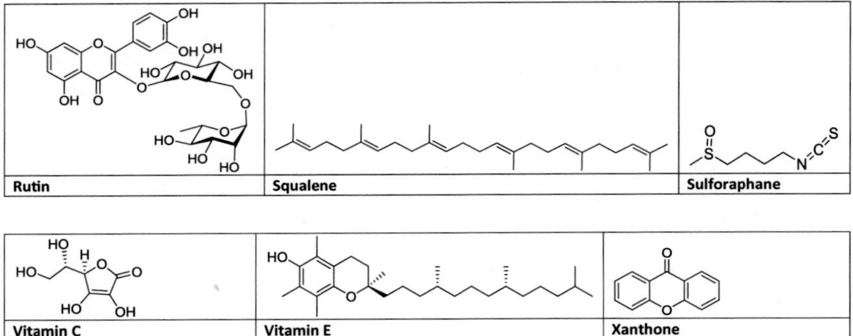

FIGURE 2.3 Cont'd.

The cause of skin cancer is not as highly associated with dietary lifestyle, but more with UV radiation, much research has shown the potential of functional foods and nutraceuticals to fight skin cancer initiation and progression. This is mainly due to the antioxidant and antiinflammatory activities that these foods and substances possess. The following sections will discuss those functional foods and nutraceuticals that have shown potential as chemopreventive agents for skin cancer (Wargovich, Woods, Hollis, & Zander, 2001).

2.3.4.1 Fruit

Apples are a common fruit that has a wide distribution across the world. Apples and apple juice have been said to be some of the most popular consumable foods among the fruit category. Apple and apple juice intake are also associated with all age groups. Investigations on the health benefits of apples have shown that apples have antimutagenic, antioxidant, and antiinflammatory activities. Apple juice has also been shown to inhibit cancer cell proliferation. The reported chemopreventive activity of apples was demonstrated in an animal model study, in which mouse skin papilloma occurrence was decreased after the mice were fed with aqueous apple peel extracts. Another animal study also displayed an increase in the survival of mice transplanted with B16 mouse melanoma cells, after apple polyphenols were added to their drinking water. Although these animal model studies showed the potential of apples for chemopreventive abilities for skin cancer in mice, human clinical trials are still necessary to advocate the effect apples have on skin cancer in humans (Gerhauser, 2008).

Grapes and grape-based products are well known throughout the world. Wine is one of the largest and most diverse grape-based products and has been vigorously studied for potential cancer prevention activity since the isolation of resveratrol from red wines. The demonstrated anticancer activity against skin cancer of red grapes and red wine is mainly attributed to the resveratrol content in the skin of red grapes and seeds of the grapes. Resveratrol and grape seed extract will be discussed later on as nutraceuticals (Kaur, Agarwal, & Agarwal, 2009).

TABLE 2.4 Functional Foods and Nutraceuticals With Demonstrated Cancer Chemopreventive Activities

Functional Food/ Nutraceutical	Known Active Phytochemical Constituents	Cancer Against Which Potential Chemoprevention Was Demonstrated	References
Fruits			
Apples	Flavonols (quercetin (Fig. 2.3)); catechins; oligomeric procyanidins; triterpenoids; anthocyanins.	Various cancers such as **skin**; breast and colon cancer.	Gerhauser (2008) and Reddy et al. (2003)
Citrus fruits (oranges; lemons; grapefruit; tangerines; lime)	Quercetin (Fig. 2.3); kaempferol (Fig. 2.3); tangerine; nobiletin (Fig. 2.3); rutin (Fig. 2.3); D-limonene (Fig. 2.3); heptamethoxyflavone; limonoids; vitamin C (Fig. 2.3); apigenin (Fig. 2.3); carotenoids.	Esophageal; thyroid; pancreatic; lung; prostate; renal; stomach; **skin** cancer.	Park and Pezzuto (2002), Reddy et al. (2003) and Chinembiri et al. (2014)
Grapes (red wine; whole grapes; grape juice; and grape seeds)	Resveratrol (Fig. 2.3); coumarin; chlorogenic acid (Fig. 2.3); ellagic acid (Fig. 2.3); catechins.	Various cancers including colon; **skin**; breast; and lung cancer.	Kaur et al. (2009), Park and Pezzuto (2002), Reddy et al. (2003) and Russo (2007)
Vegetables			
Asparagus	Vitamin E (tocopherol) (Fig. 2.3); selenium	**Skin** cancer	Chinembiri et al. (2014) and Bialy et al. (2002)
Cruciferous vegetables (including broccoli; cabbage; radish; kale; cauliflower)	Sulforaphane (Fig. 2.3)	Various types of cancer including **skin** cancer.	Chinembiri et al. (2014)

Continued

TABLE 2.4 Functional Foods and Nutraceuticals With Demonstrated Cancer Chemopreventive Activities—cont'd

Functional Food/ Nutraceutical	Known Active Phytochemical Constituents	Cancer Against Which Potential Chemoprevention Was Demonstrated	References
Herbs and Spices			
Basil	Apigenin (Fig. 2.3)	Breast; cervical; **skin**; lung; leukemia; and prostate	Patel et al. (2007)
Oregano	Apigenin (Fig. 2.3); quercetin (Fig. 2.3)	Breast; cervical; **skin**; lung; leukemia; breast; and prostate	Chinembiri et al. (2014), Patel et al. (2007) and Al-Kalaldeh, Abu-Dahab, and Afifi (2010)
Spearmint	Apigenin (Fig. 2.3)	Breast; cervical; **skin**; lung; leukemia; and prostate	Patel et al. (2007)
Turmeric	Curcumin (Fig. 2.3)	Various cancers	Khan et al. (2007)
Ginger	Gingerol (Fig. 2.3)	**Skin** cancer	Chinembiri et al. (2014) and Khan et al. (2007)
Thyme	Apigenin (Fig. 2.3)	Head and neck **squamous cell carcinoma**; **melanoma**; and liver cancer.	Chinembiri et al. (2014)
Rosemary	Carnosol (Fig. 2.3); ursolic acid	Cervical and **skin** cancer.	Berrington and Lall (2012), Bialy et al. (2002) and Johnson (2011)

Parsley	Apigenin (Fig. 2.3)	Head and neck **squamous cell carcinoma**; **melanoma**; and liver cancer.	Chinembiri et al. (2014)
Tarragon	Apigenin (Fig. 2.3)	Breast; cervical; **skin**; lung; leukemia; and prostate	Patel et al. (2007)
Sage	Carnosol (Fig. 2.3)	Various types of cancer including **skin** cancer.	Johnson (2011).
Teas			
Green tea	Epigallocatechin-3-gallate (EGCG) (Fig. 2.3); apigenin (Fig. 2.3)	Cervical; esophageal; stomach; breast; **skin** cancer.	Patel et al. (2007), Park and Pezzuto (2002) and Russo (2007).
Rooibos tea	Dihydrochalcones; aspalathin (Fig. 2.3); nothofagin (Fig. 2.3)	**Skin** cancer	Marnewick et al. (2005)
Honeybush tea	Xanthone (Fig. 2.3); mangiferin (Fig. 2.3); flavanone; hesperidin (Fig. 2.3)	**Skin** cancer	Marnewick et al. (2005)
Soybeans and Oils			
Soybeans	Genistein (Fig. 2.3); daidzein (Fig. 2.3)	**Skin** cancer	Chinembiri et al. (2014) and Khan et al. (2007)
Vegetable and seed oils	Vitamin E (Fig. 2.3)	**Skin** cancer	Reddy et al. (2003)
Extra-virgin olive oil	Vitamin E (Fig. 2.3); squalene (Fig. 2.3)	**Skin**; colon; and lung cancer	Bialy et al. (2002) and Reddy et al. (2003)

Continued

TABLE 2.4 Functional Foods and Nutraceuticals With Demonstrated Cancer Chemopreventive Activities—cont'd

Functional Food/ Nutraceutical	Known Active Phytochemical Constituents	Cancer Against Which Potential Chemoprevention Was Demonstrated	References
Flaxseed	Flaxseed derivative secoisolariciresinol diglycoside (SDG)	**Melanoma**	Bialy et al. (2002)
Tea tree oil	Terpinen-4-ol	**Skin** cancer	Chinembiri et al. (2014)
Nutraceuticals			
Curcumin	–	Skin cancer	Chinembiri et al. (2014)
Resveratrol	–	**Melanoma**	Chinembiri et al. (2014)
St. John's Wort	Hypericin (Fig. 2.3)	**Skin** cancer	Chinembiri et al. (2014)
Ginseng	Ginsenosides	**Melanoma** and **nonmelanoma skin cancers.**	Shin et al. (2000) and Bialy et al. (2002)
Aloe vera	Emodin (Fig. 2.3); Aloin (Fig. 2.3)	**Melanoma**	Chinembiri et al. (2014)
Grape seed extract	–	**Skin**; colorectal; prostate; breast cancer.	Kaur et al. (2009)

–, not available.

Citrus fruits represent a group of fruit that rivals even the popularity of apples as whole fruits and fruit juice. The best-known examples of citrus fruit include lime (*Citrus aurantiifolia* (Christm.) Swingle); lemon (*Citrus limon* L.); grapefruit (*Citrus paradisi* Macfad); tangerine (*Citrus reticulata* Blanco); and *sweet lemon* (*Citrus sinensis* (L.) Osbeck). Even though most citrus fruit consumption includes only the fleshy part of these fruits, the peel of citrus fruit has been shown to contain the most active phytochemicals. The phytochemicals, nobiletin, and heptamethoxyflavone, isolated from the peel of *Citrus* fruit, have been shown to interfere with the initiation and promotion stages of skin cancer in mice model studies. Grapefruit extracts have also demonstrated inhibitory activities of mice skin tumors. The anticancer activity of citrus fruits against skin cancer has mainly been attributed to the excellent antioxidant activity associated with these fruits (Park & Pezzuto, 2002).

2.3.4.2 Vegetables

Asparagus (*Asparagus officinalis* L.) is a commonly consumed vegetable worldwide. Asparagus shoots mainly form part of dishes such as soups; salads; and vegetable dishes. The proposed activity of asparagus against skin cancer is thought to be due to two phytochemicals present in asparagus, namely vitamin E and selenium. Vitamin E is very renowned for its antioxidant activity and has been available as a dietary supplement for years. Much controversy surrounds the efficacy of vitamin E and its various forms to prevent or reduce skin cancer occurrence. Studies by Malafa, Fokum, Mowlavi, Abusief, and King (2002) and Malafa, Fokum, Smith, and Louis (2002) have shown that vitamin E succinate can inhibit melanoma tumors in mice and demonstrate antiangiogenic activity by downregulating vascular endothelial growth factor expression, whereas a study by Gensler, Aickin, Peng, and Xu (1996) indicated that this form of vitamin E does not prevent, and in some cases enhance, the occurrence of UV-induced skin cancer in mice. Vural, Canbaz, and Selcuki (1999) have shown that patients with BCC have lower levels of vitamin C and E in their blood compared with healthy individuals. Although much controversy surrounds the chemopreventive ability of vitamin E for skin cancer, this compound has a long history of benefitting the overall health of individuals and only more research will clarify the role of vitamin E in skin cancer chemoprevention. Selenium is a trace element present in a variety of foods including asparagus. Selenium has been shown to be proficient in reducing hydroxyl free radical formation that is associated with causing damage to DNA and thereby resulting in mutations that may lead to cancer formation. Many studies have indicated a positive response to chemoprevention of skin cancer by selenium, though others pointed out that there are no significant chemopreventive effects for skin cancer. (Bialy, Rothe, & Grant-Kels, 2002; Chinembiri, du Plessis, Gerber, Hamman, & du Plessis, 2014; Clark et al., 1996; Clark, Graham, & Crouse, 1984; Gensler et al., 1996; Malafa, Fokum, Mowlavi, et al., 2002; Malafa, Fokum, Smith, et al., 2002; Pence, Delver, & Dunn, 1994; Shao et al., 1996; Vural et al., 1999).

Cruciferous vegetables are clustered together based on the presence of a group of compounds called glucosinolates found within all these vegetables. Cruciferous vegetables include, but are not limited to, broccoli; cabbage; cauliflower; and brussel sprouts. Cabbage has been reported to treat many ailments and diseases including cancer since ancient times. This led to anticipation for the whole cruciferous vegetable group to have anticancer properties. Additionally, support was found based on epidemiological studies indicating that there is some correlation between the consumption of cruciferous vegetables and a decreased risk for cancer. One of the suggested active anticancer compounds found in cruciferous vegetables, sulforaphane, is formed from a glucosinolate compound. Sulforaphane has been reported to inhibit melanoma cell proliferation and metastasis and induce apoptosis via various pathways. Unfortunately, it has been reported that sulforaphane has low bioavailability and a short half-life. These characteristics make it unbefitting to be considered for further investigation and product development (Chinembiri et al., 2014; Do, Pai, Rizvi, & D'Souza, 2010; Eylen, Oey, Hendrickx, & Loey, 2007; Fimognari & Hrelia, 2007; Hamsa, Thejass, & Kuttan, 2011; Herr & Büchler, 2010; Higdon, Delage, Williams, & Dashwood, 2007; Thejass & Kuttan, 2007; Verhoeven, Goldbohm, van Poppel, Verhagen, & van den Brandt, 1996).

2.3.4.3 Herbs and Spices

Many herbs and spices are believed to have anticarcinogenic properties based on some of the active compounds found within them. One such compound is apigenin that is found in a great variety of herbs and spices such as basil (*Ocimum basilicum*); parsley (*Petroselinum crispum*); oregano (*Origanum vulgare*); spearmint (*Mentha spicata*); tarragon (*Artemisia dracunculus*); and thyme (*Thymus vulgaris*). Apigenin has demonstrated anticancer effects on head and neck SCC cells and melanoma cells. Numerous mechanisms of action displayed by apigenin have been revealed, ranging from the inhibition of cancerous cells via cell cycle arrest to the prevention of inflammation caused by UVB radiation. Apigenin has also shown to protect the skin from UVA- and UVB-induced skin carcinogenesis in mice. As such, apigenin and plants containing this active compound can be considered for future research for the prevention and treatment of skin cancer (Chinembiri et al., 2014; Patel, Shukla, & Gupta, 2007).

Ocimum basilicum, more commonly known as sweet basil, is renowned for its medicinal properties and use as an herb in the food industry. Basil has shown potential as a skin cancer chemopreventive agent in various studies. The UV-protectant capabilities of basil oil were investigated in a study by Courreges and Benencia (2002). They found that basil oil demonstrated good sun protection factor (SPF) following addition to a sunscreen formulation. Another study, by Dasgupta, Rao, and Yadava (2004), further aided in the skin cancer chemopreventive activity of basil. In their research a freshly made hydroalcoholic leaf extract of basil was able to reduce the induced skin tumor burden in mice with 12.50%–18.75%. Because apigenin has been isolated from basil, it may represent the active compound in basil that acts against skin cancer. The possible addition of basil to sunscreen seems to be a viable outcome in the future, though human clinical trials are lacking (Birt, Mitchell, Gold, Pour, & Pinch, 1997; Courreges & Benencia, 2002; Dasgupta et al., 2004; Patel et al., 2007).

Oregano (*O. vulgare*) is another herb in which apigenin is found, suggesting that oregano should be considered for possible antiskin cancer studies. Another compound, quercetin, was also found to be present in oregano and represents another active anticancer compound. Although quercetin is one of the best-known antioxidants from plants, there are many controversial findings in research directed at its activity against skin cancer. Quercetin has shown numerous mechanisms by which it can inhibit the proliferation of melanoma cells. Such mechanisms include inducing apoptosis; preventing UVB-induced oxidative stress and DNA damage and by inhibiting proteins involved in cancer formation and progression. On the other hand, quercetin has also been reported to potentiate the activity of cancer-associated enzymes, thereby aiding skin cancer progression. The presence of the controversial quercetin compound found in plants used as foods, such as oregano, should not be seen as a cause to discontinue research on such plants because of the amount of a compound present, along with other factors, may greatly influence the bioactivity of a plant against disease. Furthermore, a study by Caltagirone et al. (2000) evaluated the combined effects of apigenin and quercetin in a B16-BL6 murine melanoma metastasis model. The results indicated that the combined effects of apigenin and quercetin have the capacity to inhibit melanoma lung tumor metastasis (Caltagirone et al., 2000; Cao et al., 2014; Chinembiri et al., 2014; Olson et al., 2010; Patel et al., 2007; Rosner, Röpke, Pless, & Skovgaard, 2006; Yin et al., 2013; Zhang, Chen, Xia, & Xu, 2005).

As with basil and oregano; rosemary (*Rosmarinus officinalis*) and sage (*Salvia carnosa*) also contain an active compound that is suggested to be a possible source of anticancer activity. This active compound was named carnosol. Carnosol has been shown to impede the migration potential of B16-F10 mouse melanoma cells and to interfere with the activation of NF-κB and the activity of metalloproteinases. To date, no studies have been found that evaluate the activity of sage on skin cancer, though a methanolic extract of rosemary has been evaluated in vivo. The study on the activity of rosemary, conducted by Huang et al. (1994), indicated that the rosemary extract has the capacity to prevent tumor initiation and inhibit inflammation and tumor promotion. Human clinical trials should be considered for future research (Huang, Ho, Lin-Shiau, & Lin, 2005; Huang et al., 1994; Johnson, 2011).

Zingiber officinale, commonly known as ginger, has been used as a spice and medicine for ~2500 years in China. Ginger has been used medicinally for chronic inflammatory condition; headaches; colds; and nausea. Today, the rhizome of ginger is used around the world as a flavoring agent (powder; paste; fresh; and in teas) and even as candy. Although many compounds have been isolated from ginger, only two of the major compounds, [6]-gingerol and [6]-paradol, have indicated to have skin-cancer preventive properties in mouse skin tumorigenesis models. The ethanolic extracts of ginger have also been found to hamper tumor promotion. In addition, a pharmacokinetic study by Zick et al. (2008) has found that [6]-gingerol has decent bioavailability when a dose of ginger is orally administrated. The findings from the studies on ginger and its active constituents suggested that ginger itself and/or the active isolated compounds should be considered as potential skin-cancer preventive agents (Katiyar, Agarwal, & Mukhtar, 1996; Park & Pezzuto, 2002; Shukla & Singh, 2007; Surh et al., 1999; Zick et al., 2008).

Turmeric (*Curcuma longa*) is a popular spice that is used globally. Since early times, turmeric has also been used by traditional medicine systems for its antiinflammatory and antioxidant activity. The bright yellow color of the spice, which is made from the rhizome part of the plant, stems from a yellow plant polyphenol named curcumin. Curcumin is also considered to be the active anticancer principle in turmeric with the commercial spice usually containing about 80% of curcumin. Today, curcumin is available as an over-the-counter nutraceutical. Thus, curcumin and its anticancerous properties will be discussed in the nutraceuticals Section 2.3.4.6 (Chinembiri et al., 2014; Limtrakul, Lipigorngoson, Namwong, Apisariyakul, & Dunn, 1997).

2.3.4.4 Soybeans and Oils

The anticancer activity of soybeans is mainly attributed to the presence of two active phytoestrogens, named daidzein and genistein. Both of these phytoestrogen compounds have been shown to possess significant photoprotection properties with a topical application to the skin. Numerous studies have also indicated the great synergistic effect between daidzein and genistein for photoprotection. Genistein has also been reported to have antiinflammatory properties and to induce apoptosis in cancers cells. Although both daidzein and genistein have photoprotective abilities, the lack of penetration into the skin by daidzein has resulted in a shift to genistein as a single agent. The use of genistein for the prevention or treatment of skin cancer and skin damage with specific regards to photodamage acquired via UV radiation has been patented, and a range of products are commercially available (Chinembiri et al., 2014; Huang, Hung, Lin, & Fang, 2008; Iovine, Iannella, Gasparri, Monfrecola, & Bevilacqua, 2011; Khan, Afaq, & Mukhtar, 2007; Wang, Zhang, Lebwohl, DeLeo, & Wei, 1998, Wang, Zhang, Xie, Yu, & Zhang, 2002; Wei, Bowen, Zhang, & Lebwohl, 1998).

Many types of oil are commercially available and are extensively used not only in the kitchen but also in the cosmetics industry. The favorable properties on the skin of a wide variety of oils are mainly attributed to the vitamin E content within the oils. The controversial findings of the activity of vitamin E on skin cancer have previously been discussed in Section 2.3.4.2. Vegetable oils, seed oils, and wheat germ oil are believed to have photoprotective characteristics, which are largely ascribed to the activity of carotenoids and vitamin E. Extra-virgin olive oil has been evaluated in a mice carcinogenesis model in a study by

Ichihashi et al. (2000). This study indicated that extra-virgin olive oil delays the onset and reduces the spread of skin cancers when topically applied to mouse skin after UVB exposure. Because most oils seem to have photoprotective abilities they can be considered as additives in sunscreen lotions, whereas extra-virgin olive oil can be considered as an after-sun treatment to possibly inhibit skin cancer initiation (Bialy et al., 2002; Ichihashi et al., 2000; Reddy, Odhav, & Bhoola, 2003; Stahl, Heinrich, Jungmann, Sies, & Tronnier, 2000).

The anticancer activity from flaxseed is believed to stem chiefly from the high lignan content found in flaxseed. Although no studies were found indicating the role of flaxseed extract on skin cancer, a study by Li, Yee, Thompson, and Yan (1999) has reported the effects of a flaxseed derivative, secoisolariciresinol diglycoside (SDG), on melanoma metastasis to the lungs. The study pointed out that the SDG derivative has the capacity to slow the initiation and promotion of metastatic melanoma cells to the lungs in mice. This study provides an opportunity to evaluate the effects of the SDG derivative in clinical human trails. Depending on the outcome, the SDG derivative can be considered as a supplementary drug for patients with melanoma in which the potential of metastasis to the lungs is a possibility (Bialy et al., 2002; Li et al., 1999).

Tea tree oil is widely available commercially as a single agent or as part of a range of cosmetic products. Tea tree oil is extracted from *Melaleuca alternifolia* and is traditionally used for skin disorders; infections; and burns. The major constituents found in tea tree oil are terpenoids, which are believed to be the dominant bioactive principles responsible for its medicinal benefits. Because tea tree oil has been extensively used throughout history, it is believed to be relatively safe and to only cause minor toxicity in rare cases. Tea tree oil has been shown to have antiinflammatory properties and antimicrobial activity for numerous pathogens, which allows it to function as an antiseptic agent. The treatment ability of the oil on wounds of the skin may also function by keeping it free from infection and hampering the detrimental effects caused by inflammation. Self-medication of BCC and SCC by the use of tea tree oil has been reported to be effective by the general public. A study by Carson, Hammer, and Riley (2006) indicated that the topical application of tea tree oil to skin tumors of mice resulted in the damage and death of the skin tumor cells. All these results indicated that tea tree oil might be a suitable and easily accessible source to support the treatment of skin cancer (Carson et al., 2006; Ireland et al., 2012; WHO Monographs, 2004; Xiao et al., 2008).

2.3.4.5 Tea

Camellia sinensis, more commonly known as tea, is the second most consumed beverage in the world, besides water. Commercially available forms of black, green, and oolong tea are manufactured from the leaves and buds of *C. sinensis*. The differences between these types of tea generally lie with the way in which each tea is processed. Western civilization mostly consumes black tea, whereas Asian countries are known to consume black, green, and oolong tea. Epidemiology studies have indicated that the risk and incidence of certain cancers are lower in Asian countries, including Japan and China. This lowered risk and incidence of specific cancer types, which have been linked to the healthier dietary lifestyle of Asian countries and the consumption of green tea on a regular basis. Although black tea and oolong tea have also shown to possess anticancer properties, attention mostly has been given to studies on the effects of green tea on cancer because it has been shown to possess greater anticancer activity. The major constituent in green tea, (−)-epigallocatechin-3-gallate (EGCG), has been indicated as the most effective chemopreventive agent present in green tea. Hence, a range of animal model studies have been done to investigate the potential of orally and topically administered EGCG on skin cancer. Both oral and topical administrations of EGCG have shown to reduce UVA- and UVB-induced skin tumor incidence, growth, multiplicity, and volume in mice. In these studies it was found that topical administration of EGCG has better activity than orally administrated EGCG, which may be attributed to the fact that topical application occurs on the skin, which is the target organ in these studies. A study by Katiyar, Matsui, Elmets, and Mukhtar (1999) further supported the use of EGCG as an agent to prevent UV damage because of its ability in interfering with the immunosuppressive effect of UVB radiation. These studies in animal models provide exceptional results for the potential of green tea and specifically EGCG for the chemoprevention of skin cancer. Therefore, human clinical trials represent the logical next step in the evaluation of the skin cancer chemopreventive effects in humans (Baliga & Katiyar, 2006; Chinembiri et al., 2014; Hara, 2001; Katiyar, 2003; Katiyar & Elmets, 2001; Katiyar et al., 1999; Katiyar & Mukhtar, 1996; Mantena, Roy, & Katiyar, 2005; Mantena, Meeran, Elmets, & Katiyar, 2005).

The excitement over green tea and its bioactivity has fueled an interest in a variety of different teas and their potential in aiding the health and well-being of individuals. Rooibos (*Aspalathus linearis*) and honeybush (*Cyclopia intermedia*) are indigenous South African herbal teas that have gained a lot of popularity over the last few years, as health beverages, globally. Because of the experimental success of green tea as a chemopreventive agent for skin cancer, interest grew in how other teas would compare regarding activity against skin cancer. In a comparative study by Marnewick et al. (2005) both rooibos and honeybush tea, in processed and unprocessed forms, were evaluated for possible chemopreventive activity in a mouse skin–induced tumor model, as compared with green tea. The results from this study indicated that no tumors occurred over a 20-week period on the mice that were topically treated with green tea extracts, prior to inducing tumors, whereas the nontreated mice (positive control mice) developed tumors the same day as when induction started. Both the processed and unprocessed honeybush and rooibos extracts significantly delayed tumor development, with tumor development only occurring after 12 and 8 weeks, respectively. The unprocessed rooibos extract–treated mice showed the lowest number of tumors, whereas the unprocessed honeybush extract–treated mice showed the best reduction in the amount of mice with tumor occurrence. Interestingly, the processed honeybush extract–treated mice had the smallest tumors, regarding volume and size. Although rooibos and honeybush extracts did not perform as well as the green tea extract, these results could indicate that rooibos extracts are more functional to prevent skin cancer occurrence, whereas honeybush extracts may be useful for the prevention and treatment of skin cancer.

A study by Petrova, Davids, Rautenbach, and Marnewick (2011) focused on investigating the effects of honeybush and its constituents on UVB-induced skin damage in mice. This study indicated that honeybush extracts and two of its major constituents, hesperidin and mangiferin, do absorb UVB light thereby preventing penetration into the skin, though honeybush extracts showed much more promise than its main compounds. The antiinflammatory activity of honeybush extracts and main compound was also evaluated, with the honeybush extracts having very good activity. Hesperidin only demonstrated moderate antiinflammatory activity and mangiferin showed no antiinflammatory activity. The results found by this study point out that the two major constituents present in honeybush are not as active as the extracts themselves. It has been suggested that this may be due to synergism among the compounds in the extract, though more research is needed to justify this hypothesis. Overall it seems as though the claims and use of rooibos and honeybush tea in various skin lotions and sunscreens that are commercially available have been scientifically proven, at least to some extent (Marnewick et al., 2005, 2009; Petrova et al., 2011).

2.3.4.6 Nutraceuticals Over the Counter for Skin Cancer

Curcumin is the main active compound found in turmeric (*C. longa*). Curcumin is currently available as a nutraceutical even though it has not been approved for the treatment of any disease. The rationale behind the commercialization of curcumin as an additive health supplement is most likely due to its relatively safe toxicity record, and the activity of the compound has exhibited against a variety of ailments, including diabetes; cancer; inflammation; and HIV. A great amount of studies have found that curcumin displays many different mechanisms of action against melanoma and SCC cells and UVB-induced skin tumor formations. A few examples of these mechanisms of action involve inducing apoptosis; stimulation of the p53 tumor suppressor protein; and the inhibition of ribosomal S6 phosphorylation and the NF-κB prosurvival pathway. Curcumin has also shown potential in the treatment of symptoms such as fatigue; depression; and pain usually associated with cancer diagnosis. Although curcumin shows great potential as a chemotherapeutic agent, it was found to have low solubility and oral bioavailability, which makes it a rather difficult compound to work with in therapeutic delivery systems. Fortunately, recent advances in transdermal microemulsion and nanoformulation formulations may represent renewed potential for curcumin in therapeutic delivery systems (Anand, Sundaram, Jhurani, Kunnumakkara, & Aggarwal, 2008; Bush, Cheung, & Li, 2001; Chinembiri et al., 2014; Dahmke et al., 2013; Gupta, Patchva, Koh, & Aggarwal, 2012; Liu, Chang, & Hung, 2011; Mimeault & Batra, 2011; Moorthi & Kathiresan, 2013; Naksuriya, Okonogi, Schiffelers, & Hennink, 2014; Phillips et al., 2011; Shehzad, Lee, & Lee, 2013; Tsai et al., 2012; Zhao et al., 2013).

Resveratrol is a polyphenolic phytoalexin found in the skin and seeds of grapes, wine, peanuts, and soy. Resveratrol is believed to possess a range of beneficial bioactivities, among which its potential anticancer activity has though roughly been evaluated. Many in vitro and in vivo studies have been done on the possible chemopreventive and chemotherapeutic activity of resveratrol on skin cancer. Many studies have shown that the topical application of resveratrol inhibited UVB-induced phototoxicity and tumors. A study by Fu, Cao, Wang, Xu, and Han (2004) also indicated that oral administration of resveratrol inhibited chemically induced skin papillomas in mice. Furthermore, significant evidence suggested that resveratrol also has the capacity to inhibit human squamous carcinoma cell proliferation through an assortment of mechanisms, including cell cycle arrest at the G1-phase. Resveratrol has also exhibited activity against melanoma cells; as an antimetastatic agent and as a radiation sensitizer. On the other hand, a study by Niles et al. (2006) showed that resveratrol does not have any significant inhibitory activity on melanoma growth and even stimulates melanoma growth at higher doses in human melanoma xenograft tumor growth. This study also indicated that piceatannol, one of the main resveratrol metabolites, significantly increased the number of lung metastases. Given all the evidence, resveratrol should be considered very carefully as a nutritional supplement, especially in patients diagnosed with cancer, and more research should be done to reveal if the beneficial aspects of resveratrol can be utilized (Adhami, Afaq, & Ahmad, 2001; Afaq, Adhami, Ahmad, & Mukhtar, 2002; Afaq, Adhami, & Ahmad, 2003; Ahmad, Adhami, Afaq, Feyes, & Mukhtar, 2001; Asensi et al., 2002; Athar et al., 2007; Aziz, Afaq, & Ahmad, 2005; Baliga & Katiyar, 2006; Chen et al., 2012; Chinembiri et al., 2014; Fang, Bradley, Cook, Herrick, & Nicholl, 2013; Fu et al., 2004; Gatouillat, Balasse, Joseph-Pietras, Morjani, & Madoulet, 2010; Jang et al., 1997; Niles et al., 2003, 2006; Soleas, Grass, Josephy, Goldberg, & Diamandis, 2002).

Aloe vera is world renowned for its beneficial effects on the skin and is used in a wide range of cosmetic products. It has been used medicinally for the treatment of constipation; wounds; and skin irritation. Saline extracts of *Aloe vera* have been evaluated for its activity on murine melanoma cells by Chandu, Kumar, Bhattacharjee, and Debnath (2012). The results obtained by the study indicated that the *Aloe vera* extracts have the capacity to reduce the cell viability of the murine melanoma cells. Two active compounds, emodin and aloin, present in *Aloe vera* have been investigated for their anticancer properties. The emodin compound from Aloe has been shown to possess antimetastatic action against murine melanoma cells and to inhibit the proliferation of human and murine melanoma cells by Tabolacci et al. (2010) and Radovic et al. (2012), respectively. Aloin has also been shown to possess antimelanoma activity by significantly inhibiting the proliferation of melanoma cells and interference with cell adhesion. Richardson, Smith, McIntyre, Thomas, and Pilkington (2005) did a study of the clinical trials in which a topical application of *Aloe vera* was evaluated for the prevention and/or reduction of skin reactions in cancer patients being treated with radiation. Their finding suggested that the topical application of *Aloe vera* is not capable of preventing or reducing the occurrence of radiation-induced skin reactions in cancer patients. Altogether, the results not only point out the potential of *Aloe vera* and its active compounds for the treatment for melanoma skin cancer but also propose that *Aloe vera* cannot compensate for some of the harmful side effects caused by radiation therapy in cancer patients (Chandu et al., 2012; Chinembiri et al., 2014; Radovic et al., 2012; Richardson et al., 2005; Tabolacci et al., 2010, 2013; Van Wyk, van Oudtshoorn, & Gericke, 1997).

Hypericum perforatum, commonly known as St. John's Wort, is widely used for its medicinal properties. The potential of this plant in the treatment of skin cancer stems from an active compound hypericin found within the plant. Hypericin is activated by UVA radiation and was therefore considered for use in photodynamic therapies. Studies have shown that hypericin, activated by UVA, has the capacity to induce cell death via apoptosis and necrosis in human melanoma cells and squamous carcinoma cells. Unfortunately, clinical trials found that the photosensitizing effects of hypericin was not strong enough to be considered for further evaluation. This lack in clinical activity was postulated to be due to the low solubility and stability of hypericin in solution. Although these properties of hypericin make it a difficult candidate for consideration, a study by Sharma and Davids (2012b) has shown that depigmentation of melanoma cells increases the efficacy of the photosensitizing hypericin compound. The extract of *H. perforatum* has also shown to inhibit human melanoma cell proliferation and show promising properties for clinical evaluation (Davids, Kleemann, Kacerovská, Pizinger, & Kidson, 2008; Menichini et al., 2013; Sharma & Davids, 2012a, 2012b; Skalkos et al., 2006).

The root of ginseng (*Panax ginseng*) has been used for many years in China, Japan, and Korea for its life-prolonging and restoration of strength effects. Today, ginseng is commercially available as fresh, white, or red ginseng. Wild ginseng is believed to have more pharmacological activity as compared with cultivated ginseng and is therefore more expensive. Although many epidemiology studies have aimed to unravel the association of ginseng consumption with lowered cancer incidence, the evidence thus far are inconclusive in human studies. However, animal studies have shown promising results of ginseng in the treatment and prevention of cancer. Red ginseng extracts have been shown to have noteworthy inhibitory effects on skin cancer formation in a two-stage mouse model. Another study indicated that an extract of processed ginseng decreased the formation of skin papillomas in mice. Many mechanisms of action on how ginseng prevents cancer formation have also been suggested. Some of these mechanisms of action include antimutagenic effects, antiangiogenesis effects, immunosurveillance, and cellular defense strategies. These animal studies and potential mechanisms of actions suggest the potential use of ginseng extracts in sunscreens as a measure to prevent skin cancer formation (Chen, Liu, & Lei, 1998; Keum et al., 2000; Shin, Kim, Yun, Morgan, & Vainio, 2000).

Grape seed extract is already available as an over-the-counter supplementary health product, mainly for its antioxidant activity. The in vitro antioxidant activity found in these extracts has been shown to be greater than that of vitamin C and E. It has been put forward that grape seed extracts may potentially have anticancer activity. The activity of grape seed extracts has been suggested to stem mainly from polyphenolic compounds, such as catechin; epicatechin; oligomeric proanthocyanidins; flavan-3-ol derivatives; and resveratrol, discovered with the seeds. Many in vivo studies have supported this claim after both topical applications and oral administration of a polyphenolic fraction from grape seed extracts showed significant preventive and inhibitory activity on the initiation and promotion stages of skin cancer in mouse models. These studies advocate that the polyphenolic fraction of grape seed extracts might be used as a preventive strategy for skin cancer by both oral supplementation and as topical applications in sunscreens (Bagchi et al., 1997; Bomser, Singletary, & Meline, 2000; Bomser, Singletary, Wallig, & Smith, 1999; Katiyar, 2008; Kaur et al., 2009; Mittal, Elmets, & Katiyar, 2003; Sharma, Meeran, & Katiyar, 2007; Zhao, Sharma, & Agarwal, 1999).

2.4 SCIENTIFIC RESULTS

There is an ongoing need for the development of new anticancer agents for use against skin cancer. Plants provide a great potential for the development of these anticancer agents. Currently there is an increase in the research for medicinal plants against various types of skin cancer cells lines such as mouse melanoma (B16F10), human SCC (A431, SCC25), and human melanoma (A375, MV-3, UACC-62, Fem-x, SK-MEL-28). More focus is being placed on the scientific research of plant extracts against melanoma because this is the most dangerous type of skin cancer. In Table 2.5 a summary of plant extracts that have been tested in vitro against various types of skin cancer from across the world has been provided. The results are depicted as 50% inhibitory concentrations (IC_{50}), which can be explained as the concentration of the extract at which 50% of the carcinoma cells are no longer viable. Ideally, a plant showing toxic effects with an IC_{50} value of lower than 30 μg/mL, selective for cancerous cells would be an ideal candidate for further studies for its anticancer activity as well as mechanistic studies. Databases including the Catalogue of Life; The Plant List; Tropicos; GBIF (Global Biodiversity Information Facility); Encyclopedia of Life; Jstor Global Plants; the IUCN (International Union for Conservation of Nature and Natural Resources) red data list; and the SANBI (South African

TABLE 2.5 Plants Tested From Across the World for Cytotoxicity Against Skin Cancer

Plant/Families	Common Names	Study Area and Distribution	Plant Parts	Extract Type	Results (IC_{50}[a] in μg/mL)	References
Albertisia delagoensis (N.E.Br) Forman/ Menispermaceae	Umgandaganda.	South Africa (Mozambique and South Africa)	Leaves	Alkaloid extract	UACC62 melanoma cells: <6.25 μg/mL	De Wet, Fouche, and Van Heerden (2009)
Aloe vera (L.) Burm. f./Xanthorrhoeaceae	Many including: *Aloe vera*; true aloe; common aloe; burn plant; Indian aloe; and Mediterranean aloe.	India (Africa; Asia and North, South and Central America)	Leaves	Gel	B16F10: 300 μg/mL	Chandu et al. (2012)
Antizoma angustifolia (Burch.) Miers ex Harv. and Sond./ Menispermaceae	Maagbitterwortel; maagwortel.	South Africa (Botswana; Namibia; and South Africa)	Leaves and rhizome	Alkaloid extract	UACC62 melanoma cells: <6.25 μg/mL (leaves), 15 μg/mL (rhizome)	De Wet et al. (2009)
Antizoma miersiana Harv./ Menispermaceae	–	South Africa (Namibia and South Africa)	Leaves and rhizome	Alkaloid extract	UACC62 melanoma cells: 12.5 μg/mL (leaves), 12.25 μg/mL (rhizome)	De Wet et al. (2009)

Continued

TABLE 2.5 Plants Tested From Across the World for Cytotoxicity Against Skin Cancer—cont'd

Plant/Families	Common Names	Study Area and Distribution	Plant Parts	Extract Type	Results (IC_{50}[a] in μg/mL)	References
Boucerosia lasiantha Wight. Accepted Scientific name: *Desmidorchis umbellata* (Haw.) Decne./Apocynaceae	–	India (India and Sri Lanka)	Not specified	Methanol	A431: >100 μg/mL A375: 7.28 ± 0.18 μg/mL	Madhuri, Murthy, Amrutha, and Siva (2014)
Boucerosia umbellata (Haw.) Wight and Arn. Accepted Scientific name: *Desmidorchis umbellata* (Haw.) Decne./Apocynaceae	–	India (India and Sri Lanka)	Not specified	Methanol	A431: 78.9 μg/mL A375: 26.53 ± 0.55 μg/mL	Madhuri et al. (2014)
Caralluma stalagmifera C.E.C. Fisch./Apocynaceae	Dark purple Caralluma	India (India)	Not specified	Methanol	A431: >100 μg/mL A375: 4.93 ± 0.14 μg/mL	Madhuri et al. (2014)

Caralluma stalagmifera C.E.C. Fisch. Var. *longipetala* Karupp. and Pull./ Apocynaceae	–	India (India)	Not specified	Methanol	A431: 93.8 μg/mL A375: 8.45 ± 0.41 μg/mL	Madhuri et al. (2014)
Carica papaya L.	Papaya and melon tree	Australia (distribution in six continents-all except Antarctica)	Leaves	Ethanol (E) Acidic ethanol (AE) Acidic water (AW)	Squamous cell carcinoma (SCC25): 172.9 μg/mL (E), 77.18 μg/mL (AE) and 57.72 μg/mL (AW)	Nguyen et al. (2016)
Centratherum anthelminticum (L.) Kuntze/Compositae	Black cumin	India	Seeds	Crude chloroform fraction (CCF)	A375:8.5 ± 1.2 μg/mL	Looi et al. (2013)
Chromolaena odorata (L.) R.M. King and H.Rob./ Compositae	Bitter-bush; butterfly weed; Jack in the Bush; and christmasbush.	India (wide spread distribution in most continents)	Leaf	Ethyl acetate Hexane	A431 epidermoid carcinoma cell line: Ethyl Acetate: 35 μg/mL Hexane: 40 μg/mL	Yajarla, Nimmanapalli, Parikapandla, Gupta, and Karnati (2014)
Cissampelos capensis (L.f.) Diels/ Menispermaceae	Davidjies; davidjieswortel; dawidjies; dawidjieswortel; fynblaarklimop.	South Africa (Madagascar and South Africa)	Leaves and rhizome	Alkaloid extract	UACC62 melanoma cells: 19 μg/mL (leaves), 18.75 μg/mL (rhizome)	De Wet et al. (2009)

Continued

TABLE 2.5 Plants Tested From Across the World for Cytotoxicity Against Skin Cancer—cont'd

Plant/Families	Common Names	Study Area and Distribution	Plant Parts	Extract Type	Results (IC_{50}[a] in μg/mL)	References
Cissampelos hirta Klotzsch/ Menispermaceae	–	South Africa (Mozambique and South Africa)	Rhizome	Alkaloid extract	UACC62 melanoma cells: 12.5 μg/mL	De Wet et al. (2009)
Cissampelos mucronata A. Rich./ Menispermaceae	Ivy-vine; davidjies; davidjieswortel and; umbombo	South Africa (widespread throughout the African continent)	Rhizome	Alkaloid extract	UACC62 melanoma cells: 6.25 μg/mL	De Wet et al. (2009)
Cissampelos torulosa E. Mey. Ex Harv. and Sond./ Menispermaceae	Davidjies; davidjieswortel; ukhalimele; ukhalimele-omkhulu; and umthombo	South Africa (Malawi; Mozambique; South Africa; Swaziland; and Zimbabwe)	Rhizome	Alkaloid extract	UACC62 melanoma cells: 12.5 μg/mL	De Wet et al. (2009)
Cocculus hirsutus (L.) Diels/ Menispermaceae	Bushveld moonseed creeper; python climber; and Bosveld-halfmaanranker	South Africa (distributed in some countries in Africa; the Middle East and Asia)	Rhizome	Alkaloid extract	UACC62 melanoma cells: <6.25 μg/mL	De Wet et al. (2009)
Collaea argentina Griseb./Fabaceae	–	Argentina (Argentina and Bolivia)	Leaf	Methanol	B16F10: >500 μg/mL; HaCat: 219 ± 4 μg/mL	Mamone et al. (2011)
Etlingera elatior (Jack) R.M.Sm./ Zingiberaceae	Torch ginger and kantan	Thailand (China; Colombia; Costa Rica; Ecuador; Honduras; India; Malaysia; Panama and Thailand)	Flowers	50% hydroglycol extracts	A431: 6.182 μg/mL	Thuncharoen, Chulasiri, Nilwarangkoon, Nakamura, and Watanapokasin (2013)

Gardenia gummifera L. f./ Rubiaceae	Cumbi-gum tree; dekamella-gum gardenia; and gummy cape jasmine	India (India)	Resin and sticks	Methanol	A375: 7.816 μg/mL	Gopalakrishna, Thimappa, Thylur, Shivanna, and Sreenivasan (2014)
Helichrysum odoratissimum (L.) Sweet/Asteraceae	Imphepho	South Africa (Europe; Great Britain; Africa and Yemen)	Leaves and stems	Ethnaol	A431: 15.5 μg/mL	Twilley and Lall (2015)
Indigofera longeracemosa Baill./Fabaceae	Niagy; hengitra; Manga Maori; Aika Manga and Aika Yoruba.	India (Comoros; India; Madagascar and Senegal)	Leaves	Silver nanoparticles of water extract	Melanoma cells (SK-MEL-28): 48 μg/mL	Suseela and Lalitha (2015)
Iochroma australe Griseb./Solanaceae	Manzanillo	Argentina (Argentina and Bolivia)	Flower	Methanol	B16F10: 61 ± 3 μg/mL; HaCat: 28 ± 5 μg/mL	Mamone et al. (2011)
Ipomoea bonariensis Hook./ Convolvulaceae	Campanilla	Argentina (Argentina; Bolivia; Brazil; Paraguay; Peru; and Uruguay)	Flower	Methanol	B16F10: 147 ± 3 μg/ mL; HaCat: 139 ± 3 μg/mL	Mamone et al. (2011)
Jacaranda mimosifolia D. Don/ Bignoniaceae	Jacaranda; Black poui; and Tarco	Argentina (native distribution: Bolivia)	Flower	Methanol	B16F10: 24 ± 5 μg/mL; HaCat: 68 ± 3 μg/mL	Mamone et al. (2011)
Juglans regia L./ Juglandaceae	English walnut; walnuss; and common walnut	India (natural distribution: Central Asia)	Leaves	Methanol (M) and distilled water (W)	A375: 304 μg/mL (M), 350 μg/mL (W) B16F10: 234 μg/mL (M), 298 μg/mL (W)	Shah, Sharma, and Shah (2015)

Continued

TABLE 2.5 Plants Tested From Across the World for Cytotoxicity Against Skin Cancer—cont'd

Plant/Families	Common Names	Study Area and Distribution	Plant Parts	Extract Type	Results (IC_{50}[a] in μg/mL)	References
Leucosidea sericea Eckl. and Zeyh./ Rosaceae	Oldwood and troutwood	South Africa (Lesotho; Mozambique; South Africa; and Zimbabwe)	Leaves	Ethanol extract	B16-F10: 55 μg/mL	Sharma, Kishore, Hussein, and Lall (2014)
Manilkara zapota (L.) P.Royen/ Sapotaceae	Chicle; naseberry; and sapodilla	India (Natural distribution: Central America; Mexico; and West Indies)	Stem bark	Ethanol	A431: >80 μg/mL	Awasare, Bhujbal, and Nanda (2012)
Nitraria retusa (Forsk.) Aschers./ Nitrariaceae	–	Tunisia (Egypt; Iraq; Israel; Jordan; Kuwait; Saudi Arabia; Sinai peninsula; Syria; Algeria; Morocco; Tunisia; Libya; and Mauritania)	Leaves	Hexane (H), Chloroform (C), Ethyl acetate (EA), Methanol (M)	B16F10: 50 μg/mL (EA), 80 μg/mL (C), 340 μg/mL (H), >1000 μg/mL (M).	Boubaker et al. (2015)
Philadelphus pekinensis Rupr./ Hydrangeaceae	Pecking mock orange	Slovakia (China and North Korea)	Leaves	Ethanol	A431: 12.5 μg/mL	Val'ko, Pravdová, Nagy, Grančai, and Ficková (2007)
Philadelphus zeyheri Schrad. Ex DC. Accepted Scientific name: *Philadelphus coronarius* L./ Hydrangeaceae	Sweet mock-orange	Slovakia (natural distribution: Southern Europe)	Leaves	Ethanol	A431: 10.4 μg/mL	Val'ko et al. (2007)

Plantago afra L./ Plantaginaceae	Sand-Wegerich; African plantain; black psyllium; and glandular plantain	Spain (wide distribution in Central Asia and Tropical Africa)	Leaves	Methanol	UACC-62: 53.83 ± 16.08 μg/mL	Gálvez, Martín-Cordero, López-Lázaro, Cortéz, and Ayuso (2003)
Plantago bellardii subsp. *bellardii*/ Plantaginaceae	–	Spain (widespread through Europe; Australia; and the Middle East)	Leaves	Methanol	UACC-62: 34.77 ± 4.15 μg/mL	Gálvez et al. (2003)
Plantago coronopus L. Accepted Scientific name: *Plantago weldenii* Rchb./ Plantaginaceae	Buck's-horn plantain; cutleaf plantain and star of the earth.	Spain (distributed in Europe; Lebanon; Syria; Israel; Jordan; Egypt; Libya; Tunisia; Algeria; Morocco; and Iran)	Leaves	Methanol	UACC-62: 40.98 ± 5.80 μg/mL	Gálvez et al. (2003)
Plantago lagopus Accepted Scientific name: Plantago lagopus subsp. *lagopus*/ Plantaginaceae	Hare's-foot plantain	Spain (widely distributed in Europe; North Africa; the Middle East; and Russia)	Leaves	Methanol	UACC-62: 66.07 ± 8.76 μg/mL	Gálvez et al. (2003)
Plantago lanceolata L. Accepted Scientific name: *Plantago lanceolata* subsp. *lanceolata*/ Plantaginaceae	Narrowleaf plantain; Ribwort plantain; English plantain and lanceleaf plantain.	Spain (widely distributed in Europe; North Africa; the Middle East; and Russia)	Leaves	Methanol	UACC-62: 50.58 ± 11.15 μg/mL	Gálvez et al. (2003)

Continued

TABLE 2.5 Plants Tested From Across the World for Cytotoxicity Against Skin Cancer—cont'd

Plant/Families	Common Names	Study Area and Distribution	Plant Parts	Extract Type	Results (IC_{50}[a] in μg/mL)	References
Plantago major L./ Plantaginaceae	Common plantain; broadleaf plantain; and greater plantain	Spain (distributed in all continents except Antarctica)	Leaves	Methanol	UACC-62: 46.5 ± 8.2 μg/mL	Gálvez et al. (2003)
Plantago serraria L./ Plantaginaceae	Pulgueira and plantain Corne-de-cerf.	Spain (Portugal; Spain; Sardinia; Malta; Sicily; Italy; Greece; Tunisia; Algeria; Morocco; and the Canary Islands)	Leaves	Methanol	UACC-62: 48.94 ± 8.17 μg/mL	Gálvez et al. (2003)
Rafflesia kerrii W. Meijer/Rafflesiaceae	Bua phut	Thailand (Malaysia and Thailand)	Flowers	50% hydroglycol extracts	A431: 0.30 μg/mL	Thuncharoen et al. (2013)
Red skinned grapes	–	India	Seeds	Petroleum ether	A431: 480 μg/mL	Mohansrinivasan, Subathra, Meenakshi, Ananya, and Jemimah (2015).
Rheum emodi Wall. Accepted Scientific name: *Rheum australe* D. Don/ Polygonaceae	Himalayan rhubarb and zang bian da huang.	India (Tibet; India; Nepal; Sikkim; Bhutan; and Burma)	Resin	Methanol	A375: 11.05 μg/mL	Gopalakrishna et al. (2014)

Rhus succedanea L. Accepted Scientific name: *Toxicodendron succedaneum* (L.) Kuntze/ Anacardiaceae	Wax tree	India (Cambodia; China; India; Japan; Korea; Laos; Thailand; and Vietnam)	Stems	Methanol	A375:13.13 μg/mL	Gopalakrishna et al. (2014)
Rosa x *damascena* Herrm./Rosaceae	Damask rose	Thailand (Canada; China; and USA)	Flowers	50% hydroglycol extracts	A431: 3.22 μg/mL	Thuncharoen et al. (2013)
Salvia leriifolia BENTH./Lamiaceae	–	Italy (Afghanistan and Iran)	Aerial parts	Methanol Hexane CH_2Cl_2 AcOEt BuOH	C32 Amelanotic melanoma: Methanol: 76.2 ± 1.3 μg/mL; A375 69.1 ± 1.6 μg/mL Hexane: 11.2 ± 0.4 μg/mL; A375 20.2 ± 1.2 μg/mL CH_2Cl_2: 13.6 ± 0.9 μg/mL; A375 88.4 ± 1.3 μg/mL AcOEt: 72.2 ± 1.4 μg/mL; A375 28.4 ± 0.6 μg/mL BuOH: 29.7 ± 1.2 μg/mL; A375 32.1 ± 0.9 μg/mL	Tundis et al. (2011)

Continued

TABLE 2.5 Plants Tested From Across the World for Cytotoxicity Against Skin Cancer—cont'd

Plant/Families	Common Names	Study Area and Distribution	Plant Parts	Extract Type	Results (IC_{50}[a] in μg/mL)	References
Simmondsia chinensis (Link) C.K. Schneid/ Simmondsiaceae	Goatnut and jojoba	Jordan (Middle and North America)	Leaves (male and female), seeds and testa	Hexane (H)—seeds oil only Ethanol (E) Methanol (M)—testa only Cold pressing extract (CP)—seeds oil only	Human melanoma (MV 3): E extracts: female leaves (71 μg/mL), male leaves (65.86 μg/mL) and seeds (82.01 μg/mL) M extract: 58.51 μg/mL H extract: 39.91 μg/mL CP extract: 100 μg/mL	Qizwini, Khateeb, Mhaidat, and Saleem (2014)
Solanum amygdalifolium Steud./Solanaceae	Jazmin de Córdoba	Argentina (Argentina; Bolivia; Brazil; Paraguay; and Uruguay)	Flower	Methanol	B16F10: 36 ± 3 μg/mL; HaCat: 30 ± 2 μg/mL	Mamone et al. (2011)
Solanum chacoense Bitter Solanaceae	Papa salvaje	Argentina (Argentina; Bolivia; Brazil; New Zealand; Paraguay; and Peru)	Leaf	Methanol	B16F10: 24 ± 6 μg/mL; HaCat: 12 ± 6 μg/mL	Mamone et al. (2011)
Solanum indicum L. Accepted Scientific name: *Solanum anguivi* Lam./ Solanaceae	Billybobs and forest bitterberry	India (Arabian Peninsula; Aldabra; Mascarene islands; Madagascar; South Africa; Tropical Africa; Comoro; USA; and Peru)	Fruit	Methanol	A375: 27.94 μg/mL	Gopalakrishna et al. (2014)

Solanum sisymbriifolium Lam./Solanaceae	Sticky nightshade; espina colorada; manacader; mullaca espinudo; and João bravo.	Argentina (Argentina; Bolivia; Canada; Chile; China; Colombia; Ecuador; Mexico; Peru; South Africa; and USA)	Flower	Methanol	B16F10: 56 ± 3 μg/mL; HaCat: 30 ± 5 μg/mL	Mamone et al. (2011)
Solanum verbascifolium L. Accepted Scientific name: *Solanum donianum* Walp./ Solanaceae	Blotgett's nightshade; mullein nightshade; Fumo bravo and Kaxicuch	Argentina (Belize; Guatemala; Mexico; USA and US. Virgin Islands)	Flower	Methanol	B16F10: 75 ± 6 μg/mL; HaCat: 155 ± 3 μg/mL	Mamone et al. (2011)
Tiliacora funifera (Miers) Oliver/ Menispermaceae	–	South Africa (widespread throughout the African continent)	Leaves and rhizome	Alkaloid extract	UACC62 melanoma cells: 7 μg/mL for leaves and rhizome respectively	De Wet et al. (2009)
Tinospora fragosa (Verdoorn) Verdoorn and Troupin/ Menispermaceae	Marvel creeper and Moses' staff.	South Africa (South Africa; Namibia; Botswana; and Zimbabwe)	Leaves	Alkaloid extract	UACC62 melanoma cells: 12.5 μg/mL	De Wet et al. (2009)
Withania somnifera (L.) Dunal	Poison gooseberry; rennet; winter cherry and Withania.	India (Afghanistan; China; India Madagascar; Pakistan; and South Africa)	Roots	Water	A375: 200 μg/mL	Halder, Sing, and Thakur (2015)

[a]*IC_{50} values in μg/mL: The concentration, in μg/mL, at which the sample inhibits 50% of the cell growth and proliferation of a particular cell line.*

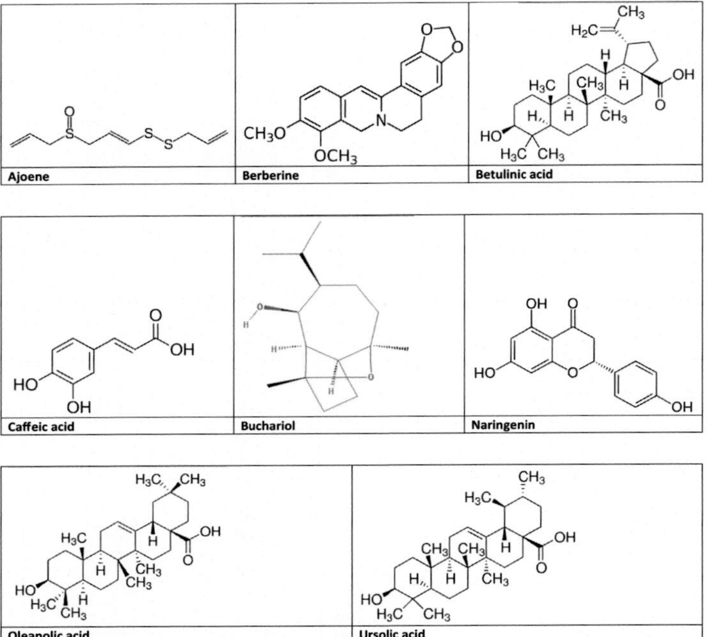

FIGURE 2.4 Chemical structures of ajoene, berberine, betulinic acid, caffeic acid, buchariol, naringenin, oleanolic acid, and ursolic acid.

National Biodiversity Institute) red data list were utilized to obtain information on the correct scientific names; vernacular names; and distributions of the plant species.

Apart from plant extracts being explored for their activity against skin cancer, there are also a variety of compounds that have been isolated from plants that have shown anticancer activity; some of these compounds are discussed below. There are also many compounds present within the "Functional Foods and Nutraceuticals" section, which have been discussed for the anticancer and chemopreventive properties; these include quercetin, resveratrol, curcumin, vitamin E, sulforaphane, apigenin, and EGCG.

Ajoene (Fig. 2.4): Ajoene is an organosulfur compound derived from allicin found in garlic, showed the reduction in cell viability of mouse melanoma (B16F10) cells and was further able to activate caspase-3. The derivative was further able to reduce the ability of the B16F10 cells to adhere to an endothelial cell monolayer by changing the integrin expression of alpha-(4)-beta-(1) expression on B16F10 cells (Ledezma, Apitz-Castro, & Cardier, 2004).

Berberine (Fig. 2.4): Berberine, a well-known alkaloid, was shown to inhibit the metastatic potential of melanoma cell (A375) and that is cell death was associated with the induction of apoptosis. The inhibition of melanoma migration was reported to be due to the inhibition of cyclooxygenase-2 (Singh, Vaid, Katiyar, Sharma, & Katiyar, 2011).

Betulinic acid (Fig. 2.4): Betulinic acid is a pentacyclic triterpenoid that has been reported to inhibit the growth of UVB-induced photocarcinogenesis and human melanoma xenografts in nude mice. It is suggested that Betulinic acid inhibits the growth of human melanoma skin cancer via inhibiting NF-κB and inducing apoptosis (Fukuda, Sakai, Matsunaga, Tokuda, & Tanaka, 2006; Honda et al., 2006; Pisha et al., 1995).

Caffeic acid (Fig. 2.4): Caffeic acid is a phenolic compound occurring naturally in a range of plants. It has been reported to possess chemopreventive activity of UV-induced skin carcinogenesis in mouse models and transformed HaCat cells by several studies. Multiple mechanisms by which caffeic acid is able to facilitate this action have also been reported. These mechanisms include: inhibition of the activator protein 1; nuclear factor-κB (NF-κB); Fyn kinase activity; inhibition of the propagation of the lipid peroxidative chain reaction and interference with the extracellular signal-regulated kinase 1 and 2 pathways and nitrogen oxides. Furthermore, caffeic acid has also shown potential as an antiskin cancer agent by inhibiting colony formation in the A431; SK-MEL-5; SK-MEL-28; and transformed HaCat cell lines and also have the capacity to inhibit metastasis of transformed HaCat cells by targeting NF-κB and the snail signaling pathway (Kang et al., 2009; Saija et al., 1999; Yang et al., 2013, 2014).

Buchariol (Fig. 2.4) and Naringenin (Fig. 2.4): The Buchariol (diacetyl triterpene) and Naringenin (flavanone) compounds have been shown to exhibit potent cytotoxic activity with IC_{50} values of 0.5 ± 0.01 and $0.6 \pm 0.09\,\mu g/mL$, respectively, against the amelanotic melanoma cell line C32. Both these compounds were isolated from *Salvia leriifolia* BENTH (Tundis et al., 2011).

Oleanolic acid (Fig. 2.4): George, Kumar, Suresh, and Kumar (2014) reported that oleanolic acid, a pentacyclic triterpene, showed antiproliferative effects against human MM (A375) cells. The compound was further able to induce apoptosis in A375 cells as noted in apoptotic-DNA fragments in ELISA and nuclear fragments in DAPI results.

Silymarin complex: Silymarin is a well-known flavanoligan complex isolated from the seeds of the milk thistle plant. This flavanoligan complex consists of six principal compounds known as silybin A, silybin B, isosilybin A, isosilybin B, silychristin A, silychristin B, and silydianin. Silymarin has been extensively evaluated for its activity against skin cancer. A study by Zi and Agarwal (1999) indicated that silymarin has the ability to induce apoptotic cell death in the A431 cell line at concentrations ranging between 75 and $150\,\mu g/mL$. Furthermore, this study showed that silymarin can induce cell cycle arrest at the G0-G1 and G2-M checkpoints. Silymarin has also been shown to inhibit inflammation via targeting COX-2 and lipoxygenase activity and can inhibit the expression of detrimental cytokines including TNF-α and IL-1α (Zhao et al., 1999; Zi, Mukhtar, & Agarwal, 1997).

Ursolic acid (Fig. 2.4): In a study by Harmand, Duval, Delage, and Simon (2005) ursolic acid (pentacyclic triterpenoid) was able to inhibit the proliferation of a human melanoma cell line (M4Beu). During cell cycle arrest an increase in sub-G1 cell population was noted, which is generally associated with the presence of apoptotic cells. It was also suggested that the induction of apoptosis could be attributed to the activation of caspase-3 and caspase-9 as well as the mitochondrial intrinsic pathway.

There are many other natural products that have shown the potential to act as an anticancer agent against various types of skin cancer. A review by Chinembiri et al. (2014) discussed the potential of various natural products for the use against skin cancer.

2.5 POTENTIAL PLANTS AND NATURAL PRODUCTS WITH CHEMOPREVENTIVE ACTIVITY AGAINST ULTRAVIOLET RADIATION

Natural substances are being considered for addition into sunscreens because of their antioxidant activity, which have been speculated to increase the photoprotective activity of a sunscreen (Ebrahimzadeh et al., 2014). Plants and compounds that have been used successfully as antioxidants include; phenolics, flavonoids, tocopherols as well as nitrogen-containing compounds. Various plant extracts have been tested for their photoprotective activity both in vitro and in vivo, and some of these include the following:

1. *Neoglaziovia variegata*: The leaves were extracted with ethanol and partitioned into hexane, chloroform, and ethyl-acetate fractions. Each extract was tested for in vitro SPF at 100 mg/L, and the SPF values obtained were 2.80, 2.97, 11.46, and 27.68, respectively (De Oliveira Junior, de Souza Araújo et al., 2013).
2. *Encholirium spectabile*: An ethanolic leaf extract was partitioned into hexane, chloroform, and ethyl-acetate fractions. Testing the in vitro SPF of the samples at 100 mg/L, SPF values of 7.22, 1.74, 8.89, and 4.60 were obtained, respectively (De Oliveira Junior, Souza et al., 2013).
3. *Buddleja cordata*: A methanolic extracts of leaves of *B. cordata* was prepared. The extract was applied to the skin (200 μL of a 2 mg/mL concentration) of female hairless mice (SKH-1) 15 min before irradiating with UVB for 10 min. The erythema was then measured after 24 h after irradiation, and it was reported that the extract was able to protect the skin from UVB irradiation (Acevedo et al., 2014).
4. *Marcetia taxifolia*: A 95% ethanolic extract of the leaves was used to determine its in vitro SPF potential. At concentrations of 250 and 125 μg/μL, SPF values of the extracts were found to be 15.52 and 8.35, respectively. The extract was also tested in a sunscreen formulation, and the results are depicted in a study by Costa, Detoni, Branco, Botura, and Branco (2015).

5. *Lippia* species: Various species of *Lippia* such as *L. brasiliensis, L. pseudothea, L. rotundifolia*, and *L. rubella.* From each plant the leaves were extracted in ethanol and added into a neutral formulation at 15% to determine the in vitro SPF potential. SPF values were determined as 9, 6, 4, and 9, respectively. The UVA protection factor was also measured for each extract and was determined as 8, 5, 3, and 6, respectively. Lastly the UVA/UVB ratio was determined for each extract and calculated as 0.91, 0.85, 0.86, and 0.77, respectively (Gonçalves, Polonini, Viccini, Brandão, & Raposo, 2015).

In our research program the antioxidant activity of three plant extracts were determined, the plants were sent for in vivo SPF clinical trials to determine whether the samples were able to boost the SPF of existing sunscreen standard (P3). Two South African plants and one exotic plant passed in vivo clinical trials as SPF boosters.

- *Helichrysum odoratissimum* (L.) Sweet was able to boost the P3 reference standard from 17.1 to 32.4
- *Syzygium jambos* (Alston.) was able to boost the P3 reference standard from 15 to 18.0
- *Greyia flanaganii* Bolus was able to boost the P3 reference standard from 16.5 to 17.3

Natural products with photoprotection properties:

1. Rutin: The SPF and protection factors UVA (PF-UVA) was measured using 10% (w/w) in a formulation. Rutin was found to have an SPF of 4.72±0.20 and a PF-UVA of 4.92±0.20. Rutin was also found to be photostable after 2h of irradiation under controlled conditions with an SPF and PF-UVA of 4.42±0.13 and 4.59±0.13, respectively, varying less than 10% of the original values. The protection increased significantly when rutin was added in combination with known inorganic filters. In combination with 10% (w/w) of titanium dioxide (TiO_2), the SPF and PF-UVA increased to 34.29±8.31 and 16.25±2.71, respectively. In combination with 10% (w/w) zinc oxide (ZnO) the SPF increased to 11.25±3.31 and the PF-UVA increased to 9.75±2.81 (Choquenet, Couteau, Paparis, & Coiffard, 2008).
2. Apigenin: The UVB protection of apigenin was determined by measuring the in vitro SPF. Apigenin can be considered as a UVB filter with a high SPF of 28.8. The ratio of UVB/UVA was determined to be 0.66 (Stevanato, Bertelle, & Fabris, 2014).
3. Silymarin: At a concentration of 10% (w/w), silymarin was able to protect against both UVA and UVB with measurements of 12.37±4.39 and 7.32±2.34 for SPF and PF-UVA, respectively. Silymarin was also photostable as determined after measuring the SPF (11.84±3.84) and PF-UVA (7.18±2.06) after 2h of irradiation (Couteau, Cheignon, Paparis, & Coiffard, 2012).

It is evident that plant extracts and natural products are not only a great source for the discovery of anticancer agents against skin cancer, as well as many other types of cancer, but that research is also moving toward using these plants and compounds as photoprotective agents in sunscreens to prevent against skin cancer.

2.6 CONCLUSION

Cancer is a devastating disease, causing millions of deaths worldwide each year. Skin cancer incidences have been increasing since 1970 because of an increase in UV radiation and the nonchalant attitude of the general public toward the negative effects of sun exposure. This increase in skin cancer cases is particularly worrying in countries with high levels of UV radiation. South Africa has high levels of UVB radiation varying between 8 and 11+ throughout most of the year, which leads to a potential increase in the risk for diseases such as skin cancer, which are related to overexposure of UV radiation. Although current treatments for cancer are extremely valuable, the general belief remains that prevention is a much better option than the treatment of cancer. Many plant-derived treatments for cancer has been discovered over the past few decades, and the potential of more such drugs are still being researched. Throughout history it has been suggested that the dietary lifestyle plays an important role in the well-being and overall health of an individual. This perspective led to a rise in research on the potential cancer chemopreventive roles that plant-based foods play that are consumed as part of the normal diet. With this increase in interest and research on functional foods with potential anticancer bioactivity, concentrated amounts of the active phytochemicals are being commercialized as over-the-counter medicines focused on the overall well-being of an individual, mainly through the antioxidant activities demonstrated by such phytochemicals. Although many functional foods and nutraceuticals have demonstrated significant activity as cancer chemopreventive agents in experimental settings, it is important to remember that most of these substances have not been approved for the treatment of cancer. It is also vital to consider that many of the phytochemicals present in plants only show cancer chemopreventive activity for specific cancer types, thus not all phytochemicals present in plants that have demonstrated anticancer activity will be active against all forms of cancer. Even though many believe that functional foods and nutraceuticals do not show a significant correlation with a decrease in the risk for the development of cancer, a healthy lifestyle is still a much better choice for overall well-being and health.

Concluding remarks:

- Plants remain an important source for the development of new treatments against skin cancer.
- It is not only important to discover new sources of anticancer drugs but it is also important to search for new plant extracts and natural products that show potential as chemopreventive agents.

- Food sources and nutraceuticals provide a great source of natural products, which are used to maintain health and function as preventative agents against cancer.
- More focus is being placed on the use of whole plant extracts for incorporation into sunscreens, due to their antioxidant activity, to prevent against skin cancer.

REFERENCES

Acevedo, J. G. A., González, A. M. E., De Maria y Campos, D. M., del Carmen Benitez Flores, J., Delgado, T. H., Maya, S. F., et al. (2014). Photoprotection of *Buddleja cordata* extract against UVB-induced skin damage in SKH-1 hairless mice. *BMC Complementary and Alternative Medicine*, *14*, 281–290.

Adhami, V. M., Afaq, F., & Ahmad, N. (2001). Involvement of the retinoblastoma (pRb)-E2F/DP pathway during antiproliferative effects of resveratrol in human epidermoid carcinoma (A431) cells. *Biochemical and Biophysical Research Communications*, *288*, 579–585.

Afaq, F., Adhami, V. M., & Ahmad, N. (2003). Prevention of short-term ultraviolet B radiation-mediated damages by resveratrol in SKH-1 hairless mice. *Toxicology and Applied Pharmacology*, *186*, 28–37.

Afaq, F., Adhami, V. M., Ahmad, N., & Mukhtar, H. (2002). Botanical antioxidants for chemoprevention of photocarcinogenesis. *Frontiers in Bioscience*, *7*, d784–d792.

Ahmad, N., Adhami, V. M., Afaq, F., Feyes, D. K., & Mukhtar, H. (2001). Resveratrol causes WAF-1/p21-mediated G1-phase arrest of cell cycle and induction of apoptosis in human epidermoid carcinoma A431. *Cells*, *7*, 1466–1473.

AIM. (2014). *FDA approved drugs for melanoma*. Retrieved from: https://www.aimatmelanoma.org/melanoma-treatment-options/fda-approved-drugs-for-melanoma/.

Al-Kalaldeh, J. Z., Abu-Dahab, R., & Afifi, F. U. (2010). Volatile oil consumption and antiproliferative activity of *Laurus nobilis*, *Origanum syriacum*, *Origanum vulgare*, and *Salvia triloba* against human breast adenocarcinoma cells. *Nutrition Research*, *30*, 271–278.

American Cancer Society. (2015a). *Signs and symptoms of basal and squamous cell skin cancers*. Retrieved from: http://www.cancer.org/cancer/skincancer-basalandsquamouscell/detailedguide/skin-cancer-basal-and-squamous-cell-signs-and-symptoms.

American Cancer Society. (2015b). *Signs and symptoms of melanoma skin cancer*. Retrieved from: http://www.cancer.org/cancer/skincancer-melanoma/detailedguide/melanoma-skin-cancer-signs-and-symptoms.

Anand, P., Sundaram, C., Jhurani, S., Kunnumakkara, A. B., & Aggarwal, B. B. (2008). Curcumin and cancer: An "old-age" disease with an "age-old" solution. *Cancer Letters*, *267*, 133–164.

Angiogenic inhibitors. (2011). *National Cancer Institute*. Retrieved from: http://www.cansa.org.za/files/2014/11/Fact-Sheet-Malignant-Melanoma-Nov-2014.pdf.

Asensi, M., Medina, I., Ortega, A., Carretero, J., Baño, M. C., Obrador, E., et al. (2002). Original contribution: Inhibition of cancer growth by resveratrol is related to its low bioavailability. *Free Radical Biology and Medicine*, *33*, 387–398.

Athar, M., Back, J. H., Tang, X., Kim, K. H., Kopelovich, L., Bickers, D. R., et al. (2007). Resveratrol: A review of preclinical studies for human cancer prevention. *Toxicology and Applied Pharmacology*, *224*, 274–283.

Awasare, S., Bhujbal, S., & Nanda, R. (2012). *In vitro* cytotoxic activity of novel oleanane type of triterpenoid saponin from stem bark of *Manilkara zapota* Linn. *Asian Journal of Pharmaceutical and Clinical Research*, *5*, 183–188.

Aziz, M. H., Afaq, F., & Ahmad, N. (2005). Prevention of ultraviolet-B radiation damage by resveratrol in mouse skin is mediated via modulation in survivin. *Photochemistry and Photobiology*, *81*, 25–31.

Bagchi, D., Garg, A., Krohn, R. L., Bagchi, M., Tran, M. X., & Stohs, S. J. (1997). Oxygen free radical scavenging abilities of vitamins C and E, and a grape seed proanthocyanidin extract *in vitro*. *Research Communications in Molecular Pathology and Pharmacology*, *95*, 179–189.

Balakrishnan, K. P., & Narayanaswamy, N. (2011). Botanicals as sunscreens: Their role in the prevention of photoaging and skin cancer. *International Journal of Cosmetic Science*, *1*, 1–12.

Baliga, M. S., & Katiyar, S. K. (2006). Chemoprevention of photocarcinogenesis by selected dietary botanicals. *Photochemical and Photobiological Sciences*, *5*, 243–253.

Berrington, D., & Lall, N. (2012). Anticancer activity of certain herbs and spices on the cervical epithelial carcinoma (HeLa) cell line. *Evidence-Based Complementary and Alternative Medicine*, 11:564927.

Bialy, T. L., Rothe, M. J., & Grant-Kels, J. M. (2002). Dietary factors in the prevention and treatment of non-melanoma skin cancer and melanoma. *Dermatologic Surgery*, *28*, 1143–1152.

Birt, D. F., Mitchell, D., Gold, B., Pour, P., & Pinch, H. C. (1997). Inhibition of ultraviolet light induced skin carcinogenesis in SKH-1 mice by apigenin, a plant flavonoid. *Anticancer Research*, *17*, 85–91.

Blanc, I. (2011). *Datura stramonium-Guixers-IB-615.JPG*. Retrieved from: https://commons.wikimedia.org/wiki/File:DATURA_STRAMONIUM_-_GUIXERS_-_IB-615.JPG.

Bomser, J., Singletary, K., & Meline, B. (2000). Inhibition of 12-O-tetradecanoylphorbol-13-acetate (TPA) induced mouse skin ornithine decarboxylase and protein kinase C by polyphenolics from grapes. *Chemico-Biological Interactions*, *127*, 45–59.

Bomser, J. A., Singletary, K. W., Wallig, M. A., & Smith, M. A. (1999). Inhibition of TPA induced tumor promotion in CD-1 mouse epidermis by a polyphenolic fraction from grape seeds. *Cancer Letters*, *135*, 151–157.

BotBln. (2010). *Merwilla plumbea KirstenboschBotGard09292010KM.JPG*. Retrieved from: https://commons.wikimedia.org/wiki/File:Merwilla_plumbea_KirstenboschBotGard09292010KM.JPG.

Boubaker, J., Bzeouich, I. M., Nasr, N., Ghozlen, H. B., Mustapha, N., Ghedira, K., et al. (2015). Phytochemical capacity of *Nitraria retusa* leaves extracts inhibiting growth of melanoma cells and enhancing melanogenesis of B16F10 melanoma. *BMC Complementary and Alternative Medicine*, *15*, 300–307.

Braxmeier, H. (2009). *Agapanthus-jewelry lilies-greenhouse*. Retrieved from: https://pixabay.com/en/agapanthus-jewelry-lilies-greenhouse-9001/.

Bush, J. A., Cheung, K. J. J., Jr., & Li, G. (2001). Curcumin induces apoptosis in human melanoma cells through a Fas receptor/caspase-8 pathway independent of p53. *Experimental Cell Research*, *271*, 305–314.

Caltagirone, S., Rossi, C., Poggi, A., Ranelletti, F. O., Natali, P. G., Brunetti, M., et al. (2000). Flavonoids apigenin and quercetin inhibit melanoma growth and metastatic potential. *International Journal of Cancer*, *87*, 595–600.

CANSA. (2010). *Fact sheet-skin cancer 2010*. Retrieved from: http://www.cansa.org.za/files/2012/05/SKIN_CANCER_Leaflet-2010.pdf.

CANSA. (2013a). *South African cancer statistics*. Retrieved from: http://www.cansa.org.za/position-statement-complementary-medicines/.

CANSA. (2013b). *Fact sheet on solar radiation skin cancer*. Retrieved from: http://www.cansa.org.za/files/2013/10/Fact-Sheet-on-Solar-Radiation-Skin-Cancer-Oct-2013.pdf.

CANSA. (2014). *Fact sheet on malignant melanoma*. Retrieved from: http://www.cansa.org.za/files/2014/11/Fact-Sheet-Malignant-Melanoma-Nov-2014.pdf.

Cao, H. H., Tse, A. K. W., Kwan, H. Y., Yu, H., Cheng, C. Y., Su, T., et al. (2014). Quercetin exerts anti-melanoma activities and inhibits STAT3 signalling. *Biochemical Pharmacology*, *87*, 424–434.

Carson, C. F., Hammer, K. A., & Riley, T. V. (2006). *Melaleuca alternifolia* (tea tree) oil: A review of antimicrobial and other medicinal properties. *Clinical Microbiology Reviews*, *19*, 50–62.

Chandu, A. N., Kumar, S. C., Bhattacharjee, C., & Debnath, S. (2012). Cytotoxicity study of plant *Aloe vera* (Linn). *Chronicles of Young Scientists*, *3*, 233–235.

Chen, M. C., Chang, W. W., Kuan, Y. D., Lin, S. T., Hsu, H. C., & Lee, C. H. (2012). Resveratrol inhibits LPS-induced epithelial-mesenchymal transition in mouse melanoma model. *Innate Immunity*, *18*, 685–693.

Chen, X., Liu, H., & Lei, X. (1998). Cancer chemopreventive and therapeutic activities of red ginseng. *The Journal of Ethnopharmacology*, *60*, 71–78.

Chinembiri, T. N., du Plessis, L. H., Gerber, M., Hamman, J. H., & du Plessis, J. (2014). Review of natural compounds for potential skin cancer treatment. *Molecules*, *19*, 11679–11721.

Chmee2. (2010). *Cannabis sativa plant (7).JPG*. Retrieved from: https://commons.wikimedia.org/wiki/File:Cannabis_sativa_plant_(7).JPG.

Choquenet, B., Couteau, C., Paparis, E., & Coiffard, L. J. M. (2008). Quercetin and rutin as potential sunscreen agents: Determination of efficacy by an in vitro method. *Journal of Natural Products*, *71*(6), 1117–1118.

Clark, L. C., Combs, G. F., Turnbull, B. W., Slate, E. H., Chalker, D. K., Chow, J., et al. (1996). Effects of selenium supplementation for cancer prevention in patients with carcinoma of the skin: A randomized controlled trial. *Journal of the American Medical Association*, *276*, 1957–1963.

Clark, L. C., Graham, G. F., & Crouse, R. G. (1984). Plasma selenium and skin neoplasms: A case controlled study. *Nutrition and Cancer*, *6*, 13–21.

Coopoosamy, R. M., & Naidoo, K. K. (2012). An ethnobotanical study of medicinal plants used by traditional healers in Durban, South Africa. *African Journal of Pharmacy and Pharmacology*, *6*(11), 818–823.

Costa, S. C. C., Detoni, C. B., Branco, C. R. C., Botura, M. B., & Branco, A. (2015). *In vitro* photoprotective effects of *Marcetia taxifolia* ethanolic extract and it's potential for sunscreen formulations. *Revista Brasileira de Farmacognosia*, *25*(4), 413–418.

Courreges, M. C., & Benencia, F. (2002). *In vitro* antiphagocytic effect of basil oil on mouse macrophages. *Fitoterapia*, *73*, 369–374.

Couteau, C., Cheignon, C., Paparis, E., & Coiffard, L. J. M. (2012). Silymarin, a molecule of interest for topical photoprotection. *Natural Product Research*, *26*(23), 2211–2214.

Cragg, G. M., & Newman, D. J. (2005). Plants as a source of anticancer agents. *Journal of Ethnopharmacology*, *100*, 72–79.

Culos, R. (2015). *Harpagophytum procumbens*. Retrieved from: https://en.wikipedia.org/wiki/Harpagophytum#/media/File:Harpagophytum_procumbens_MHNT.BOT.2005.0.1243.jpg.

Dahmke, I. N., Backes, C., Rudzitis-Auth, J., Laschke, M. W., Leidinger, P., Menger, M. D., et al. (2013). Curcumin intake affects miRNA signature in murine melanoma with mmu-miR-205-5p most significantly altered. *PLoS One*, *8*. http://dx.doi.org/10.1371/journal.pone.0081122.

Dasgupta, T., Rao, A. R., & Yadava, P. K. (2004). Chemomodulatory efficacy of Basil leaf (*Ocimum basilicum*) on drug metabolizing and antioxidant enzymes, and on carcinogen-induced skin and forestomach papillomagenesis. *Phytomedicine*, *11*, 139–151.

Davids, L. M., Kleemann, B., Kacerovská, D., Pizinger, K., & Kidson, S. H. (2008). Hypericin phototoxicity induces different modes of cell death in melanoma and human skin cells. *Journal of Photochemistry and Photobiology B: Biology*, *91*, 67–76.

de Canha, M. (2016). *Harpephyllum caffrum*. (Unpublished photograph).

De Oliveira Junior, R. G., de Souza Araújo, C., Souza, G. R., Guimarães, A. L., de Oliveira, A. P., et al. (2013). *In vitro* antioxidant and photoprotective activities of dried extracts from *Neoglaziovia variegate* (Bromeliaceae). *Journal of Applied Pharmaceutical Science*, *3*(1), 122–127.

De Oliveira Junior, R. G., Souza, G. R., Guimarães, A. L., de Oliveira, A. P., Morais, A. C. S., et al. (2013). Dried extracts of *Encholirium spectabile* (Bromeliaceae) present antioxidant and photoprotective activities *in vitro*. *Journal of Young Pharmacists*, *5*(3), 102–105.

De Wet, H., Fouche, G., & Van Heerden, F. R. (2009). *In vitro* cytotoxicity of crude alkaloidal extracts of South African Menispermaceae against three cancer cell lines. *African Journal of Biotechnology*, *8*(14), 3332–3335.

Do, D. P., Pai, S. B., Rizvi, S. A. A., & D'Souza, M. J. (2010). Development of sulforaphane-encapsulated microspheres for cancer epigenetic therapy. *International Journal of Pharmaceutics*, *386*, 114–121.

Ebrahimzadeh, M. A., Enayatifard, R., Khalili, M., Ghaffarloo, M., Saeedi, M., & Yazdani Charati, J. (2014). Correlation between sun protection factor and antioxidant activity, phenol and flavonoid contents of some medicinal plants. *Iranian Journal of Pharmaceutical Research*, *13*(3), 1041–1047.

Eylen, D. V., Oey, I., Hendrickx, M., & Loey, A. V. (2007). Kinetics of the stability of broccoli (*Brassica oleracea* cv. *italica*) myrosinase and isothiocyanates in broccoli juice during pressure/temperature treatments. *Journal of Agricultural and Food Chemistry*, *55*, 2163–2170.

Fang, Y., Bradley, M. J., Cook, K. M., Herrick, E. J., & Nicholl, M. B. (2013). A potential role for resveratrol as a radiation sensitizer for melanoma treatment. *Journal of Surgical Research*, *183*, 645–653.

Fibrich, B. (2016). *Bulbine frutescens*. (Unpublished photograph).

Fimognari, C., & Hrelia, P. (2007). Sulforaphane as a promising molecule for fighting cancer. *Mutation Research*, *635*, 90–104.

Fouche, G., Cragg, G. M., Pillay, P., Kolesnikova, N., Maharaj, V. J., & Senabe, J. (2008). *In vitro* anticancer screening of South African plants. *Journal of Ethnopharmacology*, *119*, 455–461.

Fu, Z. D., Cao, Y., Wang, K. F., Xu, S. F., & Han, R. (2004). Chemopreventive effect of resveratrol to cancer. *Chinese Journal of Cancer*, *23*, 869–873.

Fukuda, Y., Sakai, K., Matsunaga, S., Tokuda, H., & Tanaka, R. (2006). Cancer chemopreventive effect of orally administrated lupane-type triterpenoid on ultraviolet light B induced photocarcinogenesis of hairless mouse. *Cancer Letters*, *240*, 94–101.

Gálvez, M., Martín-Cordero, C., López-Lázaro, M., Cortéz, F., & Ayuso, M. J. (2003). Cytotoxic effect of *Plantago* spp. on cancer cell lines. *Journal of Ethnopharamcology*, *88*, 125–130.

Garg, J. M. (2009). *Trema orientalis (Pigeon Wood) W2 IMG 2237.jpg*. Retrieved from: https://upload.wikimedia.org/wikipedia/commons/thumb/5/5c/Trema_orientalis_%28Pigeon_Wood%29_W2_IMG_2237.jpg/800px-Trema_orientalis_%28Pigeon_Wood%29_W2_IMG_2237.jpg.

Gatouillat, G., Balasse, E., Joseph-Pietras, D., Morjani, H., & Madoulet, C. (2010). Resveratrol induces cell-cycle disruption and apoptosis in chemoresistant B16 melanoma. *Journal of Cellular Biochemistry*, *110*, 893–902.

Gensler, H. L., Aickin, M., Peng, Y. M., & Xu, M. (1996). Importance of the form of topical vitamin E for prevention of photocarcinogenesis. *Nutrition and Cancer*, *26*, 183–191.

George, V. C., Kumar, D. R. N., Suresh, P. K., & Kumar, R. A. (2014). Oleanolic acid inhibits cell growth and induces apoptosis in A375 melanoma cells. *Biomedicine and Preventive Nutrition, 4*(2), 95–99.

Gerhauser, C. (2008). Cancer chemopreventive potential of apples, apple juice, and apple components. *Planta Medica, 13*, 1608–1624.

Gonçalves, K. M., Polonini, H. C., Viccini, L. F., Brandão, M. A. F., & Raposo, N. R. B. (2015). Assessment of the photoprotective activity of *Lippia* species from Brazil and their uses as single UV filters in sunscreens. *Journal of Young Pharmacists, 7*(4), 368–372.

Gopalakrishna, S. M., Thimappa, G. S., Thylur, R. P., Shivanna, Y., & Sreenivasan, A. (2014). *In vitro* anti-cancer screening of *Solanum indicum, Rhus succedanea, Rheum emodi* and, *Gardenia gummifera* medicinal plants in cancer cells. *Journal of Pharmacy and Pharmaceutical Sciences, 3*, 22–30.

Gupta, S. C., Patchva, S., Koh, W., & Aggarwal, B. B. (2012). Discovery of curcumin, a component of golden spice, and its miraculous biological activities. *Clinical and Experimental Pharmacology and Physiology, 39*, 283–299.

Halder, B., Sing, S., & Thakur, S. S. (2015). *Withania somnifera* root extract has potent cytotoxic effect against human malignant melanoma cells. *PLoS One, 10*(9), e0137498.

Hamsa, T. P., Thejass, P., & Kuttan, G. (2011). Induction of apoptosis by sulforaphane in highly metastatic B16F-10 melanoma cells. *Drug and Chemical Toxicology, 34*, 332–340.

Hara, Y. (2001). Fermentation of tea. In Y. Hara (Ed.), *Green tea, health benefits and applications* (7th ed.) (pp. 16–21). New York: Marcel Dekker.

Harmand, P. O., Duval, R., Delage, C., & Simon, A. (2005). Ursolic acid induces apoptosis through mitochondrial intrinsic pathway and caspase-3 activation in M4Beu melanoma cells. *International Journal of Cancer, 114*(1), 1–11.

Health.24. (2015). *South Africa: 78% increase in cancer by 2030.* Retrieved from: http://www.health24.com/Medical/Cancer/Facts-and-figures/South-Africa-78-increase-in-cancer-by-2030-20120721.

Herr, I., & Büchler, M. W. (2010). Dietary constituents of broccoli and other cruciferous vegetables: Implications for prevention and therapy of cancer. *Cancer Treatment Reviews, 36*, 377–383.

Higdon, J. V., Delage, B., Williams, D. E., & Dashwood, R. H. (2007). Cruciferous vegetables and human cancer risk: Epidemiologic evidence and mechanistic basis. *Pharmacological Research, 55*, 224–236.

Hillewaert, H. (1988). *Boophone disticha.Jpg*. Retrieved from: https://commons.wikimedia.org/wiki/File:Boophone_disticha.jpg.

Hillewaert, H. (2012). *Euphorbia ingens (habitus).Jpg*. Retrieved from: https://commons.wikimedia.org/wiki/File:Euphorbia_ingens_(habitus).jpg.

Honda, T., Liby, K. T., Su, X., Sundararajan, C., Honda, Y., Suh, N., et al. (2006). Design, synthesis, and antiinflammatory activity both *in vitro* and *in vivo* of new betulinic acid analogues having an enone functionality in ring A. *Bioorganic and Medicinal Chemistry Letters, 16*, 6306–6309.

Huang, S. C., Ho, C. T., Lin-Shiau, S. Y., & Lin, J. K. (2005). Carnosol inhibits the invasion of B16/F10 mouse melanoma cells by suppressing metalloproteinase-9 through down-regulating nuclear factor-kappaB and c-Jun. *Biochemical Pharmacology, 69*, 221–232.

Huang, M. T., Ho, C. T., Wang, Z. Y., Ferraro, T., Lou, Y. R., Stauber, K., et al. (1994). Inhibition of skin tumorigenesis by rosemary and its constituents carnosol and ursolic acid. *Cancer Research, 54*, 701–708.

Huang, Z., Hung, C., Lin, Y., & Fang, J. (2008). *In vitro* and *in vivo* evaluation of topical delivery and potential dermal use of soy isoflavones genistein and daidzein. *International Journal of Pharmaceutics, 364*, 36–44.

Hummert, C. (2009). *Sutherlandia futescens 1.jpg*. Retrieved from: https://commons.wikimedia.org/wiki/File:Sutherlandia_frutescens_1.jpg.

Ichihashi, M., Ahmed, N. U., Budiyanto, A., Wu, A., Bito, T., Ueda, M., et al. (2000). Preventive effect of antioxidant on ultraviolet-induced skin cancer in mice. *Journal of Dermatological Science*, *23*, S45–S50.

Iovine, B., Iannella, M. L., Gasparri, F., Monfrecola, G., & Bevilacqua, M. A. (2011). Synergic effect of genistein and daidzein on UVB-induced DNA damage: An effective photoprotective combination. *Journal of Biomedicine and Biotechnology*. http://dx.doi.org/10.1155/2011/692846.

Ireland, D. J., Greay, S. J., Hooper, C. M., Kissick, H. T., Filion, P., Riley, T. V., et al. (2012). Topically applied *Melaleuca alternifolia* (tea tree) oil causes direct anti-cancer cytotoxicity in subcutaneous tumour bearing mice. *Journal of Dermatological Science*, *67*, 120–129.

Jang, M., Cai, L., Udeani, G. O., Slowing, K. V., Thomas, C. F., Beecher, C. W., et al. (1997). Cancer chemopreventive activity of resveratrol, a natural product derived from grapes. *Science*, *275*, 218–220.

JMK. (2013). *Lippia javanica blomme-Skeerpoort-a.jpg*. Retrieved from: https://commons.wikimedia.org/wiki/File:Lippia_javanica,_blomme,_Skeerpoort,_a.jpg.

Johnson, J. J. (2011). Carnosol: A promising anti-cancer and anti-inflammatory agent. *Cancer Letters*, *305*, 1–7.

Kang, N. J., Lee, K. W., Shin, B. J., Jung, S. K., Hwang, M. K., Bode, A. M., et al. (2009). Caffeic acid, a phenolic phytochemical in coffee, directly inhibits Fyn kinase activity and UVB-induced COX-2 expression. *Carcinogenesis*, *30*, 321–330.

Katiyar, S. K. (2003). Skin photoprotection by green tea: Antioxidant and immunomodulatory effects. *Current Drug Targets – Immune, Endocrine and Metabolic Disorders*, *3*, 234–242.

Katiyar, S. K. (2008). Grape seed proanthocyanidines and skin cancer prevention: Inhibition of oxidative stress and protection of immune system. *Molecular Nutrition and Food Research*, *52*, S71–S76.

Katiyar, S. K., Agarwal., R. Mukhtar, H. (1996). Inhibition of tumor promotion in SENCAR mouse skin by ethanol extract of *Zingiber officinale* rhizome. *Cancer Research*, *56*, 1023–1030.

Katiyar, S. K., & Elmets, C. A. (2001). Green tea polyphenolic antioxidants and skin photoprotection. *International Journal of Oncology*, *18*, 1307–1313.

Katiyar, S. K., Matsui, M. S., Elmets, C. A., & Mukhtar, H. (1999). Polyphenolic antioxidant (−)-epigallocatechin-3-gallate from green tea reduces UVB induced inflammatory responses and infiltration of leukocytes in human skin. *Photochemistry and Photobiology*, *69*, 148–153.

Katiyar, S. K., & Mukhtar, H. (1996). Tea consumption and cancer. *World Review of Nutrition and Dietetics Home*, *79*, 154–184.

Kaur, M., Agarwal, C., & Agarwal, R. (2009). Anticancer and cancer chemopreventive potential of grape seed extract and other grape-based products. *The Journal of Nutrition*, *139*, 1806S–1812S.

Keum, Y. S., Park, K. K., Lee, J. M., Chun, K. S., Park, J. H., Lee, S. K., et al. (2000). Antioxidant and anti-tumor promoting activities of the methanol extract of heat-processed ginseng. *Cancer Letters*, *150*, 41–48.

Khan, N., Afaq, F., & Mukhtar, H. (2007). Apoptosis by dietary factors: The suicide solution for delaying cancer growth. *Carcinogenesis*, *28*, 233–239.

Klug, W. S., Cummings, M. R., Spencer, C. S., & Palladino, M. A. (2009). *Concepts of genetics* (19th ed.). San Francisco: Pearson International Edition.

Koduru, S., Grierson, D. S., & Afolayan, A. J. (2007). Ethnobotanical information of medicinal plants used for treatment of cancer in the Eastern Cape Province, South Africa. *Current Science*, *92*(7), 906–908.

Köhler, F. E. (1987). *Agathosma betulina-Köhler-s Medizinal-pflanzen-020.jpg*. Retrieved from: https://commons.wikimedia.org/wiki/File:Agathosma_betulina_-_K%C3%B6hler%E2%80%93s_Medizinal-Pflanzen-020.jpg.

Lawrence, E. (2008). *Henderson's dictionary of biology* (14th ed.). London: Pearson Education Limited.

Ledezma, E., Apitz-Castro, R., & Cardier, J. (2004). Apoptotic and anti-adhesion effect of ajoene, a garlic derived compound, on the murine melanoma B16F10 cells: A possible role of caspase-3 and the alpha(4)beta(1) integrin. *Cancer Letters, 206*(1), 35–41.

Li, D., Yee, J. A., Thompson, L. U., & Yan, L. (1999). Dietary supplementation with secoisolariciresinol diglycoside (SDG) reduces experimental metastasis of melanoma cells in mice. *Cancer Letters, 142*, 91–96.

Limtrakul, P., Lipigorngoson, S., Namwong, O., Apisariyakul, A., & Dunn, F. W. (1997). Inhibitory effect of dietary curcumin on skin carcinogenesis in mice. *Cancer Letters, 116*, 197–203.

Liu, C. H., Chang, F. Y., & Hung, D. K. (2011). Terpene microemulsions for transdermal curcumin delivery: Effects of terpenes and cosurfactants. *Colloids and Surfaces B: Biointerfaces, 82*, 63–70.

Longley, D. B., Ferguson, P. R., Boyer, J., Latif, T., Lynch, M., Maxwell, P., et al. (2001). Characterization of a thymidylate synthase (TS)-inducible cell line. *Clinical Cancer Research, 7*, 3533–3539.

Looi, C. Y., Moharram, B., Paydar, M., Wong, Y. L., Leong, K. H., Mohamad, K., et al. (2013). Induction of apoptosis in melanoma A375 cells by a chloroform fraction of *Centratherum anthelminticum* (L.) seeds involves NF-kappaB, p53, and Bcl-2-controlled mitochondrial signaling pathway. *BMC Complementary and Alternative Medicine, 13*, 166–180.

Madhuri, V., Murthy, K. S. R., Amrutha, V. A., & Siva, R. K. C. (2014). Evaluation of antiproliferative properties of selected species of *Caralluma* and *Boucerosia* on skin cancer cell lines. *European Journal of Experimental Biology, 4*(1), 160–167.

Mahomoodally, M. F. (2013). Traditional medicines in Africa: An appraisal of ten potent African medicinal plants. *Evidence-Based Complementary and Alternative Medicine*, 14:617459.

Malafa, M. P., Fokum, F. D., Mowlavi, A., Abusief, M., & King, M. (2002). Vitamin E inhibits melanoma growth in mice. *Surgery, 131*, 85–91.

Malafa, M. P., Fokum, F. D., Smith, L., & Louis, A. (2002). Inhibition of angiogenesis and promotion of melanoma dormancy by vitamin E succinate. *Annals of Surgical Oncology, 9*, 1023–1032.

Mamone, L., Di Venosa, G., Valla, J. J., Rodriguez, L., Gándara, L., Batlle, A., et al. (2011). Cytotoxic effects of Argentinean plant extracts on tumour and normal cell lines. *Cell and Molecular Biology, 57*, OL1487–OL1499.

Mantena, S. K., Roy, M. A., & Katiyar, S. K. (2005). Epigallocatechin-3-gallate inhibits photocarcinogenesis through inhibition of angiogenic factors and activation of $CD8^+$ T cells in tumors. *Photochemistry and Photobiology, 81*, 1174–1179.

Mantena, S. K., Meeran, S. M., Elmets, C. A., & Katiyar, S. K. (2005). Orally administered green tea polyphenols prevent ultraviolet radiation-induced skin cancer in mice through activation of cytotoxic T cells and inhibition of angiogenesis in tumors. *Journal of Nutrition, 135*, 2871–2877.

Marnewick, J., Joubert, E., Joseph, S., Swanevelder, S., Swart, P., & Gelderblom, W. (2005). Inhibition of tumor promotion in mouse skin by extracts of rooibos (*Aspalathus linearis*) and honeybush (*Cyclopia intermedia*), unique South African herbal teas. *Cancer Letters, 224*, 193–202.

Marnewick, J. L., van der Westhuizen, F. H., Joubert, E., Swanevelder, S., Swart, P., & Gelderblom, W. C. A. (2009). Chemoprotective properties of rooibos (*Aspalathus linearis*), honeybush (*Cyclopia intermedia*) herbal and green and black (*Camellia sinensis*) teas against cancer promotion induced by fumonisin B1 in rat liver. *Food and Chemical Toxicology*, *47*, 220–229.

Massyn, A. (2006). *Eucomis autumnalis flower*. Retrieved from: https://commons.wikimedia.org/wiki/File:Eucomis_autumnalis_flower.jpg.

Menichini, G., Alfano, C., Marrelli, M., Toniolo, C., Provenzano, E., Statti, G. A., et al. (2013). *Hypericum perforatum* L. subsp. *perforatum* induces inhibition of free radicals and enhanced phototoxicity in human melanoma cells under ultraviolet light. *Cell Proliferation*, *46*, 193–202.

Mimeault, M., & Batra, S. K. (2011). Potential applications of curcumin and its novel synthetic analogs and nanotechnology-based formulations in cancer prevention and therapy. *Chinese Medicine*, *6*. http://dx.doi.org/10.1186/1749-8546-6-31.

Mittal, A., Elmets, C. A., & Katiyar, S. K. (2003). Dietary feeding of proanthocyanidins from grape seeds prevents photocarcinogenesis in SKH-1 hairless mice: Relationship to decreased fat and lipid peroxidation. *Carcinogenesis*, *24*, 1379–1388.

Mohansrinivasan, V., Subathra, D. C., Meenakshi, D., Ananya, B., & Jemimah, N. S. (2015). Exploring the anticancer activity of grape seed extract on skin cancer cells lines A431. *Brazilian Archives of Biology and Technology*, *58*(4), 540–546.

Monroe, M. M. (2015). *Cutaneous squamous cell carcinoma medication*. Retrieved from: http://emedicine.medscape.com/article/1965430-medication#4.

Moorthi, C., & Kathiresan, K. (2013). Curcumin-piperine/curcumin-quercetin/curcumin-silibinin dual drug-loaded nanoparticulate combination therapy: A novel approach to target and treat multidrug-resistant cancers. *Journal of Medical Hypotheses and Ideas*, *7*, 15–20.

Naksuriya, O., Okonogi, S., Schiffelers, R. M., & Hennink, W. E. (2014). Curcumin nanoformulations: A review of pharmaceutical properties and preclinical studies and clinical data related to cancer treatment. *Biomaterials*, *35*, 3365–3383.

National Cancer Institute. (1985). *Basal cell carcinoma*. National Institutes of Health. Retrieved from: https://commons.wikimedia.org/wiki/File:Basal_cell_carcinoma_(2).jpg.

National Cancer Institute. (1998). *Part of the ABCDs for detection of melanoma*. Retrieved from: https://commons.wikimedia.org/wiki/File:Melanoma_vs_normal_mole_ABCD_rule_NCI_Visuals_Online.jpg.

National Geographic. (2014). *Skin*. Retrieved from: http://science.nationalgeographic.com/science/health-and-human-body/human-body/skin-article/.

Nelson, K. (2012). *Squuomous cell carcinoma*. National Cancer Institute. Retrieved from: https://commons.wikimedia.org/wiki/File:Squamous_cell_carcinoma_(1).jpg.

Nguyen, T. T., Parat, M. O., Hodson, M. P., Pan, J., Shaw, P. N., & Hewavitharana, A. M. (2016). Chemical characterization and *in vitro* cytotoxicity on squamous cell carcinoma cells of *Carica papaya* leaf extracts. *Toxins*, *8*(1). http://dx.doi.org/10.3390/toxins8010007.

NIH. (2015). *Drugs approved for basal cell carcinoma*. Retrieved from: http://www.cancer.gov/about-cancer/treatment/drugs/skin.

Niles, R. M., Cook, C. P., Meadows, G. G., Fu, Y. M., McLaughlin, J. L., & Rankin, G. O. (2006). Resveratrol is rapidly metabolized in athymic (Nu/Nu) mice and does not inhibit human melanoma xenograft tumor growth. *Journal of Nutrition*, *136*, 2542–2546.

Niles, R. M., McFarland, M., Weimer, M. B., Redkar, A., Fu, Y. M., & Meadows, G. G. (2003). Resveratrol is a potent inducer of apoptosis in human melanoma cells. *Cancer Letters*, *190*, 157.

Nobili, S., Lippi, D., Witort, E., Donnini, M., Bausi, L., Mini, E., et al. (2009). Natural compounds for cancer treatment and prevention. *Pharmacological Research*, *59*, 365–378.

Olson, E. R., Melton, T., Dickinson, S. E., Dong, Z., Alberts, D. S., & Bowden, G. T. (2010). Quercetin potentiates UVB-Induced c-Fos expression: Implications for its use as a chemopreventive agent. *Cancer Prevention Research (Philadelphia, Pennsylvania)*, *3*, 876–884.

Otero, P. A. (2012). *Leonotis leonurus 001.Jpg*. Retrieved from: https://commons.wikimedia.org/wiki/File:(270112)_Leonotis_leonurus_001.jpg.

Park, E. J., & Pezzuto, J. M. (2002). Botanicals in cancer chemoprevention. *Cancer and Metastasis Reviews*, *21*, 231–255.

Patel, D., Shukla, S., & Gupta, S. (2007). Apigenin and cancer chemoprevention: Progress, potential and promise (review). *International Journal of Oncology*, *30*, 233–245.

Pence, B. C., Delver, E., & Dunn, D. M. (1994). Effects of dietary selenium on UVB induced skin carcinogenesis and epidermal antioxidant status. *Journal of Investigative Dermatology*, *102*, 759–761.

Petrova, A., Davids, L. M., Rautenbach, F., & Marnewick, J. L. (2011). Photoprotection by honeybush extracts, hesperidin and mangiferin against UVB-induced skin damage in SKH-1 mice. *Journal of Photochemistry and Photobiology B: Biology*, *103*, 126–139.

Phillips, J. M., Clark, C., Herman-Ferdinandez, L., Moore-Medlin, T., Rong, X. H., Gill, J. R., et al. (2011). Curcumin inhibits skin squamous cell carcinoma tumor growth *in vivo*. *Otolaryngology–Head and Neck Surgery*, *145*, 58–63.

Pienaar, G. (2015). *Aloe ferox 00.jpg*. Retrieved from: https://commons.wikimedia.org/wiki/File:Aloe_ferox00.jpg.

Pisha, E., Chai, H., Lee, I. S., Chagwedera, T. E., Farnsworth, N. R., Cordell, G. A., et al. (1995). Discovery of betulinic acid as a selective inhibitor of human melanoma that functions by induction of apoptosis. *Nature Medicine*, *1*, 1046–1051.

Qizwini, H. A. L., Khateeb, E. A. L., Mhaidat, N. M., & Saleem, A. (2014). Cytotoxic effects of Jordanian *Simmondsia chinensis* (link) C.K. Schnied of different cancer cell lines. *European Scientific Journal*, *10*(24), 182–195.

Radovic, J., Maksimovic-Ivanic, D., Timotijevic, G., Popadic, S., Ramic, Z., Trajkovic, V., et al. (2012). Cell-type dependent response of melanoma cells to aloe emodin. *Food and Chemical Toxicology*, *50*, 3181–3189.

Reddy, L., Odhav, B., & Bhoola, K. D. (2003). Natural products for cancer prevention: A global perspective. *Pharmacology and Therapeutics*, *99*, 1–13.

Richardson, J., Smith, J. E., McIntyre, M., Thomas, R., & Pilkington, K. (2005). *Aloe vera* for preventing radiation-induced skin reactions: A systematic literature review. *Clinical Oncology*, *17*, 478–484.

Richfield, J. (2010). *Bulbine lagopus, Uniondale, South Africa-20101017–02.jpg*. Retrieved from: https://commons.wikimedia.org/wiki/File:Bulbine_lagopus,_Uniondale,_South_Africa_-_20101017-02.jpg.

Rosner, K., Röpke, C., Pless, V., & Skovgaard, G. L. (2006). Late type apoptosis and apoptosis free lethal effect of quercetin in human melanoma. *Bioscience, Biotechnology, and Biochemistry*, *70*, 2169–2177.

Russell, P. J. (2010). *iGenetics a molecular approach* (3rd ed.). San Francisco: Pearson International Edition.

Russo, G. L. (2007). Ins and outs of dietary phytochemicals in cancer chemoprevention. *Biochemical Pharmacology*, *74*, 533–544.

Saija, A., Tomatino, A., Lo Cascio, R., Trombetta, D., Proteggente, A., De Pasquale, A., et al. (1999). Ferulic and caffeic acids as potential protective agents against photooxidative skin damage. *Journal of the Science of Food and Agriculture*, *79*, 476–480.

SAWS. (2012). *UV-B monitoring and research activities at the South African Weather Service*. Retrieved from: http://www.ehrn.co.za/sunsmart/download/symposium2012_06.pdf.

Shah, T. I., Sharma, E., & Shah, G. A. (2015). Anti-proliferative, cytotoxicity and antioxidant activity of *Juglans regia* extract. *American Journal of Cancer Prevention*, *3*(2), 45–50.

Shao, Y., Chin, C.-K., Ho, C.-T., Ma, W., Garrison, S. A., & Huang, M.-T. (1996). Anti-tumor activity of the crude saponins obtained from asparagus. *Cancer Letters*, *104*, 31–36.

Sharma, K. V., & Davids, L. M. (2012a). Hypericin-PDT-induced rapid necrotic death in human squamous cell carcinoma cultures after multiple treatment. *Cell Biology International*, *36*, 1261–1266.

Sharma, K. V., & Davids, L. M. (2012b). Depigmentation in melanomas increases the efficacy of hypericin-mediated photodynamic-induced cell death. *Photodiagnosis and Photodynamic Therapy*, *9*, 156–163.

Sharma, R., Kishore, N., Hussein, A., & Lall, N. (2014). Potential of plant *Leucosidea sericea* against *Propionibacterium acnes*. *Phytochemistry Letters*, *7*, 124–129.

Sharma, S. D., Meeran, S. M., & Katiyar, S. K. (2007). Dietary grape seed proanthocyanidins inhibit UVB-induced oxidative stress and activation of mitogen-activated protein kinases and nuclear factor-kB signaling in *in vivo* SKH-1 hairless mice. *Molecular Cancer Therapeutics*, *6*, 995–1005.

Shawka, A. (2010a). *Celtis africana tree foliage South Africa 8.JPG*. Retrieved from: https://commons.wikimedia.org/wiki/File:Celtis_africana_tree_foliage_South_Africa_8.JPG.

Shawka, A. (2010b). *Cotyledon orbiculata-pigs ear-Cape Point-SouthAfrica 3.JPG*. Retrieved from: https://commons.wikimedia.org/wiki/File:Cotyledon_orbiculata_-_pigs_ear_-_Cape_Point_-_South_Africa_3.JPG.

Shawka, A. (2011a). *Fruits of the Assegai Tree-Curtisia dentate.Jpg*. Retrieved from: https://commons.wikimedia.org/wiki/File:Fruits_of_the_Assegai_Tree_-_Curtisia_dentata.jpg.

Shawka, A. (2011b). *Juvenile Renosterbos-Elytropappus rhinocerotis-Cape Town.Jpg*. Retrieved from: https://commons.wikimedia.org/wiki/File:Juvenile_Renosterbos_-_Elytropappus_rhinocerotis_-_Cape_Town.jpg.

Shawka, A. (2011c). *Gunnera perpensa-indegenous River Pumpkin of South African wetlands ponds.Jpg*. Retrieved from: https://commons.wikimedia.org/wiki/File:Gunnera_perpensa_-_indigenous_River_Pumpkin_of_South_Africa_wetlands_ponds.jpg.

Shebs, S. (2006). *Melianthus comosus 2.jpg*. Retrieved from: https://commons.wikimedia.org/wiki/File:Melianthus_comosus_2.jpg.

Shebs, S. (2007). *Scillia natalensis 3.jpg*. Retrieved from: https://commons.wikimedia.org/wiki/File:Scilla_natalensis_3.jpg.

Shehzad, A., Lee, J., & Lee, Y. S. (2013). Curcumin in various cancers. *Biofactors*, *39*, 56–68.

Shin, H. R., Kim, J. Y., Yun, T. K., Morgan, G., & Vainio, H. (2000). The cancer-preventive potential of *Panax ginseng*: A review of human and experimental evidence. *Cancer Causes and Control*, *11*, 565–576.

Shukla, Y., & Singh, M. (2007). Cancer preventive properties of ginger: A brief review. *Food and Chemical Toxicology*, *45*, 683–690.

Singh, T., Vaid, M., Katiyar, N., Sharma, S., & Katiyar, S. K. (2011). Berberine, an isoquinoline alkaloid, inhibits melanoma cancer cell migration by reducing the expressions of cyclooxygenase-2, prostaglandin E_2 and prostaglandin E_2 receptors. *Carcinogenesis*, *32*(1), 86–92.

Skalkos, D., Gioti, E., Stalikas, C. D., Meyer, H., Papazoglou, T. G., & Filippidis, G. (2006). Photophysical properties of *Hypericum perforatum* L. extracts-novel photosensitizers for PDT. *Journal of Photochemistry and Photobiology B: Biology*, *82*, 146–151.

Skin Cancer Foundation. (2015). *The five warning signs of basal cell carcinoma*. Retrieved from: http://www.skincancer.org/skin-cancer-information/basal-cell-carcinoma/the-five-warning-signs-images#panel1–6.

Soleas, G. J., Grass, L., Josephy, P. D., Goldberg, D. M., & Diamandis, E. P. (2002). A comparison of the anticarcinogenic properties of four red wine polyphenols. *Clinical Biochemistry*, *35*, 119–124.

Stahl, W., Heinrich, U., Jungmann, H., Sies, H., & Tronnier, H. (2000). Carotenoids and carotenoids plus vitamin E protect against ultraviolet light-induced erythema in humans. *The American Journal of Clinical Nutrition*, *71*, 795–798.

Starr, F., & Starr, K. (2002). *Psidium guajava.jpg*. Retrieved from: https://commons.wikimedia.org/wiki/File:Starr_020803-0113_Psidium_guajava.jpg.

Starr, F., & Starr, K. (2007). *Pittisporum viridiflorum.jpg*. Retrieved from: https://commons.wikimedia.org/wiki/File:Starr_070404-6712_Pittosporum_viridiflorum.jpg.

Stevanato, R., Bertelle, M., & Fabris, S. (2014). Photoprotective characteristics of natural antioxidant polyphenols. *Regulatory Toxicity and Pharmacology*, *69*(1), 71–77.

Street, R. A., & Prinsloo, G. (2013). Commercially important medicinal plants of South Africa: A review. *Journal of Chemistry*, 16:205048.

Surh, Y. J., Park, K. K., Chun, K. S., Lee, L. J., Lee, E., & Lee, S. S. (1999). Anti-tumor-promoting activities of selected pungent phenolic substances present in ginger. *Journal of Environmental Pathology, Toxicology and Oncology*, *18*, 131–139.

Suseela, V., & Lalitha, G. (2015). Cytotoxic effects of green synthesized silver nanoparticles using *Indigofera longeracemosa* on skin cancer SK MEL-28 cell lines. *International Journal of Preclinical and Pharmaceutical Research*, *6*(3), 118–125.

Tabolacci, C., Lentini, A., Mattioli, P., Provenzano, B., Oliverio, S., Carlomosti, F., et al. (2010). Antitumor properties of aloe-emodin and induction of transglutaminase 2 activity in B16–F10 melanoma cells. *Life Sciences*, *87*, 316–324.

Tabolacci, C., Rossi, S., Lentini, A., Provenzano, B., Turcano, L., Facchiano, F., et al. (2013). Aloin enhances cisplatin antineoplastic activity in B16-F10 melanoma cells by transglutaminase-induced differentiation. *Amino Acids*, *44*, 293–300.

Tangopaso. (2009). *Carpobrutus edulis flower.Jpg*. Retrieved from: https://commons.wikimedia.org/wiki/File:Carpobrotus_edulis_flower.jpg.

Thejass, P., & Kuttan, G. (2007). Modulation of cell-mediated immune response in B16F-10 melanoma-induced metastatic tumor-bearing C57BL/6 mice by sulforaphane. *Immunopharmacology and Immunotoxicology*, *29*, 173–186.

Thring, T. S. A., & Weitz, F. M. (2006). Medicinal plant use in the Bredasdorp/Elim region of the Southern Overberg in the Western Cape Province of South Africa. *Journal of Ethnopharmacology*, *103*, 261–275.

Thuncharoen, W., Chulasiri, M., Nilwarangkoon, S., Nakamura, Y., & Watanapokasin, R. (2013). Apoptotic induction of skin cancer cell death by plant extracts. *Journal of the Medical Association of Thailand*, *96*, S60–S64.

Tsai, K. D., Yang, S. M., Tseng, M. J., Lin, J. C., Hsu, J. D., Lee, Y. J., et al. (2012). Curcumin protects against UVB-induced skin cancers in SKH-1 hairless mouse: Analysis of early molecular markers in carcinogenesis. *Evidence-Based Complementary and Alternative Medicine*. http://dx.doi.org/10.1155/2012/593952.

Tundis, R., Loizzoa, M. R., Menichinia, F., Bonesia, M., Colicab, C., & Menichinia, F. (2011). *In vitro* cytotoxic activity of extracts and isolated constituents of *Salvia leriifolia* Benth. Against a panel of human cancer cell lines. *Chemistry and Biodiversity*, *8*, 1152–1162.

Twilley, D., & Lall, N. (2015). *Online. Extracts and composition of Helichrysum odoratissimum for preventing and treating skin cancer*. Publication number: WO 2015049666 A1. Retrieved from: https://www.google.com/patents/WO2015049666A1?cl=en.

Val'ko, V., Pravdová, E., Nagy, M., Grančai, D., & Ficková, M. (2007). Antiproliferative activity of plant extracts from genus *Philadelphus* L.. *Acta Facultatis Pharmaceuticae Universitatis Comenianae*, *54*, 209–214.

Van Wyk, B., van Oudtshoorn, B., & Gericke, N. (1997). *Medicinal plants of South Africa* (1st ed.). Pretoria, South Africa: Briza Publications, 1–304.

Venter, P. (2012). *Hypoxis hemerocallidea 02.Jpg*. Retrieved from: https://commons.wikimedia.org/wiki/File:Hypoxis_hemerocallidea02.jpg.

Verhoeven, D. T., Goldbohm, R. A., van Poppel, G., Verhagen, H., & van den Brandt, P. A. (1996). Epidemiological studies on brassica vegetables and cancer risk. *Cancer Epidemiology, Biomarkers and Prevention*, *5*, 733–748.

Vural, P., Canbaz, M., & Selcuki, D. (1999). Plasma antioxidant defense in actinic keratosis and basal cell carcinoma. *Journal of the European Academy of Dermatology and Venereology*, *13*, 96–101.

Wang, Y., Zhang, X., Lebwohl, M., DeLeo, V., & Wei, H. (1998). Inhibition of ultraviolet B (UVB)-induced c-fos and c-jun expression *in vivo* by a tyrosine kinase inhibitor genistein. *Carcinogenesis*, *19*, 649–654.

Wang, H. Z., Zhang, Y., Xie, L. P., Yu, X. Y., & Zhang, R. Q. (2002). Effects of genistein and daidzein on the cell growth, cell cycle, and differentiation of human and murine melanoma cells. *The Journal of Nutritional Biochemistry*, *13*, 421–426.

Wargovich, M. J., Woods, C., Hollis, D. M., & Zander, M. E. (2001). Herbals, cancer prevention and health. *American Society for Nutritional Sciences*, *11*, 3034S–3036S.

Wegner, C. (2011). *Skin cancer: A shocking truth about sun damage*. Retrieved from: http://www.capechameleon.co.za/printed-version/issue-7/health2/.

Wei, H., Bowen, R., Zhang, X., & Lebwohl, M. (1998). Isoflavone genistein inhibits the initiation and promotion of two-stage skin carcinogenesis in mice. *Carcinogenesis*, *19*, 1509–1514.

WHO. (2002). *Global solar UV index*. Retrieved from: http://www.who.int/uv/publications/en/UVIGuide.pdf.

WHO. (2014). *Cancer fact sheet*. Retrieved from: http://www.who.int/mediacentre/factsheets/fs297/en/.

WHO. (2015). *Skin cancers*. Retrieved from: www.who.int/uv/faq/skincancer/en/index1.html.

WHO Monographs. (2004). *WHO monographs on selected medicinal plants – volume 2*. Retrieved from: http://apps.who.int/medicinedocs/pdf/s4927e/s4927e.pdf.

Xiao, K., Zhang, H. J., Xuan, L. J., Zhang, J., Xu, Y. M., & Bai, D. L. (2008). Stilbenoids: Chemistry and bioactivities. In A. Rahman (Ed.), *Studies in natural products chemistry 34*. (pp. 453–646). Amsterdam, The Netherlands: Elsevier.

Yajarla, V. N. G., Nimmanapalli, R. P., Parikapandla, S., Gupta, G., & Karnati, R. (2014). Anti-inflammatory and anti-proliferative properties of *Chromolaena odorata* leaf extracts in normal and skin-cancer cell lines. *Journal of Herbs, Spices and Medicinal Plants*, *20*, 359–371.

Yang, G., Fu, Y., Malakhova, M., Kurinov, I., Zhu, F., Yao, K., et al. (2014). Caffeic acid directly targets ERK1/2 to attenuate solar UV induced skin carcinogenesis. *Cancer Prevention Research (Philadelphia, Pennsylvania)*, *7*, 1056–1066.

Yang, Y., Li, Y., Wang, K., Wang, Y., Yin, W., & Li, L. (2013). P38/NF-kB/snail pathway is involved in caffeic acid-induced inhibition of cancer stem cells-like properties and migratory capacity in malignant human keratinocyte. *PLoS One*, *8*, e58915.

Yin, Y., Li, W., Son, Y. O., Sun, L., Lu, J., Kim, D., et al. (2013). Quercitrin protects skin from UVB-induced oxidative damage. *Toxicology and Applied Pharmacology*, *269*, 89–99.

Zhang, X.-M., Chen, J., Xia, Y.-G., & Xu, Q. (2005). Apoptosis of murine melanoma B16-BL6 cells induced by quercetin targeting mitochondria, inhibiting expression of PKC-α and translocating PKC-δ. *Cancer Chemotherapy and Pharmacology*, *55*, 251–262.

Zhao, Y. Z., Lu, C. T., Zhang, Y., Xiao, J., Zhao, Y. P., Tian, J. L., et al. (2013). Selection of high efficient transdermal lipid vesicle for curcumin skin delivery. *International Journal of Pharmaceutics*, *454*, 302–309.

Zhao, J., Sharma, Y., & Agarwal, R. (1999). A flavonoid antioxidant, silymarin, affords significant inhibition against 12-O-tetradecanoylphorbol 13-acetate-caused modulation of antioxidant and inflammatory enzymes, and cyclooxygenase 2 and interleukin-1α expression in SENCAR mouse epidermis: Implications in the prevention of stage I tumor. *Molecular Carcinogenesis*, *26*, 321–333.

Zi, X., & Agarwal, R. (1999). Modulation of mitogen-activated protein kinase activation and cell cycle regulators by the potent skin cancer preventive agent silymarin. *Biochemical and Biophysical Research Communications*, *263*, 528–536.

Ziarnek, K. K. (2014). *Tulbaghia violacea kx1.JPG*. Retrieved from: https://commons.wikimedia.org/wiki/File:Tulbaghia_violacea_kz1.JPG.

Zick, S. M., Djuric, Z., Ruffin, M. T., Litzinger, A. J., Normolle, D. P., Alrawi, S., et al. (2008). Pharmacokinetics of 6-gingerol, 8-gingerol, 10-gingerol, and 6-shogaol and conjugate metabolites in healthy human subjects. *Cancer Epidemiology, Biomarkers and Prevention*, *17*, 1930–1936.

Zi, X., Mukhtar, H., & Agarwal, R. (1997). Novel cancer chemopreventive effects of a flavonoid antioxidant silymarin: Inhibition of mRNA expression of an endogenous tumor promoter TNFα. *Biochemical and Biophysical Research Communications*, *239*, 334–339.

Chapter 3

Fighting the Inevitable: Skin Aging and Plants

Bianca D. Fibrich, Namrita Lall
University of Pretoria, Pretoria, South Africa

Chapter Outline

3.1 THE INEVITABLE: AGING

Every organism is affected by the inevitable passage of time and the visible changes associated with it. Before any attempt can be made to stop the aging process, or even just slow the symptoms, the question must be asked, what is aging? Harman (2003) defines aging as the accumulation of diverse and deleterious changes occurring in cells and tissues with advancing age, which are responsible for the increased risk of disease (Harman, 2003). It is a multifactorial and very complex process with many theories surrounding it (Kowald & Kirkwood, 1996; Weinert & Timiras, 2003). Although no silver bullet or fountain of youth has been discovered, the next best thing is fighting the signs of

Medicinal Plants for Holistic Health and Well-Being. http://dx.doi.org/10.1016/B978-0-12-812475-8.00003-2

aging. Because our skin is the first barrier of defense protecting us from every day stress that we are sometimes even unknowingly exposed to, it becomes subject to multiple changes, presenting a canvas of aged features. Because it is a part of the body readily seen, it becomes the target of antiaging strategies. In considering aging at this level, two types of aging exist, namely chronological (intrinsic) and photoaging (extrinsic) (Pandel, Poljšak, Godic, & Dahmane, 2013; Sahu et al., 2013).

Chronological aging occurs with the passage of time, is determined by genetic predisposition and is mainly regulated by hormones such as androgens (Jung et al., 2014; Kuno & Matsumoto, 2004). Photoaging on the other hand is determined by environmental factors such as the extent of sun exposure, the amount of melanin in the skin, repetitive facial expressions, gravity, sleeping positions as well as smoking (Fisher, Kang, & Varanti, 2002; Kligman & Kligman, 1986; Mukherjee, Maity, Nema, & Sarkar, 2011). Because the factors contributing to chronological and photoaging are somewhat distinct, they present different symptoms (Table 3.1). With time, however, the effects of

TABLE 3.1 Clinical Symptoms of Chronological and Photoaging as well as the Areas They Are Most Visible

Type of Aging	Clinical Symptoms	Areas Affected	References
Chronological aging	Laxity, fine wrinkles, benign growths, seborrheic keratosis, angiomas, no change in pigmentation, no formation of deep wrinkles, no apparent vascular damage, thin skin appearance, loss of elasticity	Entire body	Yaar, Lee, Runger, Eller, and Gilchrest (2002b), Rittié and Fisher (2002), Pandel et al. (2013), and Sahu et al. (2013)
Photoaging	Laxity, leathery appearance, solar elastosis, actinic purpura, precancerous lesions, apparent vascular damage, melanoma, skin cancer, wrinkles, mottled pigmentation, rough skin, loss of skin tone, dryness, sallowness, deep furrows, severe atrophy, telangiectasias	Sun-exposed areas; neck, face, upper chest, hands, forearms	Gilchrest (1990), Yaar, Eller, and Gilchrest (2002a), Helfrich, Sachs, and Voorhees (2008), Pandel et al. (2013), and Sahu et al. (2013)

chronological and photoaging become superimposed on one another and so the features they present are not always easily distinguishable.

The sun is the main role player in extrinsic (photo) aging as it is responsible for 90% of the structural changes that occur in the skin (Zouboulis & Boschnakow, 2001). Various types of UV radiation exist, each characterized by their wavelength and ability to penetrate the skin. UV radiations A (320–400 nm) and B (290–320 nm) are the most severe with UVA being commonly associated with cancer, penetrating up to 20% of the skin, and UVB commonly associated with sunburn, which may indirectly lead to severe photoaging (Jung et al., 2014). UVB is associated mainly with the epidermis and accounts for 70% of a person's average cumulative UV dose.

In recent years, exposure to UV radiation has increased because of increased solar UV exposures (as a consequence of stratospheric ozone depletion), use of sunscreens resulting in a false belief of being protected and hence exposures for a longer period of time, outdoor leisure activities, and prolonged life expectancy in industrialized countries. This increase in stronger, more frequent UV exposures in conjunction with the fact that our skin's ability to process these stressors have not accordingly adjusted, results in an enhanced aging process. The skin contains endogenous mechanisms of protection against this damage, which includes antioxidants as well as pigments such as melanin, however, when overwhelmed these mechanisms become no longer sufficient.

3.1.1 Aging Prevalence Worldwide

As the global population continues to grow annually, so does the life expectancy. With this increase, consumers are becoming more concerned with aging gracefully. Counting North America, Asia, Africa, and Europe, the global population aged 65 and older is expected to rise by 16% by the year 2050 that is an estimated 1511 million people (Statista, 2016). Statistically speaking, the global population of individuals aged 65 and over therefore, grows by ~8000 members daily, and is expected to continue to do so for at least the next 18 years (Transparency Market Research, 2016). The role of increased life expectancy in the increased demand for antiaging treatments and growing antiaging market was confirmed in a market research report, implicating the rising population age as a main driver of demand, specifically the baby boomer population and generations X and Y. These populations are thought to be composed of individuals exhibiting prolonged concern regarding their physical health and beauty. Of these populations, the baby boomer population comprises individuals born between 1946 and 1964 are the greatest consumers of antiaging products. The desire to age more gracefully is also seen in modern men and women from an increasingly younger age. Therefore, the market itself has expanded exponentially and is now composed of more diverse customers wanting to treat a wider range of symptoms.

3.1.2 Aging Prevalence in Africa

Projections on the African population indicate that by the year 2050, the average number of people aged 65 and older will have risen by 7% or an estimated additional 145 million people (Statista, 2016). Although geographically speaking, the antiaging market is segmented into North America, Asia Pacific, and the rest of the world, which does not implicate Africa as playing a major role in demand currently, this demand will rise as the population does, along with an important factor—disposable income.

3.1.3 Current Antiaging Approaches in Skin Care

Antiaging treatments currently include topical agents such as creams and serums, microdermabrasion, injections, and chemical peels. The ultimate goal of these treatments is to obtain healthy, blemish-free, translucent, and resilient skin after treatment has incurred by reversing the dermal and epidermal signs of aging (Vedamurthy, 2006). When selecting an antiaging strategy, many factors need to be considered including the age and health status of the patient, any previous history of procedures/treatments, type of skin, the target of the procedure, and lifestyle (Sahu et al., 2013). Although antiaging strategies are applied to reduce the signs of aging, in extensive cases they may remain visible. For this reason, preventative and protective agents should be implemented at a younger age in conjunction with the implementation of healthy lifestyle choices such as avoidance of excessive sun exposure and cessation of smoking habits where applicable.

3.1.3.1 Topical Agents

Topical agents are among the most popular as they are noninvasive and have a low risk of permanent damage (Ganceviciene, Aikaterini, Athanasios, Makrantonaki, & Zouboulis, 2012). Consequently, there is no postoperative recovery time, making topical agents much more appealing. This approach includes daily skin care regimes, and application of sun protectants that promote skin regeneration, elasticity, and smoothness (Lübbe, 2000; Tabata, O'Goshi, Zhen, Kligman, & Tagami, 2000). The mechanisms employed by topical agents can be either antioxidative or by cellular regulation (Ganceviciene et al., 2012). The goal of both these categories, however, remains unanimous: to regenerate the skins components, prevent their degradation, and target inflammatory processes (Baumann, 2007; Ganceviciene et al., 2012; Margelin, Medaisko, Lombard, Picard, & Fourtanier, 1996; Nusgens et al., 2001; Varani et al., 2000).

Antioxidants

Antioxidants are important in skin care regimes as many studies have revealed the sensitivity of the skin to the effects of reactive oxygen species (ROS) (Sander et al., 2002; Young et al., 2008). These comprise a wide array of molecules ranging from essential amino acids to vitamins and minerals (Rona, Vailati, & Berardesca, 2004). The effects of ROS on the skin are significant enough that an entire theory surrounding the aging process has been devoted to it. This theory, the free radical theory of aging, indicates that the accumulation of oxygen radicals in cells could be responsible for aging and death. The lifespan of a cell may thus be contingent on the degree of free radical damage to the mitochondria. An imbalance between the ratio of free radical production and antioxidant defenses, which diminish with time consequently results in increasing age-related oxidative stress. The free radical theory of aging is divided into several hypotheses each focusing on the exclusive role of particular organelles and types of damaged molecules in the aging process.

When the mechanisms of aging are considered, the substantial role ROS play implicates them as an important target for protective and preventative measures. Endogenous antioxidants do exist within the skin and body, however, with age the efficacy of these endogenous antioxidants becomes compromised, and thus exogenous supplementation either through topical application or within the diet through nutritional supplementation becomes necessary (Makrantonak & Zouboulis, 2007; Pandel et al., 2013; Valenzuela, Sanhuexa, & Nieto, 2003). These nutritional antioxidant supplements play roles in free radical scavenging and neutralizing, repairing oxidized membranes, and quenching iron to indirectly diminish ROS production (Baumann, 2007; Berger, 2005; Trautinger, 2001). Antioxidant agents include vitamins, polyphenols, and flavonoids (Ganceviciene et al., 2012). Vitamin C, E, and B3 are recognized as the most important antioxidants as they penetrate the skin deeply because of their low-molecular-weights (Bisset, Miyamoto, Sun, Li, & Berge, 2004). Vitamin C (L-ascorbic acid) is useful in antiaging treatments as it induces the production of collagen types I and III alongside enzymes important in the production of collagen and inhibitors of matrix metalloproteinase-1 (MMP-1) or collagenase-1 (responsible for the degradation of collagen type-I). The disadvantages, however, include that it is heat labile. Studies have also shown vitamin C protects against UV rays, corrects hypo/hyperpigmentation problems, and improves inflammatory skin conditions (Elmore, 2005; Fitzpatrick & Rostan, 2002; Nusgens et al., 2001). Vitamin E, also known as α-tocopherol, plays significant roles in reducing inflammation and proliferative properties at concentrations between 2% and 20% (Fig. 3.1) (Ganceviciene et al., 2012). It smoothes the skin and increases the ability of the stratum corneum (the outer layer of the skin) to maintain humidity, accelerate epithelialization, and contribute to photo-protection of the skin (Ganceviciene et al., 2012; Zhai et al., 2005). Although individually effective, combinations of vitamin C, E, and ferulic acid provide more extensive antioxidant protection (Ganceviciene et al., 2012; Kerscher & Buntrock, 2011; Murray et al., 2008). Vitamin B3, also known as niacinamide, is used at a concentration of 5% in antiaging products as it plays a role in regulating cell metabolism and regeneration (Draelos, 2007). After 3 months of topical

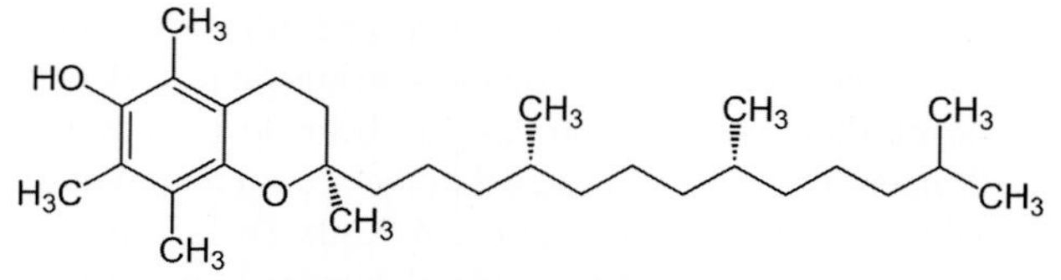

FIGURE 3.1 Vitamin E, also known as α-tocopherol plays a significant role in reducing inflammation and proliferation at concentrations between 2% and 20% and is a popular antiaging component in many formulations today.

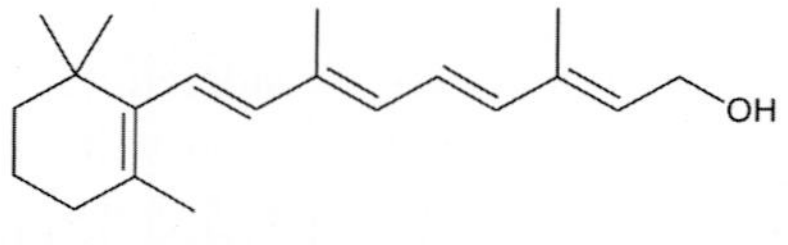

FIGURE 3.2 Retinol, a vitamin A derivative is incorporated into antiaging formulations for its ability to act as a cell regulator, speeding up cellular turnover rates to enhance the production of key structural components within the skin such as elastin and collagen.

application of antioxidants, skin elasticity, erythema, and pigmentation have been shown to improve (Bisset et al., 2004; Draelos, 2007; Kerscher & Buntrock, 2011). Procyanidins, such as Pycnogenol, are popular natural antioxidants incorporated into antiaging formulations, which act to recycle and prolong the effects of vitamin C and E, whereas exerting antioxidant effects 20 and 50 times more potent than the aforementioned antioxidants, respectively (Cossins, Lee, & Pacher, 1998; Schonlau, 2002; Rona et al., 2004). Another popular antioxidant is coenzyme Q10, also known as ubiquinone. This antioxidant acts to improve the efficiency of a cell in terms of oxygen usage and lower the production of free radicals within the cell cycle. In addition Co-Q10 has also been shown to suppress collagenase expression following UVA exposure in human dermal fibroblasts (Rona et al., 2004). Lastly, lipoic acid has also been found to be a valuable contributor to antiaging as a result of its free radical scavenging and heavy metal chelating abilities. It also exhibits excellent absorption patterns within the skin, to ensure that the effects of the application are felt (Rona et al., 2004).

Cell Regulators

These include retinol (vitamin A), retinol derivatives (retinaldehyde, tretinoin), peptides, botanicals, and growth factors (Ganceviciene et al., 2012) (Fig. 3.2). They function by directly affecting collagen and elastin metabolism to facilitate clinical and biochemical repair of photoaged skin and prevent the onset of symptoms (Ganceviciene et al., 2012; Pandel et al., 2013). Vitamin A derivatives also have strong antioxidant activity, inducing collagen biosynthesis and blocking collagenase synthesis (Ganceviciene et al., 2012; Varani et al., 2000).

Thus, it is effective in reducing the signs of UV-induced premature photoaging such as wrinkles, loss of elasticity, and pigmentation (Ganceviciene et al., 2012). Retinols are the most commonly used in antiaging formulations and when compared to its derivative, tretinoin, it provides the advantage of reduced irritation (Kafi et al., 2007; Varani et al., 2000). Tretinoin is a mild irritant and for this reason when used, it is added at a concentration of 0.05% in the United States of the retinol derivatives, only tretinoin and tazarotene are FDA approved (Rona et al., 2004).

3.1.3.2 Microdermabrasion

In this method, fine crystal particles are blasted onto the surface of the skin by an abrasion wand such that the crystals transfer their kinetic energy to the cells of the stratum corneum, facilitating detachment of dead skin cells (Bhalla & Thami, 2006). The growth of new skin cells is then stimulated because of an increased blood and nutrient supply to the skin, whereas old cells are simultaneously removed through suction (Bhalla & Thami, 2006). Thus, higher cellular turnover rates increase the elasticity of the skin as well as improve texture. Depending on the patient, the depth of the abrasion can be altered through altering vacuum pressure, speed and particulate size of the crystals, angle and speed of impaction as well as the number of passes (Bhalla & Thami, 2006; Grimes, 2005). The principle of microdermabrasion is exfoliating the skin to remove dead skin cells and promote the regeneration of new cells, and therefore, no anesthetic is needed. Side effects of treatment are short lived, lasting up to 24h posttreatment and include a temporary pink appearance in the skin accompanied by dryness and tightness. For these reasons this procedure is also considered noninvasive, as in the instance of topical agents (Bhalla & Thami, 2006).

3.1.3.3 Chemical Peels

This is used for the improvement of skin texture and tone, removal of fine lines appearing most commonly around the eyes (crow's feet) and mouth area, and to reduce the appearance of age spots (Rendon et al., 2010). In addition to being utilized as an antiaging therapy, chemical peels are often used in the rehabilitation of acne prone or severely scarred skin. These methods cause chemical ablation of defined skin layers in an attempt to induce even and tight skin as a result of the regeneration and repair mechanisms (Ganceviciene et al., 2012; Rendon et al., 2010). Chemical peels may be classified into three groups based on the depth of the peel, namely, superficial peels, medium depth peels, and deep peels (Monheit & Chastain, 2001; Rendon et al., 2010). The depth of the peel is determined by the chemical used, its concentration, pH of the solution, and the duration of the application (Fischer et al., 2009; Monheit & Chastain, 2001; Rendon et al., 2010). In some procedures the side effects after treatment may resemble that of sunburn (peeling, redness, scaling, etc.) that may last for 3 to 7 days.

Superficial peels generally utilize alpha or beta-lipohydroxy acids (such as hyaluronic acids) and trichloroacetic acid (TCA) at a concentration of 10%–30%—sufficient to exfoliate the epidermal layers without surpassing the basal layer (Gancevicìene et al., 2012). Alpha-hydroxyl acids are popular functional agents in antiwrinkle skin care and include glycolic, lactic, and citric acids. They act to thin the outer most layer of the skin, the stratum corneum, induce desquamation, stimulate collagen biosynthesis, glycosaminoglycan deposition, and promote cellular renewal, thus increasing the thickness of the dermis (Rona et al., 2004). These peels hence target the epidermis and dermal–epidermal interface. The dermal–epidermal interface is also targeted because aged skin exhibits a flattened dermal–epidermal junction (Sudel et al., 2005). The superficial nature of this peel makes it more versatile in that it can safely be applied to a wider range of skin types. Epidermal regeneration occurs quickly with the average recovery period for this peel ranging between 3 and 5 days (Rendon et al., 2010).

A medium depth peel reaches to the upper reticular dermis using TCA at a concentration above 30% but below 50%. The healing period is slightly longer than with superficial peels, with epithelialization occurring after a week. With medium depth peels, the use of sun protectants for a month posttreatment is recommended. This peel involves a higher risk than for superficial peels, even more so for patients with a darker skin (Rendon et al., 2010).

Deep peels are more commonly used in the treatment of more severely photoaged skin presenting deep wrinkles and scars, younger appearing skin is achieved through stimulating maximal collagen regeneration. Because of the strength of this peel, the recovery time is substantially longer than the two aforementioned peels, taking up to 2 months for complete recovery with continued use of sun protectants. Deep peels use a concentration of TCA above 50% in conjunction with phenol to penetrate the lower reticular dermis (Rendon et al., 2010). The phenol acts as the abrasive component and may be used with croton oil (Bensimon, 2008; Rendon et al., 2010). The phenol used in this peel is very rapidly absorbed and may potentiate cardiotoxic effects. For that reason it should be utilized with caution.

Natural acids that may be used in these peels are salicylic and glycolic acid (Kligman & Kligman, 1986). Glycolic acid is an alpha-hydroxyl acid, and salicylic acid is a beta-hydroxyl acid. The chemical differences of these acids better suite them to specific skin care therapies, for example, salicylic acid is better suited for photoaged or acne-prone skin (Fig. 3.3A). Salicylic acid, such as aspirin, beholds antiinflammatory activity that reduces inflammation and redness as a result of this procedure and thus fast-tracks the healing process (Fig. 3.3B) (Vedamurthy, 2004). Sources of salicylic acid include willow bark of the *Willow salix*, wintergreen leaves, and sweet birch (*Betula lenta*) (Vedamurthy, 2004).

The benefits of these peels can be seen several weeks posttreatment when modifications are visible. At this time, the epidermal architecture returns to

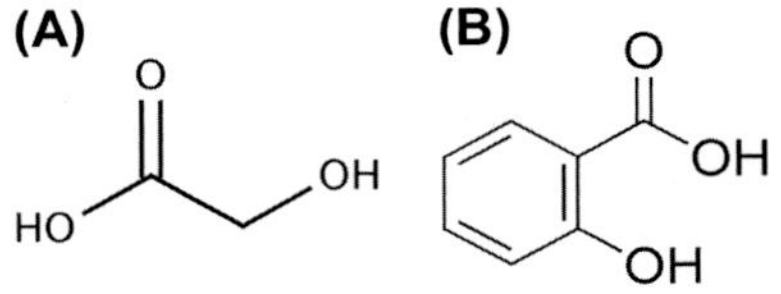

(A) O HO OH (B) O OH OH

FIGURE 3.3 Chemical structures of some natural acids used in chemical peels to fight the signs of aging. Depending on their configuration they may be better suited to different therapeutic purposes (A) Glycolic acid is an alpha-hydroxy acid (B) Salicylic acid is a beta-hydroxy acid better suited to chemical peels with the therapeutic aim of alleviating the symptoms associated with photoaged or acne-prone skin. The antiinflammatory activity associated with this acid fast-tracks the healing process associated with chemical peels as an additional benefit.

normal, melanocytes become uniformly distributed, basal cells contain small melanin grains that are homogenously distributed, and the thickness of the basal membrane is homogenous. The dermis also shows a new subepidermal band of collagen because of increased collagen production as well as a rearrangement of collagen fibers and a new network of elastin (Behin, Feuerstein, & Marovitz, 1977; Brown, Kaplan, & Brown, 1960; Vagotis & Brundage, 1995). Water and glycosaminoglycans in the dermal layer of the skin have also been shown to increase in postpeel analysis along with improvements in skin elasticity (Behin et al., 1977; Vagotis & Brundage, 1995). The effects of each peel are summarized in Table 3.2.

Although the benefits are much to be desired, they do not come without risk. The risks associated with chemical peels increase with the depth of the peel. These potential side effects include hypo/hyperpigmentation, solar lentigines, and the possibility of postoperative herpetic infections (Baker & Gordon, 1961).

3.1.3.4 Injections

Facial injections are treatments that are injected into the target site and are categorized as fillers or neuromodulators based on their function.

Dermal Fillers

Fillers are microinjected into the face and generally play no role in enhancing the synthesis or activity of endogenous structural components within the dermis (although in some cases the filler may exhibit this property as an additional benefit) but rather aim to merely take up space and by doing so plump out the appearance of wrinkles. The contents of the injection may be a single active component or a biologically compatible mixture thereof. These include vitamins, hyaluronic acid, minerals, nutrients, amino acids, growth factors, or homeopathic products (Caruso et al., 2008; Iorizzo, Padova, & Tosti, 2008). These fillers may also further be classified depending on the duration of their effect as temporary, semipermanent, or permanent. To be considered permanent, the effects of the filler are required to last for a period longer than 2 years.

TABLE 3.2 The Effects of Chemical Peels Vary Based on Their Depth, as They Penetrate the Skin Further They Are Able to Target a Wider Range of Components and Thus Result in Different Outcomes. The Depth of the Peel Is Determined by the Target of the Treatment and Hence the Desired Results

Peel	Effect
Superficial	Corneosomes are targeted, desquamation, increased epidermal activity or enzymes, epidermolysis, exfoliation.
Medium-depth	Coagulation of membrane proteins, living cells of epidermis destroyed; at high enough concentrations living cells of dermis may also be destroyed.
Deep peel	Coagulation of proteins, complete epidermolysis, basal layer restructuring, restoration of dermal architecture.

Ganceviciene, R., Aikaterini, I. L., Athanasios, T., Makrantonaki, E., & Zouboulis, C. C. (2012). Skin anti-ageing strategies. *Dermato-Endocrinology*, *4*, 308–319; Rendon, M. I., Berson, D. S., Cohen, J. L., Roberts, W. E., Starker, I., & Wang, B. (2010). Evidence and considerations in the application of chemical peels in skin disorders and aesthetic resurfacing. *Journal of Clinical Aesthetic Dermatology*, *3*, 32–43.

A semipermanent filler lasts between 1 and 2 years, and any filler lasting less than 1 year is considered temporary (Ganceviciene et al., 2012). Hyaluronic acid, a temporary "gold standard" filler, plays roles in space filling, modulation of wound healing, shock absorption, free radical scavenging, lubrication and is also implicated in regulation of cell locomotion and proliferation (Ganceviciene et al., 2012; Sherman, Sleeman, Herrlich, & Ponta, 1994). In addition, fibroblasts are activated by their presence, and collagen expression is stimulated.

Although there may be concern regarding accumulation of dermal fillers in the skin with repeated procedures, hyaluronic acid is eliminated as carbon dioxide and water by natural pathways within the body. Sources of hyaluronic acid include *Staphylococcus equine* (Ganceviciene et al., 2012). Although hyaluronic acid is also produced by plants, this native polymeric form contains no chemical modifications and so is excreted from the body at a faster rate than what is desired. For this reason many different hyaluronic acid fillers have been developed, which differ in characteristics such as particle sizes, different cross-links, mono-/biphasic structures, varying concentrations, and the presence of an anesthetic agent to produce higher retention times (Gold, 2007).

Neuromodulators

Neuromodulators are those compounds that are injected into the skin at which point they bind to the presynaptic nerve endings at specific receptor-mediated binding sites and further block signal transduction pathways to prevent involuntary or undesirable muscle contractions. By modulating nervous responses in facial muscles, wrinkles are relaxed (Cheng, 2007; Dessy et al., 2007). The effect of these injections only becomes fully apparent after a period of 10 days (Ganceviciene et al., 2012). These include Botox and Dysport, both of which are toxins isolated from the bacterium, *Clostridium botulinum.*

C. botulinum is a Gram-positive bacterium, which produces seven constitutionally similar toxins, subtypes A through G, named after their producer (Carruthers, Kiene, & Carruthers, 1996; Garcia & Fulton, 1996). Topical applications of these toxins have been proven to have no effect on the signs of aging, however, injections have been proven to reduce the signs of dynamic wrinkles (caused by excessive nerve stimulation) and exaggerated frown lines (Cheng, 2007). When injected, toxin subtype A, Botox, blocks the presynaptic release of acetylcholine at the neuromuscular junction by the action of the light and heavy protein chains, respectively, exercising the following mechanisms (Becker-Wegerich, Rauch, & Ruzicka, 2001; Ganceviciene et al., 2012):

1. The heavy chain portion binds to specific sialoglycoprotein receptors of the motor nerve end plates of cholinergic nerve endings. This binding then induces receptor-mediated endocytosis.
2. The light chain splits off from the rest of the toxin and specifically inactivates the synaptosomal-associated protein required for acetylcholine exocytosis into the neuromuscular junction.

The chemical denervation for selective muscular relaxation or paralysis produced as a result of the binding of the toxin to the presynaptic nerve endings will be visible within 24–48 h after the injection (Garcia & Fulton, 1996). Few plant-based alternatives are available for the treatment of wrinkles. This further illustrates the great potential held by plants for a more natural and safer alternative.

3.1.4 Aging Is a Complex Biological Process With Many Theories Surrounding It

More than 300 theories have been proposed to describe the aging process, including, the free radical theory of aging, the immunologic theory, the inflammation theory, and the mitochondrial theory (Table 3.3) (Medvedev, 1990). Each theory proposes a specific cause of aging to explain changes that are associated with the process of aging. Although many scientists consider each in isolation, the complexity of the aging process leaves room for suggestion that several processes operating at various levels of functional organization are actually the synergistic cause of aging, and for that reason no single aspect should be considered in isolation (Franceschi, 2000b).

The Mechanism of Aging: What Happens in Your Skin?

In the skin, epidermal cells are heavily implicated in epidermal homeostasis and regeneration of elastin and collagen fibers. With aging, the number of epidermal stem cells declines and the regenerative capacity of the skin too. Consequently, barrier disorders form and hydration and elasticity of the skin decline resulting in the formation of wrinkles. This is further aggravated by chronic exposure to sunlight (Jabs, 2012). Although extrinsic and intrinsic aging are the result of distinct factors, the molecular pathways that ensue as a result are similar and both involve telomere shortening, mitochondrial DNA mutations, oxidative stress, and a decrease in hormone levels (Sahu et al., 2013).

Mechanisms of Extrinsic Aging

Extrinsic aging follows a signal cascade stimulated by the production of ROS subsequent to UV exposure. The activation or deactivation of these pathways seems to be determined by the phosphorylation state of the proteins and enzymes involved.

As aging is mainly the result of chronic UV exposure, UV radiation provokes the production of ROS within the skin. ROS have the ability to repress activity of the enzyme protein-tyrosine phosphatase κ responsible for the maintenance of cell surface receptors in the skin. Such receptors include the epidermal growth factor, interleukin-1, keratinocyte growth factor as well as the tumor necrosis factor-α (TNF-α). These receptors are maintained in the hydrophosphorylated, inactive state (Xu, Shao, Voorhees, & Fisher, 2006). In circumstances where ROS then inhibit the function of protein-tyrosine phosphatase κ, the cell surface receptors are no longer maintained in this state but rather become activated and stimulate stress-associated mitogen-activated protein kinases (MAPK) to drive intracellular signaling (Fisher et al., 1998).

These activated kinases induce transcription of nuclear transcription complex activator protein-1 (AP-1) (Fisher et al., 2002). AP-1 may also be activated by singlet oxygen degrading the lipid membrane and releasing ceramide that leads to the formation of carbonyl groups and the accumulation of damaged oxidized dermal proteins. These oxidized membrane lipids also release arachidonic acid, which is processed to prostaglandins by cyclooxygenase. Prostaglandins recruit inflammatory cells to the area, and an inflammatory response is then launched (Yaar & Gilchrest, 2007). Activated AP-1 also increases the transcription of MMPs that are involved in collagen degradation and simultaneously downregulate the expression of procollagen I and III genes, as well as transforming growth factor-β (TGF-β) receptors. TGF-β is responsible for regulating the activity of proteases such as elastase and when downregulated cannot perform its function. The outcome of this is an overall reduced production of the dermal matrix (Fisher et al., 2000).

Collagen degradation resulting from UV radiation is generally incomplete, and these incompletely degraded collagen fibers accumulate to further reduce the structural integrity of the skin. Furthermore, UV irradiation stimulates the nuclear factor κB transcription factor, which recruits neutrophils and MMP-8 to further degrade the matrix. Neutrophils excrete degradative enzymes such as elastase

—cont'd

and collagenase (Yaar & Gilchrest, 2007). The mitochondria of each cell generates energy in the form of ATP for the cell through absorbing oxygen and is also affected by extrinsic aging, as indicated in Table 3.3 under the mitochondrial theory of aging (Harman, 1972; Miquel, Economos, Flemming, & Johnson, 1980; Tosato, Zamboni, Ferrini, & Cesari, 2007). UV exposure generates ROS that damage mitochondrial DNA by deletions, rearrangements, or insertions of DNA as a result of double-stranded DNA breaks (Prado, Cortes-Ledesma, Huertas, & Aguilera, 2003). Hence the ability of the mitochondria to produce energy is compromised with the accumulation of ROS (Sahu et al., 2013).

Telomeres are tandem repeats in a loop configuration found at the ends of chromosomes. UV radiation disrupts this loop configuration resulting in critically shortened telomeres (Table 3.3, telomere theory of aging). These exposed telomeres then have the ability to interact with the Werner protein, which activates p53, the tumor suppressor protein that damages DNA and results in cellular senescence (Yaar & Gilchrest, 2007). Upper dermal proteins accumulate ROS-induced devastation subsequent to UV exposure (Sander et al., 2002). These proteins as a result are both structurally and functionally compromised and subject to degradation by proteasomes (Shacter, 2000). Epidermal proteins are degraded by epidermal methionine sulfoxide reductases efficiently and quickly, whereas the degradation of dermal proteins is not as efficient. Accumulation of nonfunctional proteins may be toxic to the cells and induce apoptosis.

Mechanisms of Intrinsic Aging

Similarly to extrinsic aging, the activation of c-Jun by MAPK due to ROS exposures, and as a result AP-1 leads to MMP-1, MMP-3, and MMP-9 expression (Chung, Kim, Kim, & Yu, 2001). The outcome is, as with extrinsic aging, an overall decline in the levels of collagen and elastin (Sahu et al., 2013). Hormonal changes also contribute to intrinsic aging. Hormones expressed in the sex organs, pituitary glands, and adrenal glands have been shown to decline in the mid-twenties, whereas other hormones, namely, estrogens and progesterones decline later during menopause. Imbalances between estrogens and androgens result in characteristic dryness, wrinkling, epidermal atrophy, collagen breakdown, and loss of elasticity as seen in intrinsically aged skin (Kohl, Steinbauer, Landthaler, & Szeimies, 2011).

3.1.5 The Enemy Enzyme, Elastase and Its Target, Elastin

As mentioned, the dermal layer of the skin contains the structural components of the skin, which provide it with its structural integrity and elasticity. One of these structural components, elastin is an insoluble elastic fibrous protein, which forms the main component of animal connective tissues and tendons (Yan, Guon, & Lin, 1994). It is an essential part of various human tissues that depend on elasticity, including the arteries, skin, and lungs (Daamen, Verkamp, Van-Hest, & Kuppevelt, 2007; Fulop, Khalil, & Larbi, 2012). The presence of

TABLE 3.3 Theories Proposed to Describe the Process of Aging

Theory	Proposal	References
Evolutionary	Aging is the result of a decline in natural selection.	Haldane (1941) and Holliday (2006)
Free radical	The accumulation of oxygen radicals in cells could be responsible for aging and death. Lifespan may be determined by the degree of free radical damage to the mitochondria. Increasing age-related oxidative stress may be a consequence of the imbalance between free radical production and antioxidant defenses with a higher production of free radicals. The free radical theory of aging is divided into several hypotheses each focusing on the exclusive role of particular organelles and types of damaged molecules in the aging process.	Fisher et al. (2002), Harman (1972), and Weinert and Timiras (2003)
Mitochondrial	Age-related physiological decline is because of the accumulation of defects in several metabolic pathways. Mitochondrial DNA mutations accumulate progressively during life and are directly responsible for a measurable deficiency in cellular oxidative phosphorylation activity leading to enhanced ROS production. This theory of aging is considered an extended refinement of the free radical theory of aging.	Harman (1972), Miquel et al. (1980), and Tosato et al. (2007)
Gene regulation	Senescence results from changes occurring in gene expression that results in an aged phenotype.	Weinert and Timiras (2003)
Telomere	The replicative capacity of cells is limited to a certain number of replications and results in terminally arrested cells with altered physiology. In actively dividing differentiated cells a small amount of DNA is lost at the end of each chromosome with each replication cycle, hence resulting in shortened telomeres with altered structures that eventually lead to a halt in cellular proliferation. The relationship between telomeres and aging is further supported by diseases exhibiting telomeric dysfunction resulting in premature aging e.g., progeria.	Tosato et al. (2007) and Weinert and Timiras (2003)

Inflammation	Inflammation is the host's complex response to physiological and nonphysiological stress. The phases involved in stress, acute, and chronic inflammatory responses are: **1.** Intracellular activation **2.** Proinflammatory cells in tissues **3.** Vascular permeability increases **4.** Tissue damage and cell death. Individuals have stress thresholds, which determine the level of inflammatory stress they are able to cope with. When inflammation associated with this stress threshold is surpassed, unsuccessful aging occurs.	Chung et al. (2001) and Franceschi et al. (2000a)
Immune (Network)	Aging is indirectly controlled by a network of cellular and molecular defense mechanisms, which function to limit the negative effects of physical, chemical, and biological stressors. One such mechanism considered is the immune system.	Franceschi et al. (2000a)
Neuroendocrine	Aging is because of changes in neural and endocrine functions that are crucial for coordination and responsiveness of different systems to the environment, programming of physiological responses to environmental stimuli, and the maintenance of an optimal functional status for reproduction and survival.	Weinert and Timiras (2003)

elastin within these tissues is what affords them the ability to stretch and recoil and is hence the main contributor to skin elasticity and firmness (Daamen et al., 2007; Yuan & Walsh, 2006). The problem, however, is that the skin exhibits drastically declined levels of elastin with aging, which results in a loss of strength and flexibility materializing as visible wrinkles (Ndlovu, Fouche, Tselyane, Cordier, & Steenkamp, 2013). This is because the elasticity of the skin is directly influenced by the quality and quantity of elastin and collagen fibers (Jabs, 2012). These crucial parameters are increasingly perturbed with time as the activity of elastase, the key degrader of elastin, has been shown to be enhanced with age and thus undesirable degradation ensues (Nar, Werle, Bauer, Dollinger, & Jung, 2001). In addition to this, elastin is thought to be produced at a slower rate with time. The exclusivity of the elastase–elastin interaction and the decreased quality of the elastin network as a consequence of this interaction have resulted in a recent peak in the interest of enzymes such as elastase (Jabs, 2012).

Skin aging has been correlated with an increase in degradative enzymes such as elastase (Maity, Nema, Abedy, Sarkar, & Mukherjee, 2011). The elastase–elastin interaction consequently produces elastin peptides that become free to enter the bloodstream and travel to satellite locations where they act as chemotactic agents, which recruit neutrophil cells to the site, which then in turn release elastase and MMPs and further produce oxidative stress stimulating the vicious cycle (Fulop et al., 2012). The end result of this is the recruitment of inflammatory cells and hence chronic inflammation, which then also results in a reduced production of the dermal matrix and decreased structural rigidity (Jabs, 2012).

Human leukocyte elastase is a 29 kDa serine endoprotease of the proteinase S1 family capable of degrading and hydrolyzing a multitude of essential protein components found in the skin and connective tissue extracellular matrix, such as collagen, cartilage proteoglycans, fibronectin, and elastin (Sigma-Aldrich, 2015; Thring, Hili, & Naughton, 2009; Wiedow, Shroder, Gregory, Young, & Christophers, 1990; Yan et al., 1994). It exists as a peptide chain comprises 238 amino acids and 4 disulfide bonds with a preferential cleavage site for the carboxyl side of valine but will also cleave to a lesser extent after alanine. Three isoforms exist depending on the number and location of the N-linked glycans (Sigma-Aldrich, 2015).

Elastase is naturally required by the skin to degrade dysfunctional proteins that remain within the extracellular matrix after a wounding event such that they can be eliminated for repair to occur. This problem is further compounded by the fact that the number of epidermal stem cells (responsible for allowing the skin to differentiate and regenerate itself) declines with age and with it the skins ability to regenerate important structural components, such as elastin and collagen, declines as well (Jabs, 2012; Nar et al., 2001). To combat the effects of elastase on the dermis, many researchers have searched for possible inhibitors of this enzyme. One such example is ursolic acid, which is a potent antioxidant as well as inhibitor of elastase. Simultaneously it has been shown to stimulate ceramide and collagen synthesis. The downfall, however, lies in the toxicity of ursolic acid (Rona et al., 2004).

3.2 PLANTS USED TRADITIONALLY TO PREVENT THE APPEARANCE OF WRINKLES

The use of plants in medicine stemmed from the need for a cure, sometimes urgently. Skin aging is not a disease as such, for which a treatment was sought. Despite this, some literature revealed the use of plants for skin aging, as indicated in Table 3.4.

3.3 PRODUCTS AND POTENTIAL PLANTS WITH ELASTASE INHIBITION

To reduce the appearance of wrinkles, compounds such as retinol (vitamin A) and retinol derivatives are currently employed, however, they are costly and hence their application becomes somewhat limited (Im et al., 2014; Kafi et al., 2007). In the search for elastase inhibitors, tropical rain forests are an important and interesting resource, while many European plants are also becoming popular (Jabs, 2012). A study by Safayhi, Rall, Sailer, and Ammon (1997) has uncovered the inhibitory activity of boswellia acids obtained from Frankincense against the activity of human neutrophil elastase. To circumvent the damages resulting from increased elastase activity, agents that are able to inhibit their function need to permeate the protective barrier of the skin and be absorbed to the deeper layers of the skin in sufficient concentrations such that their effect can be exerted (Jabs, 2012). Plant molecules thus provide a perfect solution owing to their low-molecular-weight and small size.

A haplophyte, native to the South African dessert, used in antiwrinkle cosmetic formulations is *Mesembryanthemum crystallinum* L. This plant has been shown to accumulate flavonoids when exposed to high UV radiation (Fig. 3.4) (Ibdah, Krins, Seidlitz, Strack, & Vogt, 2002). The antiaging potential of *M. crystallinum* L. was explored by The International Research Centre for Herbal Cosmetics Yves Rocher. They have developed the antiaging skin care range, Sérum Végétal based on the antioxidant potential of this plant. Plants that have been incorporated into antiaging formulations as well as the mechanisms by which they act are mentioned in Table 3.5 and Fig. 3.5.

An increased interest has arisen in the interactions of elastase with its inhibitors because it is proposed that topical application of inhibitors to the surface of the skin may benefit UV irradiated and dry skin from the adverse effects of these exposures (Bizot-Foulon, Bouchard, Homeback, Dubert, & Bertaux, 1995). Plants have been used in cosmetics for many years and

TABLE 3.4 Plants Commonly Used to Treat the Appearance of Wrinkles

Scientific Name/ Common Name	Plant Family	Locale of Use	Traditional Use	References
Aloe vera (L.) Burm.f./ Barbados aloe	Xanthorrhoeaceae	Globally	Wrinkles, stretch marks, pigmentations, wound healing	Tobassum and Hamdani (2014)
Rosmarinus officinalis L./ Rosemary	Labiatae	Globally	Wrinkles, digestive issues, joint pain, migraines	Tobassum and Hamdani (2014)
Mentha × piperita L., pro spec. & Hylander/ Peppermint	Lamiaceae	Turkey	Indigestion, to soothe wounds, menstrual disorders, wrinkles	Ugulu (2012)
Melissa officinalis L./ Lemon balm	Lamiaceae	Globally	Flavorant, mental disorders, heart and lung disorders, antidepressant, sleeping aid, insect bites, dog bites, amenorrhea, mushroom allergies, gripe, intestinal ulcers, respiratory ailments, tumors, gripe, swelling, sore throat, emmenagogue, analgesic, scrofula, stomach complaints, tranquilizer, antispasmodic,wound disinfectant, halitosis, rabies, wrinkles toothache, earache, morning sickness, balning, crooked neck, memory, gastrointestinal ailments, nervousness, biliary ailments, hepatic ailments	Ugulu (2012) and Shakeri, Sahebkar, and Javadi (2016)

Cucumis sativus L./Cucumber	Cucurbitaceae	Turkey, India	Hepatoprotectant, wrinkles	Ugulu (2012) and Sarhadynejad et al. (2016)
Matricaria chamomilla L./Chamomile	Asteraceae	Turkey, India	Wrinkles, sedative, gastrointestinal complaints, hypnotic	Ugulu (2012) and Taher, Samud, Mezogi, and Hamza (2016)
Artemisia absinthium L./wormwood	Asteraceae	India	Colds, hypertension, laxative, vermifuge, stomachic, heart disease, carminative antiseptic, diuretic, febrifuge, antitoxic,otitis, wrinkles, cutaneous infections	Abouri et al. (2012)
Tamarindus indica L./Tamarind	Fabaceae	Sudan	Nutrition, antioxidant, antimicrobial, wrinkles, postpartum depression, measles	Abouri et al. (2012)
Daucus carota L./Carrot	Apiaceae	India	Hypotension, antifertility, hepatoprotectant, nephrosis, antibacterial, antispasmodic	Gediya, Mistry, Patel, Blessy, and Jain (2011)

FIGURE 3.4 *Mesembryanthemum crystallinum* L., commonly known as the desert vygie, has been incorporated into an antiaging formulation based on its antioxidant activity by the International Research Centre for Herbal Cosmetics, Yves Rocher.

multiple studies have revealed the usefulness of plants in inhibiting the action of degradative enzymes such as elastase (Lee et al., 1999; Ndlovu et al., 2013; Thring et al., 2009). Plants are thus an excellent source to investigate for the development of natural antioxidants involved in antiaging and inhibitors for wrinkle care as an antiaging medicine (Kim, Noh, Lee, Kim, & Lee, 1995). Active plant compounds are capable of acting as therapeutic agents because of the universal nature of amino and nucleic acids across all life forms. This then allows unrelated molecules to contain structural similarities and therefore a natural active compound produced in a plant will be able to bind to an unrelated molecule, not originally intended, and relieve or minimize its effects (Ortholand & Ganesan, 2004). *Peltophorum africanum* Sond. has traditionally been reported to treat wounds (Bizimenyera, Aderogba, Eloff, & Swan, 2007). Ndlovu et al. (2013) investigated Southern African medicinal plants for their ability to inhibit elastase as a potential antiwrinkle agent, of which *P. africanum* Sond. was one. Their results exhibited elastase inhibition of more than 80% for the stem, leaves, seeds, and bark ethanolic extracts.

3.4 SCIENTIFIC RESULTS

To incorporate a botanical component into an antiaging formulation, some scientific validation should exist to confirm the activity and benefit of its incorporation. As discussed these parameters include targeting various problems associated with aging in the skin such as loss of moisture, reduced cellular turnover rates, abnormal degradation of the structural components

TABLE 3.5 Plants Incorporated Into Products for Their Antiaging Properties

Scientific Name/ Common Names	Family	Antiaging Mechanism Employed	References
Aloe vera (L.) Burm.f./ Barbados aloe	Xanthorrhoeaceae	Reduction in expression of MMP prevents loss of hyaluronic acid promoting skin hydration	Saito et al. (2016)
Arctium lappa L./Burdock	Asteraceae	Antiinflammatory, connective tissue metabolism stimulator for regeneration of dermal structure	Knott et al. (2008)
Areca catechu L./Betel palm	Arecaceae	Elastase inhibition	
Aspalathus linearis (Burm.f.) R.Dahlgren/Rooibos	Fabaceae	Antioxidant	Joubert, Gelderblom, Louw, and Beer (2008)
Camellia japonica L./ Common Camellia	Theaceae	Inducer of collagen type 1 synthesis, moisturizing effect	Jung et al. (2007)
Camellia sinensis (L.) Kuntze/ Tea	Theaceae	Elastase inhibition, minor collagenase inhibition, antioxidant	Thring et al. (2009)
Cinnamomum verum J.Presl/ Cinnamon	Lauraceae	Induction of procollagen type-I synthesis	
Coffea arabica L./Coffee	Rubiaceae	Antioxidant, inhibition of MMPs, upregulation of procollagen type-I, down-regulation of MAPK pathway	Binic, Lazarevic, Ljubenovic, Mosja, and Sokolovic (2012)
Curcuma longa L./Tumeric	Zingiberaceae	Antioxidant	Cronin and Draelos (2010)

Continued

TABLE 3.5 Plants Incorporated Into Products for Their Antiaging Properties—cont'd

Scientific Name/ Common Names	Family	Antiaging Mechanism Employed	References
Curcuma xanthorrhiza D.Dietr./Curcumine	Zingiberaceae	Suppression of UVB induced expression of MMP-1, stimulation of procollagen-I synthesis	Oh, Shim, Gwon, Kwon, and Hwang (2009)
Cyclopia intermedia E.Mey./ Honeybush	Fabaceae	Antioxidant, reduces MMP-1 and -9 expression, upregulation of TIMP-1 mRNA expression, upregulation of genes associated with skin hydration (filaggrin, involucrin, loricrin), reduced expression and activity of inflammatory mediators	Joubert et al. (2008) and Im et al. (2014)
Dioscorea villosa L./Wild yam	Dioscoreaceae	Antiinflammatory, anticollagenase, enhances adipocyte content	Binic et al. (2012)
Glycyrrhiza glabra L./Licorice	Fabaceae	Antioxidant	Visavadiya, Soni, and Dalwadi (2009)
Hamamelis virginiana L./ Witch Hazel	Hamamelidaceae	Antioxidant, minor elastase and collagenase inhibition	Thring et al. (2009)
Illicium anisatum L./Japanese star anise	Schisandraceae	Antioxidant, antielastase, antiinflammatory	Kim et al. (2009)
Ixora parviflora Vahl/Ixora	Rubiaceae	Modulation of expression of MMPa, mitogen-associated protein kinases, cyclooxygenase-II	Wen, Fan, Tsai, Shih, and Chiang (2012)
Labisia pumila (Blume) Fern.-Vill./Kacip Fatimah	Primulaceae	Upregulation of collagen, antioxidant	Choi et al. (2010) and Chua, Lee, Abdullah, and Sarmidi (2012)
Lavandula angustifolia Moench/Lavender	Lamiaceae	Minor enzyme inhibition, antioxidant	Thring et al. (2009)

Lycium barbarum L./Goji berry	Solanaceae	Antioxidant	Nardi et al. (2016)
Morus alba L./Mulberry	Moraceae	Antioxidant	Katsube et al. (2010)
Ocimum basilicum L./Basil	Lamiaceae	Antioxidant	Bozin, Mimica-Dukic, Simin, and Anackov (2006)
Origanum vulgare L./Oregano	Lamiaceae	Antioxidant	Bozin et al. (2006)
Phlebodium aureum J.Sm./ Golden Polypody	Polypodiaceae	Inhibition of MMP-1 and -2	Philips et al. (2009)
Populus nigra L./Poplar bud	Salicaceae	Genetic modulation of antioxidant components, inflammatory responses and cell renewal	Dudonne et al. (2011)
Punica granatum L./ Pomegranate	Lythraceae	Minor enzyme inhibition, antioxidant	Zaid, Afaq, Syed, Dreher, and Mukhtar (2007), Thring et al. (2009), and Park et al. (2010)
Rosmarinus officinalis L./ Rosemary	Lamiaceae	Antioxidant	Binic et al. (2012)
Styrax japonicus Siebold & Zucc./Snowball tree	Styracaceae	MMP-1 inhibition	Moon, Lee, and Chung (2006)
Thymus caucasicus Willd. ex Benth./ *Thymus vulgaris* L./ Thyme	Lamiaceae	Antioxidant	Bozin et al. (2006)
Vitis vinifera L./Grape	Vitaceae	Antioxidant	Cronin and Draelos (2010)

MAPK, mitogen-activated protein kinases; *MMP*, matrix metalloproteinase.

FIGURE 3.5 Plants that are used for their antiaging properties. (A) *Aloe vera* (L.) Burm.f. (Zell, 2009a), (B) *Mentha × piperita* L., pro spec. & Hylander (Storch, 2007), (C) *Arctium lappa* L. (Fischer, 2008), (D) *Areca catechu* L. (Xaver, 2011), (E) *Aspalathus linearis* (Burm.f.) R.Dahlgren (Bruenkin, 2005), (F) *Coffea arabica* L., (G) *Curcuma longa* L. (Forest & Starr, 2008), (H) *Dioscorea villosa* L. (Zell, 2009b), (I) *Glycyrrhiza glabra* L. (Kojian, 2011), (J) *Illicium anisatum* L. (Kenpai, 2007), (K) *Lavandula angustifolia* Moench (Porse, 2006), (L) *Lycium barbarum* L. (Martin, 2008), (M) *Populus nigra* L. (Fischer, 2013), (N) *Punica granatum* L., (O) *Vitis vinifera* L. (Zell, 2006).

of the skin, and inflammation. Identifying botanicals that can aid in diminishing the effects of these would then be evaluated for their ability to retain moisture within the skin by retaining hyaluronic acid or supplementing the skin with it, stimulate cellular turnover, inhibit degradatory and inflammatory-associated enzymes, or supplement the skin with structural components. Table 3.6 summarizes some results pertaining to the investigation of popular botanicals for their antiaging potential regarding enzymatic inhibition (Fig. 3.6).

Despite the promising results of many experimental endeavors, few medicinal plants have been incorporated into an antiaging formulation for its ability to inhibit elastase. The field thus holds much promise and leaves much to be explored! For this resource to be explored, one needs to further investigate the various properties botanicals have to offer when considering the aging process.

TABLE 3.6 Scientific Results of Some Popular Botanicals and Their Antienzymatic Activity Conferring Their Antiaging Properties

Plant Species/ Common Name	Distribution	Part Used/Pure Compound	Experimental Parameter	Results	References
Arctium lappa L./ Burdock	From Scandinavia to the Mediterranean, the British Isles to Russia, the Middle east to China and Japan, including India	Root	Collagenase inhibition at 25 μg/mL	0%	Thring et al. (2009)
			Elastase inhibition at 25 μg/mL	51%	
Camellia sinensis (L.) Kuntze/Tea		Leaf extracts of green tea in glycerine and lyophilized powder of white tea	Collagenase inhibition	87%	Thring et al. (2009)
			Elastase inhibition at 25 μg/mL	89%	
Clerodendrum glabrum E.Mey./ Tinderwood	South Africa	Root (MeOH)[c]	Elastase inhibition at 200 μg/mL[a]	5.92%	Ndlovu et al. (2013)
		Root (EtOAc)[d]	Elastase inhibition at 200 μg/mL[a]	15.3%	
			Collagenase inhibition at 200 μg/mL[a]	31.7%	
		Stem (MeOH)[c]	Elastase inhibition at 200 μg/mL[a]	13.35%	
			Collagenase inhibition at 200 μg/mL[a]	68.09%	
		Stem (EtOAc)[d]	Elastase inhibition at 200 μg/mL[a]	89.38%	
			Collagenase inhibition at 200 μg/mL[a]	88.64%	
			Hyaluronidase inhibition at 200 μg/mL[a]	8.11%	

Continued

TABLE 3.6 Scientific Results of Some Popular Botanicals and Their Antienzymatic Activity Conferring Their Antiaging Properties—cont'd

Plant Species/ Common Name	Distribution	Part Used/Pure Compound	Experimental Parameter	Results	References
		Bark (MeOH)[c]	Elastase inhibition at 200 μg/mL[a]	13.52%	
			Collagenase inhibition at 200 μg/mL[a]	58.96%	
			Hyaluronidase inhibition at 200 μg/mL[a]	5.19%	
		Bark (EtOAc)[d]	Elastase inhibition at 200 μg/mL[a]	81.46%	
			Collagenase inhibition at 200 μg/mL[a]	81.55%	
		Fruit (MeOH)[c]	Elastase inhibition at 200 μg/mL[a]	49.38%	
			Collagenase inhibition at 200 μg/mL[a]	72.54%	
			Hyaluronidase inhibition at 200 μg/mL[a]	0.51%	
		Fruit (EtOAc)[d]	Elastase inhibition at 200 μg/mL[a]	87.95%	
			Collagenase inhibition at 200 μg/mL[a]	26.29%	
			Hyaluronidase inhibition at 200 μg/mL[a]	2.5%	

Galium aparine L./ Cleavers		Leaf and stem	Elastase inhibition at 25 μg/mL	58%	Thring et al. (2009)
			Collagenase inhibition at 25 μg/mL	7%	
Libidibia ferrea (Mart. ex Tul.) L.P.Queiroz/Juca	Brazil	Trunk bark	Elastase inhibition at 250 μg/mL[a]	35.99%	Pedrosa et al. (2016)
			Collagenase inhibition at 100 μg/mL[a]	15.2%	
			Hyaluronidase inhibition (IC_{50})[b]	8.5 μg/mL	
		Pods	Elastase inhibition at 250 μg/mL[a]	19.6%	
			Hyaluronidase inhibition (IC_{50})[b]	16.0 μg/mL	
Peltophorum africanum Sond.	South Africa	Leaves (MeOH)[c]	Elastase inhibition at 200 μg/mL[a]	55.83%	Ndlovu et al. (2013)
			Collagenase inhibition at 25 μg/mL	82.13%	
			Hyaluronidase inhibition at 200 μg/mL[a]	49.32%	
		Leaves (EtOAc)[d]	Elastase inhibition at 200 μg/mL[a]	80.20%	
			Collagenase inhibition at 200 μg/mL[a]	75.26%	
			Hyaluronidase inhibition at 200 μg/mL[a]	2.69%	
		Stem (MeOH)[c]	Elastase inhibition at 200 μg/mL[a]	64.5%	
			Collagenase inhibition at 200 μg/mL[a]	74.14%	
			Hyaluronidase inhibition at 200 μg/mL[a]	0.95%	

Continued

TABLE 3.6 Scientific Results of Some Popular Botanicals and Their Antienzymatic Activity Conferring Their Antiaging Properties—cont'd

Plant Species/ Common Name	Distribution	Part Used/Pure Compound	Experimental Parameter	Results	References
		Stem (EtOAc)[d]	Elastase inhibition at 200 μg/mL[a]	84.4%	
			Collagenase inhibition at 200 μg/mL[a]	83.25%	
			Hyaluronidase inhibition at 200 μg/mL[a]	13.78%	
		Seeds (MeOH)[a]	Elastase inhibition at 200 μg/mL[a]	85.17	
			Collagenase inhibition at 200 μg/mL[a]	80.39	
		Seeds (EtOAc)[d]	Elastase inhibition at 200 μg/mL[a]	87.09	
			Collagenase inhibition at 200 μg/mL[a]	77.34	
			Hyaluronidase inhibition at 200 μg/mL[a]	2.18	
		Bark (MeOH)[c]	Elastase inhibition at 200 μg/mL[a]	69.36	
			Collagenase inhibition at 200 μg/mL[a]	78.44	
			Hyaluronidase inhibition at 200 μg/mL[a]	48.46	
		Bark (EtOAc)[d]	Elastase inhibition at 200 μg/mL[a]	86.27	
			Collagenase inhibition at 200 μg/mL[a]	88.27	
			Hyaluronidase inhibition at 200 μg/mL[a]	23.24	

Psychotria capensis Vatke	South Africa	Leaves (MeOH)[c]	Elastase inhibition at 200 μg/mL[a]	3.28%	Ndlovu et al. (2013)
			Hyaluronidase inhibition at 200 μg/mL[a]	52.98%	
		Leaves (EtOAc)[d]	Elastase inhibition at 200 μg/mL[a]	82.84%	
			Collagenase inhibition at 200 μg/mL[a]	63.07%	
			Hyaluronidase inhibition at 200 μg/mL[a]	3.13%	
		Stem (MeOH)[c]	Elastase inhibition at 200 μg/mL[a]	42.98%	
			Collagenase inhibition at 200 μg/mL[a]	56.45%	
			Hyaluronidase inhibition at 200 μg/mL[a]	5.00	
		Stem (EtOAc)[d]	Elastase inhibition at 200 μg/mL[a]	56.58%	
			Collagenase inhibition at 200 μg/mL[a]	29.98%	
		Seeds (MeOH)[c]	Elastase inhibition at 200 μg/mL[a]	44.16	
			Collagenase inhibition at 200 μg/mL[a]	78.36	
		Seeds (EtOAc)[d]	Elastase inhibition at 200 μg/mL[a]	70.67	
			Collagenase inhibition at 200 μg/mL[a]	69.48	
		Roots (MeOH)[c]	Elastase inhibition at 200 μg/mL[a]	45.90	
			Collagenase inhibition at 200 μg/mL[a]	59.41	
			Hyaluronidase inhibition at 200 μg/mL[a]	12.86	
		Roots (EtOAc)[d]	Elastase inhibition at 200 μg/mL[a]	59.60	
			Collagenase inhibition at 200 μg/mL[a]	28.42	
			Hyaluronidase inhibition at 200 μg/mL[a]	2.28	

Continued

TABLE 3.6 Scientific Results of Some Popular Botanicals and Their Antienzymatic Activity Conferring Their Antiaging Properties—cont'd

Plant Species/ Common Name	Distribution	Part Used/Pure Compound	Experimental Parameter	Results	References
Schotia brachypetala Sond.	South Africa	Leaves (MeOH)[c]	Elastase inhibition at 200 μg/mL[a]	82.54	Ndlovu et al. (2013)
			Collagenase inhibition at 200 μg/mL[a]	81.35	
			Hyaluronidase inhibition at 200 μg/mL[a]	20.36	
		Leaves (EtOAc)[d]	Elastase inhibition at 200 μg/mL[a]	84.22%	
			Collagenase inhibition at 200 μg/mL[a]	84.70%	
			Hyaluronidase inhibition at 200 μg/mL[a]	6.29%	
		Bark (MeOH)[c]	Elastase inhibition at 200 μg/mL[a]	75.32	
			Collagenase inhibition at 200 μg/mL[a]	77.51	
			Hyaluronidase inhibition at 200 μg/mL[a]	75.13	
		Bark (EtOAc)[d]	Elastase inhibition at 200 μg/mL[a]	93.73	
			Collagenase inhibition at 200 μg/mL[a]	87.61	

[a]*Percentage inhibition at a set concentration of plant extract.*
[b]*Result given as an IC_{50}, inhibitory concentration at which 50% of the enzyme is inhibited.*
[c]*Extract prepared using methanol.*
[d]*Extract prepared using ethylacetate.*

FIGURE 3.6 A few plants discussed in Table 3.6 with promising activity for incorporation into anti-aging formulations because of their antienzymatic activity. (A) *Galium aparine* L. (B) *Libidibia ferrea* (Mart. ex Tul.) L.P.Queiroz (C) *Peltophorum africanum* Sond. (D) *Clerodendrum glabrum* E.Mey.

3.5 CONCLUSION

The world holds within itself an impressive plant species diversity, which should be explored for the ability to inhibit elastase as a prospective antiwrinkle treatment. To fully realize the potential of plants as a medicine, however, they first need to be scientifically validated. In modern times, the average life expectancy has risen, and so as people live longer, they wish to look younger. The potential of plants as antiwrinkle agents is still in its infancy and few products on the market today contain phytoconstituents specifically capable of inhibiting elastase as its mechanism of action.

REFERENCES

Abouri, M., El Mousadik, A., Msanda, F., Boubaker, H., Saadi, B., & Cherifi, K. (2012). An ethnobotanical survey of medicinal plants used in the Tata Province, Morocco. *International Journal of Medicinal Plants Research*, *1*, 99–123.

Baker, T. J., & Gordon, H. L. (1961). The ablation of rhytides by chemical means; A preliminary report. *The Journal of the Florida Medical Association*, *48*, 541.

Baumann, L. (2007). Skin and its treatment. *Journal of Pathology*, *211*, 241–251.

Becker-Wegerich, P., Rauch, L., & Ruzicka, T. (2001). Botulinum toxin A in the therapy of mimic facial lines. *Clinical and Experimental Dermatology*, *26*, 619–630.

Behin, F., Feuerstein, S. S., & Marovitz, W. F. (1977). Comparative histological study of mini pig skin after chemical peel in animal model after chemical peel and dermabrasion. *Archives of Otolaryngology*, *103*, 271.

Bensimon, R. H. (2008). Croton oil peels. *Journal of Aesthetic Surgery*, *28*, 1–15.

Berger, M. M. (2005). Can oxidative damage be treated nutritionally? *Clinical Nutrition*, *24*, 172–183.

Bhalla, M., & Thami, G. P. (2006). Microdermabrasion: Reappraisal and brief review of literature. *Dermatologic Surgery*, *52*, 809–814.

Binic, I., Lazarevic, V., Ljubenovic, M., Mosja, J., & Sokolovic, D. (2012). Skin aging: natural weapons and strategies. *Evidence Based Complementary and Alternative Medicine*, *2013*, 1–10.

Bisset, D. L., Miyamoto, K., Sun, P., Li, J., & Berge, C. A. (2004). Topical niacinamide reduces yellowing, wrinkling, red blotchiness and hyperpigmented spots in ageing facial skin. *International Journal of Cosmetic Science*, *26*, 231–238.

Bizimenyera, E. S., Aderogba, M. A., Eloff, J. N., & Swan, G. E. (2007). Potential of neuro-protective antioxidant based therapeutic from *Plectranthus africanum* Sond. (Fabaceae). *African Journal of Traditional Complementary and Alternative Medicine*, *4*, 99–106.

Bizot-Foulon, V., Bouchard, B., Homeback, W., Dubert, L., & Bertaux, B. (1995). Unco-ordinate expression of type I and III collagens, collagenase, and tissue inhibitors of matrix metalloproteinases I along *in vitro* proliferative life span of human skin fibroblasts: Regulation by all trans retinoic acids. *Cell Biology International*, *19*, 129–136.

Bozin, B., Mimica-Dukic, N., Simin, N., & Anackov, G. (2006). Characterization of the volatile composition of essential oils of some Lamiaceae species and antimicrobial and antioxidant activity of the entire oil. *Journal of Agriculture and Food Chemistry*, *8*, 1822–1828.

Brown, A. M., Kaplan, L. M., & Brown, M. E. (1960). Phenol induced histological skin changes: Hazards, technique and uses. *British Journal of Plastic Surgery*, *13*, 158–169.

Bruenkin, W. (2005). *Rooibos, Aspalanthus linearis*. In *Wikimedia commons*. Online available: https://commons.wikimedia.org/wiki/File:Rooibos_(Aspalathus_linearis).jpg.

Carruthers, A., Kiene, K., & Carruthers, J. (1996). Botulinum A exotoxin use in clinical dermatology. *Journal of the American Academy of Dermatology*, *34*, 788–797.

Caruso, M. K., Roberts, A. T., Bissoon, L., Self, K. S., Guillot, T. S., & Greenway, F. L. (2008). An evolution of mesotherapy solutions for inducing lipolysis and treating cellulite. *Journal of Plastic Reconstructive Aesthetic Surgery*, *61*, 1321–1324.

Cheng, C. M. (2007). Cosmetic use of the botulinum toxin type A in the elderly. *Clinical Interventions in Ageing*, *2*, 81–83.

Choi, H. K., Kim, D. H., Kim, J. W., Ngadiran, S., Sarmidi, M. R., & Park, C. S. (2010). *Labisia pumila* extract protects skin cells from photoageing caused by ultraviolet radiation. *Journal of Bioscience and Bioengineering*, *109*, 291–296.

Chua, L. S., Lee, S. V., Abdullah, N., & Sarmidi, M. R. (2012). Review on *Labisia pumila* (Kacip Fatima) bioactive phytochemicals and skin collagen synthesis promoting herb. *Fitoterapia*, *83*, 1322–1335.

Chung, H. W., Kim, H. J., Kim, J. W., & Yu, B. P. (2001). The inflammation hypothesis of ageing – molecular modulation by calorie restriction. *Annuals of the New York Academy of Science*, *928*, 327–335.

Cossins, E., Lee, R., & Pacher, L. (1998). Pycnogenol prolongs the lifetime of vitamin C more than other flavonoids. *Biochemistry Molecular Biology International*, *45*, 583–597.

Cronin, H., & Draelos, Z. D. (2010). Top 10 botanical ingredients in 2010 anti-aging creams. *Journal of Cosmetic Dermatology*, *9*, 218.

Daamen, W. F., Verkamp, J. H., Van-Hest, J. C. M., & Kuppevelt, T. H. (2007). Elastin as a biomaterial for tissue engineering. *Biomaterials*, *28*, 4378–4398.

Dessy, L. A., Mazzochi, M., Rubino, C., Mazzarello, V., Spissu, N., & Scuderi, N. (2007). An objective assessment of *Botulinum* toxin A effect on superficial skin texture. *Annuals of Plastic Surgery*, *54*, 469–473.

Draelos, Z. D. (2007). The latest cosmeceutical approaches for anti-ageing. *Journal of Cosmetic Dermatology*, *6*, 2–6.

Dudonne, S., Poupard, P., Coutiere, P., Woillez, M., Richard, T., Merillon, J.-M., et al. (2011). Phenolic composition and antioxidant properties of poplar bud (*Populus nigra*) extract: Individual antioxidant contribution of phenolics and transcriptional effect of skin ageing. *Journal of Agriculture and Food Chemistry*, *59*, 4527–4536.

Elmore, A. R. (2005). Final report of the safety assessment of L-ascorbic acid, calcium ascorbate, magnesium ascorbate, magnesium ascorbyl phosphate, sodium ascorbate, and sodium ascorbyl phosphate as used in cosmetics. *International Journal of Toxicology*, *24*, 51–111.

Fischer, C. (2008). *Arctium lappa* (Asteraceae). In *C. C. Wikimedia commons*. Online available: https://commons.wikimedia.org/wiki/File:ArctiumLappa2.jpg.

Fischer, C. (2013). *Populus nigra*. In *Wikimedia commons*. Online available: https://commons.wikimedia.org/wiki/Populus_nigra#/media/File:PopulusNigra2a.jpg.

Fischer, T. C., Perosino, E., Poli, F., Viera, M. S., Dreno, B., & Cosmetic Dermatology European Group (2009). Chemical peels in aesthetic dermatology: And update. *Journal of the European Academy of Dermatology and Venereology*, *24*, 281–292.

Fisher, G. J., Datta, S., Wang, Z., Li, W. Y., Quan, T., Chung, J. H., et al. (2000). c-Jun-dependent inhibition of cutaneous procollagen transcription following ultraviolet irradiation is reversed by all-trans retinoic acid. *Journal of Clinical Investigation*, *106*, 663–670.

Fisher, G. J., Kang, S., & Varanti, J. (2002). Mechanisms of photoageing and chronological skin ageing. *Archives of Dermatology*, *138*, 1462–1470.

Fisher, G. J., Talwar, H. S., Lin, J., Lin, P., McPhillips, P., Wang, Z., et al. (1998). Retinoic acid inhibits indiction of c-Jun protein by ultraviolet radiation that occurs subsequent to activation of mitogen-activated protein kinase pathways in human skin in vivo. *Journal of Clinical Investigation*, *101*, 1432–1440.

Fitzpatrick, R. E., & Rostan, E. F. (2002). Double-blind, half face value comparing topical vitamin C and vehicle for rejuvenation of photodamage. *American Society for Dermatologica Surgery*, *28*, 231–236.

Forest, F., & Starr, K. (2008). *Curcuma longa*. In *Wikimedia commons*. Online available: https://commons.wikimedia.org/wiki/File:Starr_080103-1181_Curcuma_longa.jpg.

Franceschi, C., Bonafe, M., Valensin, S., Bonafé, M., Paolisso, G., Yashin, A. I., et al. (2000b). The network and remodeling theories of ageing: Historical background and new perspectives. *Experimental Gerontology*, *35*, 879–896.

Franceschi, C., Bonafe, M., Valensin, S., Olivieri, F., De Luca, M., Ottaviani, E., et al. (2000a). Inflamm-ageing: An evolutionary perspective on immunosenescence. *Annuals of the New York Academy of Science*, *908*, 244–254.

Fulop, T., Khalil, A., & Larbi, A. (2012). The role of elastin peptides in modulating the immune response in ageing and age-related diseases. *Pathologie Biologie*, *60*, 28–33.

Ganceviciene, R., Aikaterini, I. L., Athanasios, T., Makrantonaki, E., & Zouboulis, C. C. (2012). Skin anti-ageing strategies. *Dermato-Endocrinology*, *4*, 308–319.

Garcia, A., & Fulton, J. E., Jr. (1996). Cosmetic denervation of the muscles of facial expression with botulinum toxin. A dose dependant study. *Dermatologic Surgery*, *22*, 39–43.

Gediya, S. K., Mistry, R. B., Patel, U. K., Blessy, M., & Jain, H. N. (2011). Herbal plants: Used as cosmetics. *Journal of Natural Product and Plant Resources*, *1*, 24–32.

Gilchrest, B. A. (1990). Skin ageing and photoageing. *Dermatology Nursing*, *2*, 79–82.

Gold, M. H. (2007). Use of hyaluronic acid fillers for the treatment of the ageing face. *Clinical Interventions in Ageing*, *2*, 369–376.

Grimes, P. E. (2005). Microdermabrasion. *The American Society for Dermatologic Surgery*, *31*, 1160–1165.

Haldane, J. B. C. (1941). *New paths in genetics*. London: Allen & Unwin.

Harman, D. (1972). The biologic clock: The mitochondria? *Journal of the American Geriatrics Society*, *20*, 145–147.

Harman, D. (2003). The free radical theory of ageing. *Antioxidants & Redox Signalling*, *5*, 557–561.

Helfrich, Y. R., Sachs, D. L., & Voorhees, J. J. (2008). Overview of skin ageing and photoageing. *Dermatology Nursing*, *20*, 177–184.

Holliday, R. (2006). Aging is no longer an unsolved problem in biology. *Annuals of the New York Academy of Science*, *1067*, 1–9.

Ibdah, M., Krins, A., Seidlitz, H. W., Strack, D., & Vogt, T. (2002). Spectral dependence of flavonol and betacyanin accumulation in *Mesembryanthemum crystillinum* under enhanced UV irradiation. *Plant Cell Environment*, *25*, 1145.

Im, A.-R., Song, J. H., Lee, M. Y., Yeon, S. H., Um, K. A., & Chae, S. (2014). Anti-wrinkle effects of fermented and non-fermented *Cyclopia intermedia* in hairless mice. *BMC Complementary and Alternative Medicine*, *14*, 424.

Iorizzo, M. D. E., Padova, M. P., & Tosti, A. (2008). Biorejuvination: Theory versus practice. *Clinical Dermatology*, *36*, 177–181.

Jabs, H.-U. (2012). Elastase- the target of a novel anti ageing strategy to defy skin ageing, loss of skin elasticity and wrinkle formation. *Ästhetische Dermatologie*, *6*, 38–40.

Joubert, E., Gelderblom, W. C. A., Louw, A., & Beer, B. (2008). South African herbal teas: *Aspalathus linearis,* Cyclopia spp. and *Athrixia phylicoides* – a review. *Journal of Ethnopharmacology*, *119*, 376–412.

Jung, E., Lee, J., Baek, J., Jung, K., Lee, J., Huh, S., et al. (2007). Effect of *Camellia japonica* on human type-I procollagen production and skin barrier function. *Journal of Ethnopharmacology*, *112*, 127–131.

Jung, H.-Y., Shin, J.-C., Park, S.-M., Kim, N.-P., Kwak, W., & Choi, B.-H. (2014). *Pinus densiflora* extract protects human fibroblasts against UVB-induced photoageing by inhibiting the expression of MMPs and increasing typre-I procollagen expression. *Toxicology Reports*, *1*, 658–665.

Kafi, R., Kwak, H. S., Schumacher, W. R. E., Cho, S., Hanfit, V. N., Hamilton, T. A., et al. (2007). Improvement of naturally aged skin with vitamin A (retinol). *Archives of Dermatology*, *143*, 606–612.

Katsube, T., Yamasaki, M., Shiwaku, K., Ishijima, T., Matsumoto, I., Abe, K., et al. (2010). Effect of flavonol glycoside in mulberry *(Morus alba* L.) leaf on glucose metabolism and oxidative stress in liver in diet induced obese mice. *Science of Food and Agriculture*, *90*, 2386–2392.

Kenpai, I. (2007). *Illicium anisatum*. In *Wikimedia commons*. Online available: https://commons.wikimedia.org/wiki/File:Illicium_anisatum3.jpg.

Kerscher, M., & Buntrock, H. (2011). Anti-ageing creams. What really helps? *Der Hautarzt*, *62*, 607–613.

Kim, J. Y., Kim, S. S., Oh, T. H., Baik, J. S., Song, G., Lee, N. H., et al. (2009). Chemical composition, antioxidant, anti-elastase, and anti-inflammatory activities of *Illicium anisatum* essential oil. *Acta Pharmaceutica*, *59*, 289–300.

Kim, Y. S., Noh, Y. K., Lee, G. I., Kim, Y. K., Lee, K. S., & Min, K. R. (1995). Inhibitory effects of herbal medicines on hyaluronidase activity. *Korean Journal of Pharmacognosy*, *26*, 265–272.

Kligman, L. H., & Kligman, A. M. (1986). The nature of photoageing: It's prevention and repair. *Photodermatology*, *3*, 215–227.

Knott, A., Reuschlein, K., Mielke, H., Wensorra, U., Mummert, C., Koop, U., et al. (2008). Natural *Arctium lappa* fruit extract improves the clinical signs of ageing skin. *Journal of Cosmetic Dermatology*, *7*, 281–289.

Kohl, E., Steinbauer, J., Landthaler, M., & Szeimies, R. M. (2011). Skin ageing. *Journal of the European Academy of Dermatology and Venerology*, *25*, 873–884.

Kojian, R. (2011). *Glycyrrhiza glabra*. In *Wikimedia commons*. Online available: https://commons.wikimedia.org/wiki/Glycyrrhiza_glabra#/media/File:Gardenology.org-IMG_2804_rbgs11jan.jpg.

Kowald, A., & Kirkwood, T. B. (1996). A network theory of ageing: The interactions of defective mitochondria, aberrant proteins, free radicals and scavengers in the ageing process. *Mutation Research*, *316*, 209–236.

Kuno, N., & Matsumoto, M. (2004). *Skin beautifying agent, antiageing agent for skin, whitening agent and external agent for skin. US Patent No. 6682763 B2.*

Lee, K. K., Kim, J.-H., Cho, J.-J., & Choi, J.-D. (1999). Inhibitory effects of 150 plant extracts and their anti-inflammatory effects. *International Journal of Cosmetic Science*, *21*, 71–82.

Lübbe, L. (2000). Evidence-based corneotherapy. *Dermatology*, *2000*, 285–289.

Maity, N., Nema, N. K., Abedy, K., Sarkar, B. K., & Mukherjee, P. K. (2011). Exploring *Tagetes erecta* Linn flower for the elastase, hyaluronidase and MMP-1 inhibitory activity. *Journal of Ethnopharmacology*, *11*, 1300–1305.

Makrantonak, E., & Zouboulis, C. C. (2007). Characteristics and patho-mechanisms of endogenously aged skin. *Dermatology*, *214*, 352–360.

Margelin, D., Medaisko, C., Lombard, D., Picard, J., & Fourtanier, A. (1996). Hyaluronic acid and dermatan sulphate are selectively stimulated by retinoic acid in irradiated and nonirradiated hairless mouse skin. *Journal of Investigative Dermatology*, *106*, 505–509.

Martin, J. (2008). *Lycium barbarum*. In *Wikimedia commons*. Online available: https://commons.wikimedia.org/wiki/Lycium_barbarum#/media/File:Lycium_barbarum_Flower_Closeup_Miguelturra_CampodeCalatrava.jpg.

Medvedev, Z. A. (1990). An attempt at rational classification of the theories of ageing. *Biological Reviews*, *65*, 375–398.

Miquel, J., Economos, A. C., Flemming, J., & Johnson, J. E., Jr. (1980). Mitochondrial role in cell ageing. *Experimental Gerontology*, *15*, 575–591.

Monheit, G. D., & Chastain, M. A. (2001). Chemical peels. *Facial Plastic Surgery Clinics of North America*, *9*, 239–255.

Moon, H.-I., Lee, J., & Chung, J. H. (2006). The effect of erythrodiol-3-acetate on the expression of MMP-1 and type-I procollagen caused by UV irradiated cultured primary old aged human skin fibroblasts. *Phytomedicine*, *13*, 707–711.

Mukherjee, P. K., Maity, N., Nema, N. K., & Sarkar, B. K. (2011). Bioactive compounds from natural resources against skin ageing. *Phytomedicine*, *19*, 64–73.

Murray, J. C., Burch, J. A., Streilein, R. D., Iannacchione, M. A., Hall, R. P., & Pinnell, S. R. (2008). A topical antioxidant solution containing vitamins C and E stabilized by ferulic acid provides protection for human skin against damage caused by ultraviolet irradiation. *The Journal of the American Academy of Dermatology*, *59*, 418–425.

Nardi, G. M., Januario, A. G. F., Freire, C. G., Megiolaro, F., Schneider, K., Perazzoli, M. R. A., et al. (2016). Anti-inflammatory activity of berry fruits in mice model of inflammation based oxidative stress modulation. *Pharmacognosy Research*, *8*, 42–49.

Nar, H., Werle, K., Bauer, N. M. T., Dollinger, H., & Jung, B. (2001). Crystal structure of human macrophage elastase (MMP-12), in conjunction with a hydroxamic acid inhibitor. *Journal of Molecular Biology*, *312*, 743–751.

Ndlovu, G., Fouche, G., Tselyane, M., Cordier, W., & Steenkamp, V. (2013). In vitro determination of the anti-aging potential of four Southern African medicinal plants. *BMC Complementary and Alternative Medicine*, *13*, 304–310.

Nusgens, B. V., Humbert, P., Rougier, A., Colige, A. C., Haftek, M., Lambert, C. A., et al. (2001). Topically applied vitamin C enhances the mRNAlevel of collagens I and III, their processing enzymes and tissue inhibitor of matrix metalloproteinase I in the human dermis. *Journal of Investigative Dermatology*, *116*, 853–859.

Oh, H. I., Shim, J. S., Gwon, S. H., Kwon, H. J., & Hwang, J. K. (2009). The effect of xanthorrhizol on the expression of matrix metalloproteinase-1 and type-I procollagen in ultraviolet-irradiated human skin fibroblasts. *Phytotherapy Research*, *23*, 1299–1302.

Ortholand, J., & Ganesan, A. (2004). Natural products and combinatorial chemistry: Back to the future. *Current Opinion in Chemical Biology*, *8*, 271–280.

Pandel, R., Poljšak, B., Godic, A., & Dahmane, R. (2013). Skin photoageing and the role of antioxidants in it's prevention. *ISRN Dermatology*, *2013*, 1–11.

Park, H. M., Moon, E., Kim, A. J., et al. (2010). Extract of *Punica granatum* inhibits skin photoaging induced by UVB irradiation. *International Journal of Dermatology*, *49*, 276–282.

Pedrosa, T. D., Barros, A. O., Nogueira, J. R., Fruet, A. C., Rodrigues, I. C., Calcagno, D. Q., et al. (2016). Anti-wrinkle and anti-whitening effects of jucá (*Libidibia ferrea* Mart.) extracts. *Archives in Dermatological Research*, *308*, 643–654.

Philips, N., Cante, J., Chen, Y. J., Natrajan, P., Tow, M., Keller, T., et al. (2009). Beneficial regulation of matrix metalloproteinases and their inhibitors, fibrillary collagens and transforming growth factor β by *Polypodium leucotomos*, directly or in dermal fibroblasts, UV radiated fibroblasts and melanoma cells. *Archives of Dermatological Research*, *301*, 487–495.

Porse, S. (2006). *Lavandula angustifolia*. In *Wikimedia commons*. Online available: https://commons.wikimedia.org/wiki/Lavandula_angustifolia#/media/File:Lavandula-angustifolia-flowering.JPG.

Prado, F., Cortes-Ledesma, F., Huertas, P., & Aguilera, A. (2003). Mitotic recombination in *Saccharomyces cerevisiae*. *Current Genetics*, *42*, 185–189.

Rendon, M. I., Berson, D. S., Cohen, J. L., Roberts, W. E., Starker, I., & Wang, B. (2010). Evidence and considerations in the application of chemical peels in skin disorders and aesthetic resurfacing. *Journal of Clinical Aesthetic Dermatology*, *3*, 32–43.

Rittié, L., & Fisher, G. J. (2002). UV-light induced signal cascades and skin ageing. *Research Reviews*, *1*, 705–720.

Rona, C., Vailati, F., & Berardesca, E. (2004). The cosmetic treatment of wrinkles. *Journal of Cosmetic Dermatology*, *3*, 26–34.

Safayhi, H., Rall, B., Sailer, E.-R., & Ammon, H. P. T. (1997). Inhibition by boswellic acids of human leukocyte elastase. *The Journal of Pharmacology and Experimental Therapeutics*, *281*, 460–463.

Sahu, R. K., Roy, A., Matlam, M., Deshmukh, V. K., Dwivedi, J., & Jha, A. K. (2013). Review on skin ageing compilation of scientific validated medicinal plants, prominence to flourish a better research reconnoiters in herbal cosmetics. *Research Journal of Medicinal Plants*, *7*, 1–22.

Saito, M., Tanaka, M., Misawa, E., Yao, R., Nabeshima, K., Yamauchi, K., et al. (2016). Oral administration of Aloe vera gel powder prevents UVB induced decrease in skin elasticity via suppression of overexpression of MMPs in hairless mice. *Bioscience, Biotechnology, and Biochemistry*, *80*.

Sander, C. S., Chang, H., Salzmann, S., Muller, C. S., Ekanayake-Mudiyanselage, S., Elsner, P., et al. (2002). Photoageing is associated with protein oxidation in human skin in vivo. *Journal of Investigative Dermatology*, *118*, 618–625.

Sarhadynejad, Z., Sharififar, K., Pardakhty, A., Nemtollahi, M. H., Sattaie-Mokhtari, S., & Mandegary, A. (2016). Pharmacological safety evaluation of traditional herbal medicine "Zereshk-e-Saghir" and assessment of its hepatoprotective effects on carbon tetrachloride induced hepatic damage in rats. *Journal of Ethnopharmacology*, *22*, 387–395.

Schonlau, F. (2002). The cosmeceutical Pycnogenol. *Journal of Applied Cosmetics*, *20*, 38–45.

Shacter, E. (2000). Protein oxidative damage. *Methods in Enzymology*, *319*, 428–436.

Shakeri, A., Sahebkar, A., & Javadi, B. (2016). *Melissa offocinalis* L.- A review of its traditional uses, phytochemistry and pharmacology. *Journal of Ethnopharmacology*, *188*, 204–228.

Sherman, L., Sleeman, J., Herrlich, P., & Ponta, H. (1994). Hyaluronidase receptors: Key players in growth, differentiation and tumour progression. *Current Opinions in Cellular Biology*, *6*, 726–733.

Sigma-Aldrich. (2015). *Human neutrophil elastase*. Online available: http://www.sigmaaldrich.com/south-africa.html.

Statista. (2016). Online available: www.statista.com.

Storch, H. (2007). *Mentha x piperita*. In *004.Jpg, C. C. Wikimedia commons*. Online available: https://commons.wikimedia.org/wiki/File:Pfefferminze_Mentha_%C3%97_piperita_IMG_6544.JPG.

Sudel, K. M., Venzke, K., Mielke, H., Breitenbach, U., Mundt, C., Jaspers, S., et al. (2005). Novel aspects of intrinsic and extrinsic ageing of human skin; beneficial effects of soy extract. *Phytochemistry and Photobiology*, *81*, 581–587.

Tabata, N., O'Goshi, K., Zhen, Y. X., Kligman, A. M., & Tagami, H. (2000). Biophysical assessment of persistent effects of moisturizers after their daily application; evaluation of corneotherapy. *Dermatology*, *200*, 308–313.

Taher, Y. A., Samud, A. M., Mezogi, J. S., & Hamza, A. M. (2016). CNS depressant activity of Chamomile flower methanol extract in mice. *Pakistan Journal of Pharmaceutical Research*, *2*, 98102.

Thring, T. S. A., Hili, P., & Naughton, D. P. (2009). Anti-collagenase, anti-elastase and anti-oxidant activities of extracts from 21 plants. *BMC Complemenary and Alternative Medicine*, *19*, 27.

Tobassum, N., & Hamdani, M. (2014). Plants used to treat skin diseases. *Pharmacognosy Review*, *8*, 52–60.

Tosato, M., Zamboni, V., Ferrini, A., & Cesari, M. (2007). The ageing process and potential interventions to extend life expectancy. *Clinical Interventions in Ageing*, *2*, 401–412.

Transparency Market Research. (2016). Online available: www.transparencymarketresearch/press-release/anti-aging-market.htm.

Trautinger, F. (2001). Mechanisms of photodamage of the skin and its functional consequences for skin ageing. *Clinical and Experimental Dermatology*, *26*, 573–577.

Ugulu, I. (2012). Fidelity level and knowledge of medicinal plants used to make therapeutic Turkish baths. *Ethnomedicine*, *6*, 1–9.

Vagotis, F. L., & Brundage, S. R. (1995). Histologic study of dermabrasion and chemical peel in an animal model after pre-treatment with Retin-A. *Aesthetic Plastic Surgery*, *19*, 243–246.

Valenzuela, A., Sanhuexa, J., & Nieto, S. (2003). Natural antioxidants in foods: From food safety to health benefits. *Grassas y Aceties*, *54*, 298–303.

Varani, J., Warner, R. L., Gharaee-Kermani, M., Phan, S. H., Kang, S., & Chung, J. H. (2000). Vitamin A antagonisers decreased cell growth and elevated collagen-degrading matrix metalloproteinases and stimulates collagen accumulation in naturally aged human skin. *Journal of Investigative Dermatology*, *114*, 480–486.

Vedamurthy, M. (2004). Salicylic acid peels. *Indian Journal of Dermatology, Venereology and Leprology*, *70*, 136–138.

Vedamurthy, M. (2006). Anti-aging therapies. *Indian Journal of Dermatology, Venereology and Leprology*, *72*, 183–186.

Visavadiya, N. P., Soni, B., & Dalwadi, N. (2009). Evaluation of antioxidant and anti-atherogenic properties of *Glycyrrhiza glabra* root using *in vitro* models. *International Journal of Food Sciences and Nutrition*, *60*, 135–149.

Weinert, B. T., & Timiras, P. S. (2003). Theories of ageing. *Journal of Applied Physiology*, *95*, 1706–1716.

Wen, K. C., Fan, P.-C., Tsai, S.-Y., Shih, I.-C., & Chiang, H.-M. (2012). *Ixora parviflora* protects against UVB-induced photoageing by inhibiting the expression of MMPs, MAP kinases and COX-2 and by promoting type-I procollagen synthesis. *Evidence Based Complementary and Alternative Medicine*, *2012*, 1–11.

Wiedow, O., Shroder, J. M., Gregory, H., Young, J. A., & Christophers, E. (1990). Elafin: An elastase specific inhibitor of human skin. Purification, characterization and complete amino-acid sequence. *Journal of Biological Chemistry*, *15* 2665–2671.

Xaver, F. (2011). *Areca catechu*. In *Wikimedia commons*. Online available: https://commons.wikimedia.org/wiki/File:Areca_catechu_1.jpg.

Xu, Y., Shao, Y., Voorhees, J. J., & Fisher, G. J. (2006). Oxidative inhibition of receptor-type protein-tyrosine phosphatase kappa by ultraviolet irradiation activates epidermal growth factor receptor in human keratinocytes. *Journal of Biological Chemistry*, *281*, 27389–27397.

Yaar, M., Eller, M. S., & Gilchrest, B. A. (2002a). Fifty years of skin ageing. *Journal of Investigative Dermatology Symposium Proceedings*, *7*, 51–58.

Yaar, M., & Gilchrest, B. A. (2007). Photoageing: Mechanism, prevention and therapy. *British Journal of Dermatology*, *157*, 874–887.

Yaar, M., Lee, M. S., Runger, T. M., Eller, M. S., & Gilchrest, B. (2002b). Telomere mimetic oligonucleotides protect skin cells from oxidative damage. *Annales de Dermatologie et de Venereologie*, *129*, 1–18.

Yan, Z. Y., Guon, G. X., & Lin, Z. F. (1994). Elastolytic activity from *Flavobacterium odoratum*. Microbial screening and cultivation, enzyme production and purification. *Process Biochemistry*, *29*, 427–436.

Young, C. N., Koepke, J. I., Terlecky, L. J., Borkin, M. S., Boyd, M. S., & Terlecky, S. R. (2008). Reactive oxygen species in tumour necrosis factor α-activated primary human keratinocytes: Implications for psoriasis and inflammatory skin disease. *Journal of Investigative Dermatology*, *128*, 2606–2614.

Yuan, Y. V., & Walsh, N. A. (2006). Antioxidant and antiproliferative activities of extracts from a variety of edible seaweeds. *Food and Chemical Toxicology*, *44*, 1144–1150.

Zaid, M. A., Afaq, F., Syed, D. N., Dreher, M., & Mukhtar, H. (2007). Inhibition of UVB-mediated oxidative stress and markers of photoaging in immortalized HaCaT keratinocytes by pomegranate polyphenol extract POMx. *Photochemistry and Photobiology*, *83*, 882–888.

Zell, H. (2006). *Vitis vinifera*. In *Wikimedia commons. C. C.* Online available: https://commons.wikimedia.org/wiki/Vitis_vinifera#/media/File:Vitis_vinifera_001.jpg.

Zell, H. (2009a). *Aloe vera* (L.) Burm. F. In *004.Jpg, C. C. Wikimedia commons.* Online available https://commons.wikimedia.org/wiki/File:Aloe_vera_Lanzarote.jpg.

Zell, H. (2009b). *Dioscorea villosa*. In *C. C. Wikimedia commons*. Online available: https://commons.wikimedia.org/wiki/File:Dioscorea_villosa_001.JPG.

Zhai, H., Beham, S., Villarama, C. D., Arens-Corell, M., Choi, M. J., & Maibach, H. I. (2005). Evaluation of the antioxidant capacity and preventative effects of a topical emulsion and its vehicle control on the skin in response to UV exposure. *Skin Pharmacology and Physiology*, *18*, 288–293.

Zouboulis, C. C., & Boschnakow, A. (2001). Chronological ageing and photoageing of the human sebaceous gland. *The Journal of Clinical and Experimental Dermatology*, *26*, 600–607.

FURTHER READING

Ferreira, M. R. A., & Soares, L. A. L. (2015). *Libidibia ferrea* (Mart. Ex Tul) L.P. Queriroz: A review of the biological activities and phytochemical composition. *Journal of Medicinal Plants Research*, *9*, 140–150.

Kim, Y., Uyama, H., & Kobayashi, S. (2005). Inhibition effects of (+) catechin aldehyde polycondensates on proteinases causing proteolytic degradation of extracellular matrix. *Biochemical and Biophysical Research Communities*, *320*, 256–261.

Lourens, A. C. U., Viljoen, A. M., & van Heerden, F. R. (2008). South African *Helichrysum* species: A review of the traditional uses, biological activity and phytochemistry. *Journal of Ethnopharmacology*, *119*, 630–650.

Martin, J. (2011). *Mesembryanthemum crystillinum*. Online available https://commons.wikimedia.org/wiki/File:Mesembryanthemum_crystallinum_Enfoque_2011-7-03_LagunadelaMata.jpg.

Popoola, O. K., Marnewick, J. L., Rautenbach, F., Iwuoha, E. I., & Hussein, A. A. (2015). Acylphoroglucinol derivatives from the South African *Helichrysum niveum* and their biological activities. *Molecules*, *20*, 17309–17324.

Shielde, B. (2003). The effect of sesquiterpene lactones on the release of human neutrophil elastase. *Biochemical Pharmacology*, *65*, 897–903.

United States Department of Agriculture. (2005). *Citrus fruits. Agricultural research service*. Online available https://en.wikipedia.org/wiki/Vitamin_C#/media/File:Ambersweet_oranges.jpg.

Viegas, D. A., Palmeira-de-Oliveira, A., & Salgueiro, L. (2014). *Helichrysum italicum*: From traditional use to scientific data. *Journal of Ethnopharmacology*, *151*, 54–65.

Chapter 4

Exploiting Medicinal Plants as Possible Treatments for Acne Vulgaris

Isa Anina Lambrechts, Marco Nuno de Canha, Namrita Lall
University of Pretoria, Pretoria, South Africa

Chapter Outline

Medicinal Plants for Holistic Health and Well-Being. http://dx.doi.org/10.1016/B978-0-12-812475-8.00004-4

4.1 ACNE VULGARIS

Acne vulgaris is one of the most common chronic inflammatory skin diseases affecting around 85% of the general population at some point throughout their lives (Fig. 4.1A). It is also among the top three skin disorders treated by dermatologists. Although mortality rates associated with skin disorders are low, it is a major health concern as it has both physical and psychosocial effects, which may endure long after active lesions have been treated (Chiu, Chon, & Kimball, 2003; Truter, 2009).

Acne can vary in severity, which is determined based on lesion type, size, number, scarring, and postinflammatory hyperpigmentation. Acne severity is also governed by lesion position where, with milder acne, lesions frequently appear on the face and, with more severe cases, lesions can be found on the back, shoulders, and chest (Food and Drug Administration, 2005). The mildest form of acne, known simply as acne, comprises mostly noninflammatory lesions such as blackheads and whiteheads and between 0 and 5 papules and pustules. Moderate acne comprises both noninflammatory and inflammatory lesions, such as papules and pustules. Moderate acne is identified in patients with 6–20 papules and pustules. Severe acne that often results in scarring has not only mixed lesions but also includes nodules and cysts. Severe acne can be identified in patients with between 21 and 50 papules and pustules. In very severe cases more than 50 lesions are present (Hayashi, Akamatsu, & Kawashima, 2008; Layton, 2005; Webster, 2002).

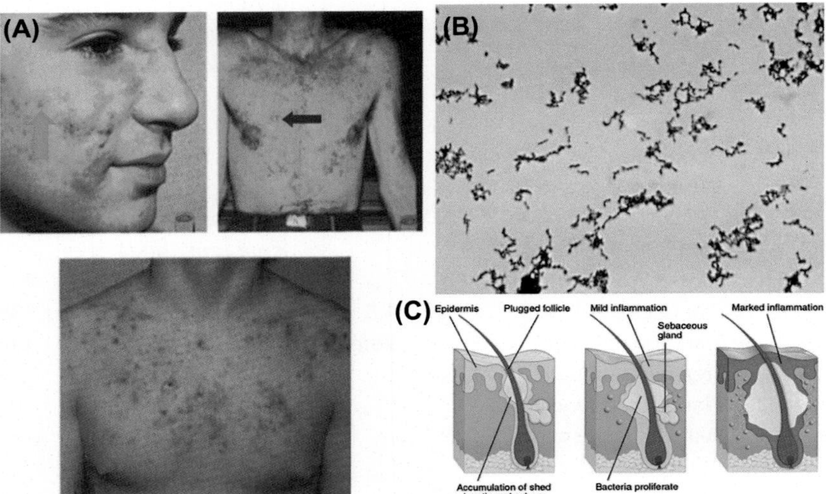

FIGURE 4.1 (A) Acne lesions of the face, trunk, neck, and shoulders (Chris73, 2005), (B) these are the result of the causative agent of acne, Gram-positive *Propionibacterium acnes*, depicted here after Gram staining (Strong, 1972), and (C) the progression of acne vulgaris that ultimately results in marked inflammation (OpenStax College, 2013).

4.1.1 *Propionibacterium acnes* and the Skin

The skin is the largest organ of the human body, acting as an ecosystem for a variety of microbiota. The skin has physical and chemical features, which select for certain sets of microorganisms that are adapted to the niche in which they live. These habitats are determined by the thickness of the skin, folds, and the density of hair follicles and glands. Skin disorders or infections often occur when there is a disruption in the delicate balance between the host and microorganism (Grice & Segre, 2011).

Propionibacterium acnes has been identified as the causative microorganism involved in the progression of acne vulgaris (Fig. 4.1B). It is a Gram-positive, aerotolerant anaerobic bacillus with the ability to produce propionic acid as a metabolic by-product. This microorganism forms part of the normal skin microflora and remains on the skin from birth until death. In 1963, *P. acnes* was injected into sterile cysts of steatocystomas (noncancerous cysts originating in the sebaceous glands), resulting in subsequent inflammation, providing evidence that this microorganism is involved in inflammatory acne (Cogen, Nizet, & Gallo 2008; Webster & Graber, 2008).

Antibiotic Treatment and *Propionibacterium acnes* Resistance

Antibiotics are a common form of treatment for acne. Macrolides (antibiotics containing a macrolide ring structure) such as erythromycin, clindamycin, and azithromycin are used for their antibacterial effect as they bind irreversibly to the 50S ribosomal subunit of the bacterium inhibiting translocation during the synthesis of proteins. The tetracycline antibiotics such as tetracycline, doxycycline, and minocycline are generally used for their ability to bind to the 30S ribosomal subunit of *P. acnes*, preventing translocation (Fig. 4.2A). These also have

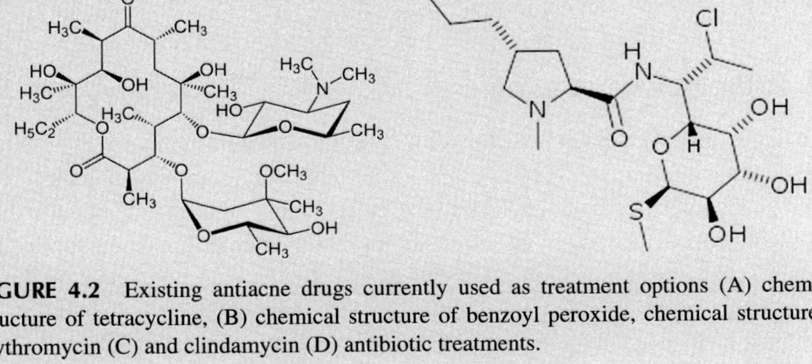

FIGURE 4.2 Existing antiacne drugs currently used as treatment options (A) chemical structure of tetracycline, (B) chemical structure of benzoyl peroxide, chemical structure of erythromycin (C) and clindamycin (D) antibiotic treatments.

Continued

—cont'd

antiinflammatory activity by inhibiting neutrophil and monocyte chemical signaling by chelating calcium preventing the accumulation of these immune cells to the follicle (Webster & Graber, 2008).

Although not classified as an antibiotic, benzoyl peroxide is used for its ability to reduce *P. acnes* numbers (Fig. 4.2B). Once applied to the skin, benzoyl peroxide is broken down into two components, benzoic acid and hydrogen peroxide. The hydrogen peroxide is responsible for the majority of the bactericidal activity and although the function of the benzoic acid is unclear, its antioxidant activity is said to be beneficial against acne (Webster & Graber, 2008).

P. acnes has developed some mechanisms of resistance to the above treatments. Among these are the point mutations in genes that encode for domain V of the peptidyl transferase loop of the 23S ribosomal ribonucleic acid (rRNA) that results in a cross-resistance phenotype that is resistance to MLS (macrolides, lincosamides, and type B streptogramins) antibiotics. Among highly resistant *P. acnes* strains, the Tn 5432 transposon that houses the erythromycin ribosome methylase (*erm*) (X) resistance gene has been found, which could account for MLS resistance.

This *erm* gene encodes for methyl transferase that is responsible for methylating the N6 position of the 23S rRNA, which is the binding site for many MLS antibiotics, preventing their ability to bind.

It is often noted that resistance to erythromycin and clindamycin are also accompanied with resistance to tetracycline (Fig. 4.2C and D). This resistance is acquired through a single G to C transition in the 16S rRNA of the small ribosomal subunit and cross-resistance to minocycline and doxycycline may also occur. It has been observed that *P. acnes* strains that are resistant to a single member of an antibacterial class are likely to show varying degrees of resistance against derivatives within that class. Bacterial biofilms aid in the pathogenesis of many human infections. Bacterial cells growing in a biofilm have high levels of resistance to antimicrobial agents due to restricted penetration of antimicrobials, decreased growth rates, expression of resistance genes, and the presence of persister cells. It has been hypothesized that *P. acnes* growing within the pilosebaceous unit grow as a biofilm. Studies have shown that some *P. acnes* strains have the ability to form biofilms on a number of biomaterials. The genome sequence of *P. acnes* also shows evidence for the ability of the bacteria to grow in a biofilm based on the presence of three gene clusters that encode for extracellular polysaccharide biosynthesis and adhesion proteins, which are key factors in the biofilm formation (Coenye, Peeters, & Nelis, 2007; Nord & Oprica, 2006).

4.1.2 Pathogenesis of Acne

Acne affects the pilosebaceous follicle, a structure comprised of the hair follicle and its associated sebaceous gland. The exact sequence and nature of these factors are not yet known, but there is a general agreement that acne vulgaris is a multifactorial disease. The etiology of acne is hypothesized to occur in four main steps, namely increased levels of sebum secretion, hypercornification (blockage and hardening) of the follicular duct, proliferation of *P. acnes*, and inflammation (Fig. 4.1C) (Falcocchio, Ruiz, Javier Pastor, Saso, & Diaz, 2006).

4.1.3 Androgens and Sebum Production

There are a number of compounds that can be found in human sebum including cholesterol, fatty acids, squalene, triglycerides, cholesteryl esters, diglycerides, and wax esters. Sebum production is considered the first step in the pathogenesis of acne vulgaris.

Increased sebum production creates an ideal environment for sustaining the colonization of the acne-causing bacterium *P. acnes*. Androgens are hormones that are associated with the pathogenesis of acne vulgaris and play an essential role in promoting the growth of the sebaceous glands, sebum production, and hyperkeratinization. There are three major enzymes involved in the production of androgen hormones and they include 3β-hydroxysteroid dehydrogenase, 17β-hydroxysteroid dehydrogenase (17β-HSD) type 2, and 5α-reductase type 1. These enzymes are involved in the production of acne-causing hormones such as testosterone and 5α-dihydrotestosterone (DHT) from adrenal androgen dehydroepiandrosterone that does not cause acne. The net activity of 17β-HSD type 2 and 5α-reductase type 1 can lead to the overproduction of acne-causing androgens, testosterone, and DHT within the sebaceous glands of areas affected by acne. These enzymes have the ability to stimulate sebaceous secretion, where the hormone testosterone is irreversibly metabolized into DHT by the enzyme 5α-reductase type 1 (Fig. 4.3). The enzyme 5α-reductase can be divided into two isoenzymes, type 1 and type 2 that differ in their distribution in the areas on the skin. Type 1 can mostly be found in the sebaceous glands of the scalp and

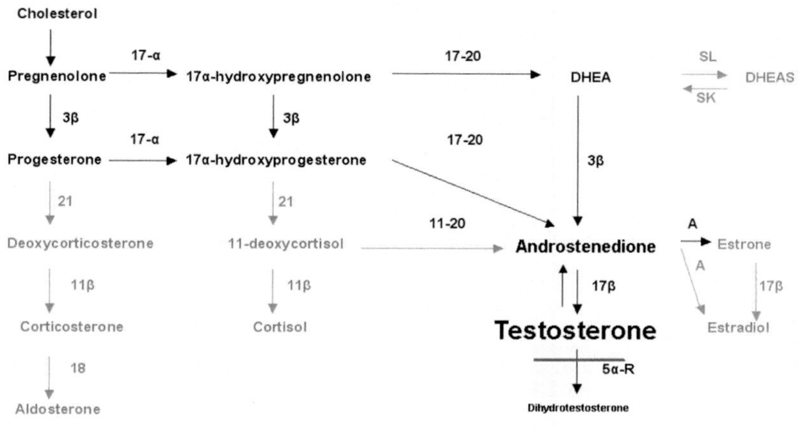

FIGURE 4.3 The biosynthesis of testosterone and the reaction catalyzed by 5α-reductase (Broere, 2007).

face, while type 2 is found predominantly in zones not affected by acne (Arora, Yadav, & Saini, 2011; Pawin et al., 2004).

Although the exact mechanism by which androgens increase sebaceous gland size and secretion is not yet identified, it is believed that testosterone and DHT bind to nuclear androgen receptors and form complexes. In the nuclei of sebaceous cells, the androgen-receptor complexes interact with DNA to regulate genes that are involved in lipid production and cell growth (Arora et al., 2011).

Nuclear receptors are involved in maturing the sebaceous gland and contribute to the initiation of inflammation associated with acne vulgaris. Peroxisome proliferator-activated receptors (PPAR) α, β, and γ and retinoid X receptors are among the most essential receptors for sebum production as they regulate sebocyte differentiation and epidermal growth. It has been suggested that PPAR found in the sebaceous glands could provide a new target for the treatment of acne (Arora et al., 2011; Pawin et al., 2004).

4.1.4 Hyperkeratinization

Follicular hyperkeratinization (abnormally rapid shedding of skin cells) in the sebaceous gland and follicular infundibulum (uppermost section of the hair follicle, near the opening of the pores) can be considered one of the crucial events in the development of acne lesions. It is believed to be the second step in the pathogenesis of acne vulgaris. There are several terms for hyperkeratinization that include ductal hypercornification and retention hyperkeratosis. Retention hyperkeratosis occurs when keratinocytes or skin cells of the follicle become cohesive and don't shed normally on the surface of the skin. This abnormal shedding of the skin leads to the formation of a microcomedone under the skin 8 weeks before the acne lesion being visible on the skin surface. Ductal hypercornification can result due to two main factors; hyperproliferation of ductal keratinocytes, which is abnormal shedding of keratinocytes or a reduced separation of ductal corneocytes leading to clumping of the cells in the follicle. Patients with acne have shown altered shedding (desquamation) of the epithelium. The stratum corneum is the outer most layer of the skin and consists of keratinized cells or dead skin cells containing keratin. Desmosomes are specialized cell junctions mostly found in the epidermis that connect two outer cell membranes together with intermediate keratin filaments also known as tonofilaments (Fig. 4.4A). Keratinocytes of follicles affected by acne have more desmosomes and tonofilaments resulting in a thicker and more cohesive stratum corneum.

Lipid droplets are present on the keratinocytes and with lamellar granules being exocytosed into the intracellular spaces in the granular cell layer, there is a decrease in the number of lamellar granules, which prevents normal shedding of the cells into the lumen. The result is a keratinous plug that occludes the follicular canal and leads to the formation of a comedone (Fig. 4.4B) (Pawin et al., 2004; Thiboutot, 2000; Toyoda & Morohashi, 2001).

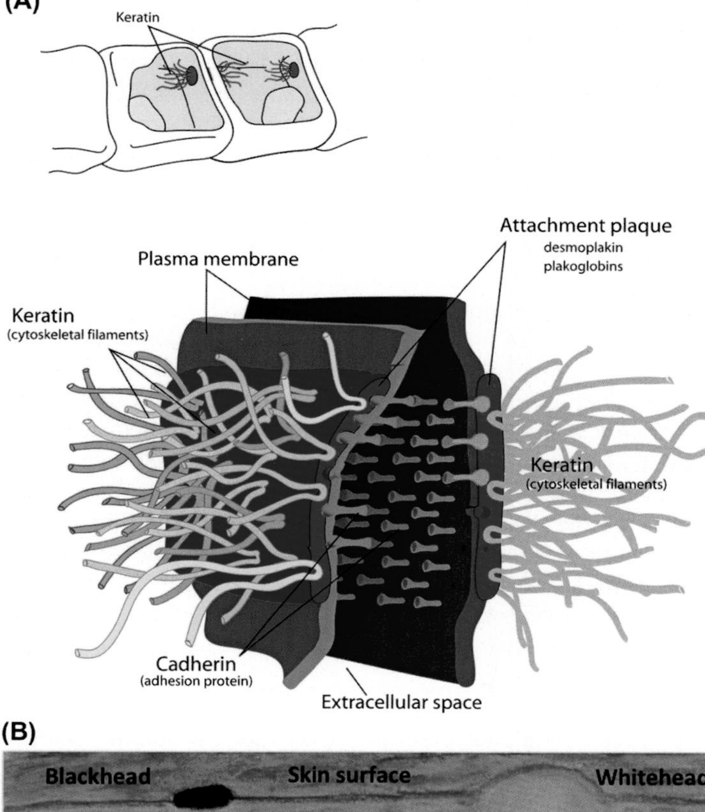

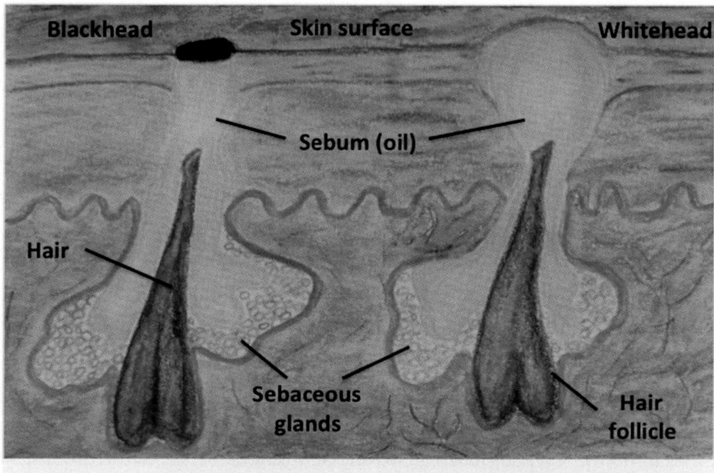

FIGURE 4.4 (A) Desmosome cell junction with tonofilaments that contribute to skin barrier function (Ruiz, 2007), and (B) blackhead and whitehead comedones (Bastian, 2013).

There are a number of factors believed to contribute to hyperkeratinization. These include a decrease in the concentration of sebaceous linoleic acid due to hyperseborrhea causing abnormal keratinocyte differentiation. Linoleic acid is an essential fatty acid and it has been found that a deficiency in animals results in scaliness. It has been hypothesized that 5α-reductase type 1 could be associated with abnormal hyperkeratinization, leading to microcomedones being formed that are associated with acne lesions. Inflammatory cytokines such as interleukin 1-α (IL-1α) are produced by keratinocytes in the epidermis and lymphocytes. Inflammation and androgen hormones have the ability to stimulate keratinocyte differentiation that is mediated by IL-1α and growth factors, leading to hyperkeratinization and comedogenesis. The acne-causing bacteria *P. acnes* has the ability to form biofilms. A biofilm is a complex structure that is a result of the aggregation of the bacteria within an extracellular polysaccharide lining after attachment to the surrounding surface. Various substances are secreted by *P. acnes* in an attempt to adhere to the follicular lining during the formation of the biofilm. The polymeric matrix provides a physical barrier for protection against the external environment and antimicrobial agents. It is hypothesized that the glycocalyx polymer secreted during biofilm formation can become incorporated in the sebum, acting as a biological glue between keratinocytes and contributing to comedone formation (Arora et al., 2011; Dessinioti & Katsambas, 2010; Toyoda & Morohashi, 2001).

4.1.5 Proliferation of *Propionibacterium acnes*

P. acnes proliferation and increases in bacterial load within the pilosebaceous unit are associated with the preceding aetiological steps in the pathogenic pathway. Sebaceous glands that are affected by hypercornification or defective keratinocyte differentiation form a microcomedone. This not only allows for the formation of an anaerobic environment but also traps sebum, creating a lipid-rich environment. *P. acnes* is therefore provided with an oxygen-deprived and nutrient-rich environment, allowing optimal growth and subsequent colonization (Williams, Dellavalle, & Garner, 2012).

4.1.6 Inflammation

Inflammation is characterized as one of the key components in acne pathogenesis. It has been shown that in inflammatory acne lesions, genes involved in the inflammatory processes are upregulated (Makrantonaki, Ganceviviene, & Zouboulis, 2011). The *P. acnes* bacterium contributes to the inflammatory process through several mechanisms. Based on the genomic sequence of the bacterium and several studies, it is known that *P. acnes* secretes lytic enzymes such as lipase, protease, and hyaluronidase (Burkhart, Burkhart, & Lehmann, 1999). The *P. acnes* lipase enzyme, glycerol-ester hydrolase A, is involved in the hydrolysis of sebum triglycerides into glycerol and free fatty acids. These free fatty acids contribute to inflammation as they act as chemotactic factors for immune cells and are highly irritating to the cells of the sebaceous follicle. These chemotactic factors act as signals for neutrophil cells. Once phagocytosed by the neutrophils, *P. acnes* cells release hydrolases that contribute to inflammation by disrupting the follicular wall of the pilosebaceous unit. Changes in lipid content due to lipase activity also affect linoleic acid levels, often decreasing its concentration. This fatty acid acts as a natural radical scavenger for reactive oxygen species (ROS) produced by neutrophils. *P. acnes* also contains genes encoding the hyaluronidase enzyme. This enzyme is involved in the degradation of hyaluronan, one of the crucial components of the extracellular matrix of connective tissues. Studies have indicated that the breakdown products of hyaluronan are commonly found at sites of inflammation and tissue injury. Matrix metalloproteinases are endopeptidase enzymes produced by keratinocytes. These enzymes contribute to inflammation by causing rupture of the pilosebaceous follicle (Bruggemann, 2010, Chap. 66; Dessinioti & Katsambas, 2010; Falcocchio et al., 2006; Portugal, Barak, Ginsburg, & Kohen, 2007; Vowels, Yang, & Leyden, 1995).

Human sebocytes are able to express cyclooxygenase isoenzymes COX-1 and COX-2 involved in inflammation. The COX-2 enzyme is responsible for the production of prostaglandins from the substrate arachidonic acid and forms part of the inflammatory response (Fig. 4.5A). This enzyme is selectively upregulated in sebaceous glands that are affected by acne (Makrantonaki et al., 2011).

Patients suffering from acne often develop an immune response to *P. acnes*. This is based on evidence of elevated levels of circulating immune complexes. These elevated levels correspond to the severity of acne observed in patients. Although the antibodies for *P. acnes* have not been fully characterized, they are most likely from the IgG class, as IgG3 has been found in higher levels in acne patients. These antibodies enhance B-cell activity. B cells can either be plasma cells or memory cells. Plasma B cells can either inactivate pathogens directly through the release of antibodies or make them more accessible to other cells of the immune system, which may result in inflammation. *P. acnes* is able to activate complement through both the classic and alternate pathways. Skin specimens of acne patients observed using immunofluorescence have indicated the presence of C3 deposits, involved in the induction of inflammation. Other studies have reported the formation of the C5a complex by *P. acnes* activation of complement (Burkhart et al., 1999; Vowels et al., 1995).

The innate immune response is also activated by *P. acnes*. This process is controlled by a set of receptors called Toll-like receptors (TLRs), which act as mechanisms of recognition for bacterial molecular structures such as lipopolysaccharides of Gram-negative bacteria and peptidoglycans in the cell wall of Gram-positive bacteria. Studies have shown that the TLR-2 receptor is responsible for the recognition of lipopeptides in Gram-positive bacteria and is responsible for the production of proinflammatory cytokines.

Studies have shown that the increased secretion of proinflammatory cytokines such as IL-1β (Interleukin-1β), IL-8 (Interleukin-8), and TNF-α (Tumor

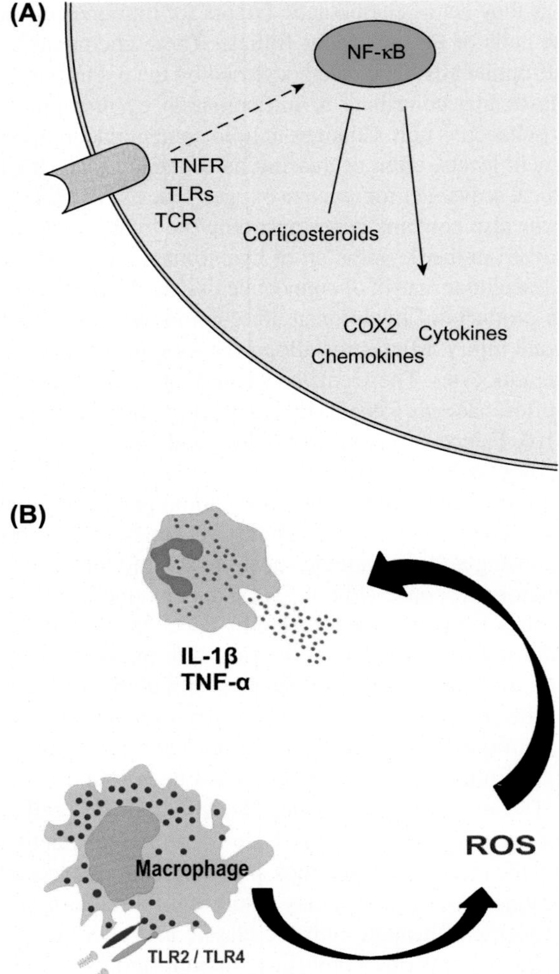

FIGURE 4.5 (A) Recognition of bacterial cell wall components by TLRs leading to cytokine release (Arne, 2012), and (B) reactive oxygen species (ROS) resulting in proinflammatory cytokine release (Campos-Rodríguez et al., 2016). *COX2*, cyclooxygenase 2; *IL-1β*, interleukin-1β; *TCR*, T-cell receptor; *TLR*, toll-like Receptor; *TNF-α*, tumor necrosis factor α; *TNFR*, tumor necrosis factor receptor.

Necrosis Factor α) by mononuclear cells (macrophages) is activated through the TLR-2 receptor pathway (Fig. 4.5A). IL-8 acts as a signal molecule for neutrophils to the site of the lesion, where they release lysosomal enzymes that ultimately lead to the rupture of the follicular epithelium and cause inflammation. IL-1β is the most studied member of the IL-1 family and has a role in mediating autoinflammatory diseases. The secretion of IL-12 is also activated through this

TLR-2 pathway in monocytes. This proinflammatory cytokine acts as an activation mechanism for the Th1 adaptive immune response. The implication of this is that the overproduction of Th1 cytokines is linked to tissue injury in some autoimmune and inflammatory diseases (Dinarello, 2009; Kim et al., 2002).

The presence of *P. acnes* cells can also enhance inflammation through the generation of ROS (Fig. 4.5B). These are short-lived small molecular structures that are produced during aerobic metabolism, often resulting in cellular damage and tissue injury. The production of ROS is related to acne pathophysiology as their production by neutrophils is enhanced with increases in acne severity as shown by ROS levels detected in crude cutaneous biopsies. *P. acnes* does not only stimulate phagocyte cells but also keratinocytes to produce ROS. The superoxide anion (O_2^-) is produced in keratinocytes in the presence of *P. acnes*, however, when the natural antioxidant mechanism of cells becomes overwhelmed, there is an overproduction of O_2^- and NO (nitric oxide). The NO produced can act as a cell signaling molecule ultimately resulting in inflammation (Grange, Weill, Dupin, & Batteux, 2010; Portugal et al., 2007).

4.1.7 Acne Vulgaris Prevalence Worldwide and in South Africa

According to the Global Burden of Disease Project, acne vulgaris is ranked as the eighth most prevalent disease worldwide, affecting ~9.4% of the world's entire population (Hay et al., 2014; Tan & Bhate, 2015). A number of population prevalence studies have also shown that the peak prevalence in terms of age is 16–20 years. Gender differences also contribute to prevalence where acne is more prevalent in females in younger age ranges when compared with males who are more likely to develop acne only once they reach puberty. Males are also said to be more prone to suffer from severe acne. Acne prevalence can also vary between different ethnic groups with percentage prevalence for African Americans (37%), Hispanic (32%), Asian (30%), Caucasian (24%), and Continental Indian (23%) being recorded. Studies have also reported that mild acne is the most prevalent form (66%) followed by moderate acne (33%) and lastly severe acne (<10%). Acne lesions commonly occur on the face but are also prevalent on the upper back (52%), upper chest (30%), lower back (22%), shoulders/arms (16%), and the neck (8%) (Tan & Bhate, 2015). Using population data from the 2004 US Census Bureau, the estimated population of South Africa was 44,448,470, which extrapolated to 2,778,029 people who would suffer from acne (Health Grades Incorporated, 2014).

4.2 WHY USE PLANTS FOR THE TREATMENT OF ACNE VULGARIS?

Although acne vulgaris and the acne-causing bacterium, *P. acnes*, have been intensively studied, little research has focused on the use of medicinal plants for the treatment of this skin disease. However, there are a number of medicinal plants used traditionally to treat acne, and this provides the impetus for scientists to explore their medicinal properties (Tables 4.1 and 4.2) (Mabona & Van Vuuren, 2013; Van Wyk, Van Oudtshoorn, & Gericke, 2009).

TABLE 4.1 Medicinal Plants With Traditional Usage and Biological Activity for the Treatment of Acne Vulgaris and Its Symptoms

Plant Species	Family	Distribution	Used Part and/or Form	Traditional Use	Reported Biological Activities	References
Aloe ferox Mill. (Fig. 4.6A)	Xanthorrhoeaceae	South Africa and Lesotho	Leaves and roots	Applied topically or ingested to treat eczema, dermatitis, and acne	Laxative effects, skin and wound healing, antioxidant, antiinflammatory, antimicrobial (MIC of 250 μg/mL on *Staphylococcus epidermidis*), antimalarial, and anticancer	Chen, Van Wyk, Vermaak, and Viljoen (2012), Coopoosamy and Naidoo (2012), Mabona and Van Vuuren (2013), and Van Wyk et al. (2009)
Aloe vera (L.) Burm.f. (Fig. 4.6J)	Xanthorrhoeaceae	The Caribbean, India, the Mediterranean, North Africa and South America	Leaf gel	Applied topically to treat skin ailments and acne vulgaris	Antiinflammatory, antiviral, antibacterial and antiaging activity	Surjushe, Vasani, and Saple (2008)
Aspalathus linearis (Burm.f.) R. Dahlgren (Fig. 4.6B)	Fabaceae	South Africa	Leaf extract	Topical applications are believed to alleviate dermatological problems associated with eczema, acne and nappy rash	Antioxidant, decreased neutrophil infiltration in *acne pustulosa* (in vivo study)	Carraz-Bernabei (2011), Joubert, Gelderblom, Louw, and de Beer (2008), and Van Wyk et al. (2009)
Bulbine frutescens Willd.	Asphodelaceae	South Africa, Lesotho, and Swaziland	Leaf juice	Treatment for acne wounds, burns, rashes, and itches	Cutaneous wound healing in vitro	Coopoosamy and Naidoo (2012), Van Wyk (2008), and Van Wyk et al. (2009)

Centella asiatica (L.) Urb. (Fig. 4.6C)	Apiaceae	South Africa, India, Sri Lanka, China, Indonesia, and Malaysia	Leaves and roots	Leprosy, wounds prevention of scar tissue and acne	Antimicrobial activity against *Propionibacterium acnes* (5 mg/mL), antioxidant activity and antiinflammatory activity	Chomnawang, Surassmo, Nukoolkarn, and Gritsanapan (2005), Ohran (2012), Oyedeji and Afolayan (2005), Van Wyk (2008), and Van Wyk et al. (2009)
Glycyrrhiza glabra L. (Fig. 4.6I)	Fabaceae	Asia, France, Italy, Iran, Iraq, Spain, Syria, and Turkey	Roots	Cysts, dermatitis, eczema, and pruritus	Antimicrobial activity against *P. acnes* (200 μg/mL ATCC 6919 strain and 100 μg/mL ATCC 11827 strain) and antiinflammatory activity	Nam, Kim, Sim, and Chang (2003) and Saeedi, Morteza-Semnani and Ghoreishi (2003)
Harpephyllum caffrum Bernh. (Fig. 4.6D)	Anacardiaceae	South Africa, southern Mozambique, Swaziland, and Zimbabwe	Bark	Applied externally to treat eczema and acne	Tyrosinase inhibitory activity, cytotoxicity, effect on melanogenesis, antimicrobial activity against various Gram-positive and Gram-negative bacteria including *P. acnes* (MIC 0.18 mg/mL DCM:MeOH and 0.5 mg/mL Aqueous extract)	Coopoosamy and Naidoo (2012), Mabona et al. (2013), Mapunya, Nikolova, and Lall (2012), and Van Wyk et al. (2009)

MIC, minimum inhibitory concentration.

TABLE 4.2 Plant Compounds Reported to Have Antiacne Activity

Compound Name	Structure	Activity	References
Aloesin		Antiinflammatory activity	Chen et al. (2012)
Anacardic acid	R = 6-[8(Z),11(Z),14-Pentadecatrienyl] salicylic acid R = 6-[8(Z),11(Z)-Pentadecadienyl] salicylic acid R = 6-[8(Z)-Pentadecenyl] salicylic acid R = 6-Pentadecylsalicylic acid	Antibacterial activity with an MIC 0.78 μg/mL against *Propionibacterium acnes* Anacardic acid analogue from *Syzygium jambos* with an MIC of 7.9 μg/mL against *P. acnes*	Kubo et al. (1994), Sharma, Kishore, Hussein, and Lall (2013)
Asiaticoside		Antibacterial, antiinflammatory has also been used in antiacne formulations	James and Dubery (2009)

Berberine	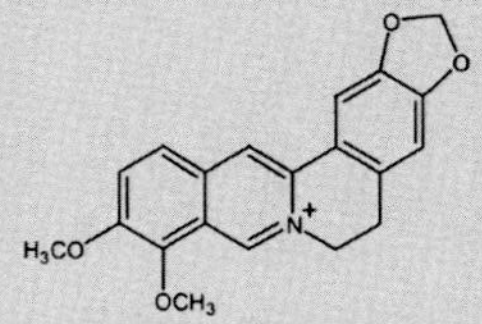 H_3CO, OCH_3, N^+, O, O	Antimicrobial activity against *P. acnes* with an MIC of 5–25 μg/mL	Slobodnikova, Alova, Labudova, Kotulva, and Kettmann (2004)
Plumbagin	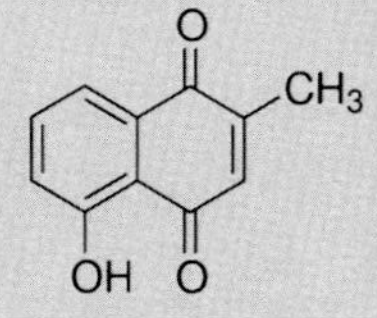 O, CH_3, OH, O	Antibacterial activity against *Staphylococcus epidermidis* with an MIC of 4.0 μg/mL (microorganism also associated with acne)	Mabona et al. (2013)

MIC, minimum inhibitory concentration.

4.2.1 Plant Extracts Traditionally Used for the Treatment of Acne

Medicinal plants have the potential to be commercialized not only for their ability to act as antimicrobial agents but also for the treatment of symptoms associated with acne vulgaris. The species with potential commercial value have been summarized in Table 4.1, which includes the traditional use and the biological activity that may support their use.

4.2.2 Plant Compounds With Antiacne and Antiinflammatory Activity

Although plants contain many constituents, the isolation of pure compounds from plants is an important part of the drug discovery process. The isolation of compounds known to be responsible for biological activity may not only enhance both the isolation process and the effectiveness but also provide valuable information of the mechanism of action. Table 4.2 summarizes some isolated plant compounds that show activity against *P. acnes* and acne symptoms.

4.3 ESSENTIAL OILS FOR THE TREATMENT OF ACNE

Aromatherapy is a term used to describe the treatment of various diseases and also to maintain general heath through the use of essential oils as inhalations, massage perfumes, and in baths. Essential oils are volatile secondary metabolites obtained from fragrant plants. They are either called essential oils, ethereal oils, or volatile oils because they have the ability to evaporate quickly. They have antibacterial, antifungal, and antiviral properties making them ideal to use in the pharmaceutical, cosmetic, medical, and food industries. The medicinal benefits of essential oils are seen when absorbed through the mucus membranes of the lungs and nose or through the skin. Essential oils have been used for centuries, and they are considered one of mankind's first medicines. Evidence of their use can be seen in Egyptian hieroglyphics and Chinese manuscripts. The Egyptians used essential oils not only for embalming but also considered them to be holy, and there were temples dedicated to the production and the mixing of these precious oils. The recipes were written in hieroglyphics on the walls of these temples, and papyrus manuscripts were found mentioning fine oils that were used to treat ailments. The San people of South Africa, also known as Bushmen, reportedly massaged themselves with aromatic plants mixed with fats and oils to treat various diseases (Higley & Higley, 1998; Van Wyk & Wink, 2015).

Essential oils can be extracted from various plant parts such as flowers, leaves, wood, stems, roots, seeds, fruits, bark, rhizomes, and gums or oleoresin exudations. There are a number of techniques to obtain essential oils, each method with its own advantages and disadvantages. The methods commonly used for extraction include the following: water distillation, water and steam distillation, steam distillation, cohobation, maceration, and enfleurage. In South

Africa the essential oil industry is still very young, despite the fact that Southern Africa is richly gifted with an immense diversity of plants including aromatic plants. This provides motivation for South African plants to be explored to their full potential (Jäger & Van Staden, 2000, Chap. 3; Yojana, 2014, pp. 1–17).

South Africa is home to a number of aromatic plant families, including the Rutaceae, Asteraceae, and the Lamiaceae families with many of these species being endemic. However, little research has focused on the use of these volatile oils in the treatment of acne vulgaris and their activity toward *P. acnes* (Table 4.3) (Van Vuuren, 2008).

4.3.1 Plant Essential Oils Traditionally Used and Commercially Available for the Treatment of Acne

Essential oils are defined as natural, complex, multicomponent systems which are mainly made up of terpenes but can include nonterpene components as well. This could explain their extensive use in the cosmetic, pharmaceutical, and food industries. Table 4.3 depicts some essential oils prepared from plants species with the potential to treat acne vulgaris.

4.4 SCIENTIFIC REPORTS ON PLANTS AGAINST *PROPIONIBACTERIUM ACNES*

Essential oils have been found to have lower antimicrobial activity compared to crude extracts and should therefore be classified differently. Currently there are no set criteria that classify essential oils as having noteworthy or insignificant activity. It has been suggested, however, that a minimum inhibitory concentration (MIC) of 2 mg/mL or lower should be considered noteworthy for an essential oil (Van Vuuren, 2008). For plant extracts and pure compounds, it has been suggested that an MIC threshold of 100 μg/mL or lower and 10 μg/mL or lower, respectively, indicate a noteworthy antibacterial activity (Sharma, Kishore, Hussein, & Lall, 2014). A significant or noteworthy MIC can be described as a concentration of plant extract or pure compound with the closest activity (or least difference) compared with known antibacterial treatments, regardless of natural or synthetic origin. Plants or pure compounds with the lowest MICs are generally considered for further research as less of the extract or compound is necessary to achieve the desired effect; however, it is always important to consider the possible toxic effects to the human host at the same concentration that may kill bacterial cultures.

The essential oil of *Maleuca alternifolia* Cheel, also known as tea tree oil, is used in many cosmeceuticals to treat acne vulgaris due to its antibacterial activity. The good antibacterial activity of tea tree oil and essential oils, in general, is believed to be due to the active compounds in the oils disrupting the cellular membrane of bacteria, resulting in potassium leakage. Terpenes found in tea tree oil have been reported to disrupt the lipid bilayer through affecting the

TABLE 4.3 Essential Oils and Their Components With the Potential to Treat Acne Vulgaris

Plant Species	Family	Distribution	Used Part	Active Compounds	Reported Biological Activities	References
Cryptomeria japonica D.Don (Fig. 4.6K)	Cupressaceae	Japan, southern China, North America, and Europe	Leaves	γ-eudesmol and sabinene	Antibacterial activity against *Propionibacterium acnes* (MIC of 0.156 μL/mL) and antiinflammatory properties	Yoon, Kim, Oh, Lee, and Hyun (2009)
Melaleuca alternifolia Cheel (tea tree oil) (Fig. 4.6E)	Myrtaceae	Australia, Europe, and North America	Leaves	terpinen-4-ol, α-terpineol, and α-pinene	Antibacterial activity against *P. acnes* (MIC of 0.05–0.63 %vol/vol) and antiinflammatory properties	Kanlayavattanakul and Lourith (2011), Carson et al. (2006), and Sinha, Srivastava, Mishra, and Yadav (2014)
Eucalyptus globulus Labill. (*Eucalyptus* oil) (Fig. 4.6L)	Myrtaceae	East and south–east Tasmania, China, South America, and sub-Saharan Africa	Leaves	γ-terpinene	Antibacterial activity against *P. acnes* (MIC of 9.38 mg/mL)	Athikomkulchai et al. (2008) and Sinha et al. (2014)
Psidium guajava L. (Guava oil) (Fig. 4.6M)	Myrtaceae	Central America and southern Mexico	Leaves	α-pinene	Antibacterial activity against *P. acnes* (MIC of 9.38 mg/mL)	Athikomkulchai et al. (2008) and Sinha et al. (2014)
Rosmarinus officinalis L. (*Rosemary* oil) (Fig. 4.6N)	Lamiaceae	Mediterranean, widely cultivated	Leaves	1,8-cineole, α-pinene, camphor, and camphene	Antibacterial activity against *P. acnes* (MIC of 0.56 mg/mL)	Fu et al. (2007) and Sinha et al. (2014)
Cymbopogon nardus (L.) Rendle (citronella oil) (Fig. 4.6O)	Poaceae	South America, west and eastern Africa, Asia	Leaves	Citronellal and geranial	Antibacterial activity on *P. acnes* (MIC of 0.005–0.3 μL/mL), antiinflammatory properties, and antioxidant activity	Lertsatitthanakorn, Taweechaisupapong, Aromdee, and Khunkitti (2006) and Sinha et al. (2014)

Cymbopogon citratus Stapf (Lemongrass oil) (Fig. 4.6P)	Poaceae	Indonesia, Africa, China, and South America	Leaves	Citral	Antibacterial activity against *P. acnes* (MIC of 0.6 μL/mL)and antioxidant activity	Lertsatitthanakorn et al. (2006) and Sinha et al. (2014)
Ocimum tenuiflorum L. (holy basil oil) (Fig. 4.6Q)	Lamiaceae	Tropical and subtropical Asia	Leaves	Eugenol	Antibacterial activity against *P. acnes* (MIC of 5.0 μL/mL)and antioxidant activity	Lertsatitthanakorn et al. (2006) and Sinha et al. (2014)
Citrus hystrix DC. (kaffir lime oil) (Fig. 4.6R)	Rutaceae	Philippines, India, Burma, Malay Peninsula, and Ceylon	Leaves	D-limonene, citronellal and geranial	Antibacterial activity against *P. acnes* (MIC of 5.0 μL/mL)	Lertsatitthanakorn et al. (2006) and Sinha et al. (2014)
Lippia javanica Spreng. (Fig. 4.6F)	Verbenaceae	Southern parts of South Africa, Mozambique, and Zimbabwe	Leaves	Monoterpenoids	Antiinflammatory activity	Van Wyk (2011) and Van Wyk et al. (2009)
Osmitopsis asteriscoides (L.) Less. (Bels)	Asteraceae	Endemic to the Western Cape	Leaves	Eucalyptol and camphor	Antibacterial activity against *Staphylococcus aureus* (MIC 32 mg/mL), antiinflammatory properties, and antioxidant activity	Van Wyk et al. (2009) and Viljoen et al. (2003)
Pelargonium graveolens L'Her. (Fig. 4.6G)	Geraniaceae	Northern and southern parts of South Africa, Mozambique, and Zimbabwe	Leaves	Geraniol, linalool and citronellol	Antibacterial activity against *P. acnes*, antiinflammatory properties, and antioxidant activity	Culbert and Olness (2009), Van Wyk et al. (2009), and Van Wyk (2011)

MIC, minimum inhibitory concentration.

packing of the lipids in the cellular membrane. This disruption increases the fluidity of the bilayer affecting cell permeability (Rai and Kon, 2013). Tea tree oil has previously been shown to inhibit *P. acnes* at a MIC between 0.05 and 0.63 (%v/v) (Carson, Hammer, & Riley, 2006). Previous studies by Enshaieh, Jooya, Siadat, and Iraji (2007), reported that a gel containing 5% tea tree oil was effective in treating mild-to-moderate acne in 60 patients in a randomized, double-blind clinical trial. Bassett, Pannowitz, and Barnetson (1990) determined that the effect of a 5% tea tree oil gel was comparable to a 5% benzoyl peroxide lotion frequently used to treat acne lesions. The clinical trials were performed on 124 patients in a single-blind, randomized clinical trial, and both treatments had a significant effect in reducing inflamed lesions. Although tea tree oil took longer to be effective, patients experienced fewer side effects such as skin irritation, itching, and redness.

Plants contain a number of different secondary metabolites that are known to have antibacterial activity and could be useful in targeting *P. acnes*. A number of known compounds isolated from plants have antibacterial activity against *P. acnes*. Phenolic compounds, such as salicylic acid isolated from *Anacardium occidentale* L. commonly known as the cashew plant, have been found to be effective in inhibiting *P. acnes* at a MIC of 50 μg/mL (Himejima & Kubo, 1991). The mechanism of action of polyphenols is believed to be due to enzyme inhibition through reacting with sulfhydryl groups and proteins associated with the bacteria. *Podocarpus nagi (Thunb.) Pilg.*, also known as tree bard, is known to contain high levels of totarol, a flavonol that has previously been found to inhibit *P. acnes* and other Gram-positive bacteria (Cowan, 1999). Kubo, Muroi, and Kubo (1994) found totarol to inhibit *P. acnes* at a MIC of 0.39 μg/mL. The activity of flavonols is believed to be due to their ability to form a complex with the extracellular proteins and cell walls of bacteria, inactivating the proteins and resulting in loss of function (Cowan, 1999).

Essential oils can also be used in combination with other plant extracts to increase the effectiveness of both the essential oil and the extract. Clinical studies done by Orafidiya et al. (2004) evaluated the synergistic effect of *Aloe vera* (L.) Burm.f. gel and *Ocimum gratissimum* Forssk. essential oil on acne lesions. A lotion of 2% *Ocimum* oil containing grading concentrations of aloe gel were tested on 84 patients against a 1% clindamycin phosphate solution, Dalacin, a known antiacne drug. They found that the addition of aloe gel enhanced the effectiveness of *Ocimum* oil on papules and pustules. A lotion of 2% *Ocimum* oil and 50% *A. vera* (L.) Burm.f. gel were more effective in reducing inflammatory lesions than Dalacin.

A study by Mabona, Viljoen, Shikanga, Marston, and Van Vuuren (2013) reported the potential of some South African plant extracts, for example, the dichloromethane:methanol (DCM:MeOH = 1:1) extracts of *Terminalia sericea* Burch. ex DC. roots and *Gunnera perpensa* L. leaves, which could be used as possible antiacne agents as both showed an MIC of 0.03 mg/mL. The aqueous extract of *T. sericea* Burch. ex DC. also showed noteworthy antimicrobial activity with a MIC of 0.25 mg/mL. Aqueous extracts are often favored, as traditional healers generally prepare remedies using water as the vehicle, although oil, honey, and fermented extracts through the addition of sugar are also used (Bodeker, Bhat, Burley, & Vantomme, 1997).

4.4.1 Antiinflammatory Activity of *Leucosidea sericea*

Leucosidea sericea Eckl. & Zeyh. a plant belonging to the Rosaceae family, is indigenous to South Africa. It is found in the Eastern Cape, Free State, and KwaZulu-Natal provinces (Fig. 4.6H). It is traditionally used against various ailments including severe inflammation of the eye. In combination with other plants, this species is used as an astringent. Astringents are generally prescribed for patients who suffer from very oily skin.

The ethanolic extract of *L. sericea* was tested for its antibacterial, antioxidant, and antiinflammatory potential. The MIC of the ethanolic extract of the leaves of *L. sericea* against *P. acnes* was 15.6 μg/mL, which was comparable to that of tetracycline with an MIC of 3.1 μg/mL. The antioxidant activity was also determined due to the role of oxidants in inflammation, and the radical scavenging capacity of the ethanolic leaf extract of *L. sericea* showed an IC_{50} of 2.01 μg/mL using the DPPH assay. The antiinflammatory activity of *L. sericea* was determined by stimulation of macrophage cells with *P. acnes* and detecting the levels of IL-8 and TNF-α. The ethanol leaf extract was able to decrease IL-8 and TNF-α levels when treated with 25, 12.5, and 6.25 μg/mL of *L. sericea* extract (Sharma et al., 2014). Inflammation is one of the major symptoms of acne, resulting in the formation of pustules, papules and cysts. In more severe cases of inflammatory acne, the formation of cysts is often accompanied with the formation of scars. This has a significant impact on psychosocial attributes, especially in teenagers who are commonly afflicted with acne. This work provides a basis for the possibility of this plant extract to be applied in a topical formulation, accompanying the use of conventional oral treatments, to improve efficiency and response to treatment. Increased efficiency could potentially be accompanied by reduced development of resistant strains of *P. acnes*.

4.4.2 Antiacne Formulations From *Leucosidea sericea*

L. sericea is one of the few plants that has passed through extensive research and development in terms of antiacne activity. The research performed at the University of Pretoria, Department of Plant Science has led to the filing of an international (PCT—Patent Cooperation Treaty) patent. *L. sericea* was investigated in a full clinical study on human volunteers. Volunteers of two different skin types were used, Fitzpatrick skin type IV and V. The Fitzpatrick skin type system was developed by dermatologist, Thomas B Fitzpatrick in 1975, as a classification system for skin types that react differently to Ultraviolet light. Type IV is characterized as a moderate brown, typically Mediterranean, which

FIGURE 4.6 Medicinal plants reported to have traditional usage and biological activity against *Propionibacterium acnes* and/or acne-related symptoms. (A) *Aloe ferox* Mill., (B) *Aspalathus linearis* (Burm.f.) R. Dahlgren., (C) *Centella asiatica* (L.) Urb., (D) *Harpephyllum caffrum* Bernh., (E) *Melaleuca alternifolia* Cheel., (F) *Lippia javanica* Spreng., (G) *Pelargonium graveolens* L'Her., (H) *Leucosidea sericea* Eckl. & Zeyh, (I) *Glycyrrhiza glabra* L., (J) *Aloe vera* (L.) Burm.f., (K) *Cryptomeria japonica* D.Don, (L) *Eucalyptus globulus* Labill., (M) *Psidium guajava* L., (N) *Rosmarinus officinalis* L., (O) *Cymbopogon nardus* (L.) Rendle, (P) *Cymbopogon citratus* Stapf, (Q) *Ocimum tenuiflorum* L., and (R) *Citrus hystrix* DC.

rarely burns and tans well. Type V is characterized as dark brown, typically Middle Eastern, which very rarely burns and tans well. These skin types are more prone to visible scarring and postinflammatory hyperpigmentation caused by acne lesions and were chosen for this study.

Volunteers of various ages were also considered. The clinical trial consisted of 25 healthy volunteers between 16 and 30 years old. For the clinical study conducted, the profile of the 25 volunteers was as follows:

- 9 subjects were characterized as Type IV Fitzpatrick skin types
- 16 subjects were characterized as Type V Fitzpatrick skin types

Volunteers returned to the facility after 14 and 28 days where physical counts of noninflammatory and inflammatory lesions were obtained.

Results showed that at 10% in a gel cream base, the extract was able to achieve the following:

- Was **effective** in reducing the number of **comedones**; after 28 days of consecutive use; when compared to the placebo control
- Was **effective** in reducing the number of **papules**; after 14 to 28 days of consecutive use; when compared to the placebo control
- Was **effective** in reducing the number of **pustules**; after 28 days of consecutive use; when compared to the placebo control
- Was **effective** in reducing the number of **whiteheads**; after 14 to 28 days of consecutive use; when compared to the placebo control

Future recommendations have been made for this active ingredient to be readily available for commercial scale use by formulators of cosmetic products.

4.5 CONCLUSION

The use of medicinal plants that are used by local communities, traditional healing practitioners, as well as those known to contain medicinally beneficial compounds, plays a vital role in the search for new and more effective treatments for acne. There is currently a strong focus on medicinal plants at a research level and therefore strong motivation exists for the use of these ingredients in cosmetic care products that need to be proven safe and effective with the aid of clinical research.

REFERENCES

Arne, L. H. (2012). *Wikimedia commons, peripheral GABA (B) receptor targets*. Online Available: https://commons.wikimedia.org/wiki/File:Peripheral_GABA(B)_receptor_targets.png.

Arora, M. K., Yadav, A., & Saini, V. (2011). Role of hormones in acne vulgaris. *Clinical Biochemistry*, *44*, 1035–1040.

Athikomkulchai, S., Watthanachaiyingcharoen, R., Tunvichien, S., Vayumhasuwan, P., Karnsomkiet, P., Sae-Jong, P., et al. (2008). The development of anti-acne products from *Eucalyptus globulus* and *Psidium guajava* oil. *Journal of Health Research*, *22*(3), 109–113.

Bassett, I. B., Pannowitz, D. L., & Barnetson, R. S. (1990). A comparative study of tea-tree oil versus benzoylperoxide in the treatment of acne. *The Medical Journal of Australia*, *153*(8), 455–458.

Bastian, H. (2013). *Wikimedia commons, comedones*. Online Available: https://commons.wikimedia.org/wiki/File:Comedones.jpg.

Bodeker, G., Bhat, K. K. S., Burley, J., & Vantomme, P. (1997). Global Initiative for Traditional Systems (GIFTS) of health. *Medicinal Plants for Forest Conservation and Health Care*, 92.

Broere, C. M. (2007). *Wikimedia commons, test biosynthesis 5-alpha reductase*. Online Available: https://commons.wikimedia.org/wiki/File:Test_biosynth_5ARD2.jpg.

Bruggemann, H. (2010). Skin: Acne and *Propionibacterium acnes* genomics. In *Handbook of hydrocarbon and lipid microbiology* (pp. 3216–3225). Springer-Verlag Berlin Heidelberg.

Burkhart, C. G., Burkhart, C. N., & Lehmann, P. F. (1999). Acne: A review on immunologic and microbiologic factors. *Postgraduate Medical Journal, 75*, 328–331.

Campos-Rodríguez, R., Gutiérrez-Meza, M., Jarillo-Luna, R. A., Drago-Serrano, M. E., Abarca-Rojano, E., Ventura-Juárez, J., et al. (2016). *Wikimedia commons, role of neutrophils in rodent amebic liver abscess*. Online Available: https://commons.wikimedia.org/wiki/File:Parasite150072-fig2_Role_of_neutrophils_in_rodent_amebic_liver_abscess.png.

Carraz-Bernabei, E. (2011). African teas. In Anonymous (Ed.), *Healing teas for your body, mind & soul* (p. 12). England: Apex Publishing Ltd.

Carson, C. F., Hammer, K. A., & Riley, T. V. (2006). *Melaleuca alternifolia* (tea tree) oil: A review of antimicrobial and other medicinal properties. *Clinical Microbiology Reviews*, *19*(1), 50–62.

Chen, W., Van Wyk, B., Vermaak, I., & Viljoen, A. M. (2012). Cape aloes – a review of the phytochemistry, pharmacology and commercialisation of *Aloe ferox*. *Phytochemistry Letters*, *5*, 1–12.

Chiu, A., Chon, S. Y., & Kimball, A. B. (2003). The response of skin disease to stress: Changes in the severity of acne vulgaris as affected by examination stress. *Archives of Dermatology*, *139*(7), 897–900.

Chomnawang, M. T., Surassmo, S., Nukoolkarn, V. S., & Gritsanapan, W. (2005). Antimicrobial effects of Thai medicinal plants against acne-inducing bacteria. *Journal of Ethnopharmacology*, *101*, 330–333.

Chris73. (2005). *Wikimedia commons, acne vulgaris USMIL*. Online Available: https://commons.wikimedia.org/wiki/File:AcneVulgarisUSMIL.jpg.

Coenye, T., Peeters, E., & Nelis, H. J. (2007). Biofilm formation by *Propionibacterium acnes* is associated with increased resistance to antimicrobial agents and increased production of putative virulence factors. *Research in Microbiology*, *158*, 386–392.

Cogen, A. L., Nizet, V., & Gallo, R. T. (2008). Skin microbiota: a source of disease or defence? *British Journal of Dermatology*, *158*(3), 442–455.

Coopoosamy, R., & Naidoo, K. (2012). An ethnobotanical study of medicinal plants used by traditional healers in Durban, South Africa. *African Journal of Pharmacy and Pharmacology*, *6*, 818–823.

Cowan, M. M. (1999). Plant products as antimicrobial agents. *Clinical Microbiology Reviews*, *12*(4), 564–582.

Culbert, T., & Olness, K. (2009). Dermatology and skin infections. In Anonymous (Ed.), *Integrative pediatrics*. Oxford University Press.

Dessinioti, C., & Katsambas, A. D. (2010). The role of *Propionibacterium acnes* in acne pathogenesis: Facts and controversies. *Clinics in Dermatology*, *28*, 2–7.

Dinarello, C. A. (2009). Immunological and inflammatory functions of the interleukin-1 family. *Annual Review of Immunology*, *27*, 519–550.

Enshaieh, S., Jooya, A., Siadat, A. H., & Iraji, F. (2007). The efficacy of 5% topical tea tree oil gel in mild to moderate acne vulgaris: A randomized, double-blind placebo-controlled study. *Indian Journal of Dermatology, Venereology, and Leprology*, *73*(1), 22.

Falcocchio, S., Ruiz, C., Javier Pastor, F. I., Saso, L., & Diaz, P. (2006). *Propionibacterium acnes* GehA lipase, an enzyme involved in acne development, can be successfully inhibited by defined natural substances. *Journal of Molecular Catalysis B: Enzymatic*, *40*, 132–137.

Food and Drug Administration. (2005). *Guidance for industry, acne vulgaris: Developing drugs for treatment*. Online Available: http://www.fda.gov/downloads/Drugs/.../Guidances/UCM071292.pdf.

Fu, Y., Zu, Y., Chen, L., Efferth, T., Liang, H., Liu, Z., et al. (2007). Investigation of antibacterial activity of rosemary essential oil against *Propionibacterium acnes* with atomic force microscopy. *Planta Medica*, *73*(12), 1275–1280.

Grange, P. A., Weill, B., Dupin, N., & Batteux, F. (2010). Does inflammatory acne result from imbalance in keratinocyte innate immune response? *Microbes and Infection*, *12*, 1085–1090.

Grice, E. A., & Segre, J. A. (2011). The skin microbiome. *Nature Reviews Microbiology*, *9*(4), 244–253.

Hayashi, N., Akamatsu, H., & Kawashima, M. (2008). Establishment of grading criteria for acne severity. *The Journal of Dermatology*, *35*(5), 255–260.

Hay, R. J., Johns, N. E., Williams, H. C., Bolliger, I. W., Dellavalle, R. P., Margolis, D. J., et al. (2014). The global burden of skin disease in 2010: An analysis of the prevalence and impact of skin conditions. *Journal of Investigative Dermatology*, *134*(6), 1527–1534.

Health Grades Incorporated. (2014). *Statistics for acne by country*. Online Available: http://www.rightdiagnosis.com/a/acne/stats-country.htm.

Higley, C., & Higley, A. (1998). *Reference guide for essential oils*. Abundant Health.

Himejima, M., & Kubo, I. (1991). Antibacterial agents from the cashew *Anacardium occidentale* (Anacardiaceae) nut shell oil. *Journal of Agricultural and Food Chemistry*, *39*(2), 418–421.

Jäger, A. K., & Van Staden, J. (2000). Salvia in southern Africa. In *The genus salvia* (p. 47).

James, J. T., & Dubery, I. A. (2009). Pentacyclic triterpenoids from the medicinal herb, *Centella asiatica* (L.) urban. *Molecules*, *14*, 3922–3941.

Joubert, E., Gelderblom, W. C. A., Louw, A., & de Beer, D. (2008). South African herbal teas: *Aspalathus linearis*, *Cyclopia* spp. and *Athrixia phylicoides* – a review. *Journal of Ethnopharmacology*, *119*, 376–412.

Kanlayavattanakul, M., & Lourith, N. (2011). Therapeutic agents and herbs in topical application for acne treatment. *International Journal of Cosmetic Science*, *33*(4), 289–297.

Kim, J., Ochoa, M., Krutzik, S. R., Takeuchi, O., Uematsu, S., Legaspi, A. J., et al. (2002). Activation of toll-like receptor 2 in acne triggers inflammatory cytokine responses. *The Journal of Immunology*, *169*(3), 1535–1541.

Kubo, I., Muroi, H., & Kubo, A. (1994). Naturally occurring antiacne agents. *Journal of Natural Products*, *57*(1), 9–17.

Layton, A. M. (2005). Acne vulgaris and similar eruptions. *Medicine*, *33*(1), 44–48.

Lertsatitthanakorn, P., Taweechaisupapong, S., Aromdee, C., & Khunkitti, W. (2006). *In vitro* bioactivities of essential oils used for acne control. *International Journal of Aromatherapy*, *16*(1), 43–49.

Mabona, U., & Van Vuuren, S. F. (2013). Southern African medicinal plants used to treat skin diseases. *South African Journal of Botany*, *87*, 175–193.

Mabona, U., Viljoen, A., Shikanga, E., Marston, A., & Van Vuuren, S. F. (2013). Antimicrobial activity of southern African medicinal plants with dermatological relevance: From an ethnopharmacological screening approach, to combination studies and the isolation of a bioactive compound. *Journal of Ethnopharmacology*, *148*(1), 45–55.

Makrantonaki, E., Ganceviviene, R., & Zouboulis, C. (2011). An update on the role of the sebaceous gland in the pathogenesis of acne. *Dermato-Endocrinology*, *3*(1), 41–49.

Mapunya, M. B., Nikolova, R. V., & Lall, N. (2012). Melanogenesis and antityrosinase activity of selected South African plants. *Evidence-Based Complementary and Alternative Medicine*, *2012*, 1–6.

Nam, C., Kim, S., Sim, Y., & Chang, I. (2003). Anti-acne effects of oriental herb extracts: A novel screening method to select anti-acne agents. *Skin Pharmacology and Physiology*, *16*(2), 84–90.

Nord, C. E., & Oprica, C. (2006). Antibiotic resistance in *Propionibacterium acnes*. Microbiological and clinical aspects. *Anaerobe*, *12*, 207–210.

Ohran, I. E. (2012). *Centella asiatica* (L.) Urban: From traditional medicine to modern medicine with neuroprotective potential. *Evidence-Based Complementary and Alternative Medicine*, *2012*, 1–8.

OpenStax College. (2013). *Wikimedia commons, acne formation*. Online Available: https://commons.wikimedia.org/wiki/File:515_Acne_formation.jpgv.

Orafidiya, L. O., Agbani, E. O., Oyedele, A. O., Babalola, O. O., Onayemi, O., & Aiyedun, F. F. (2004). The effect of aloe vera gel on the anti-acne properties of the essential oil of *Ocimum gratissimum* Linn leaf–a preliminary clinical investigation. *International Journal of Aromatherapy*, *14*(1), 15–21.

Oyedeji, O., & Afolayan, A. (2005). Chemical composition and antibacterial activity of the essential oil of *Centella asiatica*, growing in South Africa. *Pharmaceutical Biology*, *43*, 249–252.

Pawin, H., Beylot, C., Chivot, M., Faure, M., Florence, P., Revuz, J., et al. (2004). Physiopathology of acne vulgaris: Recent data, new understanding of the treatments. *European Journal of Dermatology*, *14*, 4–12.

Portugal, M., Barak, V., Ginsburg, I., & Kohen, R. (2007). Interplay among oxidants, antioxidants and cytokines in skin disorders: Present status and future considerations. *Biomedicine and Pharmacotherapy*, *61*, 412–422.

Rai, M., & Kon, K. (Eds.). (2013). *Fighting multidrug resistance with herbal extracts, essential oils and their components*. Academic Press.

Ruiz, M. (2007). *Wikimedia commons, desmosome cell junction*. Online Available: https://commons.wikimedia.org/wiki/File:Desmosome_cell_junction_en.svg.

Saeedi, M., Morteza-Semnani, K., & Ghoreishi, M. R. (2003). The treatment of atopic dermatitis with licorice gel. *Journal of Dermatological Treatment*, *14*(3), 153–157.

Sharma, R., Kishore, N., Hussein, A., & Lall, N. (2013). Antibacterial and anti-inflammatory effects of *Syzygium jambos* L. (Alston) and isolated compounds on acne vulgaris. *BMC Complement and Alternative Medicine*, *292*, 1–10.

Sharma, R., Kishore, N., Hussein, A., & Lall, N. (2014). The potential of *Leucosidea sericea* against *Propionibacterium acnes*. *Phytochemistry Letters*, *7*, 124–129.

Sinha, P., Srivastava, S., Mishra, N., & Yadav, N. P. (2014). New perspectives on antiacne plant drugs: Contribution to modern therapeutics. *BioMed Research International*, *2014*, 1–19.

Slobodnikova, L., Alova, D. K., Labudova, D., Kotulva, D., & Kettmann, V. (2004). Antimicrobial activity of *Mohania aquifolium* crude extract and its major isolated alkaloids. *Phytotherapy Research*, *18*, 674–676.

Strong, B. (1972). *Wikimedia commons, Propionibacterium acnes*. Online Available: https://commons.wikimedia.org/wiki/File:Propionibacterium_acnes.jpg.

Surjushe, A., Vasani, R., & Saple, D. G. (2008). Aloe vera: A short review. *Indian Journal of Dermatology*, *53*(4), 163.

Tan, J. K. L., & Bhate, K. (2015). A global perspective on the epidemiology of acne. *British Journal of Dermatology*, *172*(S1), 3–12.

Thiboutot, D. (2000). The role of follicular hyperkeratinization in acne. *Journal of Dermatological Treatment*, *11*, 5–8.

Toyoda, M., & Morohashi, M. (2001). Pathogenesis of acne. *Medical Electron Microscopy*, *34*, 29–40.

Truter, I. (2009). Evidence based pharmacy practice: Acne vulgaris. *SA Pharmaceutical Journal*, 12–19.

Van Vuuren, S. (2008). Antimicrobial activity of South African medicinal plants. *Journal of Ethnopharmacology*, *119*, 462–472.

Van Wyk, B. (2008). A broad review of commercially important southern African medicinal plants. *Journal of Ethnopharmacology*, *119*, 342–355.

Van Wyk, B. (2011). The potential of South African plants in the development of new medicinal products. *South African Journal of Botany*, *77*, 812–829.

Van Wyk, B., Van Oudtshoorn, B., & Gericke, N. (2009). In R. Ferreira (Ed.), *Medicinal plants of South Africa* (pp. 7–43). South Africa: Briza Publications.

Van Wyk, B. E., & Wink, M. (2015). In *Phytomedicines, herbal drugs, and poisons*. University of Chicago Press.

Viljoen, A., Van Vuuren, S., Ernst, E., Klepser, M., Demirci, B., Başer, H., et al. (2003). *Osmitopsis asteriscoides* (Asteraceae) – the antimicrobial activity and essential oil composition of a Cape-Dutch remedy. *Journal of Ethnopharmacology*, *88*, 137–143.

Vowels, B. R., Yang, S., & Leyden, J. J. (1995). Induction of pro-inflammatory cytokines by a soluble factor of *Propionibacterium acnes*: Implications for chronic inflammatory acne. *Infection and Immunity*, *36*(8), 3158–3165.

Webster, G. F. (2002). Clinical review: Acne vulgaris. *British Medical Journal*, *325*, 475–479.

Webster, G. F., & Graber, E. M. (2008). Antibiotic treatment for acne vulgaris. *Seminars in Cutaneous Medicine and Surgery*, *27*, 183–187.

Williams, H. C., Dellavalle, R. P., & Garner, S. (2012). Acne vulgaris. *The Lancet*, *379*, 361–372.

Yojana, R. K. V. (2014). *Extraction methods of natural essential oils*. TNAU Agritech Portal: Horticulture.

Yoon, W. J., Kim, S. S., Oh, T. H., Lee, N. H., & Hyun, C. G. (2009). *Cryptomeria japonica* essential oil inhibits the growth of drug-resistant skin pathogens and LPS-induced nitric oxide and pro-inflammatory cytokine production. *Polish Journal of Microbiology*, *58*(1), 61–68.

Chapter 5

Medicinal Plants as Alternative Treatments for Progressive Macular Hypomelanosis

Analike Blom van Staden, Namrita Lall
University of Pretoria, Pretoria, South Africa

Chapter Outline

Medicinal Plants for Holistic Health and Well-Being. http://dx.doi.org/10.1016/B978-0-12-812475-8.00005-6

5.1 PROGRESSIVE MACULAR HYPOMELANOSIS

Progressive macular hypomelanosis (PMH) is characterized by hypopigmented spots primarily found on the chest and the back, but in extreme cases it spread to the neck and face (Fig. 5.2). PMH appears to be a stable disease and does not cause any pain or inflammation (Halder & Rodney, 2013). A comparison of lesional and nonlesional skin of PMH patients, by means of an electron microscopic study, provided insight into the mechanisms of this hypopigmented disease. In a study conducted earlier, it was reported that both nonlesional and lesional skin contained melanosome precursors (primary components to synthesize melanin pigments) in the melanocytes, but the developed melanosomes were remarkably smaller and less melanized in hypopigmented skin areas than those of in normal skin areas (Marks & Seabra, 2001; Martínez-Martínez, Azaña-Defez, Rodríguez-Vázquez, Faura-Berruga, & Escario-Travesedo, 2012; Westerhof, Relyveld, Kingswijk, de Man, & Menke, 2004). A decrease in melanogenesis, melanin transfer as well as pathogenesis caused by *Propionibacterium acnes* is possibly the reasons for this hypopigmented disease.

5.1.1 Melanogenesis and Progressive Macular Hypomelanosis

Melanogenesis (the synthesis of melanin) takes place in the melanocytes, which are specialized pigment-producing cells, where melanin is generated as pigment granules, otherwise known as lysosome-related organelles called melanosomes (Videira, Moura, & Magina, 2013; Wasmeier, Hume, Bolasco, & Seabra, 2008). Melanin is derived from the amino acid—tyrosine—through a synthesis regulated by the enzyme tyrosinase. The final product results in a range of brown and black pigments that gives animal skin and hair its color (Lawrence, 2008a). Accordingly, melanin consists of a mosaic of eumelanin (brown–black pigments) and pheomelanin (yellow–red pigments) (Fig. 5.1) (Liu & Simon, 2005). Exploration of the melanin synthesis pathway provides an opportunity to understand the regulation of melanogenesis and how to direct the degree of pigmentation to ultimately find treatments for hypopigmented diseases.

In 2004, Westerhof et al. provided a possible explanation to what causes PMH and described three main hypotheses for the development of hypopigmentation: "firstly, *P. acnes* produces a melanogenesis inhibitory factor; secondly, a decrease in melanin synthesis due to a defect in tyrosinase's structure or tyrosinase inhibition; thirdly, melanin transfer from the melanocyte to the keratinocyte is obstructed and melanosome destruction occurs" (Fig. 5.2).

5.1.2 Causal Bacteria and Current Treatments

The hypothesis of Westerhof et al. (2004) was supported by experiments done by themselves and Relyveld, Kingswijk, and Reitsma (2006). PMH has previously been misdiagnosed as pityriasis versicolor due to the presence of small hypopigmented lesions. This misperception was resolved when lesional skin from both PMH and pityriasis versicolor skin was analyzed with Wood's lamp,

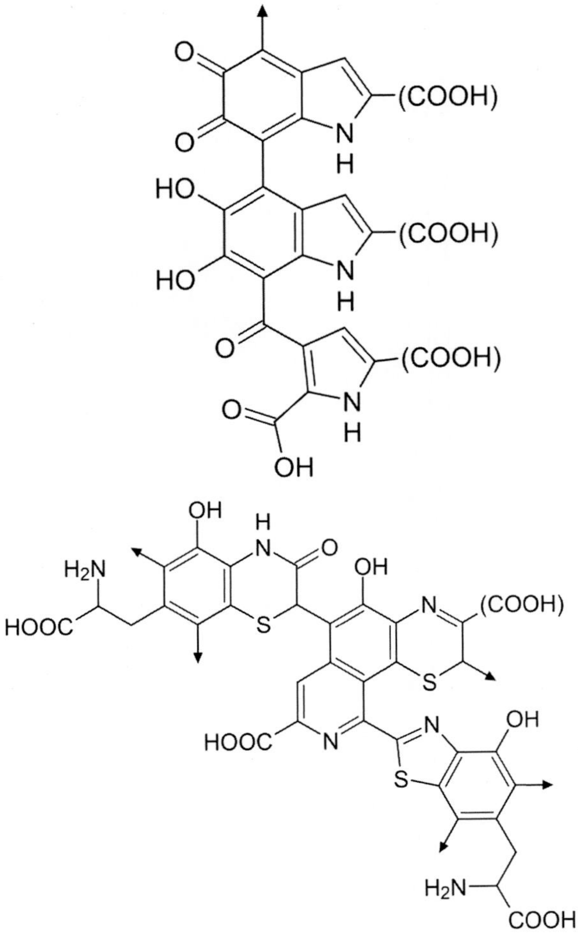

FIGURE 5.1 Structural formula of eumelanin (top) and pheomelanin (bottom), the pigments responsible for the brown–black and yellow–red color associated with animal skin and hair.

a procedure that uses transillumination (light) to investigate pigment disorders. Westerhof et al. (2004) reported that only lesional skin of PMH patients emitted red fluorescence under Wood's lamp, similar to those observed in acne skin and no red fluorescence was observed in pityriasis versicolor lesions. Red fluorescence is emitted by a porphyrin ring, which is characteristic of *P. acnes*, consequently the assumption was made that PMH is possibly caused by *P. acnes* (Westerhof et al., 2004). *P. acnes* is a rod shaped gram-positive bacteria, normally found on all human skin (Fig. 5.2; Toyoda & Morohashi, 2001).

In 2006, Relyveld executed a supporting experiment where lesional skin of different PMH patients was compared with the *Propionibacterium* genus. Forty three percent compared well with the *P. acnes* strain 6919, while 47% compared

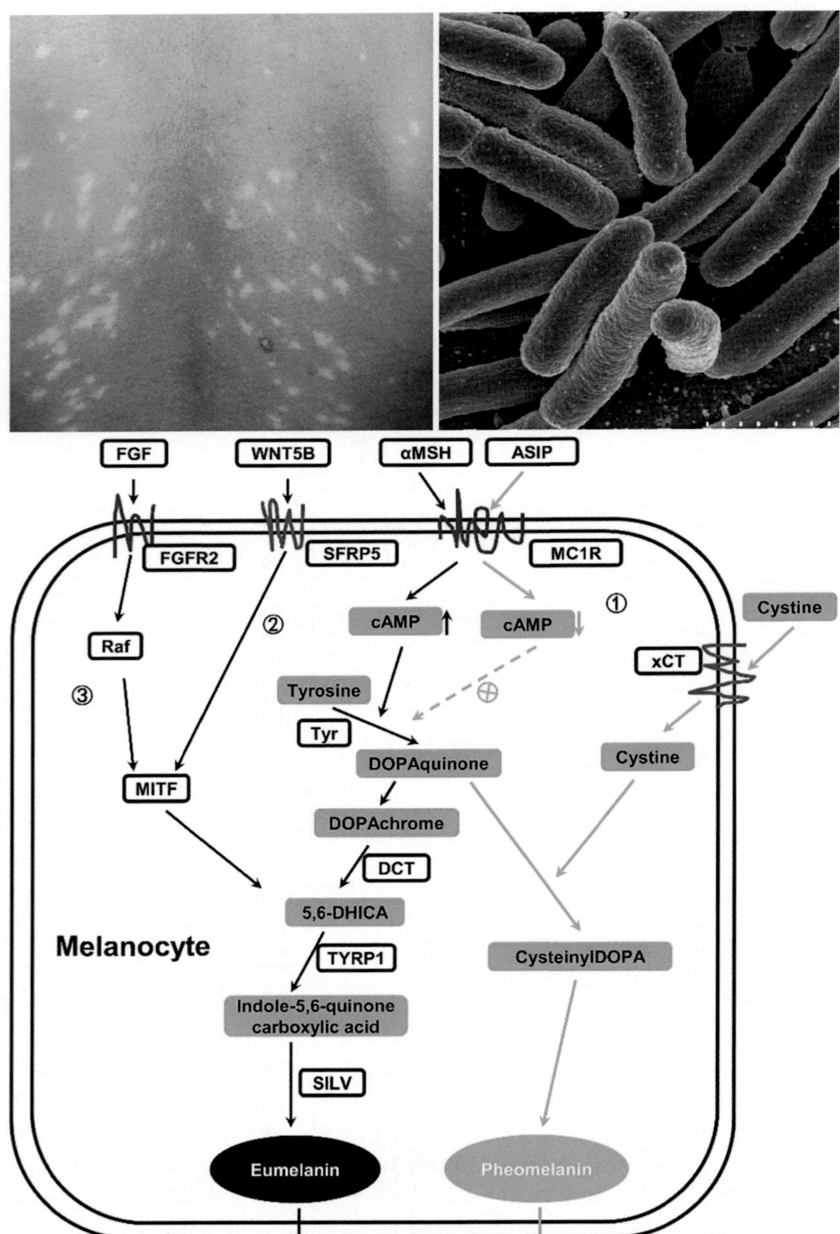

FIGURE 5.2 (Top left) hypopigmented lesions found on the back of a patient with progressive macular hypomelanosis (Koley, Mandal, Choudshary, & Bandyopadhyay, 2013), (top right) rod shaped, gram-positive bacteria (NIAID, 2005), and (bottom) melanin synthesis pathway (Jiang et al., 2014). *MITF*, microphthalmia-associated transcription factor.

well with the *Propionibacterium* genus but the specific species was unknown. Therefore, it was evident that *P. acnes* is most probably the causative bacteria of PMH (Relyveld et al., 2006).

The fact that *P. acnes* was identified as the causative bacteria of PMH led to the use of antibiotics as a treatment for PMH. The current treatment includes the application of 5% benzoyl peroxide and 1% clindamycin gel, combined with ultraviolet "A" (UVA) light therapy or psoralens combined with UVA (PUVA) (Westerhof et al., 2004). The mechanism by which PUVA leads to repigmentation in hypopigmented lesions is through stimulating the release of cytokines by the keratinocytes. These cytokines then act as signals that promote the transfer of melanosomes from the melanocytes to the keratinocytes (Kostoviæ, Nola, Buèan, & Šitum, 2003, p. 163). There are, however, some concerns about the side effects and safety of pigmentation stimulators, such as PUVA (Brown, 2001). Some of the consequences reported due to the exposure of Ultraviolet radiation (UVR) are premature aging of the skin, inflammation due to damaged keratinocytes, DNA breakage, and the depletion of antioxidants in the cell or the production of reactive oxygen species (Gilchrest & Eller, 1999; Pillai, Oresajo, & Hayward, 2005; Rhie et al., 2001; Wood et al., 1996). Consequently, an alternative treatment to UVR and antibiotics, to stimulate melanin production, is needed.

5.1.3 Melanin Production and the Role of Tyrosinase

The synthesis of eumelanin and pheomelanin starts with the substrate L-tyrosine that is converted to L-DOPA (dihydroxyphenylalanine) that is subsequently converted to dopaquinone (Fig. 5.3). The transition of L-tyrosine to L-DOPA is catalyzed by tyrosine hydroxylase and the transition of L-DOPA to dopaquinone is catalyzed by DOPA-oxidase, collectively these enzymes are known as tyrosinase (Olivares, Jimenez-Cervantes, Lozano, Solano, & Garcia-Borron, 2001). The aforementioned initial steps in melanin production are fundamental as the remainder of reactions proceed consequentially (Halaban et al., 2002). Tyrosinase that is responsible for the rate-limiting step of melanogenesis can be activated (phosphorylated) by means of two pathways, namely intrinsic and extrinsic regulation (Munemura, Eskay, Kebabian, & Long, 1980).

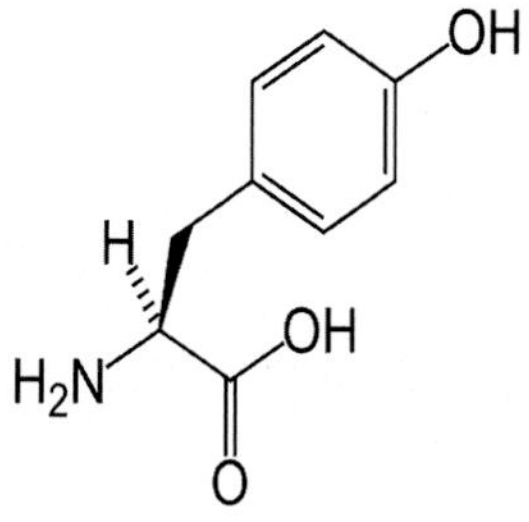

OH
H
OH
H_2N
O

FIGURE 5.3 Structural formula L-Tyrosine, the substrate from which dopaquinone is formed in melanogenesis.

Intrinsic Regulation of Melanogenesis

In the pituitary gland, located in the brain, two hormones; melanocyte stimulating hormone (α-MSH) and adrenocorticotropic hormone (ACTH), are derived from proopiomelanocortin. However, the pituitary production of the tetrapeptide His-Phe-Arg-Trp found in α-MSH and ACTH is insufficient to stimulate melanogenesis, therefore, melanin production rests on melanocytes and keratinocytes to also produce the tetrapeptide (Lin & Fisher, 2007). When α-MSH or ACTH connects to the melanocortin-1 receptor, an adenylate cyclase enzyme is activated by phosphorylation and, consequently, there is an increase in cAMP in the cell (Fig. 5.2).

An increase in cAMP activates protein kinase A (PKA) that successively activates CREB (cAMP response element-binding protein). CREB acts as a transcription factor, involving the microphthalmia-associated transcription factor (MITF) that in its activated (phosphorylated) state regulates the expression of melanogenic enzymes, for an example, tyrosinase (Park, Kosmadaki, Yaar, & Gilchrest, 2009).

The phosphorylation of MITF is reliant on mitogen-activated protein kinases (MAPKs), whose activity is stimulated by the binding of keratinocyte-produced stem cell factor to the c-kit receptor tyrosine kinase (Tsatmali, Ancans, & Thody, 2002). MITF protein expression is not only dependent on CREB but MITF can also be regulated by transcription factors and facilitators produced by keratinocytes (Yamaguchi & Hearing, 2009). Keratinocytes may also produce catecholamines from L-DOPA and bind to alpha 1 and beta 2 adrenergic receptors in melanocytes, which stimulated melanogenesis either via cAMP as described earlier or protein kinase C-β (PKC-β) (Slominski, Tobin, Shibahara, & Wortsman, 2004). PKC-β phosphorylation is essential as it gives rise to activation of tyrosinase. Activation of PKC-β occurs because of membrane-associated diacylglycerol (DAG) that is either released through adrenergic receptor activation or UVR action. Several factors are produced by keratinocytes as a response to UVR exposure, which may inhibit or stimulate melanogenesis, and is known as extrinsic regulation (Tsatmali et al., 2002).

Extrinsic Regulation of Melanogenesis

Plasma membrane lipids release DAG that activates PKC-β and sequentially tyrosinase, after the exposure to UVR (Schallreuter, Kothari, Chavan, & Spencer, 2008). UVR is the dominant external factor in the regulation of melanogenesis and which leads to "tanning" but can also cause DNA damage (Lin & Fisher, 2007). It was reported by Khlgatian et al. in 2002 that the tumor-suppressor protein, p53, affects the tanning response as it regulates the α-MSH production in keratinocytes and regulates the transcription of the tyrosinase transcription factor and hepatocyte nuclear factor 1α in melanocytes (Khlgatian et al., 2002).

5.1.4 Melanin Transfer and the Factors Involved

Following the synthesis of melanin and the formation of pigment granules, stimulated either through intrinsic factors or extrinsic factors, the melanosomes move toward the tips of melanocyte dendrites, where they are transferred to adjacent keratinocytes (Cardinali et al., 2005). Microfibrils and particular receptors within melanocyte dendrites are required for melanosomes' transfer from melanocytes to keratinocytes. Hypopigmentation therefore, occurs with a decrease in the rate of melanosome transfer (Kumarasinghe et al., 2006; Minwalla, Zhao, Poole, Wickett, & Boissy, 2001; Relyveld, Dingemans, Menke, Bos, & Westerhof, 2008; Westerhof et al., 2004). Although the biological processes involved in melanin transfer are not completely defined, three mechanisms of melanin transfer have been identified, namely cytophagocytosis, exocytosis, and direct injection of the melanosome in the keratinocyte via a filopodia tube (Singh, Kurfurst, et al., 2010; Singh, Sharma, Dudhe, & Singh, 2010).

Cytophagocytosis-Mediated Melanin Transfer

Cytophagocytosis is the uptake of cellular segments (keratinocyte in melanin transfer) by means of enveloping the plasma membrane and enclosing the melanin granule (Lawrence, 2008b). A transmission light microscopy study reported by Okazaki in 1976 described how melanin transfer is mediated by cytophagocytosis. Microscopic observations exhibited how the tip of the melanocyte dendrite (containing the pigment granules) appeared to penetrate the keratinocyte and subsequently form a small pouch. After an hour the constriction of the pouch was almost completed, although it was still linked to the melanocyte dendrite through a fine string. The melanocyte dendrite slowly retracted and the keratinocyte constrictive movement caused the pouch to be pinched off completely after 3 h. Further investigation under a phase-contrast microscope showed how the dendrite pouch membrane was broken down in the cytoplasm of the keratinocytes, and ultimately the melanosomes were surrounded only by a single membrane belonging to the keratinocyte. Cytophagocytosis was also confirmed with scanning electron microscopy and transmission electron microscopy (Okazaki, Uzuka, Morikawa, Toda, & Seiji, 1976).

It was reported that an increase in phagocytic activity of keratinocytes comes about through upregulation in the expression and activation of the proteinase-activated receptor-2 (PAR-2) (Sharlow et al., 2000). Consequently, melanin transfer could not take place when PAR-2 activation is inhibited in vitro (Paine et al., 2001). PAR-2 belongs to the G-protein coupled transmembrane receptor family, which is widely distributed in the body and is expressed by epithelial cells, for example, keratinocytes (Mari et al., 1996). PARs are permanently activated through the proteolytic cleavage of the amino acid terminus, which then exposes the amino terminus and forms a linked peptide ligand, which then acts as the receptor's activator by means of connection (Vu, Hung, Wheaton, & Coughlin, 1991). In HaCaT cells (keratinocytes) PAR-2 is activated by trypsin, mast cell tryptase and factors VII and Xa (Camerer, Huang, & Coughlin, 2000; Kong et al., 1997).

PAR-2 is also activated by a synthetic peptide (SLIRL) and a human peptide (SLIGKV), which corresponds to the cleaved amino terminus, independent of the receptor cleavage (Böhm et al., 1996; Steinhoff et al., 1999). Activation of PAR-2 also leads to the activation of Rho, which belongs to the Rho family of GTPases (guanosine triphosphatase—hydrolyzing enzymes) (Scheffzek & Ahmadian, 2005). The Rho family GTPases (Cdc42—Cell division control protein 42, Rac1—Ras-related C3 botulinum toxin substrate, and RhoA—Ras homolog gene family,

Continued

—cont'd

member A) play an important role in cytoskeletal remodeling during phagocytosis as well as regulate many properties of intracellular actin assembly (Bustelo, Sauzeau, & Berenjeno, 2007; Niedergang & Chavrier, 2004; Scott et al., 2003).

Actin assembly is also needed for phagocytosis induced in vitro by the keratinocyte growth factor (KGF) (Cardinali et al., 2005). The KFG stimulates phagocytosis by binding to the KFG receptor on the keratinocyte (Miki et al., 1992). An experiment was performed by Cardinali et al. in 2005 to determine the role of KGF in melanin transfer. Cardinali et al. (2005) found that the presence of the KGF receptors in the phagosomes increased when treated with KFG and there was an uptake in the latex beads (representing melanin) added therefore, it can be assumed that KGF not only stimulate phagocytosis but also melanin transfer.

Exocytosis-Mediated Melanin Transfer

Melanin transfer mediated by exocytosis occurs via the merging of the melanosomal membrane with the plasma membrane of the melanocyte, resulting in extracellular pigment granules. The extracellular melanin is then phagocytised by the keratinocytes (Van Den Bossche, Naeyaert, & Lambert, 2006). Exocytosis was recognized when electron microscopic observation of human skin showed the presence of extracellular melanin and subsequently the enveloping of these pigments granules by keratinocyte pseudopods (Yamamoto & Bhawan, 1994). The excretion of melanin (in vitro) in the extracellular space increased in quantity when the melanocyte was stimulated by the α-MSH and β-amyloid precursor proteins (Van Den Bossche et al., 2006). The electron microscopy analysis also showed the presence of exocytosis regulators, such as SNAREs (**SNA**P (Soluble NSF Attachment Protein) **RE**ceptor) and Rab GTPases, more specifically Rab 3. SNAREs comprise of three membrane-associated protein families that function in the later events of membrane fusion (Scott & Zhao, 2001).

Filopodia-Mediated Melanin Transfer

The hypothesis of filopodia-mediated melanin transfer was investigated through a scanning electron microscope (SEM) and reported by Singh et al. The SEM of epidermal melanocytes and keratinocytes revealed numerous filopodia extending from the melanocyte and interlinking with neighboring keratinocytes (Singh, Kurfurst, et al., 2010; Singh, Sharma, et al., 2010). Filopodia are kinetic narrow tubular membrane extensions (200–300 nm across—approximately the width of a melanosome), which contain long actin filaments and are fast growing at the end of protrusion (Faix & Rottner, 2006). Melanin granules were also identified (using SEM) within the filopodia extending from the melanocyte to the adjacent keratinocyte. The transfer of melanin was also confirmed with time-lapse video micrography (Singh, Kurfurst, et al., 2010; Singh, Sharma, et al., 2010). Actin-based motor protein MyoX (myosin X) is present throughout normal human skin but is localized to the dendrite actin-rich cell tips of melanocytes.

MyoX regulates filopodia formation and elongation, where it moves along the actin bundle from the filopodia starting point to become concentrated as the tips of the filopodia (Bohil, Robertson, & Cheney, 2006; Nagy et al., 2008; Tokuo, Mabuchi, & Ikebe, 2007). Cell division control protein 42 (Cdc 42) also induce

—cont'd

melanin transfer and filopodia development; nonetheless Cdc 42 cannot induce melanin transfer if the cells do not express MyoX too. MyoX is present downstream of Cdc 42 and guarantee melanosome transfer from the melanocyte to the keratinocyte. The loss of MyoX and Cdc 42 expression in the melanocytes lead to inhibition of melanin transfer (Singh, Kurfurst, et al., 2010; Singh, Sharma, et al., 2010). Mutations in myosins have been reported in pigmentation disorders (Ménasché et al., 2003; Mercer, Seperack, Strobel, Copeland, & Jenkins, 1991). Change in melanin transfer is of substantial therapeutic and clinical value in hypo- or hyperpigmentary conditions, therefore, upregulation or downregulation of melanocyte filopodia development or keratinocyte phagocytosis may be beneficial in the treatment of those diseases (Singh, Kurfurst, et al., 2010; Singh, Sharma, et al., 2010).

5.1.5 Prevalence of Progressive Macular Hypomelanosis

PMH has been identified worldwide but is more prevalent in dark skinned individuals, originating from tropical countries, and young females (Relyveld, Menke, & Westerhof, 2012). However, the true prevalence of PMH is still unknown, and PMH has not been fully reported in Africa or the Middle East (Hassan et al., 2014; Relyveld, Menke, & Westerhof, 2007). Other hypopigmented diseases, including Vitiligo, occur worldwide with an estimated prevalence 1% in most populations (Taïeb & Picardo, 2009).

5.2 MEDICINAL PLANTS USED FOR PROGRESSIVE MACULAR HYPOMELANOSIS AND OTHER HYPOPIGMENTED DISORDERS

There is no record of plants traditionally used specifically for PMH, however, certain plants are well documented in treating leukoderma that is also a hypopigmentary disorder. In India (Churu district—Thar Desert) a survey was carried out in 2005 to document the utilization of indigenous medicinal plants for leukoderma (Upadhyay, Roy, & Kumar, 2007). One of medicinal plants for leukoderma consisted of the bark of young branches of *Tecomella undulata* Seem., commonly known as marwar teak or desert teak (Nandwani, 2014). *T. undulata* originated in north-western India and Arabia. It grows wild in Rajasthan but is also found in Jodhpur, Bikaner, Jaisalmer, Barmer, Sikar, Naguar, Jhunjhunu, Churu, and Ganganagar (Negi et al., 2011). Globally it is found in southern Iran, eastern Afghanistan, and Pakistan (Hameed, 2016).

In another instance, it has been reported that paste made from the seeds of *Raphanus sativus* L. and *Achyranthes aspera* L., and the leaves of *Citrullus colocynthis* (L.) Schrad. were examples of plants that are applied to the hypopigmented areas in leukoderma patients in India (Fig. 5.4A and B) (Upadhyay et al., 2007). *R. sativus*, or better known as radish, is grown worldwide for its edible taproot. Radish originated in the eastern parts of the Mediterranean. In 2000 BC, radish was a very important food crop grown in Egypt.

In 500 BC it was spread to China and AD 700 reached Japan. Radish was only introduced in Britain during the mid-16th century (KEW, 2016a). *A. aspera*, commonly known as the Devil's horsewhip, on the other hand originated in Southeast Asia and Africa but is now distributed all over the world; from Taiwan in Asia to South Africa and Mexico to the Bahamas and Australia (Fig. 5.4B; CABI, 2014). *C. colocynthis* is commonly known as bitter cucumber, bitter apple, or vine of Sodom (Herbcyclopedia, 2014) and is endemic to

FIGURE 5.4 A few medicinal plants reported to show efficacy for hypopigmented diseases. (A) *Raphanus sativus* L. (Jengod, 2009), (B) *Achyranthes aspera* L. (Jose, 2010) used in for the treatment of leukoderma (Zell, 2009), (C) *Withania somnifera*, (D) *Terminalia prunioides* (Dupont, 2011), (E) *Psoralea corylifolia* L. (Ganguly, 2013), (F) *Gloriosa superba* L., (G) *Xanthium strumarium* L., traditionally used in Indian medicine (Mohlenbrock, 2005), (H) *Hypericum revolutum* (Rae, 2010), (I) *Olea europaea* subs. *africana* (Mill.) P.S. Green, and (J) *Angelica archangelica* L. (Kat, 2007). *(C, F, I) Photo: Analike Blom van Staden.*

southern Tunisia but is found in the desert areas of Punjab and Sindh provinces of Pakistan (Marzouk et al., 2010).

Another Indian plant used for leukoderma is *Xanthium strumarium* L. that was recorded in a book entitled Indian medicinal plants (Fig. 5.4G; Khare, 2007). Extracts of the whole plant, including the fruits, seeds, leaves, and roots are applied to the affected areas (Kamboj & Saluja, 2010). *X. strumarium*, commonly known as cocklebur or burweed, originated in Central and South America but is now found in the subtropical and Mediterranean parts of Asia, Africa, North America, Europe, and several islands (CABI, 2016c). The name *Xanthium strumarium*, stemmed from the Greek words "xanthos" and "strumarium," which means "yellow" and "cushion like swelling," due to its round seedpods that turn from green to yellow as they ripen (Dharmananda, 2003).

The utilization of Indian plants, such as *Telosma pallida* (Roxb.) Craib and *Launaea asplenifolia* Hook.f. for leukoderma is normally using the latex of fruit and fruit paste that is applied externally on the localized white patches (in the initial stage of the disease). After application of the paste, the patient is requested to sit in sunlight for 20–30 min. Another application of the paste is given at night and left for the whole night. The paste is then washed with water the following morning. The treatment is continued until some improvement is observed. During observation, it was found that the white patches on the skin of patient was reduced almost 90% after 6 months (Singh & Narain, 2010). *T. pallida* is not only found in India but also in Nepal, Pakistan, Jammu and Kashmir, Thailand, Sikkim, and Burma (GBIF, 2016). *L. asplenifolia* is mostly distributed in India, in the Aurangabad, Osmanabad, and Nanded districts and in the Uttar Pradesh region of India, but it has been found in the rice fields of Southeast Asia, where it is considered a weed (Moody, 1989; Singh & Narain, 2010).

Several plants in literature have been reported to stimulate melanogenesis and/or upregulate tyrosinase activity in the melanocytes. The *Piper nigrum's* L., commonly known as black pepper, fruit extract exhibited growth-stimulatory activity toward cultured melanocytes. Black pepper originated in the south western regions of India, where it was later cultivated for trading. From India, black pepper was taken to Malaysia, Indonesia, and 27 other countries, including Vietnam, China, Cambodia, Fiji, Brazil, Mexico, Madagascar, Zimbabwe, and Kenya (Ravindran, 2003). The aqueous extract of only 0.1 mg/mL *P. nigrum* L. instigated nearly 300% stimulation of the growth of cultured mouse melanocytes, within 8 days. The growth stimulating effect and dendrite elongation of *P. nigrum* L. fruit is probably due to the presence of the alkaloid piperine (1-piperoyl piperidine) (Lin, Hoult, Bennett, & Raman, 1999). A synthetic derivative of piperine was later developed and is used in "Kalawalla," an herbal treatment used for vitiligo (Tahir, Pramod, Ansari, & Ali, 2010). The synthetic piperine compound stimulates pigmentation in the melanocytes and results in a light brown color after 6 weeks of treatment, however, the compound works best when combined with UVR (Gutierrez & Writer, 2014).

Plants such as *Psoralea corylifolia* L., *Angelica archangelica* L., *Olea europaea subs. africana* (Mill.) P.S.Green and *Gloriosa superba* L. have been used to restore melanin production in the hypopigmented areas (Fig. 5.4E and J; Chandak, Devdhe, & Changediya, 2009; Chopra, Nayar, & Chopra, 1956, p. 81; Khandel, Ganguly, Bajaj, & Khan, 2012; Vaidya, 2006). *P. corylifolia*, commonly known as Babchi, is a traditional Chinese Ayurvedic medicine. It is mostly used against skin diseases but has been studied for its antioxidant and antiinflammatory activity (Khushboo, Jadhav, Kadam, & Sathe, 2010). *P. corylifolia*, is found all over India and has spread to the tropical and subtropical regions, including Southern Africa and parts of China (Krishnamurthi, 1969; Sharma, Yelne, Dennis, Joshi, & Billore, 2000). *A. archangelica* L. is believed to be native to Syria from where it spread to northern countries, such as Scotland, Lapland, and Iceland. *A. archangelica*, also known as *Angelica officinalis*, but commonly known as the garden Angelica, is cultivated for its medicinal use in Thuringia and the roots are exported to Spain (Grieve, 2016). *O. europaea subs. africana* and *G. superba* are both South African plants, distributed all over South Africa, but is also found in other African countries, India, and Southeastern Asia (Fig. 5.4F and I; Joffe, Notten, & Mutshinyalo, 2001).

5.3 PLANT PRODUCTS FOR HYPOPIGMENTED DISORDERS FROM MEDICINAL PLANTS

"Anti-vitiligo" is a traditional herbal formulation, which contributes to the repigmentation in the hypopigmented skin patches of vitiligo patients and can be used either orally or in the form of topical preparations (Tahir et al., 2010). An increase in the melanin context and the amount of melanocytes in the affected epidermal skin areas is due to the upregulated activity of tyrosinase by the ethanol extracts of *P. corylifolia*, the main ingredient in "Anti-vitiligo" (Fig. 5.4E; Li, Zhu, & Xia, 2001). *P. corylifolia* is a rich source of psoralens, which belong to the furacoumarins phytochemical group (Song & Tapley, 1979). Psoralen-containing plants have been used for many years in the treatment of vitiligo (Scott, Pathak, & Mohn, 1976). Psoralens was first isolated by Fahmy and Abu-Shady in 1947 and in the same year, they initiated the treatment of vitiligo patients with 8-methoxypsoralen and 5-methoxypsoralen combined with sun exposure—describing it as "phototherapy" (Fahmy & Abu-Shady, 1947; Roelandts, 2002). Other plants containing psoralens, used in other traditional Chinese medicine, are *Paeonia veitchii*, *Ligusticum chuanxiong*, *Cuscuta chinensis*, and *Tribulus terrestris*, which could possibly be used for vitiligo (Li et al., 2001).

The current treatment for PMH includes the application of 1% clindamycin and 5% benzoyl peroxide gel, combined with UVA light therapy or psoralens combined with UVA (Fig. 5.5; Westerhof et al., 2004). There are, however, no herbal-based treatments specifically for PMH.

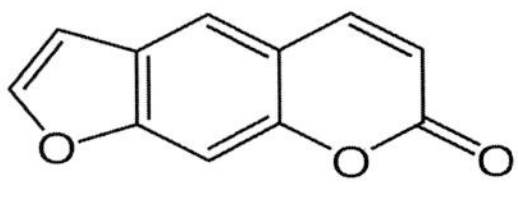

FIGURE 5.5 Skeletal formula of psoralen that is used in conjunction with ultraviolet A for the treatment of progressive macular hypomelanosis.

5.4 COMPOUNDS STIMULATING MELANIN PRODUCTION

Melanogenesis is stimulated and regulated by many different plant metabolites. These components regulating melanogenesis could in turn be influenced by factors such as plant extracts. Several compounds isolated from plants have been reported to possess a stimulatory effect on melanin synthesis. The following section gives the synopsis of some of these compounds.

5.4.1 Terpenoids

Lupane triterpenes have variable effects on melanogenesis depending on their structure (Fig. 5.6). According to earlier reports the ketone formation at carbon 3, as in the case of lupeol, enhances its stimulating effect on melanogenesis. It has been reported earlier that lupeol stimulated melanin production by inducing dendrite growth, mediated by the rearrangement of the actin cytoskeleton (Hata et al., 2006). It was also previously found that the redisposition of cytoskeletal components led to dendrite outgrowth of B16 cells (Bertolotto et al., 1998). Lupeol activated cofilin, an actin depolymerizing factor, which caused actin fiber disassembly in B16 2F2 cells. Western blot analysis by Hata et al. in 2006 indicated that lupeol, at a concentration of 10 mM, induced the activation (phosphorylation) of p38 MAPK in B16 2F2 cells, when compared to untreated cells (Hata et al., 2006). It was also determined that activation of p38 MAPK by lupeol is prevented by the inhibition of PKA, which indicated that p38 MAPK is a downstream target of the cAMP-PKA pathway (Lee, Kim, & Park, 2007). Conversely, lupane triterpenes, that have a carbonyl group at carbon 17, inhibited melanocyte proliferation and also induced apoptotic effects on B16 2F2 cells. These lupane triterpenes that have a carboxyl group at carbon 17 showed

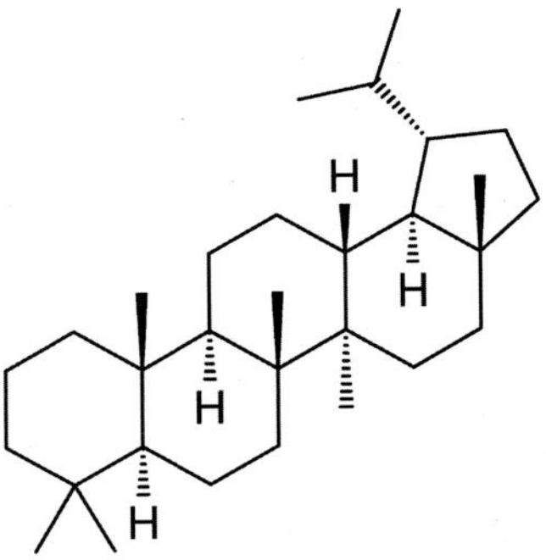

FIGURE 5.6 Chemical structure of the lupane triterpene that has variable effects on melanogenesis.

severe cytotoxic effects on B16 2F2 cells (Hata et al., 2006). Therefore, triterpenes with different structures have different kinds of effect on melanogenesis.

Glycyrrhizin, another triterpene, found in the rhizomes and roots of *Glycyrrhiza glabra* (liquorice), is composed of glycyrrhetinic acid and two glucuronic acids (Andersen, 2007). *G. glabra* is native to Europe, Asia, Northern Africa and Western Asia but is now found in USA and cultivated in Spain, Russia, and the Middle East (KEW, 2016a). A recent report has shown that glycyrrhizin is involved in melanogenesis (Jung, Yang, Song, & Park, 2001). Glycyrrhizin induces melanin production in B16 melanoma cells by upregulating both tyrosinase messenger RNA (mRNA) and tyrosinase-related protein-2 (TRP-2) mRNA. Glycyrrhizin induces melanogenesis through cAMP signaling, which was indicated by the reduction in stimulated melanogenesis through the inhibition of PKA and blocking of cAMP response element (CRE) activation by H-89. Therefore, glycyrrhizin activates CREB and also showed an increase in the production of cAMP. Glycyrrhizin operates upstream of PKA as the inhibition of protein kinase by H-89 did not affect the stimulation of cAMP production by glycyrrhizin. Furthermore, glycyrrhizin activated cyclic response filament "CRE" promoters and activator protein-1 (AP-1) but not the nuclear factor-κB promoter. Together with the activation of AP-1 promoter by glycyrrhizin, MAP, p42/44 MAPK, was also stimulated but does not have a direct effect on the amount of melanin content as MAPK was not blocked by PD98059, an MEK1 inhibitor (Lee et al., 2005).

5.4.2 Phenolics

Phenolics can be classified in two groups, namely simple phenols and polyphenols, based on the number of phenol subunits. Simple phenols include phenolic acids or phenols with the carboxyl group determining the function of the compound. Polyphenols, such as flavonoids, contain at least two phenol rings (Marinova, Ribarova, & Atanassova, 2005). These secondary compounds play an important role in melanin synthesis, which is described in the following section.

5.4.2.1 Flavonoids

Quercetin (3,3′,4′,5,7-pentahydroxyflavone) is a diphenylpropanoid (a phenylalanine derivative) and forms part of the flavonoid secondary metabolites (Takeyama, Takekoshi, Nagata, Osamura, & Kawana, 2004). Quercetin induced melanogenesis in both natural human epidermal melanocytes and human melanoma cell line VII at concentrations of 1 and 20 μM, respectively (Fig. 5.7; Nagata, Takekoshi, Takeyama, Homma, & Yoshiyuki Osamura, 2004). Melanogenesis was augmented through the upregulation of tyrosinase and the inhibition of melanogenic inhibitors (Nagata et al., 2004; Takeyama et al., 2004). Phase-contrast microscopy also confirmed the increase in melanin granules at the tips of the dendrites and an increase of dendrite protrusions toward

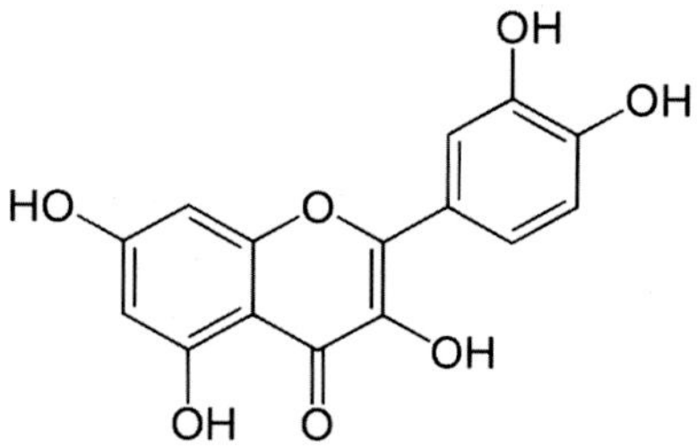

FIGURE 5.7 Chemical structure of quercetin, a compound known to induce melanogenesis.

the adjacent keratinocytes after the 10 μM quercetin treatment. Conversely, the melanocytes in the control group had short dendrites and less melanized melanosomes (Takeyama, et al., 2004).

Tyrosinase activation by quercetin was analyzed by DOPA activity staining and by investigating the effect of actinomycin D (transcription inhibitor) and cyclohexamide (protein synthesis inhibitor) on the quercetin-activated tyrosinase (Nagata et al., 2004; Sobell, 1985). Both actinomycin D and cyclohexamide hindered tyrosinase activation, which suggests that both transcriptional and translational events are responsible for the increase in melanogenesis by quercetin. However, immune-blot analysis showed a poor correlation between the DOPA reactivity and the tyrosinase protein levels, with no statistically significant difference between the treated cells and the control. Quercetin also didn't affect the expression of tyrosinase mRNA (Nagata et al., 2004). Another mode of tyrosinase activation by quercetin could be through the activation of wild type p53, which is known to activate both tyrosinase and tyrosinase-related protein 1 (TRP-1) (Nylander et al., 2000; Plaumann, Fritsche, Rimpler, Brandner, & Hess, 1996). Quercetin suppresses cytotoxicity that is caused by hydrogen peroxide (H_2O_2), through activating glutathione peroxides, responsible for converting H_2O_2 into glutathione disulphide and water. The reduction of H_2O_2 by quercetin may play a key role in the modulation of melanogenesis (Nagata, Takekoshi, Takagi, Honma, & Watanabe, 1999).

Naringenin (4′,5,7-trihydroxyflavanone), a citrus flavanone is chemically related to quercetin and also increases melanogenesis in mouse B16 melanoma cells (Chiang, Lin, Hsiao, Tsai, & Wen, 2011; Ohguchi, Akao, & Nozawa, 2006). Increase in tyrosinase activity was measured with western blotting, which indicated that MITF is mainly responsible for the increase in melanin content through the addition of naringenin (Ohguchi et al., 2006).

5.4.2.2 Phenolic Acid

Rosmarinic acid (a-o-caffeoyl-3,4-dihydroxyphenyl lactic acid) is found in perilla, sweet basil, and rosemary (Baba et al., 2005; Erkan, Ayranci, & Ayranci, 2008; Jayasinghe, Gotoh, Aoki, & Wada, 2003). Lee et al. in 2007 established that rosmarinic acid increases the melanin content and tyrosinase expression in

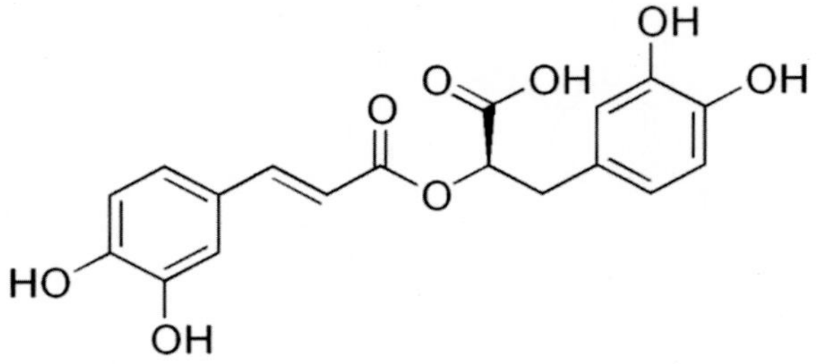

FIGURE 5.8 Chemical structure of rosmarinic acid that has been shown to increase melanin content as well as the expression of tyrosinase in B16 melanoma cells.

B16 melanoma cells (Fig. 5.8). The tyrosinase expression was determined by Western blot analysis with α-MSH and forskolin as the positive controls. The mechanism linking PKA to the increased tyrosinase expression (stimulated by rosmarinic acid) was selected based on (1) the reduction of tyrosinase expression—when H-89 and KT 5720 (PKA inhibitors) were added—and (2) the inconsequential effect when SB203580 (p38 MAPK inhibitor) or Ro-32-0432 (PKC inhibitor) was added (Lee et al., 2007).

Further investigation into the mechanism determined whether the production of cAMP—the activator of PKA—was also affected by rosmarinic acid (Das et al., 2007). Rosmarinic acid activated the CRE reporter activity but did not have any effect on cAMP production. Consequently, the upregulation of melanogenesis by rosmarinic acid occurred downstream from cAMP production. Rosmarinic acid activates the transcription factor—CREB—and PKA but have no effect on the phosphorylation of p38 MAPK, serine-threonine protein kinase (Akt) and glycogen synthase kinase (GSK) 3β (Lee et al., 2007). Accordingly MITF (which contains a CRE element) is promoted by cAMP when CREB and PKA have been activated by rosmarinic acid, which then leads to the increase in melanogenesis (Hamid, Sarmidi, & Park, 2012). As yet, the activation of PKA by rosmarinic acid may be facilitated by a cAMP-dependent pathway, but research conducted by other authors indicated that TGF β and TNF α-induced reactive oxygen species could also activate PKA (Levy, 1992; Zhang et al., 2004). Consequently, PKA could also be activated by rosmarinic acid through interactions.

5.4.2.3 Acetophenones

Xanthoxylin is a phenolic ketone ($C_{10}H_{12}O_4$) present in plants such as the Brazilian plant *Sebastiania schottiana*, the Asian plant *Blumea balsamifera*, and the Japanese plant *Zanthoxylum piperitum* (Moleephan, 2012; Wang et al., 2014). Xanthoxylin showed similar result to quercetin by increasing melanin content and dendrite elongation explained earlier. After a Western blot analysis, where H-89 and KT 5720 (PKA inhibitors) and SB203580 (p38 MAPK inhibitor) or Ro-32-0432 (PKC inhibitor) were added to the xanthoxylin-stimulated mouse B16F10 melanoma cells, it was concluded that PKA was the main signaling pathway mechanism as explained for rosmarinic acid. Moleephan et al. in 2012 showed that xanthoxylin induce melanin synthesis by means of PKC, PKB, and MAPK-signaling pathways to some extent. The mechanism involved in the upregulation of melanin could be determined through investigating the effect of PKA, PKC, PKB, and MEK1 inhibitors on the melanin content stimulated by xanthoxylin. The xanthoxylin-stimulated melanin content in the control cell group was 100% in comparison to the cells containing the PKA, PKC, PKB, and MEK1 inhibitor resulting in 44.66%, 67.25%, 84.39%, and 86.28% melanin content, respectively. Xanthoxylin also significantly induced MITF expression at all concentrations used in the study (Moleephan, 2012).

5.4.3 Plant Hormones

Cytokinins are an important group of plant growth regulatory substances especially 6-benzylaminopurine (6-BAP), which promotes plant growth. It has been found that the cytokinin, 6-BAP, stimulated melanogenesis by increasing the level of tyrosinase, TRP-1, TRP-2, and MITF protein in the cells. Western blot analysis showed that 6-BAP upregulated the level of CREB leading to increased melanin synthesis. The addition of several inhibitors such as H-89 (PKA inhibitor), SB203580 (p38 MAPK inhibitor), and Ro-32-0432 (PKC inhibitor) showed that 6-BAP stimulated melanin production through protein kinase 1 but it did not increase the intracellular cAMP level (Kim et al., 2009). For that reason, 6-BAP activates PKA via a cAMP-independent pathway. Consequently, PKA (stimulated by 6-BAP) activation could take place via interactions with other cellular signaling pathways, such as with transforming growth factor α (TGF- α), which is directly regulated by the smad3/4 complex, or sphingosine that activates PKA via a unique cAMP-independent mechanism (Ma et al., 2005; Zhang et al., 2004). Endothelin-1, a vasoactive peptide, stimulates PKA activity by involving the degradation of IκB independent of cAMP (Dulin, Niu, Browning, Richard, & Voyno-Yasenetskaya, 2001).

5.5 ANTIBACTERIAL COMPOUNDS

Pathogenicity of *P. acnes* is primarily due to its ability to produce exocellular enzymes or proteins that influence the immune system and give rise to inflammation (Perry & Lambert, 2006). Several studies on the antibacterial activity of plant extracts have been conducted and many compounds have been isolated, which have been dealt within the following section.

5.5.1 Terpenes

Essential oils found in most aromatic plants are a good source of antibacterial qualities (Prabuseenivasan, Jayakumar, & Ignacimuthu, 2006). It was reported that the essential oil of *Abies koreana*, commonly known as Korean fir and

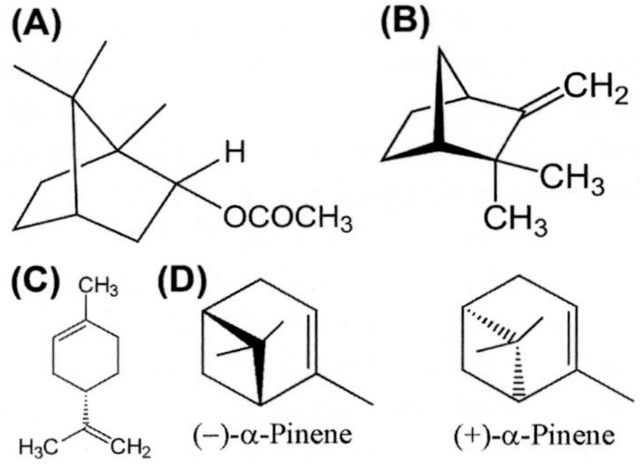

FIGURE 5.9 Chemical compounds isolated from *Abies koreana*, the Korean fir, responsible for the antibacterial activity of the plant against *Propionibacterium acnes*. (A) bornyl acetate, (B) camphene, (C) limonene, and (D) α-pinene.

endemic to South Korea, exhibited good antibacterial activities against *P. acnes* (IUCN, 2016a). The active components were identified as bornyl acetate (ester), camphene (monoterpene), limonene (cyclic terpene), and a-pinene (terpene) (Fig. 5.9). The minimum inhibitory concentrations of bornyl acetate, camphene, limonene, and a-pinene against the *P. acnes* (ATCC 3314 strain) were 5.000, 0.156, 0.625 and 0.500, and 20.000, 0.156, 0.625 and 0.125 μL/mL against the *P. acnes* SKA-7 strain, respectively (Yoon, Kim, Oh, Lee, & Hyun, 2009). The antibacterial activity of cyclic terpenes is ascribed to their initiation of bacterial membrane integrity loss and the failing of the bacteria's proton motive force (Sikkema, De Bont, & Poolman, 1995).

The terpenes—α-terpineol and linalool—and monoterpene—sabinene—form the main constituents of *Zingiber cassumunar*, Cheonyahagyul and Geumgamja oils. Alpha-terpineol and linalool showed to inhibit *P. acnes* growth with minimum inhibitory concentration (MIC) of 1.250 and 0.625 μL/mL, respectively (Kim et al., 2008; Pithayanukul, Tubprasert, & Wuthi-Udomlert, 2007). The antibacterial activity of basil oils from *Ocimum basilicum* (sweet basil) and *Ocimum sanctum* (holy basil) against *P. acnes* were established by using an agar dilution assay. MIC values of sweet basil and holy basil oils were 2.0% and 3.0% v/v, respectively. Although α-bergamotene and γ-caryophyllene are present in the highest concentrations, the antibacterial activity of plant extracts with high contents of eugenol has been reported in the past (Himejima & Kubo, 1992; Kubo, Muroi, & Kubo, 1994). As a result, eugenol may be the major antimicrobial compound of basil oil (Viyoch et al., 2006).

Biological activities such as antimicrobial activity of diterpenoids is linked to the α,β-unsaturated ketone group in the D-ring (characteristic of Rabdosia diterpenoids) which acts as alkylation agents, given that the corresponding dihydro derivatives typically do not exhibit any activity (Fujita & Node, 1984; Kubo, Kamikawa, & Kubota, 1974; Yamaguchi, Taniguchi, Kubo, & Kubota, 1977). The abietane diterpene—2b-acetoxyferruginol—isolated from *Prumnopitys andina*, endemic to Chile (IUCN, 2016a, 2016b), and four diterpenes—rosthornins, with an α,β-unsaturated ketone in the D-ring—isolated from Chinese *Rabdosia rosthornii* exhibited MIC of 4 μg/mL and between 3.17 and 25 μg/mL against *P. acnes*, respectively (Kubo, Xu, & Shimizu, 2004; Smith, Wareham, Zloh, & Gibbons, 2008). The four rosthornins also exhibited minimum bactericidal concentrations (MBC) ranging from 3.17 to 25.00 μg/mL. Therefore, no difference in their MIC and MBC was observed, suggesting no enduring bacteriostatic activity (Kubo et al., 2004). A possible explanation for the bactericidal effect may be that the highly lipophilic rosthornin enters the lipid-bilayer portions of the bacterial cytoplasmic membranes. Various enzymes, energy converting systems such as electron transport chains and ATPases are embedded in the bacterial membrane lipid bilayers. Once inside the lipid bilayer, rosthornin may inhibit the oxidative phosphorylation as was found in the mitochondria isolated from rat liver with Rabdosia diterpenoids (Kubo et al., 2004; Yamaguchi et al., 1977).

5.5.2 Alpha Acids

The flower extracts of *Humulus lupulus*, commonly known as hops, contained humulone, lupulone, and xanthohumol, which have shown moderate to strong anticollagenase inhibitory activities and exhibited bactericidal activity against *P. acnes*. Humulone, lupulone, and xanthohumol exhibited MIC at 10, 0.1, and 3.0 μg/mL, respectively (Fig. 5.10). Lupulone and xanthohumol showed to be bactericidal at 0.3 and 3.0 μg/mL, respectively (Yamaguchi, Satoh-Yamaguchi, & Ono, 2009). *H. lupulus* originates in Europe, and the major crops are produced in the United States, Germany, England, and Czech Republic (Snyder & Conway, 2008).

5.5.3 Phenolics

Phenolic compounds have similar mechanisms concerning their antibacterial activity, which includes the disruption of bacterial cell membranes, but the capacity to do this is dependent on the compounds' ability to cross the bacterial cell membrane, which is determined by the compounds' lipophilicity (Smith et al., 2008). Consequently, the phenolic compound rhinacanthin-C (a naphthoquinone ester), isolated from *Rhinacanthus nasutus* and the crude extract itself showed antibacterial activity against *P. acnes* with MIC values of 8 and 16 μg/mL, respectively. The crude extract also exhibited effective bactericidal activity with an MBC value of 32 μg/mL, but rhinacanthin-C did not exhibit bactericidal effects that could be due to synergy between other compounds found in *R. nasutus* (Puttarak, Charoonratana, & Panichayupakaranant, 2010). *R. nasutus* is indigenous to Bengal, but its origin is both India and China (Das, 2006).

Other phenolic compounds with antibacterial activity against *P. acnes* have been reported in the *Magnolia* species. Polyphenols namely honokiol and magnolol showed to have their MIC of 4 and 9 μg/mL, respectively (Fig. 5.11). The rate of bacterial killing was also determinant at different concentrations of magnolol or honokiol and resulted in more than 10^5 organisms/mL being destroyed

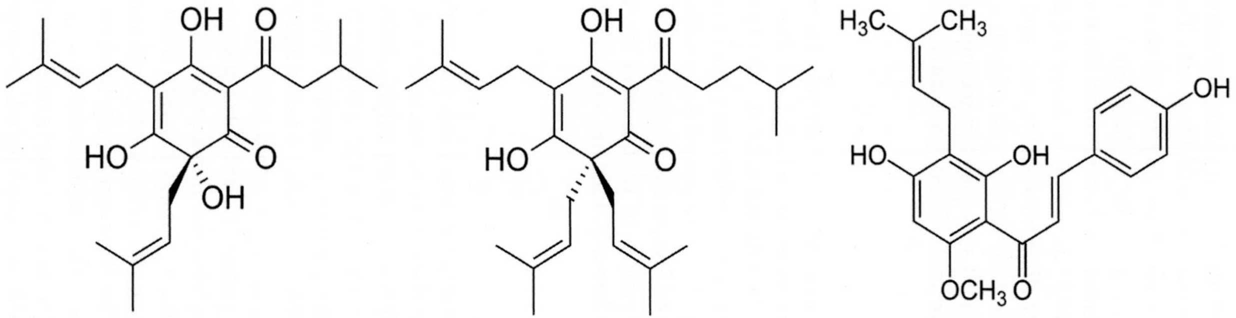

FIGURE 5.10 Compounds isolated from flower extracts of *Humulus lupulus* or hops with antibacterial activity against *P. acnes*. (Left) humulone, (middle) lupulone, and (right) xanthohumol.

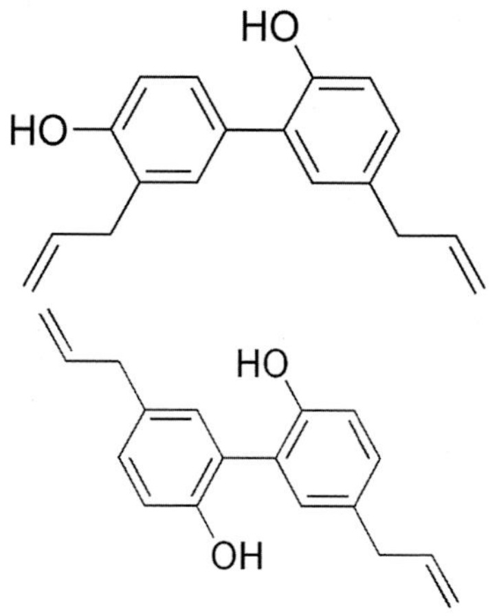

HO
HO
HO
OH

FIGURE 5.11 Polyphenolic compounds isolated from *Magnolia* species that have been shown to have antibacterial activity against *P. acnes.* (Top) honokiol and (bottom) magnolol.

(reduction to <4 CFU/mL) within 2 h at the highest concentration tested. The bacterial cultures were examined for 24 h and no regrowth was detected (Park et al., 2004).

5.5.4 Chalcones and Flavonoids

Several research have shown the antibacterial activity against gram-positive bacteria due to the presence of flavonoids and compounds derived from flavonoids such as 2′,6′-Dihydroxy-3′-methyl-4′-methoxy-dihydrochalcone, eucalyptin and 8-desmethyleucalyptin (Kubo et al., 1974; Lim, Kim, & Seo, 2007; Roopashree, Dang, Rani, & Narendra, 2008; Takahashi, Kokubo, & Sakaino, 2004; Vijayalakshmi, Tripura, & Ravichandiran, 2011). Flavonoids are known to mutilate bacterial cell walls and membranes, take apart the cytoplasmic proteins from the cytoplasm and reduce protein synthesis such as heat shock proteins and lipases (Fu et al., 2009; Kim, Kim, & Lim, 2010; Pan, Chen, Lin, & Lin, 2009). The structures of the flavonoids also influence the activity against bacteria. According to the MIC and MBC values obtained by Wang et al. in 2013, quercetin was found to have lower antibacterial activity against *P. acnes* than retinoic acid and erythromycin but exhibited greater activity than isoquercetin, rutin, and azelaic acid. The antibacterial activity of quercetin against *P. acnes* was 16 times greater than isoquercetin, while rutin did not demonstrate any inhibitory activity. The structural differences between quercetin, isoquercetin, and rutin are at the carbon 3 position, where quercetin has a free OH group on carbon 3, whereas isoquercetin has a glycoside at carbon 3 resulting from C3-glycosylation of quercetin, and rutin is formed from the rhamnosylation of isoquercetin (Barber & Behrman, 1991). For that reason, the findings by Wang et al. in 2013 support the hypothesis that flavonoid's aglycones exhibited more antibacterial activity than when present as glycosides and that the structure of

C3—OH is essential for significant antibacterial activity of flavonoids (Mori, Nishino, Enoki, & Tawata, 1987; Saising, Hiranrat, Mahabusarakam, Ongsakul, & Voravuthikunchai, 2008; Tsuchiya et al., 1996). Glycosylation of quercetin lowers the lipid-solubility and it is also larger in size, leading to more steric hindrance in membrane penetration (Wang, Yang, Qin, Shan, & Ren, 2013).

5.5.5 Tannins

Tannins found in *Garcinia mangostana*, *Momordica charantia*, and *Terminalia arjuna* exhibited antibacterial activity against *P. acnes* with MIC ranging between 0.425 and 0.625 μg/mL, respectively (Pothitirat, Chomnawang, Supabphol, & Gritsanapan, 2009; Roopashree et al., 2008; Vijayalakshmi et al., 2011). *G. mangostana* is a popular fruit, commonly known as mangosteen, of Southeast Asia but originated in Peninsular Malaysia (Nazre, 2014). *M. charantia*, known as bitter gourd, is a common vegetable in Asia but native to African and Australian continents. It is currently cultivated in North, South and Central America, the West Indies and islands found in the Pacific ocean (CABI, 2016b). *T. arjuna* is commonly known as arjuna and is an ancient Indian remedy; the stem bark powder was used for heart ailments (Dwivedi & Chopra, 2014). The standardized pomegranate rind extract of *Punica granatum* fruits contained several hydrolyzable tannins of the ellagitannin group. The fruit extract exhibited antibacterial activity against *P. acnes*, with an MIC of 15.6 μg/mL, respectively (Panichayupakaranant, Tewtrakul, & Yuenyongsawad, 2010). The pomegranate tree originated in India and the Middle East. It has been used centuries for medicinal purposes and is now cultivated in United States, Israel, Spain, and India (Mertens-Talcott, Jilma-Stohlawetz, Rios, Hingorani, & Derendorf, 2006).

5.5.6 Xanthonoid

Xanthone and its derivatives have shown antibacterial activities against several gram-positive bacteria but the mechanism of action is still unidentified (Chomnawang, Surassmo, Nukoolkarn, & Gritsanapan, 2005; Pothitirat et al., 2009). Mangostin, a xanthone derivative produced by guttiferaeous plants, such as *G. mangostana* had an MIC of 1.95 μg/mL and was also bactericidal with an MBC of 1.95 μg/mL, respectively (Pothitirat et al., 2009).

5.5.7 Phenylpropanoids

Eugenol and methyl eugenol, the two phenylpropenes, which are also present in the antibacterial basil oils from *O. sanctum* and *O. basilicum* (described at the terpenes), showed inhibitory activity toward *P. acnes* growth (Viyoch et al., 2006). Resveratrol (3,40,5-trihydroxystilbene) showed 50% inhibitory concentrations (IC_{50}) after 24 h of treatment at 82, 63, and 73 mg/L for the three

P. acnes strains ATCC 25746, ATCC 29399, and ATCC 33179, respectively. The concentration 50, 100, or 200 mg/L of resveratrol tested were found to be bactericidal against the ATCC strain 25746 after 24 h of treatment, whereas the ATCC 29399 and ATCC 33179 strains still produced a small number of colonies after 24 h of treatment. Docherty et al. in 2007 found that there is a link between *P. acnes'* generation time and their susceptibility toward resveratrol, as ATCC strain 25746 that replicated the fastest with a generation time of 3.3 h, was the most subjected to resveratrol, while ATCC strain 29399 with the slowest generation time was the least responsive to resveratrol treatment (Docherty, McEwen, Sweet, Bailey, & Booth, 2007). *O. basilicum*, known as sweet basil, originated in the tropical Indo-Malayan and Indo-Pakistan regions but is now available on shelf worldwide (Khair-ul-Bariyah, Ahmed, & Aujla, 2012). *O. sanctum*, known as Holy Basil or Tulsi leaves, is grown in Nepal as a religious plants but can be found in people's gardens all over the world (Kayastha, 2014).

5.5.8 Alkaloids

Alkaloids, such as berberine and harmine, have also shown inhibition of *P. acnes* growth by intercalating with AT base pairs in the bacteria's DNA (Fig. 5.12) (Hopp, Cunningham, Bromel, Schermeister, & Khalil, 1975; Iwasa et al., 2001; Kumar et al., 2007). Intercalation of berberine, an isoquinoline alkaloid, in the bacterial DNA results in the impairment of both reproduction and transcriptional activity of the bacterial cells (Kumar et al., 2007). The crude extract of *Mahonia aquifolium*, commonly known as holly leaved barberry or Oregon grape, together with the alkaloids berberine and jatrorrhizine displayed effective inhibitory activity against 20 clinical isolates of *P. acnes* with MIC between 5 and 50 μg/mL, but berberine was the most active with an MIC of 5 μg/mL, respectively (Slobodníková, KoSt'álová, Labudová, Kotulová, & Kettmann, 2004). *M. aquifolium* is native to the western states of United States and is

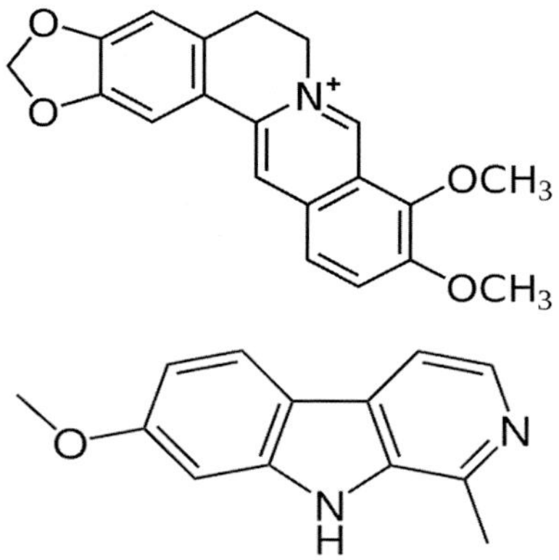

O
O
N^+
OCH_3
OCH_3
O
N
N
H

FIGURE 5.12 Chemical structures of alkaloids exhibiting antibacterial action against *P. acnes*. (Top) berberine and (bottom) harmine.

cultivated in Europe. *M. aquifolium* is seen as an alien shrub in Germany as it invades seminatural habitats (CABI, 2016a, 2016b, 2016c).

5.5.9 Fatty Acids

Naturally occurring fatty acids, such as azeliac acids, linoleic, and lauric acids, have shown to inhibit comedone formation and their antiinflammatory properties were found to inhibit bacterial activity of *P. acnes* (Dweck, 1997; Gibson, 1996; Holland & Bojar, 1989). Seed oils from plants such as pumpkin (*Cucurbita pepo*), sunflower (*Helianthus annuus*), and flax (Linum sp.) have a high fatty acid content that includes fatty acids such as azeliac acid, linolenic acid, and linoleic acid and could form part of topical treatments for acne (Kanlayavattanakul & Lourith, 2011; Tolkachev & Zhuchenko, 2004).

5.6 SCIENTIFIC RESULTS

The research on PMH has recently been initiated; therefore, there is still a lot of research to be done, including pathogenesis of the disease and finding alternative treatments. Most experiments currently conducted focused either on the bacterial aspect or on the melanin production aspect, but none investigated if plant extracts could have a dual effect; i.e., an inhibitory effect on *P. acnes* and stimulatory effect on melanin production. Chiang et al. in 2011 looked at the effect of increased naringenin content in citrus extract on melanogenesis, which could be applied in suntan lotions. They found that not only did the increased level of naringenin induce cellular melanogenesis but it also increased tyrosinase expression (Chiang et al., 2011).

Naringenin belongs to the flavanone phytochemical group and are known as phytoestrogens (Fig. 5.13; Erlund, Meririnne, Alfthan, & Aro, 2001). As discussed previously glycyrrhizin, which is also a phytoestrogen, stimulated melanin production in a similar way. Phytoestrogens mimic the biological activities of human estrogen (Malaivijitnond, 2012). Estrogen has previously shown to regulate melanogenesis and skin proliferation in normal human skin culture, therefore it is expected that phytoestrogens, which mimics the behavior of estrogen, will also result in increased melanogenesis (McLeod, Ranson, & Mason, 1994). This impersonating behavior was confirmed with molecular docking

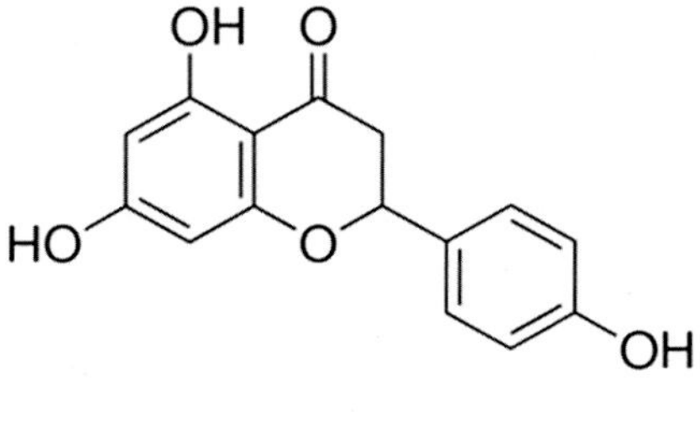

FIGURE 5.13 Chemical structure of Naringenin, a phytoestrogen capable of stimulating melanogenesis.

analysis, which showed that genistein, another phytoestrogen, perfectly fit the active site of the estrogen receptor alpha and formed hydrogen bonds between the receptor and ligand (Yingngam & Rungseevijitprapa, 2012).

Matsuda et al. isolated several compounds from *Piper methysticum* in 2006 that have been reported to stimulate melanogenesis in murine B16 melanoma cells to prevent the development of gray hair. Theophylline, the positive control, which has been isolated from plant extracts was found to increase both gamma-glutamyl transpeptidases and tyrosinase reactive cells, which led to increased pigmentation (Hu, 1982). The rhizome extracts of *P. methysticum* (Kava) exhibited significant increase in melanogenesis activity with a slight increase in cell proliferation at a concentration of 10 μg/mL, respectively. Three kavalactones, namely yangonin (1), (+)-methysticin (2), and 7,8-Epoxyyangonin (3), were isolated from the acetone-methanol extract (Fig. 5.14). Compound 1 and 2 at a concentration of 10 μg/mL showed only a slight increase in total melanin produced when compared to the control. Compound 3, however, resulted in a 62% increase in total melanin content when compared to the control and increased cell proliferation, by 16.9% greater than the control (Matsuda et al., 2006).

A similar outcome was obtained for *Pyrostegia venusta* Miers, a woody vine native to Brazil and is commonly known as "flame vine." Moreira et al. established in 2012 that both the leaf and flower extracts of *P. venusta* stimulated melanogenesis in melanoma cells after 4 days of incubation. The leaf extract at a concentration of 3 μg/mL, increased melanogenesis by 33.3 ± 3%, while the flower extract increased melanogenesis by 23.4 ± 3% when treated with a concentration of 0.1 μg/mL. However, neither extract had any effect on tyrosinase activity (Moreira et al., 2012).

B16F1 melanoma cells treated with the leaf extracts of *G. mangostana*, at concentrations of 4–32 μg/mL, showed not only an increase in melanogenesis but also an increase in tyrosinase activity, in a dose-dependent manner. The

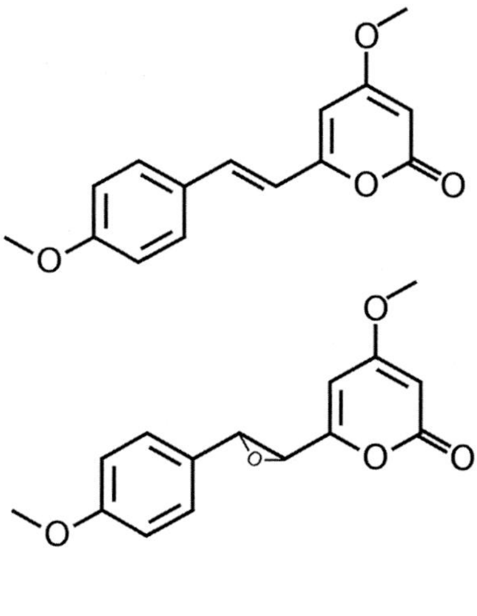

FIGURE 5.14 Kavalactone structures of compounds identified from the rhizome extract of *Piper methysticum* that have been shown to stimulate melanogenesis. (Top) yangonin and (bottom) 7,8-epoxyyangonin.

increase in tyrosinase activity is due to the upregulation of the tyrosinase gene expression, which is activated by an increase in MITF protein levels. The treatment with *G. mangostana* extract resulted in increased levels of both tyrosinase and MITF expressions. *G. mangostana* L. is a tree native to Southeast Asia but is also found in China, Thailand, Taiwan, Indonesia, Cambodia, Malaysia, Singapore, and the Philippines. *G. mangostana* is commonly known as mangosteen and its fruit is consumed either fresh or as preserves (Hamid et al., 2012).

Fallopia multiflora (Thunb.) Haraldson is a traditional Chinese herb used mainly for antiaging and hair blackening. In a study conducted by Liu and Ma in 2015, the *F. multiflora* ethanol extract showed a significant ($P < .05$) increase in tyrosinase activity. Four active compounds isolated from *F. multiflora*, including gallic acid, emodin, physcion, and 2,3,4′,5-tetrahydroxystilbene 2-O-β-D-glucoside (THSG) were analyzed for their effect on tyrosinase activity. The results obtained indicated that mushroom tyrosinase activity was increased to 286.2%, 140.5%, 123.9%, and 111.3% for THSG, gallic acid, physcion, and emodin, respectively. Treatment of B16 cells with *F. multiflora* extract for 3 days greatly ($P < .05$) increased the melanin production when compared to the control cells and resulted in 53% increase in pigmentation (Liu & Ma, 2015).

Nelumbo nucifera, Gaertn. commonly known as lotus and identified by its yellow flowers, is known for its variety in medical application in Eastern Asia. In China the seeds are mostly used for conditions such as cancer, tissue inflammation, poisoning, and leprosy, but Jeon et al. established in 2009 that lotus essential oil stimulate melanogenesis in human melanocytes. The lotus flower's essential oil including the petals, at a concentration of 10 μg/mL, not only stimulated melanin synthesis by 2.1-fold, compared to the control, and resulted in an increase in tyrosinase activity but also induced the expression of the tyrosinase gene, MITF, and TRP-2 proteins. The different constituents of the essential oil were determined as palmitoleic acid methyl ester (7.55%), palmitic acid methyl ester (22.66%), linoleic acid methyl ester (11.16%), and linolenic acid methyl ester (5.16%). However, palmitoleic acid methyl ester was the only constituent that induced melanogenesis and increased tyrosinase expression (Fig. 5.15; Jeon, Kim, Koo, Kim, & Lee, 2009).

Hypericum revolutum Vahl subsp. revolutum is commonly known as the curry bush and is native to Zimbabwe but is also found in south west Arabia, Eastern Africa, Cameroon, and Madagascar (Fig. 5.4H; FOZ, 2009). The *Hypericum* genus is traditionally used for skin infections and leukoderma, a hypopigmentation disease, caused by exposure to chemicals, inflammation etc (Capitanio, Cappelletti, & Filippini, 1989; SANBI, 2014; Zofou et al., 2011). *H. revolutum* contains coumarins that have been shown to induce melanin

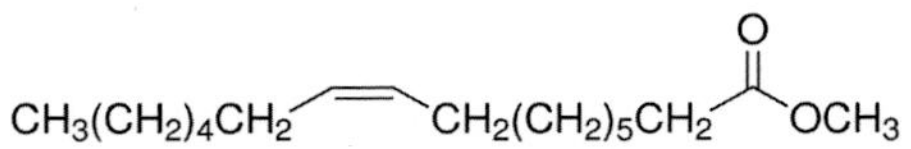

FIGURE 5.15 Chemical structure of palmitoleic acid methyl ester found in lotus essential oil, which was found to induce both melanogenesis and tyrosinase expression.

production (Matsuda et al., 2005). *Withania somnifera* (L.) Dunal is commonly known as the winter cherry (English), bofepha (Sotho) and ubuvuma (Xhosa) and is traditionally used for leukoderma and skin diseases (Fig. 5.4C; Basha, Anjaneyulu, Krishna, Parveen, & Sudarsanam, 2014; Prakash, Gupta, & Dinda, 2002; SANBI, 2014; Singh, Kurfurst, et al., 2010; Singh, Sharma, et al., 2010). *W. somnifera* grows wild in South Africa, India, Afghanistan, Philistine, Egypt, Pakistan, Jordan, Spain, Morocco, Sri Lanka, Eastern Africa, Canary Island, Madagascar and in the Congo (Fig. 5.4C; Kumar, 2012). Extracts containing withanone—as is the case with *W. somnifera*—has shown to increase hair melanin in clinical trials conducted by Bone and Morgan (1996) (Fig. 5.16; Widodo et al., 2009). It has been reported earlier that Withaferin A and quercetin—both found in *W. somnifera*—stimulated melanin dispersion mediated by cyclic AMP that led to skin darkening (Fig. 5.17; Ali & Meitei, 2011, 2012; Novales, 1972).

Terminalia prunioides is commonly known as sterkbos (Afrikaans); purple pod terminalia (English), hareri (Somali), and mwalambe (Swahili) and is traditionally used for bacterial infections (gram-positive organisms) and skin diseases (Fig. 5.4D; Eloff, Katerere, & McGaw, 2008). Traditional application is by means of making a powder of the leaves and mixing it with water to make a paste, which is then applied to the affected skin areas (Ayyanar & Ignacimuthu, 2005; Chouhan & Singh, 2011).

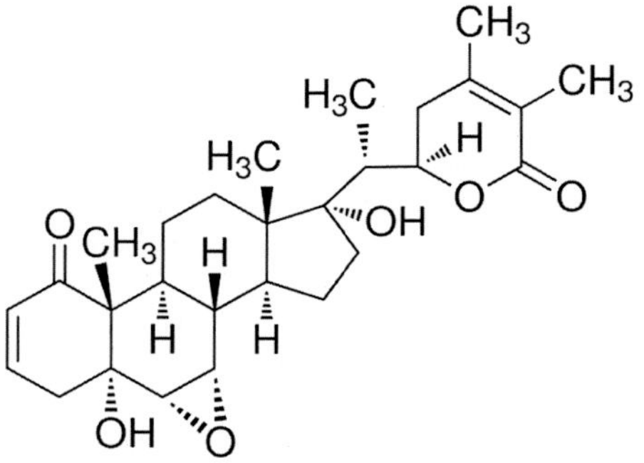

FIGURE 5.16 Chemical structure of withanone, isolated from *Withania somnifera*, which has been shown to increase hair melanin.

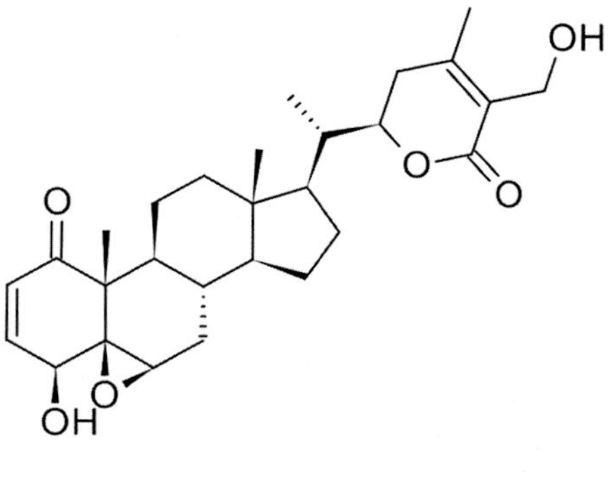

FIGURE 5.17 Chemical structure of Withaferin A, a compound that stimulates the dispersion of melanin, ultimately leading to darkening of the skin.

5.7 CONCLUSION

PMH is a disorder affecting people's (especially dark skinned individuals) self-esteem and outlook on life. There are several plants that are used for hypopigmentary diseases, such as leukoderma and vitiligo, but no herbal treatments or plant extracts are specifically used for PMH; this should encourage researchers to find an alternative from natural sources.

REFERENCES

Ali, S. A., & Meitei, K. V. (2011). On the action and mechanism of withaferin-A from *Withania somnifera*, a novel and potent melanin dispersing agent in frog melanophores. *Journal of Receptor and Signal Transduction Research*, *31*(5), 359–366.

Ali, S. A., & Meitei, K. V. (2012). *Withania somnifera* root extracts induce skin darkening in wall lizard melanophores via stimulation of cholinergic receptors. *Natural Product Research*, *26*(17), 1645–1648.

Andersen, A. (2007). Definition and structure. *International Journal of Toxicology*, *26*(2), 79–112.

Ayyanar, M., & Ignacimuthu, S. (2005). Traditional knowledge of Kani tribals in Kouthalai of Tirunelveli hills, Tamil Nadu, India. *Journal of Ethnopharmacology*, *102*(2), 246–255.

Baba, S., Osakabe, N., Natsume, M., Yasuda, A., Muto, Y., Hiyoshi, T., et al. (2005). Absorption, metabolism, degradation and urinary excretion of rosmarinic acid after intake of *Perilla frutescens* extract in humans. *European Journal of Nutrition*, *44*(1), 1–9.

Barber, G., & Behrman, E. (1991). The synthesis and characterization of uridine 5′-(β-l-rhamnopyranosyl diphosphate) and its role in the enzymic synthesis of rutin. *Archives of Biochemistry and Biophysics*, *288*(1), 239–242.

Basha, S., Anjaneyulu, E., Krishna, S., Parveen, D., & Sudarsanam, G. (2014). Plants used in the treatment of leucoderma by the tribals of Yerramalai forest of Kurnool district, Andhra Pradesh, India. *The Journal of Ethnobiology and Traditional Medicine. Photon*, *121*, 761–766.

Bertolotto, C., Abbe, P., Englaro, W., Ishizaki, T., Narumiya, S., Boquet, P., et al. (1998). Inhibition of Rho is required for cAMP-induced melanoma cell differentiation. *Molecular Biology of the Cell*, *9*(6), 1367–1378.

Bohil, A. B., Robertson, B. W., & Cheney, R. E. (2006). Myosin-X is a molecular motor that functions in filopodia formation. *Proceedings of the National Academy of Sciences*, *103*(33), 12411–12416.

Böhm, S. K., Khitin, L. M., Grady, E. F., Aponte, G., Payan, D. G., & Bunnett, N. W. (1996). Mechanisms of desensitization and resensitization of proteinase-activated receptor-2. *Journal of Biological Chemistry*, *271*(36), 22003–22016.

Bone, K., Morgan, M. (1996). *Clinical applications of ayurvedic and Chinese herbs: monographs for the western herbal practitioner*. Phytotherapy Press.

Brown, D. A. (2001). Skin pigmentation enhancers. *Journal of Photochemistry and Photobiology B: Biology*, *63*(1), 148–161.

Bustelo, X. R., Sauzeau, V., & Berenjeno, I. M. (2007). Gtp-binding proteins of the Rho/Rac family: Regulation, effectors and functions in vivo. *Bioessays*, *29*(4), 356–370.

CABI. (2014). *Achyranthes aspera (devil's horsewhip)*. Invasive Species Compendium. http://www.cabi.org/isc/datasheet/2664.

CABI. (2016a). *Mahonia aquifolium (Oregongrape)*. Invasive Species Compendium. http://www.cabi.org/isc/datasheet/32269.

CABI. (2016b). *Momordica charantia (bitter gourd)*. Invasive Species Compendium. http://www.cabi.org/isc/datasheet/34678.

CABI. (2016c). *Xanthium strumarium (common cocklebur)*. Invasice Species Compendium. http://www.cabi.org/isc/datasheet/56864.

Camerer, E., Huang, W., & Coughlin, S. R. (2000). Tissue factor-and factor X-dependent activation of protease-activated receptor 2 by factor VIIa. *Proceedings of the National Academy of Sciences*, *97*(10), 5255–5260.

Capitanio, M., Cappelletti, E., & Filippini, R. (1989). Traditional antileukodermic herbal remedies in the mediterranean area. *Journal of Ethnopharmacology*, *27*(1), 193–211.

Cardinali, G., Ceccarelli, S., Kovacs, D., Aspite, N., Lotti, L. V., Torrisi, M. R., et al. (2005). Keratinocyte growth factor promotes melanosome transfer to keratinocytes. *Journal of Investigative Dermatology*, *125*(6), 1190–1199.

Chandak, R., Devdhe, S., & Changediya, V. (2009). Evaluation of anti-histaminic activity of aqueous extract of ripe olives of Olea europea. *Journal of Pharmacy Research*, *2*(3), 416–420.

Chiang, H. M., Lin, J. W., Hsiao, P. L., Tsai, S. Y., & Wen, K. C. (2011). Hydrolysates of citrus plants stimulate melanogenesis protecting against UV-induced dermal damage. *Phytotherapy Research*, *25*(4), 569–576.

Chomnawang, M. T., Surassmo, S., Nukoolkarn, V. S., & Gritsanapan, W. (2005). Antimicrobial effects of Thai medicinal plants against acne-inducing bacteria. *Journal of Ethnopharmacology*, *101*(1), 330–333.

Chopra, R., Nayar, S., & Chopra, I. (1956). *Glossary of Indian medicinal plants*. New Delhi: CSIR Publication.

Chouhan, H. S., & Singh, S. K. (2011). A review of plants of genus *Leucas*. *Journal of Pharmacognosy and Phytotherapy*, *3*(2), 13–26.

Das, N. (2006). Propagation prospects of dye yielding plant *Rhinacanthus nasutus* (Linn.) Kurz. *Natural Product Radiance*, *5*(1), 42–43.

Das, R., Esposito, V., Abu-Abed, M., Anand, G. S., Taylor, S. S., & Melacini, G. (2007). cAMP activation of PKA defines an ancient signaling mechanism. *Proceedings of the National Academy of Sciences*, *104*(1), 93–98.

Dharmananda, S. (2003). *Safety issues affecting Chinese herbs: The case of Xanthium*. ITM.

Docherty, J. J., McEwen, H. A., Sweet, T. J., Bailey, E., & Booth, T. D. (2007). Resveratrol inhibition of *Propionibacterium acnes*. *Journal of Antimicrobial Chemotherapy*, *59*(6), 1182–1184.

Dulin, N. O., Niu, J., Browning, D. D., Richard, D. Y., & Voyno-Yasenetskaya, T. (2001). Cyclic AMP-independent activation of protein kinase A by vasoactive peptides. *Journal of Biological Chemistry*, *276*(24), 20827–20830.

Dupont, B. (2011). Terminalia prunoides. In *File:Purple-pod cluster-leaf Terminalia prunoids (8390500976).jpg*. Wikimedia commons. https://commons.wikimedia.org/wiki/File:Purple-pod_Cluster-leaf_Terminalia_prunoides_%288390500976%29.jpg.

Dweck, A. C. (1997). Skin treatment with plants of the Americas: Indigenous plants historically used to treat psoriasis, eczema, wounds and other conditions. *Cosmetics and Toiletries*, *112*(10), 47–66.

Dwivedi, S., & Chopra, D. (2014). Revisiting *Terminalia arjuna*: An ancient cardiovascular drug. *Journal of Traditional and Complementary Medicine*, *4*(4), 224–231.

Eloff, J., Katerere, D., & McGaw, L. (2008). The biological activity and chemistry of the southern African Combretaceae. *Journal of Ethnopharmacology*, *119*(3), 686–699.

Erkan, N., Ayranci, G., & Ayranci, E. (2008). Antioxidant activities of rosemary (*Rosmarinus Officinalis* L.) extract, blackseed (*Nigella sativa* L.) essential oil, carnosic acid, rosmarinic acid and sesamol. *Food Chemistry*, *110*(1), 76–82.

Erlund, I., Meririnne, E., Alfthan, G., & Aro, A. (2001). Plasma kinetics and urinary excretion of the flavanones naringenin and hesperetin in humans after ingestion of orange juice and grapefruit juice. *The Journal of Nutrition*, *131*(2), 235–241.

Fahmy, I., & Abu-Shady, H. (1947). *Ammi majus* Linn.; pharmacognostical study and isolation of a crystalline constituent, ammoidin. *Quarterly Journal of Pharmacy and Pharmacology*, *20*(3), 281.

Faix, J., & Rottner, K. (2006). The making of filopodia. *Current Opinion in Cell Biology*, *18*(1), 18–25.

FOZ. (2009). *Hypericum revolutum Vahl*. Flora of Zimbabwe. http://www.zimbabweflora.co.zw/speciesdata/species.php?species_id=140380.

Fu, Y., Chen, L., Zu, Y., Liu, Z., Liu, X., Liu, Y., et al. (2009). The antibacterial activity of clove essential oil against *Propionibacterium acnes* and its mechanism of action. *Archives of Dermatology*, *145*(1), 86–88.

Fujita, E., & Node, M. (1984). Diterpenoids of *Rabdosia* species. *Progress in the Chemistry of Organic Natural Products*, *46*, 77–157.

Ganguly, B. (2013). *Psoralea corylifolia*. Wikimedia commons. https://commons.wikimedia.org/wiki/File:Psoralea_corylifolia_-_Agri-Horticultural_Society_of_India_-_Alipore_-_Kolkata_2013-01-05_2282.JPG.

GBIF. (2016). *Telosma pallida (Roxb.) Craib*. Global Biodiversity Information Facility. http://www.gbif.org/species/3582593.

Gibson, J. (1996). Rationale for the development of new topical treatments for acne vulgaris. *Cutis*, *57*(1 Suppl.), 13–19.

Gilchrest, B. A., & Eller, M. S. (September 1999). DNA photodamage stimulates melanogenesis and other photoprotective responses. *Journal of Investigative Dermatology Symposium Proceedings*, *4*(1), 35–40. Nature Publishing Group.

Grieve, M. (2016). *Angelica*. Botanical.com. http://www.botanical.com/botanical/mgmh/a/anegl037.html.

Gutierrez, D., & Writer, S. (2014). *Skin pigment disease reversed with piperine nutrient from black pepper*. Natural news. http://www.naturalnews.com/025302_disease_piperine_ultraviolet_radiation.html.

Halaban, R., Patton, R. S., Cheng, E., Svedine, S., Trombetta, E. S., Wahl, M. L., et al. (2002). Abnormal acidification of melanoma cells induces tyrosinase retention in the early secretory pathway. *Journal of Biological Chemistry*, *277*(17), 14821–14828.

Halder, R. M., & Rodney, I. J. (2013). Disorders of hypopigmentation. In A. F. Alexis, & V. H. Barbosa (Eds.), *Skin of color: A practical guide to dermatologic diagnosis and treatment*. New York, USA: Springer.

Hameed, S. (2016). *Genetic diversity and variations in the endangered tree [Tecomella undulata] in Rajasthan*. Flora of Pakistan. http://www.efloras.org/florataxon.aspx?flora_id=5&taxon_id=250071457.

Hamid, M. A., Sarmidi, M. R., & Park, C. S. (2012). Mangosteen leaf extract increases melanogenesis in B16F1 melanoma cells by stimulating tyrosinase activity in vitro and by up-regulating tyrosinase gene expression. *International Journal of Molecular Medicine*, *29*(2), 209–217.

Hassan, A. M., El-Badawi, M. A., Abd-Rabbou, F. A., Gamei, M. M., Moustafa, K. A., & Almokadem, A. H. (2014). Progressive macular hypomelanosis pathogenesis and treatment: A randomized clinical trial. *Journal of Microscopy and Ultrastructure*, *2*(4), 205–216.

Hata, K., Mukaiyama, T., Tsujimura, N., Sato, Y., Kosaka, Y., Sakamoto, K., et al. (2006). Differentiation-inducing activity of lupane triterpenes on a mouse melanoma cell line. *Cytotechnology*, *52*(3), 151–158.

Herbcyclopedia. (2014). *Citrullus colocynthis (Bitter apple)*. Herbcyclopedia. http://www.herbcyclopedia.com/item/citrullus-colocynthis-bitter-apple-2.

Himejima, M., & Kubo, I. (1992). Antimicrobial agents from *Licaria puchuri*: Major and their synergistic effect with polygodial. *Journal of Natural Products*, *55*(5), 620–625.

Holland, K., & Bojar, R. (1989). The effect of azelaic acid on cutaneous bacteria. *Journal of Dermatological Treatment*, *1*(Suppl. 1), 17–19.

Hopp, K., Cunningham, L., Bromel, M., Schermeister, L., & Khalil, S. (1975). In vitro antitrypanosomal activity of certain alkaloids against *Trypanosoma lewisi*. *Lloydia*, *39*(5), 375–377.

Hu, F. (1982). Theophylline and melanocyte-stimulating hormone effects on gamma-glutamyl transpeptidase and DOPA reactions in cultured melanoma cells. *Journal of Investigative Dermatology*, *79*(1), 57–62.

IUCN. (2016a). *Abies koreana IUCN red list of threatened species*. http://www.iucnredlist.org/details/31244/0.

IUCN. (2016b). *Prumnopitys andina IUCN red list of threatened species*. http://www.iucnredlist.org/details/35934/0.

Iwasa, K., Moriyasu, M., Yamori, T., Turuo, T., Lee, D., & Wiegrebe, W. (2001). In vitro cytotoxicity of the protoberberine-type alkaloids. *Journal of Natural Products*, *64*(7), 896–898.

Jayasinghe, C., Gotoh, N., Aoki, T., & Wada, S. (2003). Phenolics composition and antioxidant activity of sweet basil (*Ocimum basilicum* L.). *Journal of Agricultural and Food Chemistry*, *51*(15), 4442–4449.

Jengod. (2009). Raphanus sativus. In *File:Radish 3371103037 4ab07db0bf o.jpg*. Wikimedia commons. https://commons.wikimedia.org/wiki/File:Radish_3371103037_4ab07db0bf_o.jpg.

Jeon, S., Kim, N.-H., Koo, B.-S., Kim, J.-Y., & Lee, A.-Y. (2009). Lotus (*Nelumbo nuficera*) flower essential oil increased melanogenesis in normal human melanocytes. *Experimental and Molecular Medicine*, *41*(7), 517–524.

Jiang, Y. Z., Songhao, Xu, J., Feng, J., Mahboob, S., Al-Ghanim, K. A., et al. (2014). *Diagram of putative gene pathways in the common carp skin pigmentation process*. Figshare. https://dx.doi.org/10.1371/journal.pone.0108200.g004.

Joffe, P., Notten, A., & Mutshinyalo, T. (2001). *Gloriosa superba L. and Olea europaea L. subsp. africana (Mill.) P.S.Green*. SANBI. http://pza.sanbi.org/olea-europaea-subsp-africana.

Jose, J. (2010). Achyranthes aspera. In *File:Achyranthes kadavoor.jpg*. Wikimedia commons. https://commons.wikimedia.org/wiki/File:Achyranthes.

Jung, G., Yang, J., Song, E., & Park, J. (2001). Stimulation of melanogenesis by glycyrrhizin in B16 melanoma cells. *Experimental and Molecular Medicine*, *33*(3), 131–135.

Kamboj, A., & Saluja, A. (2010). Phytopharmacological review of *Xanthium strumarium* L. (Cocklebur). *International Journal of Green Pharmacy*, *4*(3), 129.

Kanlayavattanakul, M., & Lourith, N. (2011). Therapeutic agents and herbs in topical application for acne treatment. *International Journal of Cosmetic Science*, *33*(4), 289–297.

Kat. (2007). Angelica archangelica. In *File:Angelica archangelica (1118596627).jpg*. Wikimedia commons. https://commons.wikimedia.org/wiki/File:Angelica_archangelica_%281118596627%29.jpg.

Kayastha, B. L. (2014). Queen of herbs tulsi (*Ocimum sanctum*) removes impurities from water and plays disinfectant role. *Journal of Medicinal Plants Stud*, *2*(2).

KEW. (2016a). *Glycyrrhiza glabra (liquorice)*. Kew Royal Botanic Gardens. http://www.kew.org/science-conservation/plants-fungi/glycyrrhiza-glabra-liquorice.

KEW. (2016b). *Raphanus sativus (radish)*. KEW Royal Botanic Gardens. http://www.kew.org/science-conservation/plants-fungi/raphanus-sativus-radish.

Khair-ul-Bariyah, S., Ahmed, D., & Aujla, M. I. (2012). Comparative analysis of *Ocimum basilicum* and *Ocimum sanctum*: Extraction techniques and urease and alpha-amylase inhibition. *Pakistan Journal of Chemistry*, *2*(3), 134–141.

Khandel, A. K., Ganguly, S., Bajaj, A., & Khan, S. (2012). New records, ethno-pharmacological applications & indigenous uses of *Gloriosa superba* L. (Glory lily) practices by Tribes of Pachmarhi Biosphere Reserve, Madhya Pradesh, Central India. *Nature and Science*, *10*(5), 23–48.

Khare, C. (2007). *Xanthium strumarium* Linn. In *Indian medicinal plants*. New York: Springer.

Khlgatian, M. K., Hadshiew, I. M., Asawanonda, P., Yaar, M., Eller, M. S., Fujita, M., et al. (2002). Tyrosinase gene expression is regulated by p53. *Journal of Investigative Dermatology*, *118*(1), 126–132.

Khushboo, P., Jadhav, V., Kadam, V., & Sathe, N. (2010). *Psoralea corylifolia* Linn.-"Kushtanashini". *Pharmacognosy Reviews*, *4*(7), 69.

Kim, S., Baik, J. S., Oh, T., Yoon, W.-J., Lee, N. H., & Hyun, C. (2008). Biological activities of Korean *Citrus obovoides* and *Citrus natsudaidai* essential oils against acne-inducing bacteria. *Bioscience, Biotechnology and Biochemistry*, *72*(10), 2507–2513.

Kim, J., Kim, N., & Lim, Y. (2010). Evaluation of the antibacterial activity of rhapontigenin produced from rhapontin by biotransformation against *Propionibacterium acnes*. *Journal of Microbiology and Biotechnology*, *20*(1), 82–87.

Kim, S., Lee, J., Jung, E., Lee, J., Huh, S., Hwang, H., et al. (2009). 6-Benzylaminopurine stimulates melanogenesis via cAMP-independent activation of protein kinase A. *Archives of Dermatological Research*, *301*(3), 253–258.

Koley, S., Mandal, R. K., Choudhary, S., & Bandyopadhyay, A. (2013). Post-kala-azar dermal leishmaniasis developing in miltefosine-treated visceral leishmaniasis. *Indian Journal of Dermatology*, *58*(3), 241.

Kong, W., McConalogue, K., Khitin, L. M., Hollenberg, M. D., Payan, D. G., Böhm, S. K., et al. (1997). Luminal trypsin may regulate enterocytes through proteinase-activated receptor 2. *Proceedings of the National Academy of Sciences*, *94*(16), 8884–8889.

Kostoviæ, K., Nola, I., Buèan, Ž., & Šitum, M. (2003). *Treatment of vitiligo: Current methods and new approaches*. LOGICA.

Krishnamurthi, A. (1969). *The wealth of India: Raw materials* (Vol. VIII). . Ph-Re.

Kubo, I., Kamikawa, T., & Kubota, T. (1974). Studies on constituents of Isodon *Japonicus hara*: The structures and absolute stereochemistry of isodonal, trichodonin and epinodosin. *Tetrahedron*, *30*(5), 615–622.

Kubo, I., Muroi, H., & Kubo, A. (1994). Naturally occurring antiacne agents. *Journal of Natural Products*, *57*(1), 9–17.

Kubo, I., Xu, Y., & Shimizu, K. (2004). Antibacterial activity of ent-kaurene diterpenoids from *Rabdosia rosthornii*. *Phytotherapy Research*, *18*(2), 180–183.

Kumar, A. (2012). *Pharmacognosy of Ashwagandha*. Science 2.0. http://www.science20.com/humboldt_fellow_and_science/blog/pharmacognosy_ashwwagandha-93560.

Kumarasinghe, S. P. W., Tan, S. H., Thng, S., Thamboo, T. P., Liang, S., & Lee, Y. S. (2006). Progressive macular hypomelanosis in Singapore: A clinico-pathological study. *International Journal of Dermatology*, *45*, 737–742.

Kumar, G., Jayaveera, K., Kumar, C., Sanjay, U. P., Swamy, B., & Kumar, D. (2007). Antimicrobial effects of Indian medicinal plants against acne-inducing bacteria. *Tropical Journal of Pharmaceutical Research*, *6*(2), 717–723.

Lawrence, E. (2008a). Definitions. In E. J. Lawrence, & J. M. Jackson (Eds.), *Henderson's Dictionary of biology: (Vol. 8)*. England: Pearson Education Limited.

Lawrence, E. (2008b). Melanin. In *Henderson's Dictionary of biology* (Vol. 14). England: Pearson Education Limited.

Lee, J., Jung, E., Park, J., Jung, K., Park, E., Kim, J., et al. (2005). Glycyrrhizin induces melanogenesis by elevating a cAMP level in B16 melanoma cells. *Journal of Investigative Dermatology*, *124*(2), 405–411.

Lee, J., Kim, Y. S., & Park, D. (2007). Rosmarinic acid induces melanogenesis through protein kinase A activation signaling. *Biochemical Pharmacology*, *74*(7), 960–968.

Levy, S. B. (1992). Dihydroxyacetone-containing sunless or self-tanning lotions. *Journal of the American Academy of Dermatology*, *27*(6), 989–993.

Lim, Y., Kim, I., & Seo, J. (2007). In vitro activity of kaempferol isolated from the *Impatiens balsamina* alone and in combination with erythromycin or clindamycin against *Propionibacterium acnes*. *Journal of Microbiology (Seoul, Korea)*, *45*(5), 473–477.

Lin, J. Y., & Fisher, D. E. (2007). Melanocyte biology and skin pigmentation. *Nature*, *445*(7130), 843–850.

Lin, Z., Hoult, J., Bennett, D. C., & Raman, A. (1999). Stimulation of mouse melanocyte proliferation by *Piper nigrum* fruit extract and its main alkaloid, piperine. *Planta Medica*, *65*(7), 600–603.

Liu, S.-H., & Ma, L.-J. (2015). Effects of Chinese herbal extracts on tyrosinase activity and melanogenesis. *Natural Products Chemistry and Research*, *3*, 183.

Liu, Y., & Simon, J. D. (2005). Metal–ion interactions and the structural organization of Sepia eumelanin. *Pigment Cell Research*, *18*(1), 42–48.

Li, H., Zhu, W., & Xia, M. (2001). Melanogenic effects of ethanol extracts obtained from 5 Traditional Chinese Medicines on shape and properties of melanocytes from skin of brownish guinea pigs. *Journal of Clinical Dermatology*, *30*(2), 69–71.

Malaivijitnond, S. (2012). Medical applications of phytoestrogens from the Thai herb *Pueraria mirifica*. *Frontiers of Medicine*, *6*(1), 8–21.

Ma, Y., Pitson, S., Hercus, T., Murphy, J., Lopez, A., & Woodcock, J. (2005). Sphingosine activates protein kinase A type II by a novel cAMP-independent mechanism. *Journal of Biological Chemistry*, *280*(28), 26011–26017.

Mari, B., Guerin, S., Far, D. F., Breitmayer, J., Belhacene, N., Peyron, J., et al. (1996). Thrombin and trypsin-induced Ca (2+) mobilization in human T cell lines through interaction with different protease-activated receptors. *The FASEB Journal*, *10*(2), 309–316.

Marinova, D., Ribarova, F., & Atanassova, M. (2005). Total phenolics and total flavonoids in Bulgarian fruits and vegetables. *Journal of the University of Chemical Technology and Metallurgy*, *40*(3), 255–260.

Marks, M. S., & Seabra, M. C. (2001). The melanosome: Membrane dynamics in black and white. *Nature Reviews Molecular Cell Biology*, *2*(10), 738–748.

Martínez-Martínez, M. L., Azaña-Defez, J. M., Rodríguez-Vázquez, M., Faura-Berruga, C., & Escario-Travesedo, E. (2012). Progressive macular hypomelanosis. *Pediatric Dermatology*, *29*(4), 460–462.

Marzouk, B., Marzouk, Z., Haloui, E., Fenina, N., Bouraoui, A., & Aouni, M. (2010). Screening of analgesic and anti-inflammatory activities of *Citrullus colocynthis* from southern Tunisia. *Journal of Ethnopharmacology*, *128*(1), 15–19.

Matsuda, H., Hirata, N., Kawaguchi, Y., Naruto, S., Takata, T., Oyama, M., et al. (2006). Melanogenesis stimulation in murine B16 melanoma cells by Kava (*Piper methysticum*) rhizome extract and kavalactones. *Biological and Pharmaceutical Bulletin*, *29*(4), 834.

McLeod, S. D., Ranson, M., & Mason, R. S. (1994). Effects of estrogens on human melanocytes in vitro. *The Journal of Steroid Biochemistry and Molecular Biology*, *49*(1), 9–14.

Ménasché, G., Ho, C. H., Sanal, O., Feldmann, J., Tezcan, I., Ersoy, F., et al. (2003). Griscelli syndrome restricted to hypopigmentation results from a melanophilin defect (GS3) or a MYO5A F-exon deletion (GS1). *Journal of Clinical Investigation*, *112*(3), 450.

Mercer, J. A., Seperack, P. K., Strobel, M. C., Copeland, N. G., & Jenkins, N. A. (1991). Novel myosin heavy chain encoded by murine dilute coat colour locus. *Nature*, *349*(6311), 709–713.

Mertens-Talcott, S. U., Jilma-Stohlawetz, P., Rios, J., Hingorani, L., & Derendorf, H. (2006). Absorption, metabolism, and antioxidant effects of pomegranate (*Punica granatum* L.) polyphenols after ingestion of a standardized extract in healthy human volunteers. *Journal of Agricultural and Food Chemistry*, *54*(23), 8956–8961.

Miki, T., Bottaro, D. P., Fleming, T. P., Smith, C. L., Burgess, W. H., Chan, A., et al. (1992). Determination of ligand-binding specificity by alternative splicing: Two distinct growth factor receptors encoded by a single gene. *Proceedings of the National Academy of Sciences*, *89*(1), 246–250.

Minwalla, L., Zhao, Y., Le Poole, I. C., Wickett, R. R., & Boissy, R. E. (2001). Keratinocytes play a role in regulating distribution patterns of recipient melanosomes in vitro. *Journal of Investigative Dermatology*, *117*(2), 341–347.

Mohlenbrock, R. H. (2005). Xanthium strumarium. In *File:Xanthium strumarium L..jpg*. Wikimedia commons. https://commons.wikimedia.org/wiki/File:Xanthium_strumarium_L..jpg.

Moleephan, W. (2012). Effect of xanthoxylin on melanin content and melanogenic protein expression in B16F10 melanoma. *Asian Biomedicine*, *6*(3).

Moody, K. (1989). *Weeds reported in rice in South and Southeast Asia*. International Rice Research Institute.

Moreira, C. G., Horinouchi, C. D., Souza-Filho, C. S., Campos, F. R., Barison, A., Cabrini, D. A., et al. (2012). Hyperpigmentant activity of leaves and flowers extracts of *Pyrostegia venusta* on murine B16F10 melanoma. *Journal of Ethnopharmacology*, *141*(3), 1005–1011.

Mori, A., Nishino, C., Enoki, N., & Tawata, S. (1987). Antibacterial activity and mode of action of plant flavonoids against *Proteus vulgaris* and *Staphylococcus aureus*. *Phytochemistry*, *26*(8), 2231–2234.

Munemura, M., Eskay, R., Kebabian, J., & Long, R. (1980). Release of α-melanocyte-stimulating hormone from dispersed cells of the intermediate lobe of the rat pituitary gland: Involvement of catecholamines and adenosine 3′,5′-monophosphate. *Endocrinology*, *106*(6), 1795–1803.

Nagata, H., Takekoshi, S., Takagi, T., Honma, T., & Watanabe, K. (1999). Antioxidative action of flavonoids, quercetin and catechin, mediated by the activation of glutathione peroxidase. *The Tokai Journal of Experimental and Clinical Medicine*, *24*(1), 1–11.

Nagata, H., Takekoshi, S., Takeyama, R., Homma, T., & Yoshiyuki Osamura, R. (2004). Quercetin enhances melanogenesis by increasing the activity and synthesis of tyrosinase in human melanoma cells and in normal human melanocytes. *Pigment Cell Research*, *17*(1), 66–73.

Nagy, S., Ricca, B. L., Norstrom, M. F., Courson, D. S., Brawley, C. M., Smithback, P. A., et al. (2008). A myosin motor that selects bundled actin for motility. *Proceedings of the National Academy of Sciences*, *105*(28), 9616–9620.

Nandwani, D. (2014). *Sustainable horticultural systems: Issues, technology and innovation*. Springer International Publishing.

Nazre, M. (2014). New evidence on the origin of mangosteen (*Garcinia mangostana* L.) based on morphology and ITS sequence. *Genetic Resources and Crop Evolution*, *61*(6), 1147–1158.

Negi, R., Sharma, M., Sharma, K., Kshetrapal, S., Kothari, S., & Trivedi, P. (2011). Genetic diversity and variations in the endangered tree [*Tecomella undulata*] in Rajasthan. *Indian Journal of Fundamental and Applied Life Sciences*, *1*(1), 50–58.

NIAID. (2005). Escherichia coli. In *File:Escherichia coli NIAID.jpg*. Wikimedia commons. https://commons.wikimedia.org/wiki/File:EscherichiaColi_NIAID.jpg.

Niedergang, F., & Chavrier, P. (2004). Signaling and membrane dynamics during phagocytosis: Many roads lead to the phagos (R) ome. *Current Opinion in Cell Biology*, *16*(4), 422–428.

Novales, R. R. (1972). Recent studies of the melanin-dispersing effect of MSH on melanophores. *General and Comparative Endocrinology*, *3*, 125–135.

Nylander, K., Bourdon, J. C., Bray, S. E., Gibbs, N. K., Kay, R., Hart, I., et al. (2000). Transcriptional activation of tyrosinase and TRP-1 by p53 links UV irradiation to the protective tanning response. *The Journal of Pathology*, *190*(1), 39–46.

Ohguchi, K., Akao, Y., & Nozawa, Y. (2006). Stimulation of melanogenesis by the citrus flavonoid naringenin in mouse B16 melanoma cells. *Bioscience, Biotechnology, and Biochemistry*, *70*(6), 1499–1501.

Okazaki, K., Uzuka, M., Morikawa, F., Toda, K., & Seiji, M. (1976). Transfer mechanism of melanosomes in epidermal cell culture. *Journal of Investigative Dermatology, 67*(4), 541–547.

Olivares, C., Jimenez-Cervantes, C., Lozano, J., Solano, F., & Garcia-Borron, J. (2001). The 5, 6-dihydroxyindole-2-carboxylic acid (DHICA) oxidase activity of human tyrosinase. *Biochemistry Journal, 354*, 131–139.

Paine, C., Sharlow, E., Liebel, F., Eisinger, M., Shapiro, S., & Seiberg, M. (2001). An alternative approach to depigmentation by soybean extracts via inhibition of the PAR-2 pathway. *Journal of Investigative Dermatology, 116*(4), 587–595.

Pan, C., Chen, J., Lin, T., & Lin, C. (2009). In vitro activities of three synthetic peptides derived from epinecidin-1 and an anti-lipopolysaccharide factor against *Propionibacterium acnes*, *Candida albicans*, and *Trichomonas vaginalis*. *Peptides, 30*(6), 1058–1068.

Panichayupakaranant, P., Tewtrakul, S., & Yuenyongsawad, S. (2010). Antibacterial, anti-inflammatory and anti-allergic activities of standardised pomegranate rind extract. *Food Chemistry, 123*(2), 400–403.

Park, H., Kosmadaki, M., Yaar, M., & Gilchrest, B. (2009). Cellular mechanisms regulating human melanogenesis. *Cellular and Molecular Life Sciences, 66*(9), 1493–1506.

Park, J., Lee, J., Jung, E., Park, Y., Kim, K., Park, B., et al. (2004). In vitro antibacterial and anti-inflammatory effects of honokiol and magnolol against *Propionibacterium* sp. *European Journal of Pharmacology, 496*(1), 189–195.

Perry, A., & Lambert, P. A. (2006). Propionibacterium acnes. *Letters in Applied Microbiology, 42*(3), 185–188.

Pillai, S., Oresajo, C., & Hayward, J. (2005). Ultraviolet radiation and skin aging: Roles of reactive oxygen species, inflammation and protease activation, and strategies for prevention of inflammation-induced matrix degradation–a review. *International Journal of Cosmetic Science, 27*(1), 17–34.

Pithayanukul, P., Tubprasert, J., & Wuthi-Udomlert, M. (2007). In vitro antimicrobial activity of *Zingiber cassumunar* (Plai) oil and a 5% Plai oil gel. *Phytotherapy Research, 21*(2), 164–169.

Plaumann, B., Fritsche, M., Rimpler, H., Brandner, G., & Hess, R. D. (1996). Flavonoids activate wild-type p53. *Oncogene, 13*(8), 1605–1614.

Pothitirat, W., Chomnawang, M. T., Supabphol, R., & Gritsanapan, W. (2009). Comparison of bioactive compounds content, free radical scavenging and anti-acne inducing bacteria activities of extracts from the mangosteen fruit rind at two stages of maturity. *Fitoterapia, 80*(7), 442–447.

Prabuseenivasan, S., Jayakumar, M., & Ignacimuthu, S. (2006). In vitro antibacterial activity of some plant essential oils. *BMC Complementary and Alternative Medicine, 6*(1), 39.

Prakash, J., Gupta, S. K., & Dinda, A. K. (2002). *Withania somnifera* root extract prevents DMBA-induced squamous cell carcinoma of skin in Swiss albino mice. *Nutrition and Cancer, 42*(1), 91–97.

Puttarak, P., Charoonratana, T., & Panichayupakaranant, P. (2010). Antimicrobial activity and stability of rhinacanthins-rich *Rhinacanthus nasutus* extract. *Phytomedicine, 17*(5), 323–327.

Rae, A. (2010). Hypericum revolutum. In *File:Hypericum revolutum 1.jpg*. Wikimedia commons. https://commons.wikimedia.org/wiki/File:Hypericum_revolutum_1.jpg.

Ravindran, P. N. (2003). *Black pepper: Piper nigrum*. CRC Press.

Relyveld, G., Dingemans, K., Menke, H., Bos, J., & Westerhof, W. (2008). Ultrastructural findings in progressive macular hypomelanosis indicate decreased melanin production. *Journal of the European Academy of Dermatology and Venereology, 22*(5), 568–574.

Relyveld, G. N., Kingswijk, M., & Reitsma, J. B. (2006). Benzoyl peroxide/clindamycin/ultraviolet A is more effective than fluticasone/ultraviolet A in progressive macular hypomelanosis: A randomized study. *Journal of Academic Dermatology, 55*(5), 836–843.

Relyveld, G. N., Menke, H. E., & Westerhof, W. (2007). Progressive macular hypomelanosis. *American Journal of Clinical Dermatology*, *8*(1), 13–19.
Relyveld, G. N., Menke, H. E., & Westerhof, W. (2012). Progressive macular hypomelanosis. *American Journal of Clinical Dermatology*, *8*(1), 13–19.
Rhie, G., Shin, M. H., Seo, J. Y., Choi, W. W., Cho, K. H., Kim, K. H., et al. (2001). Aging-and photoaging-dependent changes of enzymic and nonenzymic antioxidants in the epidermis and dermis of human skin in vivo. *Journal of Investigative Dermatology*, *117*(5), 1212–1217.
Roelandts, R. (2002). The history of phototherapy: Something new under the sun? *Journal of the American Academy of Dermatology*, *46*(6), 926–930.
Roopashree, T., Dang, R., Rani, R. S., & Narendra, C. (2008). Antibacterial activity of antipsoriatic herbs: *Cassia tora*, *Momordica charantia* and *Calendula officinalis*. *International Journal of Applied Research in Natural Products*, *1*(3), 20–28.
Saising, J., Hiranrat, A., Mahabusarakam, W., Ongsakul, M., & Voravuthikunchai, S. P. (2008). Rhodomyrtone from *Rhodomyrtus tomentosa* (Aiton) Hassk. as a natural antibiotic for staphylococcal cutaneous infections. *Journal of Health Science*, *54*(5), 589–595.
SANBI. (2014). *Olea europeae*. SA National Biodiversity Institute. http://www.plantzafrica.com.
Schallreuter, K. U., Kothari, S., Chavan, B., & Spencer, J. D. (2008). Regulation of melanogenesis–controversies and new concepts. *Experimental Dermatology*, *17*(5), 395–404.
Scheffzek, K., & Ahmadian, M. R. (2005). GTPase activating proteins: Structural and functional insights 18 years after discovery. *Cellular and Molecular Life Sciences*, *62*(24), 3014–3038.
Scott, G., Leopardi, S., Parker, L., Babiarz, L., Seiberg, M., & Han, R. (2003). The proteinase-activated receptor-2 mediates phagocytosis in a Rho-dependent manner in human keratinocytes. *Journal of Investigative Dermatology*, *121*(3), 529–541.
Scott, B. R., Pathak, M. A., & Mohn, G. R. (1976). Molecular and genetic basis of furocoumarin reactions. *Mutation Research/Reviews in Genetic Toxicology*, *39*(1), 29–74.
Scott, G., & Zhao, Q. (2001). Rab3a and SNARE proteins: Potential regulators of melanosome movement. *Journal of Investigative Dermatology*, *116*(2), 296–304.
Sharlow, E., Paine, C., Babiarz, L., Eisinger, M., Shapiro, S., & Seiberg, M. (2000). The protease-activated receptor-2 upregulates keratinocyte phagocytosis. *Journal of Cell Science*, *113*(17), 3093–3101.
Sharma, P., Yelne, M., Dennis, T., Joshi, A., & Billore, K. (2000). *Database on medicinal plants used in Ayurveda*.
Sikkema, J., De Bont, J., & Poolman, B. (1995). Mechanisms of membrane toxicity of hydrocarbons. *Microbiological Reviews*, *59*(2), 201–222.
Singh, S. K., Kurfurst, R., Nizard, C., Schnebert, S., Perrier, E., & Tobin, D. J. (2010). Melanin transfer in human skin cells is mediated by filopodia—a model for homotypic and heterotypic lysosome-related organelle transfer. *The FASEB Journal*, *24*(10), 3756–3769.
Singh, U., & Narain, S. (2010). Traditional treatment of leucoderma by Kol tribes of Vindhyan region of Uttar Pradesh. *Indian Journal of Traditional Knowledge*, *9*(1), 173–174.
Singh, G., Sharma, P., Dudhe, R., & Singh, S. (2010). Biological activities of *Withania somnifera*. *Annals of Biology Research*, *1*(3), 56–63.
Slobodníková, L., KoSt'álová, D., Labudová, D., Kotulová, D., & Kettmann, V. (2004). Antimicrobial activity of *Mahonia aquifolium* crude extract and its major isolated alkaloids. *Phytotherapy Research*, *18*(8), 674–676.
Slominski, A., Tobin, D. J., Shibahara, S., & Wortsman, J. (2004). Melanin pigmentation in mammalian skin and its hormonal regulation. *Physiological Reviews*, *84*(4), 1155–1228.
Smith, E. C., Wareham, N., Zloh, M., & Gibbons, S. (2008). 2β-Acetoxyferruginol—a new antibacterial abietane diterpene from the bark of *Prumnopitys andina*. *Phytochemistry Letters*, *1*(1), 49–53.

Snyder, R., & Conway, S. (2008). Humulus lupulus. In *Food for thought: The science, culture and politics of food*.

Sobell, H. M. (1985). Actinomycin and DNA transcription. *Proceedings of the National Academy of Sciences*, *82*(16), 5328–5331.

Song, P. S., & Tapley, K. J. (1979). Photochemistry and photobiology of psoralens. *Photochemistry and Photobiology*, *29*(6), 1177–1197.

Steinhoff, M., Corvera, C., Thoma, M., Kong, W., McAlpine, B., Caughey, G., et al. (1999). Proteinase-activated receptor-2 in human skin: Tissue distribution and activation of keratinocytes by mast cell tryptase. *Experimental Dermatology*, *8*(4), 282–294.

Tahir, M. A., Pramod, K., Ansari, S., & Ali, J. (2010). Current remedies for vitiligo. *Autoimmunity Reviews*, *9*(7), 516–520.

Taïeb, A., & Picardo, M. (2009). Vitiligo. *New England Journal of Medicine*, *360*(2), 160–169.

Takahashi, T., Kokubo, R., & Sakaino, M. (2004). Antimicrobial activities of Eucalyptus leaf extracts and flavonoids from *Eucalyptus maculata*. *Letters in Applied Microbiology*, *39*(1), 60–64.

Takeyama, R., Takekoshi, S., Nagata, H., Osamura, R. Y., & Kawana, S. (2004). Quercetin-induced melanogenesis in a reconstituted three-dimensional human epidermal model. *Journal of Molecular Histology*, *35*(2), 157–165.

Tokuo, H., Mabuchi, K., & Ikebe, M. (2007). The motor activity of myosin-X promotes actin fiber convergence at the cell periphery to initiate filopodia formation. *The Journal of Cell Biology*, *179*(2), 229–238.

Tolkachev, O., & Zhuchenko, A. (2004). Biologically active substances of flax: Medicinal and nutritional properties (a review). *Pharmaceutical Chemistry Journal*, *34*(7), 360–367.

Toyoda, M., & Morohashi, M. (2001). Pathogenesis of acne. *Medical Electron Microscopy*, *34*(1), 29–40.

Tsatmali, M., Ancans, J., & Thody, A. J. (2002). Melanocyte function and its control by melanocortin peptides. *Journal of Histochemistry and Cytochemistry*, *50*(2), 125–133.

Tsuchiya, H., Sato, M., Miyazaki, T., Fujiwara, S., Tanigaki, S., Ohyama, M., et al. (1996). Comparative study on the antibacterial activity of phytochemical flavanones against methicillin-resistant *Staphylococcus aureus*. *Journal of Ethnopharmacology*, *50*(1), 27–34.

Upadhyay, B., Roy, S., & Kumar, A. (2007). Traditional uses of medicinal plants among the rural communities of Churu district in the Thar Desert, India. *Journal of Ethnopharmacology*, *113*(3), 387–399.

Vaidya, A. (2006). Reverse pharmacological correlates of ayurvedic drug actions. *Indian Journal of Pharmacology*, *38*(5), 311.

Van Den Bossche, K., Naeyaert, J. M., & Lambert, J. (2006). The quest for the mechanism of melanin transfer. *Traffic*, *7*(7), 769–778.

Videira, I. F. D.S., Moura, D. F. L., & Magina, S. (2013). Mechanisms regulating melanogenesis*. *Anais Brasileiros de Dermatologia*, *88*(1), 76–83.

Vijayalakshmi, A., Tripura, A., & Ravichandiran, V. (2011). Development and evaluation of anti-acne products from *Terminalia arjuna* bark. *International Journal ChemTech Research*, *3*, 320–327.

Viyoch, J., Pisutthanan, N., Faikreua, A., Nupangta, K., Wangtorpol, K., & Ngokkuen, J. (2006). Evaluation of in vitro antimicrobial activity of Thai basil oils and their micro-emulsion formulas against *Propionibacterium acnes*. *International Journal of Cosmetic Science*, *28*(2), 125–133.

Vu, T.-K. H., Hung, D. T., Wheaton, V. I., & Coughlin, S. R. (1991). Molecular cloning of a functional thrombin receptor reveals a novel proteolytic mechanism of receptor activation. *Cell*, *64*(6), 1057–1068.

Wang, Y., Shi, L., Wang, A., Tian, H., Wang, H., & Zou, C. (2014). Preparation of high-purity (−)-Borneol and Xanthoxylin from leaves of *Blumea balsamifera* (L.) DC. *Separation Science and Technology*, *49*(10), 1535–1540.

Wang, L., Yang, X., Qin, P., Shan, F., & Ren, G. (2013). Flavonoid composition, antibacterial and antioxidant properties of tartary buckwheat bran extract. *Industrial Crops and Products*, *49*, 312–317.

Wasmeier, C., Hume, A. N., Bolasco, G., & Seabra, M. C. (2008). Melanosomes at a glance. *Journal of Cell Science*, *121*(24), 3995–3999.

Westerhof, W., Relyveld, G. N., Kingswijk, M. M., de Man, P., & Menke, H. E. (2004). *Propionibacterium acnes* and the pathogenesis of progressive macular hypomelanosis. *Archives of Dermatology*, *140*(2), 210–214.

Widodo, N., Shah, N., Priyandoko, D., Ishii, T., Kaul, S C., & Wadhwa, R. (2009). Deceleration of senescence in normal human fibroblasts by withanone extracted from ashwagandha leaves. *Journals of Gerontology Series A: Biomedical Sciences and Medical Sciences*, *64*(10), 1031–1038.

Wood, L. C., Elias, P. M., Calhoun, C., Tsai, J. C., Grunfeld, C., & Feingold, K. R. (1996). Barrier disruption stimulates interleukin-1α expression and release from a pre-formed pool in murine epidermis. *Journal of Investigative Dermatology*, *106*(3), 397–403.

Yamaguchi, Y., & Hearing, V. J. (2009). Physiological factors that regulate skin pigmentation. *Biofactors*, *35*(2), 193–199.

Yamaguchi, N., Satoh-Yamaguchi, K., & Ono, M. (2009). In vitro evaluation of antibacterial, anticollagenase, and antioxidant activities of hop components (*Humulus lupulus*) addressing acne vulgaris. *Phytomedicine*, *16*(4), 369–376.

Yamaguchi, M., Taniguchi, M., Kubo, I., & Kubota, T. (1977). Inhibitory effect of antibacterial and antitumor diterpenoids on oxidative phosphorylation in mitochondria isolated from rat liver. *Agricultural and Biological Chemistry*, *41*(12), 2475–2477.

Yamamoto, O., & Bhawan, J. (1994). Three modes of melanosome transfers in Caucasian facial skin: Hypothesis based on an ultrastructural study. *Pigment Cell Research*, *7*(3), 158–169.

Yingngam, B., & Rungseevijitprapa, W. (2012). Molecular and clinical role of phytoestrogens as anti-skin-ageing agents: A critical overview. *Phytopharmacology*, *3*, 227–244.

Yoon, W.-J., Kim, S.-S., Oh, T.-H., Lee, N. H., & Hyun, C.-G. (2009). *Abies koreana* essential oil inhibits drug-resistant skin pathogen growth and LPS-induced inflammatory effects of murine macrophage. *Lipids*, *44*(5), 471–476.

Zell, H. (2009). Citrullus colocynthis. In *File:Citrullus colocynthis 004.jpg*. Wikimedia commons. https://commons.wikimedia.org/wiki/File:Citrullus_colocynthis_004.JPG.

Zhang, L., Duan, C. J., Binkley, C., Li, G., Uhler, M. D., Logsdon, C. D., et al. (2004). A transforming growth factor β-induced Smad3/Smad4 complex directly activates protein kinase A. *Molecular and Cellular Biology*, *24*(5), 2169–2180.

Zofou, D., Kowa, T. K., Wabo, H. K., Ngemenya, M. N., Tane, P., & Titanji, V. P. (2011). *Hypericum lanceolatum* (Hypericaceae) as a potential source of new anti-malarial agents: A bioassay-guided fractionation of the stem bark. *Malaria Journal*, *10*(167), 17.

Chapter 6

The Role of Medicinal Plants in Oral Care

Dikonketso Bodiba, Karina Mariam Szuman, Namrita Lall
University of Pretoria, Pretoria, South Africa

Chapter Outline

Medicinal Plants for Holistic Health and Well-Being. http://dx.doi.org/10.1016/B978-0-12-812475-8.00006-8

6.1 INTRODUCTION

The term "oral cavity" is used to describe the inner portions of the mouth (Fig. 6.1A). The human oral cavity is home to the various structures that help play a role in the first stages of digestion, speech, and taste and include the lips, cheeks, palate, gums, teeth, and tongue. These various structures of the oral cavity are exposed to a variety of external sources including the air when we speak and food or liquids when we eat or drink. The regular exposure of certain structures to external factors make the oral cavity and its contents a perfect surface to which various Gram-positive and Gram-negative bacteria, fungi, and certain yeasts can attach, making it one of the most complex microbial habitats found in humans.

6.1.1 Teeth, Gums, and Alveolar Bones

Each tooth in our mouth is composed of a crown and root covered by protective calcified tissues; dentin (covers the bulk of the tooth), enamel (covers

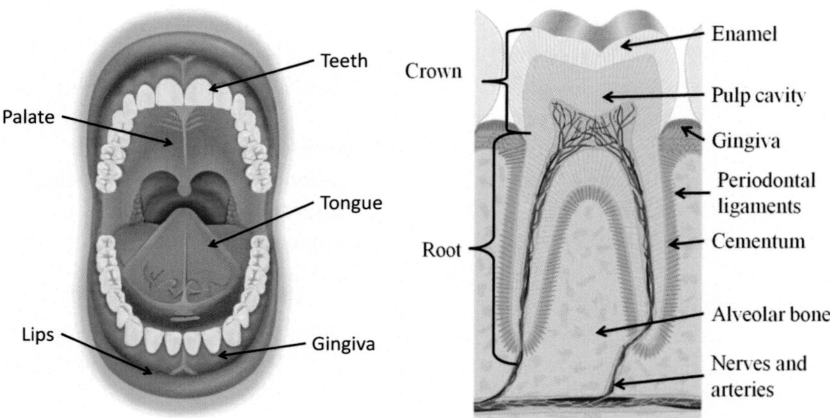

FIGURE 6.1 (A) The major parts of the human oral cavity; and (B) A cross section of a tooth. *(A) Adapted from Wikimedia Commons. (2016). Human mouth. Online Available: https://commons.wikimedia.org/wiki/File:Human_mouth.jpg. (B) Adapted from Goran, G. (2014). Cross section of teeth. Online Available: https://commons.wikimedia.org/wiki/File:Cross_sections_of_teeth.svg.*

the crown), and cementum (covers the root). Each tooth root is surrounded by the alveolar bone that helps to keep each tooth stabilized via the periodontal ligament (connective tissue that connects the root of each tooth to the alveolar bone), whereas the surrounding gums (gingiva) protect the roots from decay while keeping the teeth in place. A cross section of the tooth would reveal the pulp cavity, which runs from the crown to the root of each tooth, supplying the tooth with blood and nerves (Walker, 1990) (Fig. 6.1B).

The surfaces of the teeth are the perfect place on which dental plaque can attach. Dental plaque is produced through the release of the various types of bacterial metabolic products that accumulate on the surface of each tooth (Henley-Smith, Botha, & Lall, 2013). Plaque accumulates either supragingival (in between teeth or on the tooth crown above the gingival tissue) or subgingival (below the gingiva within the surrounding crevice of the teeth) and its role in the development of periodontal diseases is highlighted further on in this chapter.

The gaps between teeth and the crevices within the gingiva are also hotspots for the initiation of periodontal diseases. The small spaces that naturally occur between teeth make it easy for small particles of food to get lodged. Without regular flossing and the difficulty of reaching between the gaps of teeth during regular brushing, the growth of bacteria increases rapidly overtime within these areas.

6.1.2 Saliva

Saliva is a clear watery liquid produced by the six salivary glands present within the oral cavity. The presence of saliva within the oral cavity plays an important role in digestion, moistening the mouth for speaking, chewing, and swallowing as well as washes the mouth to prevent gum diseases and cavities (via enamel remineralization) due to its several functioning components. Although saliva naturally contains antibacterial agents such as, lysozymes (enzymes that break down bacterial cell walls) and lactoperoxidases (a natural occurring antibacterial enzyme with a broad spectrum of activity), it does have a neutral pH. The neutral pH nature of saliva coupled with the average human body and oral cavity temperature of 37°C makes the saliva found within the oral cavity the perfect breeding ground (microbial habitat) for various oral pathogens in the presence of food particles and shredded epithelial cells within the mouth (Devine & Cosseau, 2008).

6.2 PERIODONTAL DISEASES

6.2.1 Periodontitis

Within the oral cavity, the term "periodontium" is used. The periodontium is a collective term used to describe the specialized tissues that keep the tooth within its socket in the gums (Berthelot & Le Goff, 2010). Each tooth is held up in the alveolar socket by collagen fibers known as periodontal ligaments, which are embedded within the alveolar bone and cementum. The supporting structures include the periodontal ligament, tooth root, and alveolar bone socket, which make up the specialized tissues known as the periodontium. The tissues which cover these supporting structures include the gingival and alveolar mucosa (Williams, 1990). Periodontal diseases are divided into two major classes depending on which tissues of the oral cavity they disturb. A disease associated with the gingiva implies that a person has gingivitis. Gingivitis is characterized by swelling and reddening of the gums, a change in the gingival margin (usually a reduction) and bleeding of the gingiva during brushing (Fig. 6.2A). As the disease progresses into severity, and increased pain, it is no longer known as just gingivitis but rather acute necrotizing ulcerative gingivitis (Williams, 1990). If treated with correct and regular oral hygiene, gingivitis is a reversible event and can be completely eradicated, however, if left untreated, it can progress to periodontitis (Pizzo, Guiglia, Russo, & Campisi, 2010) (Fig. 6.2B).

When the disease extends into the destruction of the periodontium, it is termed periodontitis. Several factors including smoking, diabetes, and poor oral hygiene lead to the development of these periodontal diseases (Albander, 2005). Increased inflammation due to gingivitis extends into the periodontal ligament and alveolar bone causing bone degradation and destruction of the supporting structures, leading to the development of periodontitis (Williams, 1990). Along with their periodontal effect, periodontal diseases have shown numerous associated relationships with systemic diseases. The pathogens associated with gingivitis and periodontitis as well as their metabolic products and inflammatory

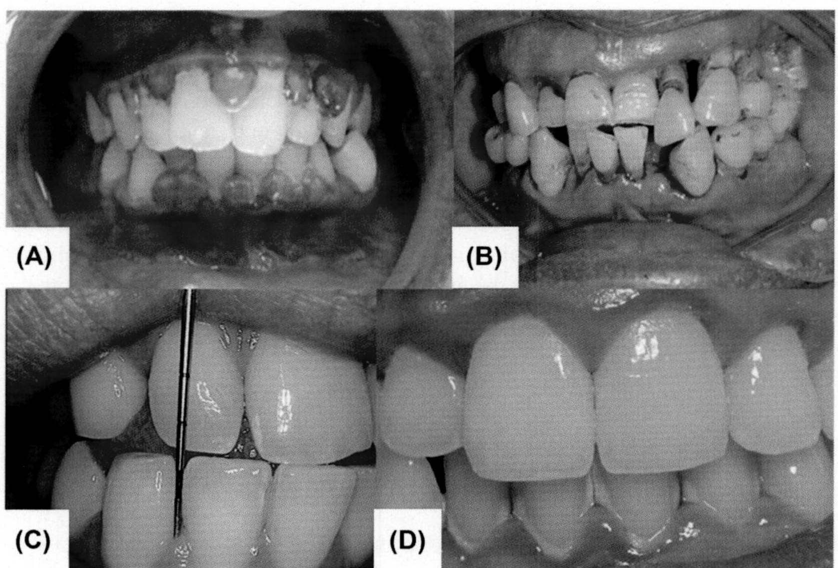

FIGURE 6.2 (A) Symptomatic observations of gingivitis include swollen, red and bleeding gums (Rosenbach, 2010); (B) Severe chronic periodontitis that has caused degradation of the supporting structures around the teeth exposing the roots (Zeron, 2010); (C) A periodontal probe is used to measure the pocket depth of inflamed gums (Strong, 2012); and (D) Plaque forms on the surface of teeth along the gum margin when irregular oral hygiene occurs (Wikimedia Commons, 2013).

mediators enter the bloodstream contributing to cardiovascular diseases (coronary heart disease, strokes, and high blood pressure) (Pizzo et al., 2010).

6.2.1.1 Causes of Periodontitis

As periodontitis develops, inflammation of the gums occurs and chronic gingival inflammation slowly degrades the periodontium overtime leading to the loss of teeth. Although symptomatic of an established disease (gingivitis), the initial sign is bleeding of the gums. The initial step of a dentist, when dealing with a patient who has inflamed gums, is to measure the depth of their periodontal pockets (Albander, 2005).

When the probe is inserted into the periodontal pocket of a patient with gingivitis, the initial response is bleeding (Fig. 6.2C). A pocket depth greater than 5 mm (stage II) indicates that separation has occurred between the gingiva and the tooth allowing entry for dental plaque, food, and oral bacteria. All these additional factors that colonize the periodontal pocket increases inflammation causing the gums to swell and recede from the teeth, causing it to bleed easily when touched. As the gums recede from the tooth it exposes the root. Exposure of the root surface to external factors, such as hot and cold food and beverages, causes severe pain and discomfort for the patient. As the disease worsens, the alveolar bone undergoes resorption (stage III, 6 mm pocket depth) that leads to the detachment of the periodontal ligament causing teeth to become very mobile, which ultimately ends in the loss of teeth (Berthelot & Le Goff, 2010).

The main cause of these periodontal inflammatory conditions is due the occurrence of dental plaque (Pizzo et al., 2010) (Fig. 6.2D). Dental plaque forms due to the attachment of early colonizing bacteria to the tooth surface after brushing. In cases of irregular brushing or poor oral hygiene, these early colonizing bacteria grow and change the environment in such a way so as to make it more suitable for the more pathogenic anaerobic bacteria to grow. As these organisms attach and metabolize, they form biofilms that act not only as a scaffold for additional bacteria to colonize but also help to keep them within the plaque. Eventually overtime, the biofilm becomes a habitat for a diverse community of interacting microorganisms (Marsh, 2010).

In healthy patients, the oral cavity is colonized by predominately Gram-positive bacteria but as the colonization shifts to an increased population of anaerobic Gram-negative bacteria, periodontal disease progress (Devine & Cosseau, 2008). This shift in equilibrium may be due to either one or all of the following factors (Pizzo et al., 2010):

- Changes in the environment due to the interaction of various bacterial communities or the accumulation of dental plaque
- Uses of systemic antibiotic, which reduce the populations of beneficial bacteria causing a reduction in inhibitory substances that would usually outcompete the pathogenic bacteria
- Compromised host immune system

This shift in microbial populations can be seen in the increased population difference in healthy individuals (10^2–10^3) and patients with gingivitis (10^4–10^8) or periodontitis (10^5–10^8) (Dixon, Bainbridge, & Darveau, 2004). As mentioned above, an increase in the depth of the periodontal pockets allows for the colonization of food, dental plaque, and oral bacteria (Berthelot & Le Goff, 2010). Bacteria are always present in the human mouth and even healthy gingival tissues possess inflammatory cells, but when the plaque builds up on the tooth surface, these inflammatory cells increase in number causing the gingival tissues to not only swell but also bleed and become edematous. When this occurs, usually the first colonizers are *Streptococci* bacteria. *Streptococci* colonization causes an increase in the amount of bleeding and tissue inflammation, resulting in the release of nutrients that are then readily available and required by other microbes, including *Prevotella intermedia*, assisting in its proliferation (Loesche & Grossman, 2001). Among the human oral bacteria within the oral microbiota there is a vast amount of communication through cell recognition and exchange of genetic and metabolic material. This communication is essential for the colonization and formation of plaque on the tooth enamel. The oral bacteria communicate with one another and the environment by adhering to the surfaces of teeth and forming a community of mixed oral bacterial species. As each cell adheres to the tooth surface, it provides an additional surface for more bacterial cells to adhere to. This adherence is important for the survival of oral bacteria as without it, the bacteria would be flushed out of the mouth via the host's saliva. On the other hand, when too much bacteria is adhered to the surface of teeth, the sudden increase in anaerobic bacteria within the mouth leads to aggregation of the bacteria and dominance in *P. intermedia* populations. This aggregation and dominance causes periodontitis and associated gum diseases (Loesche & Grossman, 2001).

6.2.1.2 *Prevotella intermedia*

P. intermedia is a black pigmented, anaerobic, Gram-negative oral pathogen (Dorn, Leung, & Progulske-Fox, 1998). These pathogens are found living within the periodontal pockets in between human teeth where they coexist with other oral microorganisms forming part of the human oral microbiota; however, when *P. intermedia* is found at higher bacterial populations than the other oral bacteria, periodontitis and acute necrotizing ulcerative gingivitis are the symptomatic diseases often observed (Haffajee & Socransky, 1994; Marcotte & Lavoie, 1998).

The formation of plaque stimulates the growth of *P. intermedia* leading to an increase of gingival crevicular fluid within the mouth. Gingival crevicular fluid is an inflammatory exudate allowing one to determine at which stage of development the periodontal disease is at within the host through a noninvasive biochemical analysis of its constituents (Subrahmanyam & Sangeetha, 2003). The presence of large amounts of gingival crevicular fluid, according to Marcotte and Lavoie (1998), stimulates the growth of *P. intermedia* as it contains nutrients including hemin and vitamin K, which are both essential for the growth of the pathogen. The available nutrients together with natural body conditions, such as a slightly basic pH and a constant temperature between 34 and 36°C, provides the perfect environment allowing for a rapid increase in the *P. intermedia* populations within the human mouth.

Immunoglobulins (Ig): Also known as antibodies that are produced by white blood cells and have an important role in immune response. They recognize and bind to specific molecules known as antigens produced by bacteria or viruses, further degrading them.
Proteases: Enzymes that degrade proteins.

P. intermedia is more evasive than most oral bacteria. The initial defense response of the host is to excrete salivary Ig namely, IgA1 and IgA2. *P. intermedia* is able to dominate the oral microbiota as it is able to produce IgA proteases that efficiently degrade the host's IgA defenses, allowing *P. intermedia* to evade initial defense responses and become successful in establishing themselves on teeth surfaces and within the oral mucosa (Marcotte & Lavoie, 1998). As the bacteria continue to proliferate within the gingival crevices it produces additional virulence factors to further establish a disease. Bacterial infections of this nature lead to *P. intermedia* secreting proteolytic enzymes including proteases. In addition to degrading host defenses, proteases such as collagenases have the ability to break down tissues and bone. In a previous study conducted by Kornman and Loesche (1982) on the effects of increased levels of steroid hormones (estrogen and progesterone) in the host's saliva, it was found that these hormones act as growth factors for the pathogen thus during pregnancy and puberty, when these hormones are at high production levels, there is a high risk of it being accompanied by gingivitis.

The onset of periodontitis is a complex process that not only causes damage to the oral cavity of humans but has also been associated with more severe diseases such as coronary heart disease. Research conducted over the years has illustrated the severe health risk that periodontitis may have on the association and development of coronary heart disease. *P. intermedia* was found to colonize not only the oral epithelial cells but also human coronary artery endothelia and coronary artery smooth muscle cells resulting in the association between periodontitis and heart disease in patients. Thus treatment of diseases caused by this pathogen is very important as patients are at a risk of not only gum disease but heart disease too (Beck, Garcia, Heiss, Vokonas, & Offenbacher, 1996; DeStefano, Anda, Kahn, Williamson, & Russell, 1993; Dorn, Dunn, & Progulske-Fox, 1999; Mattila et al., 1989).

6.2.2 Dental Caries

Dental caries/tooth decay is a preventable disease that is characterized by the progressive demineralization of the surface of the tooth, eventually leading to tooth loss (Fig. 6.3A). This disease is one of the most common and widespread diseases that affect the majority of the population, particularly those from poor backgrounds. The impact that dental caries has is a profound one and poses a challenge to the general health of the population.

Tooth decay has a negative effect on the quality of life of the people who suffer from it. This is due to the pain and discomfort that accompanies cavities and tooth loss. This disease can also have a negative effect on the self-esteem, confidence, and communication skills of the people who suffer from it.

- According to the World Health Organization, many school and work hours are lost annually due to the high prevalence of absenteeism as a result of the discomfort and pain associated with caries and cavities.
- People often overlook oral health as an important part of their general well-being because the mouth is generally seen as a separate part from the rest of the body.
- The cost of oral health and oral disease treatment is very expensive, with it being the fourth most expensive disease treatment in some countries.

6.2.2.1 Causes of Dental Caries

Dental caries and tooth decay is primarily caused by the Gram-positive cocci (spherical) bacterium *Streptococcus mutans*, which is vertically transmitted from mother to child. This bacterium is a facultative anaerobe (capable of switching to fermentation or *anaerobic* respiration if oxygen is absent) that

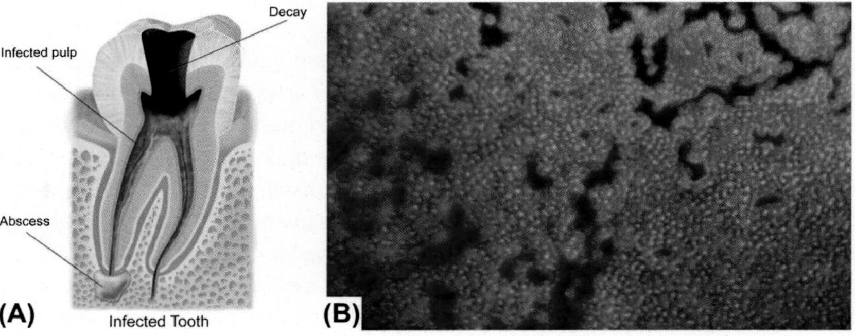

FIGURE 6.3 (A) Cross section illustration of a tooth depicting a tooth infected with *Streptococcus mutans* resulting in tooth decay (Blaus, 2013); and (B) Gram stain of the Gram-positive bacterium *S. mutans*.

grows at an optimal temperature of 37°C, which is the same temperature as that of a normal human host (Fig. 6.3B).

The development of this disease is dependent on a number of other factors including the presence of sugars and the presence of teeth. The bacterium is able to directly adhere to the surface of the tooth and metabolizes sugars that are consumed in the diet to produce acid as a by-product. The acid then causes the breakdown of the tooth enamel.

S. mutans initiates the onset of dental caries by using the host's saliva to produce sticky like substances known as pellicles, which aid in its direct attachment to the enamel of the tooth. Once the bacterium attaches, it then begins to produce an aggregate of cells around that area known as biofilm. Once the bacterium establishes a conducive environment for infection in the biofilm, it secretes organic acids (by-products of metabolism) that decrease the pH in the mouth, leading to the demineralization of tooth tissues and eventual decay of the tooth, forming cavities.

The presence of biofilm in the mouth also allows for the attachment of other oral pathogens and provides a suitable environment for them to establish themselves. These bacteria are known as secondary colonizers and include *Fusobacterium* and anaerobic *Actinomyces*.

- The oral cavity has natural flora present that causes no harm until there is a shift in the environment that makes it conducive for disease development.
- Poor oral hygiene as well as environmental factors can lead to the development of dental caries.
- Decreased saliva production increases the chances of developing caries as it leads to a decrease in pH in the oral cavity.
- The action of brushing teeth can reduce the amount of bacteria present in the oral cavity thereby decreasing the chances of further development of the disease.

6.2.2.2 Mechanisms Employed by Streptococcus mutans *to Cause Disease*

- Glucosyltransferases (GTFs) are enzymes that are encoded by the *gtf* genes.
- Antigen I/II is another adhesin that *S. mutans* produces to aid in attachment, encoded for by the *spaP* gene.

The success of *S. mutans* to cause disease is dependent on its ability to effectively attach onto the surface of the tooth. The development of dental caries can be divided into three distinct stages. The first stage of dental caries is one of the most important steps, which involves the presence of sucrose and GTF enzymes, required for the initial accumulation of *S. mutans*. Saliva found in the oral cavity forms a thin film on the surface of the tooth. The bacterium uses several proteins found in these films to form pellicles on the surface of the tooth. The second stage involves the bacteria interacting with the proteins in the saliva to continue producing these pellicles that allow for the accumulation of *S. mutans*, eventually forming a biofilm. The last step involves the metabolism of dietary sugars to produce acidic by-products.

6.2.3 Oral Candidiasis

Oral candidiasis, or more commonly known as oral thrush, is a fungal infection caused by various species of *Candida*, predominantly *Candida albicans*. This disease has become an increasingly problematic health concern, particularly in individuals with compromised immune systems such as those with HIV/AIDS or diabetes. There are four types of this oral disease, namely, acute pseudomembranous candidiasis, chronic erythematous candidiasis, acute erythematous candidiasis, and chronic hyperplastic candidiasis. Some of the signs and symptoms of this disease include white patches on the tongue or other regions of the oral cavity including the throat (Fig. 6.4A).

6.2.3.1 Causes of Oral Candidiasis

C. albicans is an opportunistic fungal pathogen; a small component of the naturally occurring microflora present in the oral cavity and other parts of the body (Brown et al., 2006; Cannon & Chaffin, 1999; Inglis et al., 2012) (Fig. 6.4B). It is normally a commensal microorganism whose occurrence is not an indication of disease, except in those people with a compromised immune system. Immune-compromised individuals include those with HIV/AIDS and diabetes, in such cases *C. albicans* populations increase to levels where it leads to the development of oral candidiasis. It has also been shown to mainly cause disease

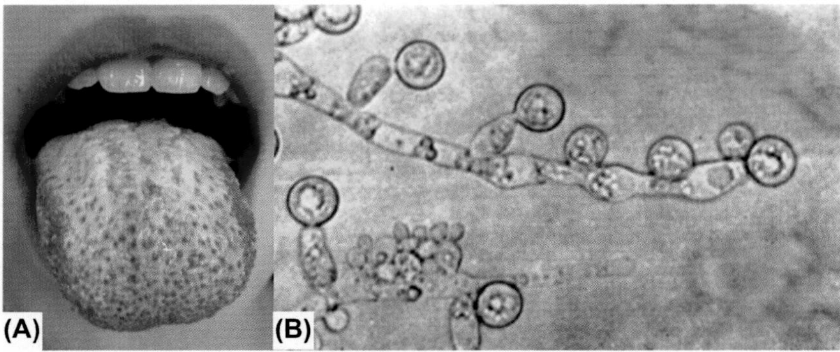

FIGURE 6.4 (A) Human tongue infected with *Candida albicans*, showing the fungal symptoms of oral candidiasis (Heilman, 2015); (B) *C. albicans* under a transmission electron microscope (Colm, 2010).

in the elderly and the very young. Infections of this pathogen can be both superficial and systemic. The treatment of diseases caused by this fungus is through the use of fungicides, however it is difficult due to the potential toxicity to cells (Cao et al., 2005; Williams & Lewis, 2011).

6.2.3.2 *Mechanisms Employed by* Candida albicans *to Cause Disease*

Strategies Used by *C. albicans* to Cause Disease

Virulence factor	Effect
• Adherence	Promotes retention in the mouth
• Phenotypic switching	Antigenic modification
• Hyphal development	Promotes invasion of oral epithelium
• Hydrolytic enzymes	Host cell damage

C. albicans does not have a principal virulence factor because it is found as a safe commensal organism in humans. It does, however, have putative virulence factors. These factors are important in the successful tissue damage in the host as well as the continued infectious stage of *C. albicans*. The hyphal production has been associated with an increased expression of putative virulence factors, some of which are not involved in the process of hyphal production itself (Williams & Lewis, 2011). This fungus has the special ability to undergo a phenotypic switch, from planktonic form to hyphal/filamentous form. This morphological plasticity plays an important role in its virulence. The fungus uses glucose as a trigger or signal to convert itself from yeast form to a hyphal form. The yeast form (unicellular) is important in the initial stages of infection where the pathogen needs to disseminate (disperse/spread) in the host via body fluids. The hyphal form is vital for the formation of mycelial biofilms or the infiltration into tissues (filamentous growth). This pathogen is able to produce extracellular enzymes that damage host structures (Brown et al., 2006; Cannon & Chaffin, 1999; Williams & Lewis, 2011).

6.3 PREVALENCE OF ORAL DISEASES

Chronic diseases have been attributed as the leading cause of death and disability across the world. Periodontitis and dental caries are two of the most important oral diseases that contribute to the burden of chronic diseases globally. These diseases are very common across all populations and thus can be seen as a major public health problem to many countries (Petersen & Ogawa, 2012; van Wyk & van Wyk, 2004). Oral candidiasis although not prevalent across all populations, is a major threat to around 40%–50% of individuals infected with HIV/AIDS (World Health Organisation, 2012).

6.3.1 Prevalence of Oral Diseases Worldwide

Initially presenting itself as gingivitis, periodontitis progresses when gum diseases are left untreated leading to severe tooth loss and bone destruction. Progression into periodontitis is estimated to present itself in around 15% of the global population. By the age of 65–75 years, 30% of individuals have lost all teeth, with periodontitis being the main cause (FDI World Dental Federation, 2015).

Dental caries is said to be one of the most common childhood diseases that breaches into adulthood. Statistics revealed that the untreated decay of permanent teeth has a global prevalence of around 40% for all ages combined, listing dental caries as the most prevalent conditions out of the 291 diseases included in the Global Burden Disease Study (FDI World Dental Federation, 2015).

The prevalence of HIV-infected individuals globally is estimated to be around 35 million people in 2013 with almost 50% of HIV-positive individuals developing oral diseases such as oral candidiasis during the course of infection development (FDI World Dental Federation, 2015).

6.3.2 Prevalence of Periodontal Diseases in South Africa

South Africa is regarded as a developing country with majority of the population falling into the low to middle income class. This classification causes of the vast majority of the population experience the highest burden of oral diseases due to the disadvantages in access to oral health care facilities. A total of three national oral health surveys have been conducted in South Africa, with the most recent one during the periods of 1999–2002. The results of the survey revealed that 45%–60% of South African children require treatment for dental decay or caries, while 32% of children required orthodontic treatment due to premature tooth extractions. Individuals between the ages of 35 and 44 years had the highest prevalence of dental caries and periodontal-related diseases at 95%, with 68% remaining untreated. Because of the high occurrence of HIV/AIDS infections within South Africa a number of HIV-positive individuals also suffer from associated oral candidiasis. The survey found that only 16.7% of individuals aged between 15 and 44 years were completely free of any form of periodontal disease. It was also found that about 80% of oral disease remains untreated in South Africa across all age groups (Ramphoma, 2016; van Wyk & van Wyk, 2004).

6.4 CURRENT TREATMENTS

Dental care is of great importance both in the prevention of oral diseases and the maintenance of oral hygiene to ensure proper growth and strength of teeth. Fluoride, antibiotics, and chlorhexidine gluconate have been used in the successful treatment of oral bacterial infections; however, their overuse has shown some dramatic adverse effects within the oral cavity.

The introduction of fluoride into water systems, toothpaste, and mouthwash has resulted in major reductions in the occurrence of dental caries and tooth decay. Although effective on adults, the excessive use of fluoride impacts developing teeth in children by causing enamel fluorosis. Enamel fluorosis results in the enamel of developing becoming porous. Symptoms of mild fluorosis include white opaque striations across the tooth surface, while in more severe cases, the porous regions increase in size resulting in pitting of the enamel and discoloration (Bronckers, Lyaruu, & DenBesten, 2009).

Antibiotics are used widely in the treatment of severe bacterial infections such as periodontitis. Penicillin, erythromycin, cephalosporin, clindamycin, and tetracycline have shown to be effective in reducing the populations of bacterial pathogens within the oral cavity. However their uses are limited because of allergies, cost, and more importantly, resistance development (Montgomery & Kroeger, 1984).

Chlorhexidine gluconate is the active ingredient in most commercially available antibacterial oral rinses. It is a broad spectrum antimicrobial agent that is generally effective in preventing many oral diseases; however, chlorhexidine gluconate has been reported to be inactivated by food and saliva. The adverse effects of the use of chlorhexidine gluconate include alterations of taste, mucosal irritation and stained teeth and tongue (Henley-Smith et al., 2013).

Therefore, there is a need to develop alternative preventative options and treatments that are safe, effective, and economical for everyday use.

6.5 PLANTS TRADITIONALLY USED FOR ORAL CARE

The plants mentioned in Table 6.1 have been documented to be used traditionally across the globe to treat the various symptoms associated with oral diseases.

6.6 PLANT PRODUCTS PRODUCED FROM MEDICINAL PLANTS

6.6.1 Toothpastes

Many commercially available toothpastes have incorporated the naturally active properties of plants within their formulations. The addition of tea tree oil (TTO), mint, cloves, aniseed, and olive leaf extracts enhance the activity of the toothpaste by providing effective antimicrobial activity. Adding neem leaf extracts has been shown to be effective in the inhibition of several caries forming bacteria (Wolinsky, Mania, Nachnani, & Ling, 1996). Pure extract of *Salvadora persica* is incorporated into herbal toothpastes as it helps and prevents dental caries due to the plant's anticarcinogenic activities (Gupta, Agarwal, Anup, Manujunath, & Bhalla, 2012). *Aloe ferox* is combined with other herbal extracts within a toothpaste formulation for individuals with sensitive teeth, ulcers, and bleeding gums (Meena, Figueiredo, & Trivedi, 2013). Xylitol and stevia are used in some toothpaste formulations instead of fluoride and are antimicrobial

TABLE 6.1 Reported or Documented Uses of Plants That Have Been Used Traditionally to Treat Oral Ailments

Plants						
Scientific Name	Common Name	Location of Plant	Part of Plant Used	Traditional Use	Preparation	References
Acokanthera oppositifolia (Lam.) Codd	The Bushman's poison tree	South Africa to Democratic Republic of Congo and Tanzania	Combination of plant parts including leaves, stems, and roots	Toothache	Weak infusions	Hankey (2001)
Albizia adianthifolia W. Wight	Rough-bark flat-crown	Tropical Africa, Southern Africa, and eastern Madagascar	Leaves and roots	Toothache	Paste	Krige (2007)
Allium cepa L.	Onion	Worldwide	Leaves	Sore throats, toothaches, and dental abscess	A paste is made and applied to treat toothaches, while a decoction is made and used as a gargle for sore throats and dental abscesses	Benkeblia (2004)
Aloe vera (L.) Burm.f.	*Aloe vera*	Africa and the Mediterranean	Leaves	Gingivitis	The leaves produce a gel that is then rubbed onto the infected areas of the mouth	Meena et al. (2013)

Berula erecta (Huds.) Coville	Water parsnip	Northern Hemisphere (North, South and Central America, Canada, from Europe to China), eastern Africa and South Africa.	Rhizomes	Toothache	Poultice	Grow Wild Nursery (2017)
Camellia sinensis (L.) Kuntze	Green tea	Worldwide	Leaves	Dental caries and cavities	Tea decoctions	Goenka et al. (2013)
Capsicum frutescens L.	Chilli pepper	Central and South America	Fruits	Toothache and oral candida	Paste is made from the fruits and used as a mouth rinse	Ashu Agbor and Naidoo (2015)
Carpobrotus edulis (L.) L. Bolus	The sour fig	South Africa	Leaves	Throat infections and oral thrush	Leaf juice	Henley-Smith et al. (2013)
Citrus sinensis Pers.	Sweet orange	Worldwide	Leaves	Gingivitis	Decoction	Ashu Agbor and Naidoo (2015)
Cotyledon orbiculata L.	Pig's ear	South Africa	Leaves	Toothache and inflammation associated with some gum diseases	Leaf juice is used as drops	Harris (2004)
Dichrostachys cinerea (L.) Wight & Arn.	Sickle bush	Africa, Madagascar, India, Indonesia, and Australia	Leaves	Toothache and oral candidiasis	Decoction	Cheek (2009)

Continued

TABLE 6.1 Reported or Documented Uses of Plants That Have Been Used Traditionally to Treat Oral Ailments—cont'd

Plants						
Scientific Name	**Common Name**	**Location of Plant**	**Part of Plant Used**	**Traditional Use**	**Preparation**	**References**
Dicoma anomala Sond.	Fever bush	Zimbabwe, Democratic Republic of Congo, Uganda, Tanzania to South Africa, Lesotho, and Angola	Roots	Toothache	Decoction	Maroyi (2013)
Dodonaea viscosa Jacq. var. *angustifolia*	The sand olive	South Africa, Mozambique, and Namibia	Leaves	Sore throat and oral thrush	Decoction	Teffo, Aderogba, and Eloff (2010)
Garcinia mannii Oliv.	Chewing stick	Cameroon and Nigeria	Stem	Dental caries	Chewed directly	Ashu Agbor and Naidoo (2015)
Heteropyxis natalensis Harv.	Lavender tree	South Africa, Swaziland, and Zimbabwe	Leaves and twigs	Toothache, oral and gum infections	Oral rinse (infusion)	

Ludwigia adscendens (L.) H. Hara	Water primrose	Southern Africa	Whole plant	Antiseptic for the treatment of oral ulcers	Poultice	Selim (2003)
Melianthus major L.	Giant honey flower	South Africa and India	Leaves	Used a gargle or mouthwash for the treatment of oral ulcers, gum diseases and sore throats	Warm aqueous leaf infusions	van der Walt (2000)
Piper guineense Thonn.	Pepper bush	West Africa	Fruits	Dental caries	A paste is made using water and this is either applied directly on infected teeth or used as a gargle	Idu, Obaruyi, and Erhabor (2009)
Zanthoxylum capense Harv.	Small Knobwood	South Africa, Zimbabwe, and Mozambique	Root and bark	Toothache	Powered roots and bark are used to make an infusion that can be administered as an oral rinse	Kondlo (2012)
Zea mays subsp. *mays* L.	Maize	Worldwide (used mainly in Burkina Faso)	Flowers	Toothache	Decoction	Tapsoba and Deschamps (2006)

sweeteners that help to control plaque, as well as the strengthening of teeth and gums (Riley, Moore, Ahmed, Sharif, & Worthington, 2015; Visser, 2016).

6.6.2 Oral Rinses and Herbal Mouthwash

Oral rinses are able to reach the places in the mouth that a toothbrush is not able to. This makes them a useful means of administering antibacterial ingredients that help fight against the development of dental caries.

Many commercially available oral rinses contain several plant ingredients including green tea, garlic juice, thymol, eucalyptol, liquorice root, and rooibos. Green tea extract has been incorporated into mouthwashes as not only does it provide a main antimicrobial constituent but it is also used as an alcohol replacement. By replacing the alcohol, it eliminates any irritation to the mouth, gums, and teeth that may occur. Green tea has also been shown to have extremely high antioxidant activity (higher than Vitamin C), which is an added health benefit. Garlic juice can also be used as a rinse to help reduce plaque formation and aid in the control of dental caries, while the other plant extracts used in oral rinses have known antibacterial properties.

6.6.3 Toothbrush Sticks and Chewing Sticks

The act of brushing teeth and physically removing excess plaque in the oral cavity forms part of the treatment of dental caries. In many parts of the world, particularly in Africa and Asia, plants are used to make toothbrushes and chewing sticks to maintain good oral health. The sticks are made from roots, stems, and twigs. The sticks are normally chewed or tapered at one end until they become ragged into a brush. They are soaked in water for a couple of hours to soften the natural fibers. When left in the mouth, they can stimulate saliva production and will therefore give a better cleansing effect. The sticks are sold as bundles in local markets. Some medicinal plants that are used to make these sticks include *Diospyros lycioides* (blue bush/bloubos) (Fig. 6.5A), *S. persica* (miswak/toothbrush tree) (Fig. 6.5B and C), *Acacia mellifera* (black thorn/swarthaak) (Fig. 6.5D), *Jasminum fluminense* (starry wild jasmine) (Fig. 6.5E), and *Azadirachta indica* (neem) (Fig. 6.5F).

6.7 REPORTED ANTIBACTERIAL AND ANTIFUNGAL ACTIVITY OF MEDICINAL PLANTS AND PHYTOCHEMICAL COMPOUNDS

The world is home to a diverse number of plant species, which have been traditionally used for oral care by various communities. These traditionally used plant species have been largely collected from terrestrial ecosystems; therefore, the medicinal properties of terrestrial plants have been extensively researched, while aquatic and wetland plants have remained a relatively understudied field.

FIGURE 6.5 Medicinal plants that are traditionally used as toothbrush or chewing sticks (A) *Diospyros lycioides* (JMK, 2012), (B) the sticks of *Salvadora persica* (miswak) (Osman, 2012), (C) *S. persica* leaves (Garg, 2009), (D) *Acacia mellifera* (JMK, 2013), (E) *Jasminum fluminense* (Forest & Kim Starr, 2006), (F) *Azadirachta indica* (Halady, 2012); Medicinal plants which have been reported to have growth inhibitory activities against oral pathogens (G) *Heteropyxis dehniae*, (H) *Syzygium aromaticum* commonly known as cloves (Barrios, 2007), (I) *Vaccinium vitis-idaea* (Bergsten, 2006), (J) *Cinnamomum verum* commonly known as cinnamon (Mugraby, 2007), (K) *Melaleuca alternifolia* commonly known as tea tree (Kojian, 2011), (L) Euclea natalensis (JMK, 2014), (M) *Mentha longifolia*, and (N) *Zantedeschia aethiopica*.

In both terrestrial and aquatic ecosystems, the defining factor between a normal plant versus a medicinal plant is within their ability to produce secondary metabolites (phytochemicals). These specific phytochemicals range in chemical structures, however, depending on which classes are available in the plant. These compounds provide the plant with the ability to inhibit the growth of certain oral pathogens in a variety of mechanisms.

The activity of each plant species to inhibit *S. mutans* and *P. intermedia* is commonly tested using a microtiter dilution assay that allows one to determine the lowest concentration at which the plant extract inhibits the bacteria. This assay involves growing the bacteria along with the plant extract at various concentrations to determine if the extract has any inhibitory effect against the bacteria. A growth reagent is used to determine which cells treated with extract continue to grow and which ones are inhibited by the extract. The minimum inhibitory concentration (MIC) of a particular sample can then be determined. However, depending on the organism and plant extract various other methods can also be used, such as a disc diffusion assay (measures the zone of inhibition (in mm) of a plant extract at a certain concretion toward the growth of either bacteria or fungi), to assess the antibacterial activity of a plant sample.

A number of plants found globally that have been reported to show good activity in inhibiting the growth of pathogens, which are related to causing oral diseases are discussed in the following section:

6.7.1 *Heteropyxis dehniae* Suess.

Heteropyxis dehniae is one of three plants found within the *Heteropyxis* genus; its traditional Ndebele name is *Inkiza Yentaba* (Fig. 6.5G). It is a deciduous tree and is very similar to another plant of the *Heteropyxis* genus, *Heteropyxis natalensis*. It occurs predominantly in the bushveld as well as rocky places. The tree naturally occurs in Zimbabwe (Sibanda et al., 2004).

This species contains a number of essential oils in the leaf, which largely contain 78.2% oxygenated monoterpenoids, 9.5% sesquiterpene hydrocarbons, and 6.7% oxygenated sesquiterpenoids. Linalool (58.3%), 4-terpineol (9.8%), terpineol (3.6%), and caryophyllene oxide (3.1%) are the most abundant essential oil components. The leaf oils have been shown to have cytotoxicity activity against human melanoma (SK-MEL-28) and human bladder carcinoma (5637) cells in vitro (Sibanda et al., 2004).

H. dehniae has been traditionally used by the Zulu people as tea for blood purification and is also used as a perfume to scent tobacco. In addition to this, the dried fresh leaves are smoked to relieve headaches. The leaf oils isolated from *H. dehniae* showed good activity against a number of microorganisms including *Staphylococcus aureus*, *Pseudomonas aeruginosa*, *Escherichia coli*, and *C. albicans* using the disc diffusion method with the most significant value against *C. albicans* (0.625 mg/mL) (Sibanda et al., 2004).

6.7.2 *Syzygium aromaticum* (L.) Merr. & L.M. Perry

Although traditionally used for a wide variety of ailments, the oil extracted from *Syzygium aromaticum* (cloves), is used in clinical dentistry during root canal therapy and to temporary fix tooth fillings (Fig. 6.5H). The use of this plant in clinical dentistry is attributed to its antimicrobial activity against the pathogens associated with the onset of dental caries and periodontitis (Cai & Wu, 1996).

According to Cai and Wu (1996), the methanol extract of *S. aromaticum* (clove) exhibited preferential growth-inhibitory activity against *P. intermedia* at an MIC of 156 μg/mL. Using bioassay guided chromatographic fractionation, eight active compounds were isolated from this methanol extract. These compounds were identified using spectroscopic methods as 5,7-dihydroxy-2-methylchromone 8-C-β-D-glucopyranoside, biflorin, kaempferol, rhamnocitrin, myricetin, gallic acid, ellagic acid, and oleanolic acid. These isolated compounds were then tested for their antibacterial activity against *S. mutans* and *P. intermedia*, among other oral pathogens. The flavones, kaempferol, and myricetin were observed to have the best growth-inhibitory activity against the periodontal pathogen, *P. intermedia*, with an MIC of 20 μg/mL for each compound (Fig. 6.6A and B). Other notable MICs of *S. aromaticum* compounds inhibiting the growth of *P. intermedia* included gallic acid (78 μg/mL), 5,7-dihydroxy-2-methylchromone 8-C-β-D-glucopyranoside (156 μg/mL), biflorin (625 μg/mL), and rhamnocitrin (625 μg/mL), while rhamnocitrin, ellagic acid, and

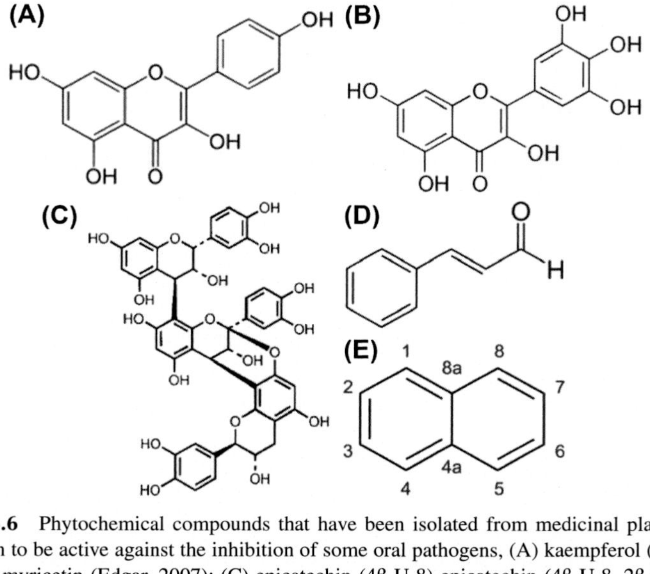

(A) (B) (C) (D) (E)

FIGURE 6.6 Phytochemical compounds that have been isolated from medicinal plants, which have shown to be active against the inhibition of some oral pathogens, (A) kaempferol (Yikrazuul, 2008); (B) myricetin (Edgar, 2007); (C) epicatechin-(4β U 8)-epicatechin-(4β U 8, 2β U O U 7)-catechin (Ho et al., 2001); (D) cinnamaldehyde (Neurotiker, 2007); and (E) the precursor from which naphthoquinones are derived, naphthalene (Edgar, 2012).

oleanolic acid all inhibited *S. mutans* at an MIC of 1250 μg/mL (which was the lowest observed).

6.7.3 *Vaccinium vitis-idaea* L.

Vaccinium vitis-idaea is a well-known berry plant found widely across the Northern Hemisphere (Fig. 6.5I). This plant is traditionally used in the treatments for gonorrhea, dysuria, and diarrhea as the leaves and berries have been reported to possess antiviral and antiinflammatory activity due to the presence of arbutin, hyperin, hydroquinone, isoquercetin, and tannins (Fokina, Roikhel, Frolova, Frolova, & Pogodina, 1993; Mitsuhashi, 1988; Perry, 1980).

Six tannins (procyanidin B-1, procyanidin B-3, proanthocyanidin A-1, cinnamtannin B1, epicatechin-(4β U 8)-epicatechin-(4β U 8, 2β U O U 7)-catechin, and epicatechin-(4β U 6)-epicatechin-(4β U 8, 2β U O U 7)-catechin) were isolated from *V. vitis-idaea* due to their good antioxidant activity as there is a correlation between antioxidant responses and the treatment of periodontitis. By decreasing the antioxidants within the oral cavity it initiates a reduction in inflammatory cellular responses which in turn reduces the inflammation observed during the initial symptoms of the development of gingivitis and periodontists (Ho et al., 2001).

The antibacterial activity of the tannins was tested against *P. intermedia* using the microtiter dilution assay method. The results indicated that only epicatechin-(4β U 8)-epicatechin-(4β U 8, 2β U O U 7)-catechin had any antimicrobial activity against *P. intermedia* with an MIC of 25 μg/mL (Ho et al., 2001) (Fig. 6.6C).

6.7.4 *Cinnamomum verum* J. Presl

Cinnamomum verum, also referred to as the "true cinnamon," is a small evergreen tropical tree belonging to the Lauraceae (laurel family) that originated in Sri Lanka (Fig. 6.5J). This tree is one of several *Cinnamomum* species that produce the commercially important spice known as cinnamon. This spice is obtained from the inner bark of the trees. The term "cinnamon" also refers to its midbrown color. Both *C. verum* and another species *Cinnamomum cassia* Blume are collectively called *Cortex Cinnamomi* for their medicinal cinnamon bark. *C. verum* is popular all over the world, whereas *C. cassia* is a well-known traditional Chinese medicine (Chaudhari, Jawale, Sharma, Kumar, & Kulkarni, 2012; Choi, Cho, Kim, Park, & Kim, 2016).

Although cinnamon is most widely used as a spice to flavor and color food, the cinnamon oils that are derived from the bark and leaves are widely used in the treatment and prevention of oral diseases such as dental caries. Toothpicks soaked in cinnamon oil can be used to effectively treat many kinds of aches and pains related to toothache. This oil has also been shown to inhibit the growth of bacteria that cause oral diseases. Volatile oils such as cinnamaldehyde, transcinnamic acid and eugenol have been previously isolated from *C. verum* and were tested for their antibacterial activity against *S. mutans*. Cinnamaldehyde showed very good activity against *S. mutans* with an MIC value of 0.02% (v/v) (Choi et al., 2016) (Fig. 6.6D).

6.7.5 *Melaleuca alternifolia* Cheel

Melaleuca alternifolia, commonly known as the narrow leaved paperbark or narrow leaved tea tree, is a tall tree or shrub endemic to Australia (Fig. 6.5K). This tree has been traditionally used as medicine for over 227 years. The oil form the tree, known as TTO, was first commercially produced using steam distillation about 95 years ago. From then, the oil has been the main constituent from the tree that is vigorously used medicinally. The primary uses of TTO have historically exploited the antiseptic and antiinflammatory properties of the oil (Hammer, Carson, & Riley, 2003). This includes treatment in ailments affecting the oral cavity, ranging empyema (pus occurring in a body cavity), catarrh (inflammation of a mucous membrane with increased flow of mucus), and gingivitis to oral thrush (Carson, Hammer, & Riley, 2006).

TTO is composed of terpene hydrocarbons, mainly monoterpenes, sesquiterpenes, and their associated alcohols. The main constituents are mostly made up of 41% terpinen-4-ol and 2.1% 1,8-cineole among 12 other components. TTO has been tested against several microorganisms and has been shown to have very good activity against *S. mutans*; with MIC values as low as 0.012% v/v (Carson et al., 2006).

6.7.6 *Azadirachta indica* A. Juss.

A. indica, commonly known as neem, nimtree or Indian lilac, is a fast growing evergreen tree belonging to the mahogany family Meliaceae (Fig. 6.5F). This tree is native to India and the Indian subcontinents but has been introduced to many other areas in the tropics. Its fruits and seeds are the source of neem oil (Tapsoba & Deschamps, 2006). Neem oil is derived by pressing the seed kernels of the neem tree. It is very bitter with a garlic/sulfur smell. This oil has been formulated in several toothpastes and has been shown to be effective against cavities and plaque formation (Hodges, 2017). Neem has been shown to have antibacterial activity against *S. mutans*, where 100 μg of neem extract had zones of inhibition of 8.2 mm (Salam, Khokon, & Mussa, 2014).

6.7.7 *Euclea natalensis* A.DC

Euclea natalensis, commonly known as the Natal ebony or large-leaved guarri, is a small bushy tree that is widely distributed across many countries in Africa (Fig. 6.5L). The twigs are commonly used for toothbrushes, with the roots and bark used as ingredients in a wide range of remedies for oral problems including for the treatment of toothache. The roots of *E. natalensis* contain naphthoquinones that are bactericidal (Fig. 6.6E). Samples of fresh roots were verified against *S. mutans*, bacterial growth was suppressed (Stander & Van Wyk, 1991). The leaves have been reported to have an MIC of 6.3 mg/mL against *S. mutans* (More, Tshikalange, Lall, Botha, & Meyer, 2008).

6.7.8 *Diospyros lycioides* Desf.

D. lycioides commonly referred to as the bluebush, star apple, or monkey plum tree, is found growing naturally in central Africa, southern Tanzania and Southern Africa (including South Africa, Lesotho, and Swaziland) (Fig. 6.5A). In Australia it has also naturalized and is currently categorized as a weed. Pieces of roots and stems of *D. lycioides* are sometimes chewed and extensively used as toothbrushes (Mujuru, 2011). Root and stem methanol extracts of *D. lycioides* showed activity against *S. mutans* at concentrations ranging from 0.3 to 11.2 mg/mL (Mbanga & Magumura, 2013).

6.7.9 *Mentha longifolia* L.

Although not fully aquatic, *Mentha longifolia* is found growing along the edges of rivers and dams due to its high affinity for water-drenched soils (Fig. 6.5M). It is one of many wetland plants that has been found to be traditionally used by communities in the treatment of various human ailments. *M. longifolia* is traditionally used as a treatment for colic, menstrual disorders, indigestion, flatulence, pulmonary infections and congestion, headaches, fever, coughs, colds, and urinary tract infections. *M. longifolia* is also used to relieve swelling and treat sores and minor wounds of the skin. The leaves and stems can be added to boiling water to release a vapor that can be inhaled to relieve nasal and bronchial congestion (van der Walt, 2004).

The essential oil obtained by hydrodistillation from the South African variety of *M. longifolia* was analyzed using gas chromatography and mass spectrometry methods. The analysis of the essential oil reveals the presence of 31 components (accounting for 99.7% of the oil) of which menthone (50.9%), pulegone (19.3%), and 1,8-cineole (11.9%) were found to be the major constituents (Oyedeji & Afolayan, 2006).

The essential oil of *M. longifolia* has been found to exhibit an MIC of 15.6 μg/mL against the oral pathogen *S. mutans* (Al-Bayati, 2009).

6.7.10 *Zantedeschia aethiopica* (L.) Spreng.

Found growing along the edges of rivers, the leaves of *Zantedeschia aethiopica* is heated and used as a dressing for wounds, boils, minor burns, insect bite, and sores (Fig. 6.5N). Patients suffering from gout or rheumatism also use the warmed leaves as poultice to reduce the pain. Traditional communities located in the southern part of South Africa, powder the rhizome of *Z. aethiopica* and use it as a poultice for inflamed wounds. The plant leaves are boiled and eaten by mixing it with honey or syrup as a treatment for asthma and bronchitis; it can also be gargled for the relief of sore throats. The plant should be boiled or cooked in some way as the raw plant material causes swelling of the throat due to the presence of microscopic calcium oxalate crystals (Roberts, 1990; Rood, 2008; Watt & Breyer-Brandwijk, 1962; Wink & van Wyk, 2008).

At present, a number of chemical constituents present within *Z. aethiopica* have been identified using spectroscopy, including two cycloartane triterpenes, 10 sterols, three lignans, and 10 phenylpropanoids (Della Greca, Ferrara, Fiorentino, Monaco, & Previtera, 1998). However, none of these compounds have been reported to possess antimicrobial activity against oral pathogens.

Antifungal studies on the efficacy of traditionally used South African plants have been performed to inhibit the growth of the oral fungal pathogen *C. albicans* (Motsei, Lindsey, van Staden, & Jager, 2003). *C. albicans* affects patients with suppressed immune systems due to HIV and AIDS as it is an opportunistic pathogen (Calderone & Fonzi, 2001). *C. albicans* like *S. mutans* colonizes the root canal eventually breaking it down. Ethanolic extracts of the leaves of *Z. aethiopica* were used in antifungal studies against *C. albicans*. The results concluded that *Z. aethiopica* had an MIC greater than 25 mg/mL (Motsei et al., 2003). This suggests that *Z. aethiopica* did not exhibit any antifungal activity at the tested concentrations. The significance of this result indicates that oral pathogens are relatively difficult to inhibit, whether they are bacterial or fungal. The concentrations, at which the antimicrobial activity was tested, were not high enough to inhibit the growth and proliferation of these resistant oral pathogens. The persistence of these oral pathogens that colonize the root canals could be explained by their ability to produce biofilms and evade dental hygiene apparatus as they colonize within and between teeth (Botelho et al., 2007).

6.8 CONCLUSION

The role that dental hygiene plays in daily routine is crucial in the prevention of infectious oral diseases including periodontitis, dental caries, and oral candidiasis. The most important thing to remember is that the causal agent of all three of these aforementioned diseases is the increased population of the unique oral pathogens, *P. intermedia*, *S. mutans*, and *C. albicans*, respectively.

Many plants used traditionally in the treatment of oral ailments have led to the emergence of many plant-based oral care products within the commercial market. Results obtained through literature reviews provide substantial evidence for the use of plant-based products for everyday oral hygiene. Expanding the search of medicinal plants from terrestrial ecosystems to include aquatic environments, could increase the use of potential natural resources within the dental care industry.

Traditional knowledge systems of plant-based treatments provide modern society with an opportunity to reduce the adverse effects such as resistance to antibiotics and staining of teeth, which are sometimes associated with chemical-based products, while still giving indigenous plants much recognition and purpose.

REFERENCES

Al-Bayati, F. A. (2009). Isolation and identification of antimicrobial compound from *Mentha longifolia* L. leaves grown wild in Iraq. *Annuals of Clinical Microbiology and Antimicrobials, 12*(8), 20.

Albander, J. M. (2005). Epidemiology and risk factors of periodontal diseases. *Dental Clinics of North America, 49*(3), 517–532.

Ashu Agbor, M., & Naidoo, S. (2015). Ethnomedicinal plants used by traditional healers to treat oral health problems in Cameroon. *Evidence-based Complementary and Alternative Medicine, 2015*.

Barrios, J. (2007). *Syzygium aromaticum*. Online Available: https://commons.wikimedia.org/wiki/Syzygium_aromaticum#/media/File:Cloves.JPG.

Beck, J. D., Garcia, R., Heiss, G., Vokonas, P. S., & Offenbacher, S. (1996). Periodontal disease and cardiovascular disease. *Journal of Periodontology, 67*, 1123–1137.

Benkeblia, N. (2004). Antimicrobial activity of essential oil extracts of various onions (*Allium cepa*) and garlic (*Allium sativum*). *LWT – Food Science and Technology, 37*(2), 263–268.

Bergsten, J. (2006). *Vaccinium vitis-idaea*. Online Available: https://commons.wikimedia.org/wiki/Vaccinium_vitis-idaea#/media/File:Vaccinium_vitis-idaea_20060824_003.jpg.

Berthelot, J., & Le Goff, B. (2010). Rheumatoid arthritis and periodontal disease. *Joint Bone Spine, 77*(6), 537–541.

Blaus, B. (2013). *Tooth decay*. Online Available: https://commons.wikimedia.org/w/index.php?search=tooth+decay&title=Special:Search&go=Go&uselang=en&searchToken=9or6m3chycfnyc10gum33gcdv#/media/File:Blausen_0864_ToothDecay.png.

Botelho, M. A., Nogueira, N. A., Bastos, G. M., Fonseca, S. G., Lemos, T. L., Matos, F. J., et al. (2007). Antimicrobial activity of the essential oil from *Lippia sidoides*, carvacrol and thymol against oral pathogens. *Brazilian Journal of Medical and Biological Research, 40*, 349–356.

Bronckers, A. L., Lyaruu, D. M., & DenBesten, P. K. (2009). The impact of fluoride on ameloblasts and the mechanisms of enamel fluorosis. *Journal of Dental Research, 88*(10), 877–893.

Brown, J. A., Sherlock, G., Myers, C. L., Burrows, N. M., Deng, C., Wu, H. I., et al. (2006). Global analysis of gene function in yeast by quantitative phenotypic profiling. *Molecular Systems Biology, 2*(1).

Cai, L., & Wu, C. D. (1996). Compounds from *Syzygium aromaticum* possessing growth inhibitory activity against oral pathogens. *Journal of Natural Products, 59*(10), 987–990.

Calderone, R. A., & Fonzi, W. A. (2001). Virulence factors of *Candida albicans*. *Trends in Microbiology, 9*(7), 327–335.

Cannon, R. D., & Chaffin, W. L. (1999). Oral colonization by *Candida albicans*. *Critical Reviews in Oral Biology and Medicine, 10*(3), 359–383.

Cao, Y. Y., Cao, Y. B., Xu, Z., Ying, K., Li, Y., Xie, Y., et al. (2005). cDNA microarray analysis of differential gene expression in *Candida albicans* biofilm exposed to farnesol. *Antimicrobial Agents and Chemotherapy, 49*(2), 584–589.

Carson, C. F., Hammer, K. A., & Riley, T. V. (2006). *Melaleuca alternifolia* (tea tree) oil: A review of antimicrobial and other medicinal properties. *Clinical Microbiology Reviews, 19*(1), 50–62.

Chaudhari, L. K. D., Jawale, B. A., Sharma, S., Kumar, H. S. M., & Kulkarni, P. A. (2012). Antimicrobial activity of commercially available essential oils against *Streptococcus mutans*. *The Journal of Contemporary Dental Practice, 13*(1), 71–74.

Cheek, M. (2009). *Dichrostachys cinerea*. Online Available: http://plantzafrika.com.

Choi, O., Cho, S. K., Kim, J., Park, C. G., & Kim, J. (2016). In vitro antibacterial activity and major bioactive components of *Cinnamomum verum* essential oils against cariogenic bacteria, *Streptococcus mutans* and *Streptococcus sobrinus*. *Asian Pacific Journal of Tropical Biomedicine*, *6*(4), 308–314.

Colm, G. (2010). *Candida albicans*. Online Available: https://commons.wikimedia.org/wiki/Candida_albicans#/media/File:Candida_albicans_2.jpg.

Della Greca, M., Ferrara, M., Fiorentino, A., Monaco, P., & Previtera, L. (1998). Antialgal compounds from *Zantedeschia aethiopica*. *Phytochemistry*, *49*(5), 1299–1304.

DeStefano, F., Anda, R. F., Kahn, H. S., Williamson, D. F., & Russell, C. M. (1993). Dental disease and risk of coronary heart disease and mortality. *British Medical Journal*, *306*, 688–691.

Devine, D. A., & Cosseau, C. (2008). Host defence peptides in the oral cavity. *Advances in Applied Microbiology*, *63*, 281–288.

Dixon, D. R., Bainbridge, B. W., & Darveau, R. P. (2004). Modulation of the innate response within the periodontium. *Periodontology*, *2000*(35), 53–74.

Dorn, B., Dunn, W., & Progulske-Fox, A. (1999). Invasion of human coronary artery cells by periodontal pathogens. *Infection and Immunity*, *67*, 5792–5798.

Dorn, B., Leung, K., & Progulske-Fox, A. (1998). Invasion of human oral epithelial cells by Prevotella Intermedia. *Infection and Immunity*, *66*(12), 6054–6057.

Edgar. (2007). *Myricetin*. Online Available: https://en.wikipedia.org/wiki/Myricetin#/media/File:Myricetin.png.

Edgar. (2012). *Napthalene*. Online Available: https://en.wikipedia.org/wiki/Naphthalene#/media/File:Naphthalene_numbering.svg.

FDI World Dental Federation. (2015). *The challenge of oral disease – a call for global action. The oral heath atlas* (2nd ed.). Geneva: FDI World Dental Federation.

Fokina, G. I., Roikhel, V. M., Frolova, M. P., Frolova, T. V., & Pogodina, V. V. (1993). The antiviral action of medicinal plant extracts in experimental tick-borne encephalitis. *Voprosy Virusologii*, *38*, 170–173.

Forest, & Kim Starr. (2006). *Jasminum fluminense*. Online Available: https://commons.wikimedia.org/wiki/File:Starr_061116-9912_Jasminum_fluminense.jpg.

Garg, J. M. (2009). *Salvadora persica (Peelu)*. Online Available: https://commons.wikimedia.org/wiki/File:Salvadora_persica_(Peelu)_W_IMG_6940.jpg.

Goenka, P., Sarawgi, A., Karun, V., Nigam, A. G., Dutta, S., & Marwah, N. (2013). *Camellia sinensis* (tea): Implications and role in preventing dental decay. *Pharmacognosy Reviews*, *7*(14), 152.

Goran, G. (2014). *Cross section of teeth*. Online Available: https://commons.wikimedia.org/wiki/File:Cross_sections_of_teeth.svg.

Grow Wild Nursery. (2017). *Berula erecta*. Online Available: http://growwild.online-community.co.za/trees/berula-erecta.

Gupta, P., Agarwal, N., Anup, N., Manujunath, B. C., & Bhalla, A. (2012). Evaluating the antiplaque efficacy of meswak (*Salvadora persica*) containing dentifrice: A triple blind controlled trial. *Journal of Pharmacy and Bioallied Sciences*, *4*(4), 282.

Haffajee, A. D., & Socransky, S. S. (1994). Microbial etiological agents of destructive periodontal diseases. *Periodontology*, *5*, 78–111.

Halady, S. (2012). *Azadirachta indica*. Online Available: https://en.wikipedia.org/wiki/Azadirachta_indica#/media/File:Tender_Neem_leaves_in_Karnataka,_India.JPG.

Hammer, K. I., Carson, C., & Riley, T. (2003). Antifungal activity of the components of *Melaleuca alternifolia* (tea tree) oil. *Journal of Applied Microbiology*, *95*(4), 853–860.

Hankey, A. (2001). *Acokanthera oppositifolia*. Online Available: http://plantzafrika.com.

Harris, S. (2004). *Cotyledon orbiculata L.* Online Available: https://www.plantzafrica.com/plantcd/cotyledorbic.htm.

Heilman, J. (2015). *Oral candidiasis.* Online Available: https://commons.wikimedia.org/wiki/Category:Oral_candidiasis#/media/File:Human_tongue_infected_with_oral_candidiasis.jpg.

Henley-Smith, C. J., Botha, F. S., & Lall, N. (2013). The use of plants against oral pathogens. In A. Mendez-Vilas (Ed.), *Microbial pathogens and strategies for combating them: Science, technology and education* (pp. 1375–1784). Formatex Research Center.

Hodges, D. G. (2017). *Neem seed oil.* Online Available: http://neemseedoil.com/tag/neem-for-toothache.

Ho, K. Y., Tsai, C. C., Huang, J. S., Chen, C. P., Lin, T. C., & Lin, C. C. (2001). Antimicrobial activity of tannin components from *Vaccinium vitis-idaea* L. *Journal of Pharmacy and Pharmacology, 53*(2), 187–191.

Idu, M., Obaruyi, G. O., & Erhabor, J. O. (2009). Ethnobotanical uses of plants among the binis in the treatment of ophthalmic and ENT (ear, nose and throat) ailments. *Ethnobotanical leaflets, 2009*(4), 9.

Inglis, D. O., Arnaud, M. B., Binkley, J., Shah, P., Skrzypek, M. S., Wymore, F., et al. (2012). *The Candida genome database incorporate.*

JMK. (2012). *Diospyros lycioides.* Online Available: https://en.wikipedia.org/wiki/List_of_Southern_African_indigenous_trees_and_woody_lianes#/media/File:Acokanthera_oppositifolia,_vrugte,_Ou_Fort,_Durban.jpg.

JMK. (2013). *Acacia mellifera, Phalandingwe.* Online Available: https://commons.wikimedia.org/wiki/File:Acacia_mellifera,_Phalandingwe,_b.jpg.

JMK. (2014). *Leaves of a Natal guarri at eMakhosini.* Online Available: https://en.wikipedia.org/wiki/Euclea_natalensis#/media/File:Euclea_natalensis_subsp_natalensis,_blare,_eMakhosini.jpg.

Kojian, R. (2011). *Melaleuca alternifolia (tea tree).* Online Available: https://commons.wikimedia.org/wiki/Category:Melaleuca_alternifolia#/media/File:Gardenology.org-IMG_2664_rbgs11jan.jpg.

Kondlo, M. (2012). *Zanthoxylum capense.* Online Available: http://plantzafrika.com.

Kornman, K., & Loesche, W. (1982). Effects of estradiol and progesterone on *Bacteroides melaninogenicus* and *Bacteroides gingivalis. Infection and Immunity, 35*, 256–263.

Krige, A. (2007). *Albizia adianthifolia.* Online Available: http://plantzafrika.com.

Loesche, W., & Grossman, N. (2001). Periodontal disease as a specific, albeit chronic, infection: Diagnosis and treatment. *Clinical Microbiology Reviews, 14*, 727–752.

Marcotte, H., & Lavoie, M. (1998). Oral microbial ecology and the role of salivary immunoglobulin A. *Microbiology and Molecular Biology Reviews, 62*, 71–109.

Maroyi, A. (2013). Traditional use of medicinal plants in south-central Zimbabwe: Review and perspectives. *Journal of Ethnobiology and Ethnomedicine, 9*(1), 31.

Marsh, P. D. (2010). Controlling the oral biofilm with anti-microbials. *Journal of Dentistry, 38*(SI), 11–15.

Mattila, K. J., Nieminen, M. S., Valtonen, V. V., Rasi, V. P., Kesaniemi, Y. A., Syrjala, S. L., et al. (1989). Association between dental health and acute myocardial infartion. *British Medical Journal, 298*, 279–282.

Mbanga, J., & Magumura, A. (2013). Antimicrobial activity of *Euclea undulata, Euclea divinorum* and *Diospyros lycioides* extracts on multi-drug resistant *Streptococcus mutans. Journal of Medicinal Plants Research, 7*(37), 2741–2746.

Meena, M., Figueiredo, N. R., & Trivedi, K. (2013). Aloe vera – an update for dentistry. *Journal of Dentofacial Sciences, 2*(4), 1–4.

Mitsuhashi, H. (1988). In M. Okada, S. Nunome, S. Terabayashi, E. Miki, T. Fujita, T. Yamauchi, et al. (Eds.), *Illustrated medicinal plants of the world in colour*. Japan: Hokuryukan Co., Ltd. 383.

Montgomery, E. H., & Kroeger, D. C. (1984). Use of antibiotics in dental practice. *Dental Clinics of North America*, *28*(3), 433–453.

More, G., Tshikalange, T. E., Lall, N., Botha, F., & Meyer, J. J. M. (2008). Antimicrobial activity of medicinal plants against oral microorganisms. *Journal of Ethnopharmacology*, *119*(3), 473–477.

Motsei, M. L., Lindsey, K. L., van Staden, J., & Jager, A. K. (2003). Screening of traditionally used South African plants for antifungal activity against *Candida albicans*. *Journal of Ethnopharmacology*, *86*(2–3), 235–241.

Mugraby, S. (2007). *Cinnamon sticks*. Online Available: https://commons.wikimedia.org/wiki/Category:Cinnamon_sticks#/media/File:Cinnamon-other.jpg.

Mujuru, L. (2011). *Diospyros lycioides Desf. PROTA (plant resources of tropical Africa)*. Online Available: http://www.prota4u.org/search.asp.

Neurotiker. (2007). *Cinnamaldehyde*. Online Available: https://en.wikipedia.org/wiki/Cinnamaldehyde#/media/File:Zimtaldehyd_-_cinnamaldehyde.svg.

Osman, I. (2012). *A pack of miswak sticks*. Online Available: https://commons.wikimedia.org/wiki/File:Miswak2.jpg.

Oyedeji, A. O., & Afolayan, A. J. (2006). Chemical composition and antibacterial activity of the essential oil isolated from South African *Mentha longifolia* (L.) L. subsp. *capensis* (Thunb.) Briq. *Journal of Essential Oil Research*, *18*.

Perry, L. M. (1980). *Medicinal plants of east and Southeast Asia: Attributed properties and uses*. In L. M. Perry, & J. Metzger (Eds.). Cambridge, MA: The MIT press. 134.

Petersen, P. E., & Ogawa, H. (2012). The global burden of periodontal disease: Towards integration with chronic disease prevention and control. *Periodontology 2000*, *60*(1), 15–39.

Pizzo, G., Guiglia, R., Russo, L., & Campisi, G. (2010). Dentistry and internal medicine: From the focal infection theory to the periodontal medicine concept. *European Journal of Internal Medicine*, *21*(6), 496–502.

Ramphoma, K. J. (2016). Oral health in South Africa: Exploring the role of dental public health specialists. *South African Dental Journal*, *71*(9), 402–403.

Riley, P., Moore, D., Ahmed, F., Sharif, M. O., & Worthington, H. V. (2015). Xylitol-containing products for preventing dental caries in children and adults. *The Cochrane Library*, *3*.

Roberts, M. (1990). *Indigenous healing plants*. South Africa: Southern Book Publishers.

Rood, B. (2008). *Uit Die Veldepteek*. Pretoria: Protea Boekhuis.

Rosenbach, D. (2010). *Gingivitis*. Online Available: https://commons.wikimedia.org/wiki/File:Gingivitis_before_and_after.jpg.

Salam, R., Khokon, J. U., & Mussa, S. B. M. (2014). Effect of neem and betel leaf against oral bacteria. *International Journal of Natural and Social Sciences*, *1*, 52–57.

Selim, M. S. (2003). *Phytochemical and pharmacological screening of Ludwigia adscendens L.B.* Bangladesh: Khulna University.

Sibanda, S., Chigwada, G., Poole, M., Gwebu, E. T., Noletto, J. A., Schmidt, J. M., et al. (2004). Composition and bioactivity of the leaf essential oil of *Heteropyxis dehniae* from Zimbabwe. *Journal of Ethnopharmacology*, *92*, 107–111.

Stander, I., & Van Wyk, C. W. (1991). Tooth brushing with the root of *Euclea natalensis*. *Journal de biologie buccale*, *19*(2), 167–172.

Strong, W. L. (2012). *Wikimedia commons*. Online Available: https://commons.wikimedia.org/wiki/File:Dr_ScharpCam_IntraOral_Camera_Perio_Probe.jpg.

Subrahmanyam, M. V., & Sangeetha, M. (2003). Gingival crevicular fluid a marker of the periodontal disease activity. *Indian Journal of Clinical Biochemistry*, *18*(1), 5–7.

Tapsoba, H., & Deschamps, J. P. (2006). Use of medicinal plants for the treatment of oral diseases in Burkina Faso. *Journal of Ethnopharmacology*, *104*(1), 68–78.

Teffo, L. S., Aderogba, M. A., & Eloff, J. N. (2010). Antibacterial and antioxidant activities of four kaempferol methyl ethers isolated from *Dodonaea viscosa* Jacq. var. *angustifolia* leaf extracts. *South African Journal of Botany*, *76*(1), 25–29.

Visser, S. (2016). *Oral hygiene – homepage of nature fresh*. Online Available: http://naturefresh.co.za/category/sue-research/.

Walker, W. B. (1990). The oral cavity and associated structures. In H. K. Walker, W. D. Hall, & J. W. Hurst (Eds.), *Clinical methods: The history, physical and laboratory examinations* (3rd ed.). Boston: Butterworths.

van der Walt, L. (2000). *Melianthus major L.* Online Available: https://www.plantzafrica.com/plantklm/melianthusmajor.htm.

van der Walt, L. (2004). *Mentha longifolia L.* Online Available: http://www.plantzafrica.com/plantklm/mentlong.htm.

Watt, J. M., & Breyer-Brandwijk, M. G. (1962). *The medicinal and poisoness plants of southern and eastern Africa* (2nd ed.). London: Livingstone.

Wikimedia Commons. (2013). *Gingivitis*. Online Available: https://commons.wikimedia.org/wiki/File:Gingivitis-before.JPG.

Wikimedia Commons. (2016). *Human mouth*. Online Available: https://commons.wikimedia.org/wiki/File:Human_mouth.jpg.

Williams, R. C. (1990). Periodontal disease. *New England Journal of Medicine*, *322*(6), 373–382.

Williams, D., & Lewis, M. (2011). Pathogenesis and treatment of oral candidosis. *Journal of Oral Microbiology*, *3*.

Wink, M., & van Wyk, B. E. (2008). *Mind altering and poisonous plants of the world*. Pretoria: Briza.

Wolinsky, L. E., Mania, S., Nachnani, S., & Ling, S. (1996). The inhibiting effect of aqueous *Azadirachta indica* (neem) extract upon bacterial properties influencing in vitro plaque formation. *Journal of Dental Research*, *75*(2), 816–822.

World Health Organisation. (2012). *World health statistics 2012*. Online Available: http://apps.who.int/iris/bitstream/10665/44844/1/9789241564441_eng.pdf.

van Wyk, P. J., & van Wyk, C. (2004). Oral health in South Africa. *International Dental Journal*, *54*(S6), 373–377.

Yikrazuul. (2008). *Kaempferol*. Online Available: https://en.wikipedia.org/wiki/Kaempferol#/media/File:Kaempferol.svg.

Zeron, A. (2010). *Periodontitis*. Online Available: https://commons.wikimedia.org/wiki/File:Periodontitis_Cr%C3%B3nica_Severa.JPG.

FURTHER READING

Kamatou, G., & Viljoen, A. (2007). Indigenous South African medicinal plants: Part 6: Salvia chamelaeagnea ('Afrikaansesalie'): Medicinal plants. *SA Pharmaceutical Journal*, *74*(10), 49.

Chapter 7

Can Medicinal Plants Provide an Adjuvant for Tuberculosis Patients?

Carel B. Oosthuizen, Anna-Mari Reid, Namrita Lall
University of Pretoria, Pretoria, South Africa

Chapter Outline

7.1 TUBERCULOSIS

7.1.1 The Casual Organism

Mycobacterium tuberculosis (Mtb), seen in Fig. 7.1, is part of the *Mycobacterium tuberculosis* complex that consists of *M. tuberculosis*, *Mycobacterium bovis*,

Medicinal Plants for Holistic Health and Well-Being. http://dx.doi.org/10.1016/B978-0-12-812475-8.00007-X

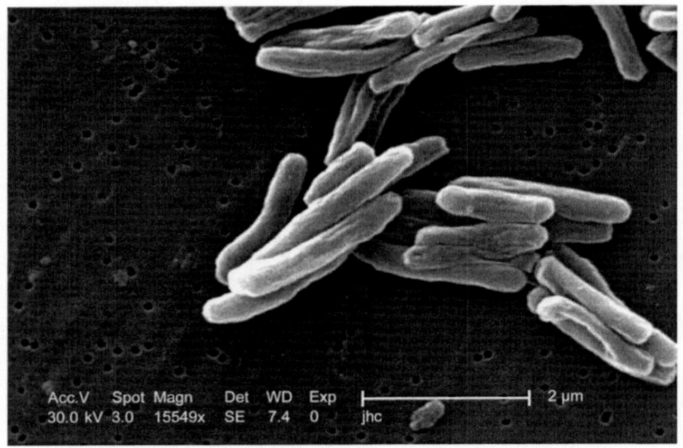

FIGURE 7.1 *Mycobacterium tuberculosis,* the causal organism of TB (Carr, 2006).

Mycobacterium microti, *Mycobacterium africanum*, *Mycobacterium pinnipedii*, *Mycobacterium avium*, *Mycobacterium intracellulare*, *Mycobacterium scrofulaceum* and *Mycobacterium caprae*.

The *M. tuberculosis* complex is a clonal progeny of a single ancestor that resulted from an evolutionary bottleneck about 30,000 years ago (Brosch et al., 2002; Gutacker et al., 2002; Hughes, Friedman, & Murray, 2002). *Mycobacterium* forms part of the Actinomycetes but has been placed in a family of its own, the Mycobacteriaceae. The *Mtb* bacterium is a Gram-positive, rod-shaped, facultative aerobic, and slow-growing organism that was first isolated by Robert Koch in 1882. The causative agent of tuberculosis (TB) is *Mtb*. It is adapted to withstand weak disinfectants and relatively long periods in the open air due to the thick lipid-rich bacterium cell wall. This cell wall is one of the most complex structures, comprising of peptidoglycan covalently attached via a linker unit to a linear galactofuran, which in turn is attached to several strands of a highly branched arabinofuran that are attached to mycolic acids. This multilayered cell wall structure is ~20 nm thick (Brennan, 2003). This bacterium normally attacks the lungs as it is a facultative aerobe and needs oxygen to survive, but it can spread easily through the air from an infected person. Due to its intracellular and pathogenic nature, it infects macrophages. In 75% of cases, the bacteria infect the lungs, termed pulmonary TB. In the other 25%, the bacteria infect other sites throughout the body. These sites include bones, joints, the central nervous system, and the lymphatic system. In these cases, it is termed miliary TB or extrapulmonary TB (Fitzgerald, Sterling, & Haas, 2009, chap. 250). Mycolic acids give mycobacteria a unique distinguishing factor that is their acid fastness, as they can only be stained by acid-fast staining methods. They are impermeable to basic dyes such as those used in Gram staining but have been detected by the use of the Ziehl–Neelsen staining procedure. With this staining method, organisms can be distinguished by their red color in between the surrounding blue tissues.

7.1.2 The Global Effect of Tuberculosis

Since its discovery, TB as a disease has received several names. One among these names is "The Great White Plague." This name was given to the infection during the time when a TB-epidemic devastated the European population around the 17th century (Microbiologybytes, 2014).

TB has had a devastating impact worldwide with 10.4 million new infections, and a mortality rate reaching 1.5 million for the year 2015. It is the second leading cause of death in the world due to an infectious disease. TB is the third leading cause of death in women aged 15–49 years (WHO, 2016). Together with this high number of TB disease another infamous epidemic is claiming its fair share with 430,000 of the infected individuals finding themselves coinfected with human immunodeficiency virus/acquired immunodeficiency syndrome (HIV/AIDS). According to the World Health Organization (WHO) report (WHO, 2016), TB comprises the heaviest disease burden geographically in Africa and Asia. Due to the lack of proper sanitation and poverty, which is evident in these continents, it becomes the perfect setting for the undesirable spread of this disease. Globally there has been a drop in the number of TB incidents from previous years, but this drop is less than 1.5% per year, indicating the lack of complete control by health authorities (WHO, 2016). Despite all the treatment plans and strategies that have been implemented by a wide variety of governmental and nongovernmental organizations to combat TB, multidrug-resistant (MDR)- and extensively drug resistant (XDR)-TB are still found to be prevalent. In most countries, it was observed that only a small number of the newly and previously infected individuals were tested for MDR-TB in 2010 (WHO, 2016).

South Africa is one of 22 countries with the highest rate of TB infections; with KwaZulu-Natal province having the highest TB prevalence in the country (SoulCity, 2015). Fortunately, TB mortality has dropped by 40% since 1990 but this has not made a huge difference when compared to the population of people still infected with TB. This means that there has been an improvement in controlling TB in the last two decades. Approximately 5%–10% of the population infected with TB will develop active TB at a certain point in their lives. During latent infections, individuals have no symptoms and cannot infect other people. Communities, the health sector, and many families have been negatively affected by TB, financially and emotionally.

7.1.3 The Treatment

Streptomycin, the first antibiotic agent used to treat infections with *M. tuberculosis,* was introduced in 1946 and isoniazid (originally an antidepressant medication) was the second, introduced in 1952 (Tomlinson, 2011) (Fig. 7.2).

Presently, the first line of defense against TB, or first-line drugs, consists of a combination of drugs including streptomycin, isoniazid, rifampicin,

FIGURE 7.2 Streptomycin, one of the pioneering compounds used against tuberculosis infections (Mills, 2008).

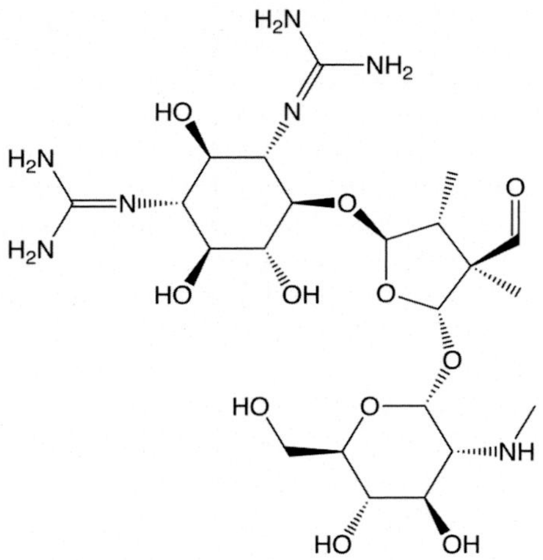

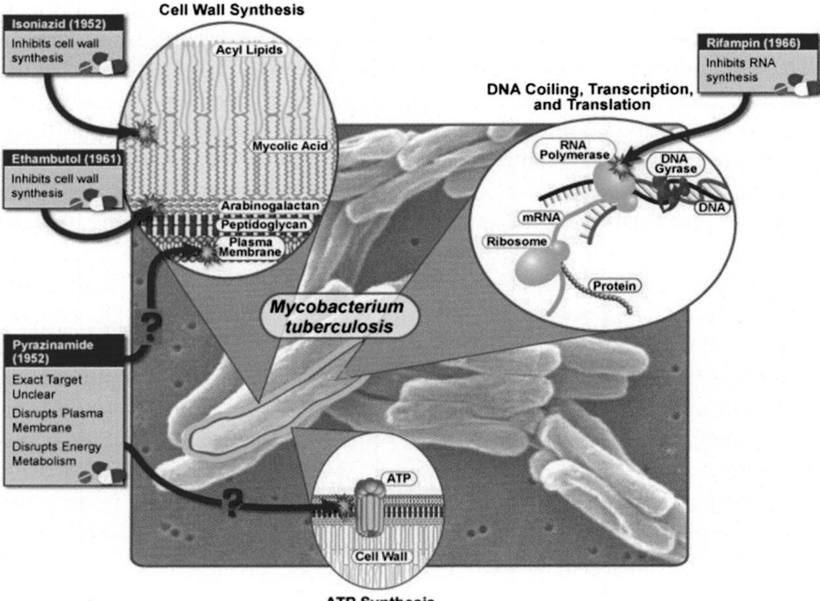

FIGURE 7.3 First-line drug treatment for tuberculosis infected individuals (NIAID, 2007).

ethambutol, and pyrazinamide at different dosages for a period of 6–9 months, depending on the severity of the infection (Fig. 7.3) (CDC, 2012). Isoniazid and ethambutol inhibit cell wall synthesis. Pyrazinamide disrupts the plasma membrane and energy metabolism. Rifampin inhibits RNA synthesis. Although these first-line drugs are very effective, they are only effective against certain strains of *Mtb*. A different drug regime needs to be administered if the patient is infected with a drug-resistant strain. TB drugs are also accompanied by hepatotoxicity (liver damage) that causes a loss of appetite; this effect is highly undesirable in HIV/AIDS and TB coinfected patients (Tostmann et al., 2008). The length of treatment results in two problems. Firstly, the patients do not comply with the time schedule of treatment, thus allowing the bacteria to develop resistance to the first-line drugs. Secondly, untreated patients and treated patients with active TB will increase the spread and rate of infection of healthy individuals (CDC, 2012). Thus, a drug regime requiring a shorter time period could be a more effective treatment than that of the one currently in place.

7.1.4 Human Immunodeficiency Virus Coinfection and Resistant Bacteria

7.1.4.1 Tuberculosis and Human Immunodeficiency Virus Coinfection

There are 34 million people living with HIV worldwide and one-third of them are infected with TB bacilli. The association with HIV, which is also an immunosuppressant disease, has a large part to play in the statistics of incidence and mortalities with an additional 0.4 million deaths occurring in HIV-positive individuals (WHO, 2016). HIV is not only just associated with additional mortalities but also increases an individual's chances of contracting TB due to the suppressed immune system. People who are HIV-positive and infected with TB are 21–34 times more prone to develop active TB than those of people who are HIV-negative and infected with TB. With HIV and TB coinfection the need arises to take drugs for both diseases together. In HIV, antiretrovirals (ARVs) form part of the prescribed drugs that need to be taken. These ARVs are mainly prescribed to inhibit the multiplication of the virus and positively control the $CD4^{+}$ cell count. In individuals that are HIV-infected, the use of ARVs at an early stage is known to decrease the chances of contracting TB, although the optimum time to use ARVs has not yet been noted (Lienhardt, Vernon, & Raviglione, 2010). In 2012 a meta-analysis was conducted that showed that taking ARVs reduces an individual's risk of contracting TB by 65% (Suthar et al., 2012). From these studies the WHO is now recommending that TB treatment is initiated first, to be followed by ARVs within 8 weeks of starting the TB treatment. However, those patients with a profound suppression of the immune system should start taking ARVs within the first 2 weeks of the TB treatment (WHO, 2016). The difficulty that is evident during HIV and TB coinfection is the observation of drug interactions between the rifampin and selected ARVs drugs such as the protease inhibitors. The presence of the immune reconstitution inflammatory syndrome, which is characterized by the appearance of new TB lesions, is also very common (Lienhardt et al., 2010).

7.1.4.2 Drug-Resistant Tuberculosis

After all the various treatment plans and strategies that have been implemented by a wide variety of governmental and nongovernmental organizations to combat TB, the prevalence of MDR- and XDR-TB still play a critical role in the numbers of infected individuals and mortalities suffered. MDR-TB involves strains that are resistant to two of the first-line drugs, namely isoniazid and rifampicin. Cases of MDR complicate the treatment of TB, especially with its ability to be resistant to the first-line drugs and the lack of testing initiatives by TB patients. To treat MDR-TB second line of drugs should be used, but they are less effective and have more side-effects. It is stated that less than 50% of the people with MDR-TB can be cured, and about 30% may die before the completion of the treatment (Fig. 7.4) (Western Cape Government, 2012).

XDR-TB is an emergent form of TB that has been reported from most parts of the world (WHO, 2016). XDR-TB is resistant to at least rifampicin, isoniazid, injectable second-line drugs such as capreomycin, kanamycin, or amikacin as well as fluoroquinolone, aminoglycosides, cycloserine, terizidone, ethionamide, prothionamide, and aminosalicylic acid (Ma, Lienhardt, McIlleron, Nunn, & Wang, 2010; McGaw, Lall, Meyer, & Elof, 2008).

To confirm XDR-TB an acid-fast stain needs to be conducted as well as a sputum culture with antibiotic susceptibility testing. In countries where funding is limited, solid media are used for *Mtb* cultures, instead of liquid culture

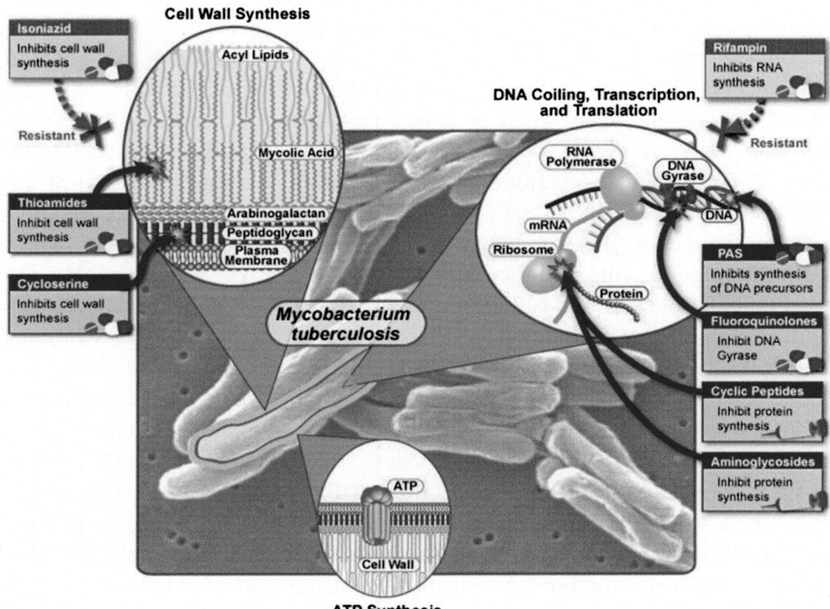

FIGURE 7.4 Second-line treatment drugs for tuberculosis infected individual (NIAID, 2007).

methods such as Mycobacteria Growth Indicator Tubes systems. This can delay the diagnosis for up to a month and therefore, is insufficient in providing timely detection. The efflux for funding, therefore, also seems to fuel this infection; and large-scale upgrading of treatment centers and clinics, especially in developing countries, needs to be undertaken (Madariaga, Lalloo, & Swindells, 2008). A noticeable outbreak of XDR-TB occurred in 2006 in the province of Gauteng and this made the health institutions, as well as the government of South Africa, focus on the fact that this is a serious disease that just cannot be ignored.

A new form of TB has also made its abysmal mark on the world, the totally drug-resistant tuberculosis (TDR-TB). This form of TB has not yet formally been introduced by the WHO but notice has since been taken as seen in their Annual Tuberculosis Report of 2014. Individuals suffering from TDR-TB show resistance to all first- and second-line drugs. Two cases of TDR-TB were reported from Italy; cases were also reported from India. These cases were found to be resistant to all drugs (Migliori, De Laco, Besozzi, Centis, & Cirillo, 2007; Udwadia, Amale, Ajbani, & Rodrigues, 2012). These cases mostly originated in very impoverished and isolated areas, but experts believe that there could be many undocumented cases. Both cases in Italy suggested initial mismanagement and noncompliance to the regimen (Migliori et al., 2007). TB is a serious infection that needs to be managed in an optimum way so that it can be ensured that such cases as these and new forms do not emerge due to human error.

All these resistant forms of TB may have developed due to noncompliance of the infected individuals; along with a high number of drugs they also involve a long treatment plan (Korbel, Schneider, & Schaible, 2008). It has been observed that in situations where poverty is evident, there are high levels of coinfection with HIV/AIDS and poor access to high-quality treatment (Ma et al., 2010). These limit the efficacy of the control programs. Only 1% of the MDR-TB cases in 2008 received proper treatment that was in agreement with the standards recommended by WHO (Ma et al., 2010). The drugs that are currently available for treatment are incapable of solving the multiple challenges and problems concerning resistance and effective treatment plans. To prevent the spreading of MDR-, XDR-, and TDR-TB, shorter treatment plans are needed that are safer, less expensive, and more effective compared with the existing treatment regimes. This emphasizes the need for new drugs, drug targets, and new regimes to be put in place to control the rise in the number of resistant strains.

7.1.5 Pathogenesis

TB can spread through a myriad of ways. It only requires one infected person with an active infection of the lungs and throat to sneeze, cough, sing, or spit to infect the individuals in the nearby vicinity (Fitzgerald et al., 2009). The droplets that are expelled into the air may remain stable in the outside environment for a couple of hours until inhaled by an unsuspecting individual. A person with active TB can infect 10–15 people in a year. The symptoms of TB may include the following (Fig. 7.5) (CDC, 2012):

- A chronic cough that lasts for several weeks
- Pain in the chest area
- When coughing, the appearance of blood or sputum
- Constant weakness or fatigue
- Considerable weight loss without any explanation
- Loss of appetite
- Chills and fever
- Night sweats

The TB bacterium has evolved to cause infection in many patients but disease only in a few. The majority (90%–95%) of infected individuals will never develop any clinical illness or infectious sign. The human innate and acquired immunity is capable and well adapted to combat this pathogen. There are several medical conditions that increase the risk of progression to TB disease; these include HIV/AIDS (as mentioned above), diabetes mellitus, renal failure, etc. TB can also develop in a healthy person who does not have these risk factors.

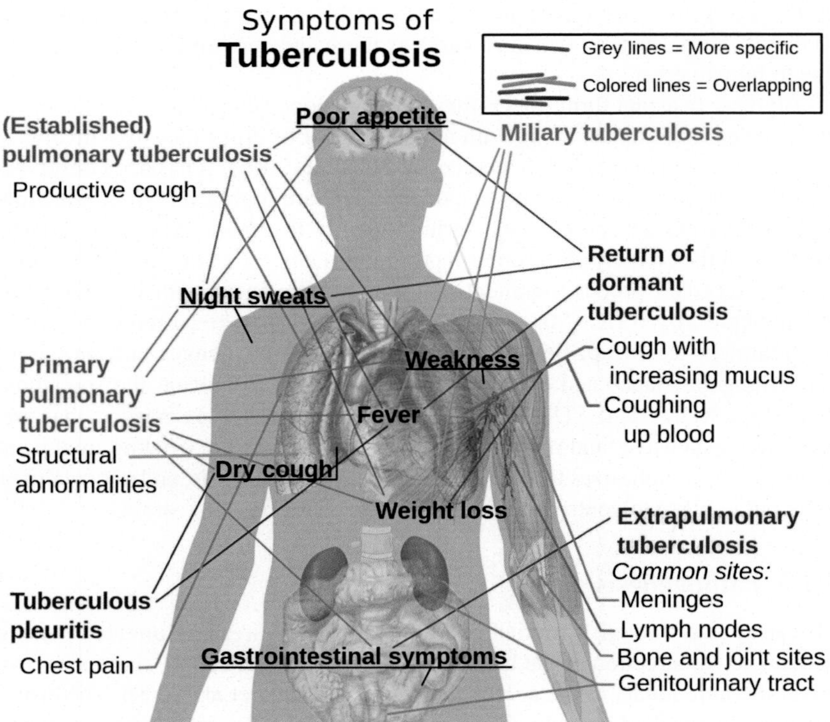

FIGURE 7.5 Symptoms of tuberculosis infected individual (Häggström, 2014).

This is probably due to a genetic susceptibility of the infected person (Antas et al., 2006; Young & Cole, 1993). *Mtb* is an airborne disease, and the bacilli can be present in the air as droplets or nuclei that are discharged by a person with an active infection. These bacilli can be stable and active in the atmosphere for several hours. This causes a major problem as the infection dose is as low as 1–10 bacilli. Thus, transmission is an extremely efficient and effective process (Bloom, 1994). Unless an infected person receives prophylactic drugs, symptomatic disease will ultimately occur in 5%–10% of infected patients. This depends on the ability of the *Mycobacterium* to survive within the macrophages; and in doing so, it is camouflaged within the host cell and evades the extracellular host defense system.

Pulmonary TB can be divided into four stages (Lurie, 1964). The infection is initiated after the inhaled *Mycobacterium* bacillus is phagocytized by the alveolar macrophages. The primary site of infection, in most cases, is the middle or lower zones of the lung, although exceptions do occur. The destruction of the bacteria that have been phagocytized by the macrophages depends on the inherent microbicidal ability of the host macrophage as well as the virulence factor of the bacilli (van Crevel, Ottenhoff, & van der Meer, 2002). Once the bacterium has been phagocytized, it modulates the behavior of its phagosome, not to merge with acidic, hydrolytically active lysosomes, and in the process evades defense by the host (Armstrong & Hart, 1971; Rohde, Yates, Purdy, & Russell, 2007). The end of stage one culminates in the disruption of the macrophages and the attraction of monocytes and other inflammatory cells to the infection site (van Crevel et al., 2002). In the second stage, the attracted monocytes will differentiate into macrophages that still ingest bacilli but do not kill the bacteria. While more blood-derived macrophages accumulate at the infection site, the bacteria continue to multiply. At this stage, little or no tissue damage occurs (van Crevel et al., 2002). The third stage begins 2–3 weeks after infection with the development of T-cell immunity. Antigen-specific T lymphocytes arrive and proliferate within the tubercles (early lesions). These antigen-specific cells activate macrophages to destroy intracellular bacteria. The logarithmic growth of the bacteria also comes to a halt in stage three (van Crevel et al., 2002). The fourth stage in pathogenesis involves necrotic lesion formation and the development of the granuloma, which inhibits extracellular growth of Mycobacteria, resulting in stationary or dormant latent infection (Fig. 7.6).

The granuloma is formed from macrophages differentiating into several specialized cell types: multinucleated giant cells, foamy macrophages, and epithelioid macrophages. A macrophage layer called fibrous cuff can form from the extracellular matrix material that results in the stratification of the granuloma. The disease is caused by the loss of blood vessels, increased necrosis, and the buildup of caseum in the granuloma center. The granuloma is ruptured in the lungs and the infectious bacilli are released into the airways (Russell, Barry, & Flynn, 2010). TB may progress under decreased immune surveillance and response. Progressive disease can happen directly after primary infection or

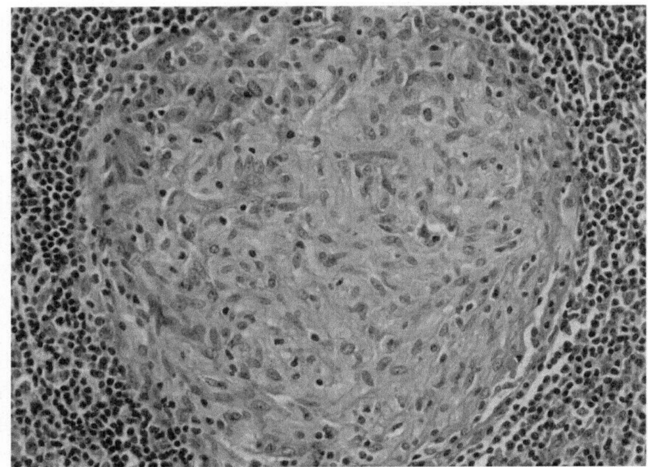

FIGURE 7.6 Granuloma structure of an infected tuberculosis individual (Freebase Data Dumps, 2013).

after months to years designated as postprimary infection. Extracellular growth can also take place within liquefied caseous foci. Rupture of bronchi can take place, allowing the bacilli to spread within the lung as well as to the outside environment (van Crevel et al., 2002). The amount of tissue damage in the form of necrosis and fibrosis, as well as the regeneration of tissue, also influences the progression of the infection (van Crevel et al., 2002).

7.1.6 Immune Response

The outcome of infection with *Mycobacterium tuberculosis* (*Mtb*) depends on the interactions that occur between the bacteria and the host's innate and acquired immune responses. These interactions can lead to control of infection, advanced disease, or the establishment of latent infection (Torrado, Robinson, & Cooper, 2011). After exposure to *Mtb*, 90% of those who develop cellular immune responses will remain healthy. This is known as latent *Mtb* infection and reflects a successful immune response and containment of *Mtb* (Grotzke & Lewinsohn, 2005).

M. tuberculosis is transmitted through aerosol droplets and, when inhaled, will result in a primary infection in 95% of cases. Cell-mediated immunity consisting of a T-cell response will clear this infection or cause containment of the tubercle bacilli within a granulomatous structure where latent *Mtb* disease is initiated. This form of TB is normally not associated with any visible symptoms as, although the tubercle bacilli are persistent, they remain in a dormant state within the unsuspecting host. This dormant state of this disease may last a lifetime without the individual developing any form of the active disease. Progression to active TB by activation or reactivation occurs in 5%–10% of infected individuals (Kumar et al., 2011). Failure of the immune system by HIV causes many individuals with a latent infection to progress to active TB infection. Other factors that may compromise the immune system include pollution and tobacco smoke, as well as a genetic predisposition (Kumar et al., 2011).

Dendritic cells and local bronchoalveolar cell subsets as well as locally active soluble effector molecules such as nitric oxide take part in the early innate immunity and then go on to initiate the adaptive immune response (Schwander & Dheda, 2011). *Mtb* infection activates two pattern recognition receptors, the Tolllike receptors and the nucleotide oligomerization domain-like receptors. It is the engagement of the Tolllike receptors that can either be immunostimulatory or immunosuppressive. Stimulation of the Tolllike receptors initiates innate immunity and takes part in the chemokine release, resulting in the accumulation of the multiple cell types that are present. T-helper 1 (Th-1)-mediated autophagy also forms part of the innate immunity and is mediated by interleukin 4 (IL-4) (Schwander & Dheda, 2011).

An important feature of the adaptive immunity is the presence of the $CD4^+$ T-cells, especially when coinfection with HIV/AIDS is a factor. HIV-coinfection results in depletion of the $CD4^+$ T-cells that give rise to increased susceptibility to reinfection (Schwander & Dheda, 2011). Adaptive immunity is mainly mediated by these $CD4^+$ T-cells but is also supported by the $CD8^+$ T-cells (Boom et al., 2003). The local innate immunity that is mediated by the alveolar macrophages may fail to control the replicating bacilli and as a result, the immune system of the infected host is exposed to increasing amounts of mycobacterial antigens, which then result in the development of the adaptive immune response. This response is mediated by the T-cells and after this, acute adaptive immunity develops. This is followed by a chronic memory immune phase. This phase is important for the control of the bacilli that are persistent as well as for the surveillance for possible reinfections. Failure of the acute or chronic adaptive immunity phase may then result in clinical TB, which makes it possible for the transmission of *Mtb* to a new host (Boom et al., 2003).

Due to such a high number of the global population that has latent TB infection, it is indicative that *Mtb* has developed many mechanisms that enable this pathogen to modulate or avoid detection by the host (Flynn & Chan, 2003). Several strategies have been identified that *Mtb* may implement during latent infection and these can be summarized as follows:

- Preventing the recognition of the infected macrophages by T-cells by inhibiting major histocompatibility complex (MHC) Class II processing and presentation.
- Evading macrophage killing mechanisms such as those mediated by nitric oxide and the reactive nitrogen intermediates (Flynn & Chan, 2003; Gupta, Kaul, Tsolaki, Kishore, & Bhakta, 2012).

A Th-1 response consists of a coordinated interaction between CD4$^+$ T-cells and the macrophages and induces cytokines such as interferon-gamma (IFN-γ) and IL-2. This interaction consists of *Mtb*-derived antigens that are processed and presented via the macrophage-associated MHC II. Antigen recognition occurs by CD4$^+$ T-cells and leads to the release of proinflammatory cytokines such as IFN-γ and TNF-α, which would then limit the intracellular growth of *Mtb* (Grotzke & Lewinsohn, 2005). *Mtb*, therefore, leads to the induction of IFN-γ secretion. It also results in higher levels of Th-2 cytokine production, especially IL-4 and IL-13. The low ratio of IFN-γ/IL-4 steers the immune system away from a protective response and causes impaired bacterial functioning. The Th-2 response is dependent on the involvement of Th-2 cytokines such as IL-4, IL-5, IL-10, and IL-13, which are known to be immunosuppressive and inhibit the macrophage activity (Ghadimi, De Vrese, Heller, & Schrezenmeir, 2010).

Controlling infection caused by *Mtb* needs elimination through the process of autophagy, which forms part of the innate immune response. Th-1 cytokines such as IFN-γ have been negatively associated with host susceptibility due to the stimulation of macrophage activity, which is also accompanied by the initiation of autophagy together with autophagy markers to the mycobacterial phagosome that gets rid of the intracellular *Mtb*. Th-2 cytokines, however, act antagonistically toward Th-1 and inhibit the process of autophagy through the production of IL-4 and IL-13. The inhibition induced by these Th-2 cytokines, therefore, prevents the elimination of intracellular mycobacteria.

The major problem with current adjuvants is that they induce both the Th-1 autophagy-promoting cytokine IFN-γ and the immunosuppressive Th-2 autophagy-inhibiting cytokines IL-4 and IL-13. This leads to a mixture of Th-1 and Th-2 that ultimately leads to undesirable effects such as reduced *Mtb* elimination (Ghadimi et al., 2010). According to Ghadimi et al. (2010) and Grotzke and Lewinsohn (2005) a skewing toward a Th-1 phenotype may not be enough for infection elimination and that it may prove to be necessary for the total suppression of major Th-2 cytokines. Current vaccines and adjuvants lead to high levels of IL-4 secretion, and therefore, successful vaccines and adjuvants should be able to suppress the subversive action of the Th-2 components.

Immunosuppressive cytokines are not only able to inhibit autophagy needed for infection elimination but they also suppress the production of nitric oxide and have an inhibitory effect on IFN-γ-induced mycobacterial phagosome maturation (Ghadimi et al., 2010). Polarization of the immune system toward a Th-1 phenotype seems to be dependent on the production of IL-12, naïve T-cells and natural killer cells. IFN-γ and IL-12 form an interdependent positive feedback loop that places increasing importance on the production of both IFN-γ and IL-12 for the in vivo suppression of infection (Ghadimi et al., 2010). As shown, cytokines are primary parts of the innate and adaptive immune responses and are secreted by almost all the nucleated cell types through an inducible response (Spelman et al., 2006). Cytokines are responsible for the phenotypes of the immune response and can, therefore, be targeted as part of therapeutic immunomodulation. Through the alteration of cytokine expression and targeting of characteristic receptors of the cytokines the potential for new therapeutics has been shown (Spelman et al., 2006). Therapeutic application of cytokines in medicine in a clinical application includes therapies such as a TNF-antagonist for the treatment of rheumatoid arthritis, inhibition of Th-2-derived cytokine expression in the treatment of asthma as well as IL-2 and IL-12 used in combination for the potential treatment of neuroblastomas. As with most therapeutic treatments, adverse reactions may form part of the treatment; and in the case of cytokine treatment with synthetic cytokines, these adverse effects, which include transient lymphopenias and monocytopenias, may hinder its future use. The immune system provides effective control of the mycobacterial infection and plays a much-needed role during infection from any type of pathogen invasion. It is during the suppression of the immune system that the protective and important role of an individual's immune system can be seen.

7.1.7 Latent Infection and Biofilm Formation

7.1.7.1 Latent Infection

The latent form of TB is prevalent in approximately one-third of the world population and an estimated 10% of this section moves on to develop active TB. Latent TB is defined as the protective immunity process that inhibits the growth of the pathogen. It is an ongoing immune cycle that consists of immune activation and suppression, which prevents the bacilli from spreading and replicating (Korbel et al., 2008). After protective immunity sets in, the *Mycobacterium* lowers its metabolism and goes over to a silent state with slow or nonreplication and therefore a persistent dormant state (Korbel et al., 2008). This dormant state may be the status of a few individuals for a couple of years or even a lifetime, but for other individuals, reactivation of the tubercle bacilli may occur due to an immune-compromising disease (Korbel et al., 2008). The objective of most of the research in TB has been to reduce the risk of reinfection in individuals who are in high-risk situations, such as those who may be in close contact with individuals having active TB or any other immune-compromising diseases. Presently isoniazid monotherapy for 6–9 months seems to be an effective treatment for reducing the risk by as much as 60%, especially for those in contact for over 2 years, with individuals having active TB (Lienhardt et al., 2010).

New drug candidates have recently been developed, such as PA-824 and TMC-207, the last of which is the only successful drug candidate that completed phase III clinical stage development and was projected to move forward to the next stage by the fourth quarter of 2014 (Klein et al., 2014; WGND, 2014). TMC-207, also known as bedaquiline, is a diarylquinoline, which forms part of the synthetically produced antibiotics. It belongs to the quinoline class of compounds with a novel mechanism of action, an ATP synthase (adenosine triphosphate) inhibitor. This compound showed effective activity against nonreplicating *Mtb* for individuals with TB infection, as well as inhibition against *Mtb* strains that are resistant against the first- and second-line drugs (Ma et al., 2010; Matteelli, Carvalho, Dooley, & Kritski, 2010; Tasneen et al., 2011). Together with known drugs such as rifapentine, TMC-207 boasts an innovative shorter treatment regime (Tasneen et al., 2011).

PA-824 was synthetically produced as a derivative of the known nitroimidazole antibiotics and forms a part of the group known as the nitroimidazo [2, 1-*b*] oxazines (Barry, Boshoff, & Dowd, 2004). PA-824 has an unknown mechanism of action but a recent study conducted by Manjunatha, Boshoff, and Barry (2009) showed the possible involvement of respiratory poisoning through the release of nitric oxide as the possible mechanism of action. It is through ongoing research that people are trying to develop drugs active against latent *Mtb* and that are also safe, short-term compliant and cost-effective.

7.1.7.2 Biofilm

A biofilm can be defined as a sessile microbial structure, characterized by the attachment of cells to a surface or each other, embedded in a matrix of extracellular polysaccharides and that demonstrates an altered phenotype with respect to growth, gene expression, and protein production (Fig. 7.7) (Donlan & Costerton, 2002). The structure of the biofilm can differ significantly between different bacteria, ranging from a single cell layer to a community of cells embedded in a thick polymeric matrix with an intricate channel network to provide the community with environmental nutrients and water (Costerton, Lewandowski, Caldwell, Korber, & Lappin-Scott, 1995; Shirtliff, Mader, & Camper, 2002). By growing in a biofilm matrix the microbes enjoy a number of advantages over the single planktonic bacterial cells. Firstly, a biofilm growth habit enables the

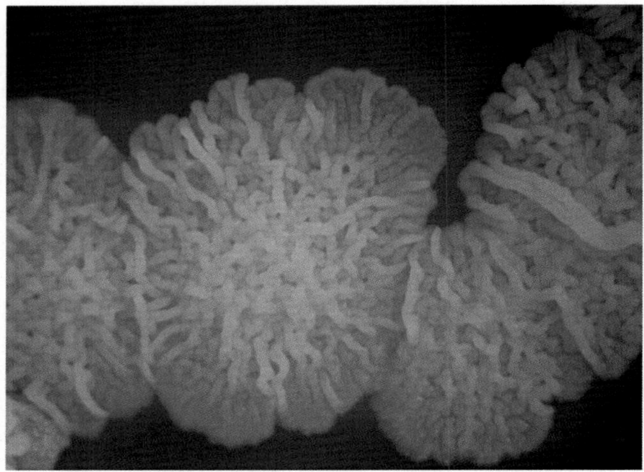

FIGURE 7.7 *Mycobacterium smegmatis* biofilm (Oosthuizen, 2015).

bacteria to overcome several removal strategies. These strategies include resistance/tolerance to antimicrobial and antifouling agents, shear stress, evading immune phagocytic clearance, as well as protection against oxygen radical and protease defenses. Secondly, the poly matrix biofilm has the ability to capture and concentrate nutrients. Thirdly, this growth habit has the potential for dispersion through detachment for the repopulation and colonization of new sites after the termination of drug treatments (Boyd & Chakrabarty, 1994).

The resistance/tolerance of the biofilm to antibiotics has been attributed to different factors; these include the low metabolic levels of bacteria in biofilms and the drastically reduced rate of cell division (Brown, Allison, & Gilbert, 1988). Biofilms have been shown to possess the ability to act as diffusion barriers and limit the penetration of some antibiotics (Islam, Richards, & Ojha, 2012). It is important to understand the mechanisms involved in the initial attachment, development, and maturation of the biofilm to elucidate the impact of the structure and identify new drug targets to overcome the problem of persistence and resistance in biofilms (Shirtliff et al., 2002). It has been shown that these communities have a tremendous impact on the local environment. An interesting aspect of microbial communities is that they can be both beneficial and devastating for humankind and clinical environment, respectively. Biofilms can develop into an almost impossible to eradicate reservoir due to the variation in physiological status of the microorganisms. Their phenotypes can result in chronic bacterial infections such as *Pseudomonas aeruginosa* in the lungs of cystic fibrosis patients, which are thought to persist largely due to the formation of biofilms (Davey & O'Toole, 2000; Namasivayam & Roy, 2013; Parsek & Singh, 2003; Zambrano & Kolter, 2005).

There is evidence of biofilm formation early in the fossil record, particularly in hydrothermal environments. Recent technology has shown that biofilms are not simple passive assemblages of cells stuck together but are rather structurally and dynamically complex biological systems. The initial attachment of bacteria to the surface is often mediated by filaments that extend from the bacteria. There are proteinaceous appendages that act as anchors, transiently fixing the bacteria on the location. Once the bacterium is attached to the surface, biofilm-associated bacteria initiate the synthesis of an extracellular matrix (Branda, Vik, Friedman, & Kolter, 2005). The milieu near the surface of the bacteria undergoes changes during biofilm formation. There can be variability in biofilm structure, which could be due to numerous conditions such as constituents of the microbial community, nutrient availability, and surfaces (Biswas et al., 2011).

There are different kinds of biofilm that can form, for instance, floating biofilm (also referred to as a pellicle) that forms on the surface of liquid culture media; the extracellular matrix constitutes secreted polysaccharides (exopolysaccharides), proteins, and sometimes even nucleic acids. An interesting feature is that mycobacteria lack polysaccharides but are still able to form remarkably robust biofilms that either attach to hydrophobic solid surfaces or float as pellicles on the surface of liquid culture media. It has been shown that the glycopeptidolipids of *Mycobacterium smegmatis* are important for initial surface attachment during biofilm formation (Branda et al., 2005; Recht, Martinez, Torello, & Kolter, 2000; Zambrano & Kolter, 2005).

A class of fatty acids has been identified in playing a pivotal role in the development of biofilm architecture in *M. smegmatis*. It has been indicated that the synthesis of these fatty acids specifically involves GroEL1, a member of the chaperone proteins found in mycobacteria. Mycolic acids that are fatty acids with a very long chain (C70-C90) have been identified in the mycobacterial cell envelope and they are usually anchored to the envelope through covalent linkage. These mycolic acids provide a permeability barrier that is largely responsible for the abilities of these organisms to resist many common therapeutic agents (Zambrano & Kolter, 2005). A study conducted by Ojha et al. (2005) shows that mycolic acid profiles of *M. smegmatis* differ greatly between free-living bacteria and bacteria associated with biofilms. A new class of mycolic acids, much shorter (C56–C68) in length, was formed by biofilm-associated bacteria. In recent years, antibacterial research has taken an antibiofilm formation approach due to most bacteria using biofilm formation as a form of defense against drugs, leading to bacterial resistance against the drugs.

7.2 MEDICINAL PLANTS AGAINST *MYCOBACTERIUM* SPECIES

Considerable effort has been made globally in the fight against TB. Financial support, however, has been lacking, jeopardizing the TB elimination goal set out by the WHO. According to the Treatment Action Group (TAG), an activist group established in South Africa to increase research for treatment of TB and HIV-coinfections, TB research funding has been underinvested in. The annual report for 2013 released by TAG states that in 2012, TB Research and Development dropped by an estimated \$30.4 million (TAG, 2013). Even though funding largely remains an issue, several institutions across the globe have been looking into the relationship of medicinal plants and their activity against several *Mycobacterium* spp. Table 7.1 contains a list of several plants investigated by researchers worldwide with significant activity against *M. tuberculosis* (only plants that exhibited an minimum inhibitory concentration (MIC) $\leq$100 μg/mL have been added to the list).

7.3 SOUTH AFRICAN MEDICINAL PLANTS (FIG. 7.8)

7.3.1 Background

In South Africa where poverty is rife and the healthcare system insufficient in providing basic health care, traditional medicine is practiced by as much as 80% of the black population (York, 2012). This is a network of knowledge strongly supported by many individuals. It is therefore, very valuable for the information that is known to be passed on for many generations. Many modern Western medicines owe their existence to the development of the active

TABLE 7.1 Plants Tested for Their Antimycobacterial Activity

Plant Name	Plant Parts Used	Solvent System	Activity Against *Mycobacterium tuberculosis* H37Rv. (MIC in μg/mL)	References
Achillea biebersteinii Afan.	Inflorescence	Aqueous	32	Orhan et al. (2012)
Anogeissus leiocarpus (DC.) Guill. & Perr.	Root bark and stem bark	Methanol	78	Mann et al. (2008)
Anthemis tinctoria var. *tinctoria* L.	Inflorescence	Aqueous	32	Orhan et al. (2012)
Anthemis pseudocotula Boiss.	Inflorescence	Aqueous	64	Orhan et al. (2012)
Artemisia austriaca Jacq.	Inflorescence	Aqueous	16	Orhan et al. (2012)
Asclepias fruticosa L.	Roots	Acetone	>100 (H37Ra)	Green et al. (2010)
Berchemia discolor (Klotzsch) Hemsl.	Bark	Acetone	12.5 (H37Ra)	Green et al. (2010)
	Leaves	Acetone	>100 (H37Ra)	Green et al. (2010)
Carica papaya L.	Leaves	Water	>100 (H37Ra)	Green et al. (2010)
		Methanol	>100 (H37Ra)	Green et al. (2010)
Cassia petersiana Bolle.	Bark	Acetone	50 (H37Ra)	Green et al. (2010)
Cedrela fissilis Vell.	Leaves	Ethanol	3.12	Ramos et al. (2008)
Chaetacme aristata Planch.	Leaves	Acetone	39 (TB 8104)	Dzoyem, Aro, McGaw, and Eloff (2016)

Continued

TABLE 7.1 Plants Tested for Their Antimycobacterial Activity—cont'd

Plant Name	Plant Parts Used	Solvent System	Activity Against *Mycobacterium tuberculosis* H37Rv. (MIC in μg/mL)	References
Chelidonium majus L.	Aerial parts	Ethanol	50 (H37Ra)	Tosun, Kizilay, Sener, Vural, and Palittapongampim (2004)
Chrysactinia mexicana A. Gray.	Roots	Ethyl ether	62.5	Molina-Salinas et al. (2007)
Cydonia oblonga Mill.	Leaves	Aqueous	16	Orhan et al. (2012)
Ficus cordata Thunb.	Stem bark	Methanol	39.06	Kuete et al. (2008)
Mixture of *Ficus chlamydocarpa* Hutch. and *Ficus cordata* Thunb	Root bark & stem bark	Methanol	78.12	Kuete et al. (2008)
Flourensia cernua DC.	Leaves	Hexane	50	Molina-Salinas et al. (2006)
Grewia villosa Willd.	Roots	Acetone	>100 (H37Ra)	Tosun et al. (2004)
Hedera helix L.	Leaves	Aqueous	32	Orhan et al. (2012)
Hypericum triquetrifolium Turra.	Aerial parts	Ethanol	100 (H37Ra)	Tosun et al. (2004)
Lantana camara L. 2949 (orange flowers)	Aerial parts	Aqueous	16	Orhan et al. (2012)
L. camara L. 2950 (orange and pink flowers)	Aerial parts	Aqueous	32	Orhan et al. (2012)

Lantana trifolia L.	Leaves	Hexane	100	Leitao et al. (2006)
		Dichloromethane	100	
Lippia lacunosa Mart & Schauer	Leaves	Hexane	100	Leitao et al. (2006)
		Dichloromethane	100	
Lippia rotundifolia Cham.	Leaves	Hexane	100	Leitao et al. (2006)
		Dichloromethane	100	
Nepeta italica L.	Aerial parts	Aqueous	16	Orhan et al. (2012)
Ononis spinosa L.	Roots	Aqueous	16	Orhan et al. (2012)
Paliurus spina-christi Mill.	Flowers	Aqueous	16	Orhan et al. (2012)
Peschiera affinis (Mull. Arg.)	Leaves	Ethanol	≤0.20	Ramos et al. (2008)
Pinus brutia (Medw.)	Aerial parts	Ethanol	50 (H37Ra)	Tosun et al. (2004)
Plantago lanceolata L.	Leaves	Aqueous	16	Orhan et al. (2012)
Plantago major L.	Flowers	Aqueous	16	Orhan et al. (2012)
P. major L.	Leaves	Aqueous	32	Orhan et al. (2012)
Plathymenia foliolosa Benth.	Leaves	Ethanol	0.78	Ramos et al. (2008)
Pouteria filipes Eyma	Leaves	Ethanol	0.78	Ramos et al. (2008)
Primula vulgaris Huds.	Leaves	Aqueous	32	Orhan et al. (2012)
Psychotria vellosiana Benth.	Leaves	Ethanol	≤0.20	Ramos et al. (2008)
Rhoicissus tridentata (L.f.) Wild & R. B. Drumm.	Leaves	Acetone	50 (H37Ra)	Green et al. (2010)

Continued

TABLE 7.1 Plants Tested for Their Antimycobacterial Activity—cont'd

Plant Name	Plant Parts Used	Solvent System	Activity Against *Mycobacterium tuberculosis* H37Rv. (MIC in μg/mL)	References
Rhynchosia precatoria (Willd.) DC.	Roots	Hexane	15.6	Coronado-Aceves et al. (2016)
Rosa canina L.	Leaves	Aqueous	16	Orhan et al. (2012)
R. canina L.	Shoot	Aqueous	32	Orhan et al. (2012)
Salvia aethiopis L.	Aerial parts	Ethanol	50 (H37Ra)	Tosun et al. (2004)
Salvia verticillata L.	Leaves	Aqueous	16	Orhan et al. (2012)
Searsia rogersii (Schonland) Moffett	Bark	Acetone	50 (H37Ra)	Green et al. (2010)
Stachys sylvatica L.	Aerial parts	Ethanol	50 (H37Ra)	Tosun et al. (2004)
Terminalia avicennioides Guill. & Perr.	Mistle toes, Root bark and fruit	Methanol	78	Mann et al. (2008)
Teucrium polium L.	Aerial parts	Aqueous	32	Orhan et al. (2012)
Ulmus glabra Huds.	Leaves	Ethanol	50 (H37Ra)	Tosun et al., 2004
Urtica dioica L.	Leaves	Aqueous	32	Orhan et al. (2012)
	Aerial parts	Ethanol	50 (H37Ra)	Tosun et al. (2004)
Vitex cooperi Standl.	Bark	methanol:water 1:1	100	Leitao et al. (2006)

MIC, minimum inhibitory concentration.

FIGURE 7.8 South African medicinal plants that showed antimycobacterial activity. (A) *Pelargonium sidoides* (Peganum, 2014), (B) *Aspalathus linearis* (Amada44, 2014), (C) *Hypoxis hemerocallidea* (Reid, 2016a), (D) *Euclea natalensis* (Reid, 2016b), (E) *Cryptocarya latifolia* (Reid, 2016c), (F) *Thymus vulgaris* (Reid, 2016c), and (G) *Ekebergia capensis* (Reid, 2016c).

substances that form part of medicinal plants. This supports the significance of plants with medicinal value and the traditional healer with whom they are associated. Ethnopharmacology encompasses all of this and can be defined as the observing, identifying, describing, and experimental investigation of ingredients of native drugs in an area and their effect (Wauthoz, Balde, Balde, Van Damme, & Duez, 2007).

South Africa has a high diversity of plants and it is reported that about 25% of all higher plants in the world are concentrated in the southern part of Africa (Van Wyk, 2008). As a result of this concentration of higher plants and the richness in diversity there is the potential of many plants being used medicinally. Of all the plant taxa present on the African continent, it is listed that there are ~5400 taxa of medicinal plants, and associated with that number, thousands more medicinal uses (Van Wyk, 2008). With these high numbers of taxa and the prevalence of traditional and indigenous knowledge, it is not surprising to find that ~3000 of those medicinal plants are used in Southern Africa for medicinal purposes (Van Wyk, 2008). Several collaborative efforts between research organizations and traditional healers have found hundreds of plant species with the potential for drug development. Several plants in South Africa have been used to detect activity against *Mycobacterium* species. Many medicinal plants that are indigenous to South Africa have been investigated for their inhibitory activity against TB. This often involves using antimycobacterial assays to investigate the inhibitory effects of crude medicinal plant extracts or pure compounds in different solvents against the *M. tuberculosis* H37Rv strain. This chapter contains a list of several plants from South Africa and their noted MIC values.

Medicinal plants have played a major role in the traditions and lives of many people all over the world and from all walks of life. Plants which were used medicinally by various people and tribes assisted with initiating the development of traditional healthcare systems such as Ayurveda and Unani, which have formed a key part of mankind for thousands of years (Gurib-Fakim, 2006). Today the medicinal importance of higher plants is still recognized with natural products and their derivatives having a 50% share of all drugs in clinical use (Gurib-Fakim, 2006). Natural and herbal medicines have shown their potential over synthetic drugs by mostly having fewer side effects and lower levels of toxicity. Pharmacologically natural medicines have also shown their importance in their use as starting material for drug synthesis or directly as therapeutics. Natural medicines can also function as models for pharmacologically active compounds that may possess higher activity and less toxicity than their synthetic counterparts (Verma & Singh, 2008). Plants have been used in many instances to treat infectious diseases, especially when these infectious diseases build up resistance against synthetic antibiotics. MDR forms of TB are a major threat to infected individuals and to the overall cure rate for South Africa. In many instances medicinal plants have been used to help with the treatment of diseases as with the case of Linctagon. *Pelargonium sidoides* forms the major component of this over the counter drug and has been successfully shown to combat respiratory infections associated with colds and flu (Nativa, 2014).

Liv52 is a polyherbal formulation, from Himalyan Herbal Healthcare, consisting of several plants such as *Capparis spinosa*, *Cichorium intybus*, Mandura Bhasma, *Solanum nigrum*, *Terminalia arjuna*, *Cassia occidentalis, Achillea millefolium*, and *Tamarix gallica* (Himalaya wellness, 2014) and is known as a potent hepatoprotectant especially against chemically-induced hepatotoxicity. It is a globally prescribed adjuvant for patients suffering from TB side-effects in countries such as Russia (ExpressPharma, 2014). Adjuvants play a significant role in combination with Western medicines against infection and diseases. There are many South African plants that are still unexplored and may harbor the potential, in a similar manner to the examples above, as adjuvants or direct treatment for infections and diseases.

It has already been mentioned that in South Africa ~80% of its population is dependent on medicinal plants for its primary healthcare needs (Street & Prinsloo, 2013). This places a large burden on the indigenous plants of South Africa. The controversial subject of biodiversity concerns is still a very important aspect that needs to be considered by individuals interested in the market that involves healthcare products from natural origin. Most of the African continent contains generations of families with rich traditional and knowledgeable information with regards to medicines of natural origin. For many individuals, medicinal plants as their main health care option is the only choice, mainly due to the high cost of primary healthcare as well as the low availability of clinics in close proximity. Traditional knowledge is important in searching for medicinal plants of interest. South Africa contains many plant species that are commercially important (Van Wyk, 2008). Some of these plants include *Aspalathus linearis*, also known as Rooibos tea, and *Hypoxis hemerocallidea*, the African potato.

A wide array of chemically and structurally different compounds can be found in plants, and some of these may have potential medicinal value. Several South African plants have been reported to have good antimycobacterial and antioxidant activities. *Euclea natalensis* a South African plant with rich traditional knowledge linking it to the treatment of chest ailments. A monomer of diospyrin known as 7-methyljuglone, which had effective activity against *Mtb* comparable to that of the anti-TB drugs, streptomycin and ethambutol were isolated from the species (Mahapatra et al., 2007).

As a developing country, South Africa has many areas that are underdeveloped and poverty stricken. These areas have thousands of people living in overcrowded and unsanitary conditions. These conditions are favorable for the pathogen *Mtb* as is evident in the number of infected individuals seen in these areas. Together with unsanitary conditions, low living standards and inadequate healthcare, the medicinal plants in the area are useful as a remedy for the many ailments that plague these areas. Medicinal plants identified by the traditional healers in these communities can be potential remedies for TB and can also aid in the symptomatic relief of this disease. By investigating higher plants, as part of the rich biodiversity in Southern Africa, phytomedicines from these plants may be identified, which could have beneficial properties such as higher biological activity, as well as more potent immunomodulatory and hepatoprotective effects than their synthetically produced counterparts (Lakshmanan et al., 2011). Selected plants may be identified by making use of plants that have been known through traditional usage in South Africa. These indigenous plants of Southern Africa may possess antimycobacterial properties, immunomodulatory, and hepatoprotective potential that could make them useful in treatment against TB disease or as supportive agents during the course of treatment to counter the side-effects that may occur.

7.3.2 Antimycobacterial Activity

Many plant species have been used traditionally to treat the symptoms of TB. According to previous research, some plants from the southern parts of Africa have shown significant activity against *Mycobacterium* in in vitro studies, with a MIC lower than 50 µg/mL. These plants include *Berchemia discolor* with MIC of 12.5 µg/mL, *Warburgia salutaris* (25 µg/mL), *Terminalia sericea* (25 µg/mL), *Bridelia micrantha* (25 µg/mL), *and E. natalensis* (8 µg/mL) (Green, Samie, Obi, Bessong, & Ndip, 2010; Lall & Meyer, 2001; Mahapatra et al., 2007). Pure compounds 7-methyljuglone and diospyrin, isolated from *E. natalensis*, showed significant inhibitory properties, exhibiting MIC of 0.5 and 8.0 µg/mL, respectively (Mahapatra et al., 2007). The inhibitory activity of these compounds against *Mtb* is significant and comparable to the existing conventional anti-TB drugs. Considering South Africa's rich plant diversity, it is imperative to explore other plants indigenous to South Africa for anti-TB activity.

Additional South African plants that have been studied for antimycobacterial activity against *M. tuberculosis* include *Cryptocarya latifolia*, *Thymus vulgaris*, *Nidorella anomala*, *Helichrysum melanacme*, *Croton pseudopulchellus*, *Ekebergia capensis*, *Polygala myrtifolia*, *Dodonaea angustifolia*, *Galenia africana*, and *E. natalensis*, which inhibit the growth of *M. tuberculosis* (Lall & Meyer, 1999, 2001; Mahapatra et al., 2007; Mativandlela et al., 2008).

These plants exhibited MIC values lower than 50 µg/mL. *E. natalensis* and *H. melanacme* were the most effective extracts against *Mtb* with MIC of 0.5 µg/mL. The high antimycobacterial activity of *E. natalensis* is because of the presence of the binaphthoquinoid compound diospyrin and a naphthoquinone, 7-methyljuglone (5-hydroxy-7-methyl-1, 4-naphthoquinone), which are active constituents responsible for the inhibitory properties of the *E. natalensis* plant. Aqueous and acetone extracts of the roots showed activity with MIC values of 100 and 1.0 µg/mL, respectively. *Chenopodium ambrosioides*, *E. capensis*, *E. natalensis*, *H. melanacme*, *N. anomala*, and *P. myrtifolia* were found to inhibit a *Mtb* resistant strain at 0.1 mg/mL (Lall & Meyer, 1999, 2001; Mahapatra et al., 2007; Mativandlela et al., 2008). A list of the reviewed plants tested at the Department of Plant and Soil Sciences, can be found in Table 7.2 with the associated activity.

TABLE 7.2 Medicinal Plants Tested for Their Antimycobacterial Activity

Plant Name	Plant Part Used	Solvent System	Activity Against *M. tb* H37Rv (MIC in μg/mL)	Activity Against Other *Mycobacterium* spp. (MIC in μg/mL)	References
Acacia karroo Hayne.	Leaves	Methanol		625.00 (*Microsporum audouinii*)	Nielsen, Kuete, Jager, Meyer, and Lall (2012)
	Stems	Methanol		78.12 (*M. audouinii*)	Nielsen et al. (2012)
Acokanthera oppositifolia (Lam.) Codd.	Stems	Methanol		312.50 (*M. audouinii*)	Nielsen et al. (2012)
Artemisia afra Jacq. ex Willd.	Leaves	Aqueous		≥25.00% (*Mycobacterium aurum*)	Ntutela et al. (2009)
		Methanol		<25.00% (*M. aurum*)	Ntutela et al. (2009)
		Dichloromethane		41.40% (*M. aurum*)	Ntutela et al. (2009)
Bridelia micrantha (Hoscht.) Baill.	Bark	Acetone	25 (H37Ra)		Green et al. (2010)
Bulbine latifolia (L.f.) Roem. et Schult.	Leaves	Methanol		312.50 (*M. audouinii*)	Nielsen et al. (2012)
Carissa edulis Vahl.	Roots	Acetone	>100 (H37Ra)		Green et al. (2010)
		Methanol	>100 (H37Ra)		Green et al. (2010)

Continued

TABLE 7.2 Medicinal Plants Tested for Their Antimycobacterial Activity—cont'd

Plant Name	Plant Part Used	Solvent System	Activity Against *M. tb* H37Rv (MIC in μg/mL)	Activity Against Other *Mycobacterium* spp. (MIC in μg/mL)	References
Combretum erythrophyllum (Burch.) Sond.	Leaves	Water	>100 (H37Ra)		Green et al. (2010)
		Methanol	>100 (H37Ra)		Green et al. (2010)
Combretum molle R. Br. ex G. Don.	Bark	Acetone	1000		Lall and Meyer (1999)
Croton pseudopulchellus Pax	Aerial parts	Acetone	1000		Lall and Meyer (1999)
Dodonaea angustifolia (L.f.) Benth.	Leaves	Ethanol	5000	3125 (*Mycobacterium smegmatis*)	Mativandlela et al. (2008)
Diospyros mespiliformis Hochst. ex A. DC.	Leaves	Hexane	100 (H37Ra)		Green et al. (2010)
Ekebergia capensis Sparrm	Bark	Acetone	100		Lall and Meyer (1999)

Euclea natalensis A. DC.	Root	Water	500		Lall and Meyer (1999)
		Acetone	500	5.70 (*M. smegmatis*), 58.50 (*Mycobacterium fortuitum*), 260.38 (*Mycobacterium bovis* BCG), 26.01 (*M. bovis* ATCC)	Lall and Meyer (1999) and McGaw et al. (2008)
		Chloroform		7.33 (*M. smegmatis*), 65.81 (*M. fortuitum*), 468.75 (*M. bovis* BCG), 48.76 (*M. bovis* ATCC)	McGaw et al. (2008)
		Methanol		42.25 (*M. smegmatis*), 469.00 (*M. fortuitum*), 664.06 (*M. bovis* BCG), 504.48 (*M. bovis* ATCC)	McGaw et al. (2008)
Euclea undulata (Thunb.)	Root	Acetone		625.00 (*M. smegmatis*), 469.00 (*M. fortuitum*), 468.63 (*M. bovis* BCG), 126.75 (*M. bovis* ATCC)	McGaw et al. (2008)
Helichrysum odoratissimum (L.) Sweet	Whole plant	Acetone	500		Lall and Meyer (1999)
Knowltonia vesicatoria (L.f.) Sims.	Aerial parts	Methanol		156.25 (*M. audouinii*)	Nielsen et al. (2012)
Lippia javanica (Burm.f.) Spreng	Leaves	Methanol	>100 (H37Ra)		Green et al. (2010)

Continued

TABLE 7.2 Medicinal Plants Tested for Their Antimycobacterial Activity—cont'd

Plant Name	Plant Part Used	Solvent System	Activity Against *M. tb* H37Rv (MIC in μg/mL)	Activity Against Other *Mycobacterium* spp. (MIC in μg/mL)	References
Pelargonium reniforme Curtis	Root	Hexane		64.00 (*M. aurum*), 512.00 (*M. smegmatis*)	Seidel and Taylor (2004)
		Ethyl acetate		512.00 (*M. aurum*), >512.00 (*M. smegmatis*)	Seidel and Taylor (2004)
		Ethanol		512.00 (*M. aurum*)	Seidel and Taylor (2004)
Pelargonium sidoides DC.	Root	Hexane		512.00 (*M. aurum*), >512.00 (*M. smegmatis*)	Seidel and Taylor (2004)
		Ethyl acetate		>512.00 (*M. aurum*), >512.00 (*M. smegmatis*)	Seidel and Taylor (2004)
		Ethanol		>512.00 (*M. aurum*)	Seidel and Taylor (2004)
Peltophorum africanum Sond.	Bark	Acetone	>100 (H37Ra)		Green et al. (2010)
Piper capense L.f.	Roots	Acetone	100 (H37Ra)		Green et al. (2010)
Polygala myrtifolia L.	Aerial parts	Acetone			Lall and Meyer (1999)

Ptaeroxylon obliquum (Thund.) Radlk	Stems	Methanol		312.50 (*M. audouinii*)	Nielsen et al. (2012)
	Bark	Methanol		78.12 (*M. audouinii*), 156.25 (*M. smegmatis*)	Nielsen et al. (2012)
Salvia africana-lutea L.	Aerial parts	Methanol		39.06 (*M. audouinii*), 312.50 (*M. smegmatis*)	Nielsen et al. (2012)
	Bark	Methanol		156.25 (*M. audouinii*)	Nielsen et al. (2012)
	Aerial parts	methanol:chloroform (1:1, v/v)	500 (H37Ra)		Kamatou, Van Vuuren, Van Heerden, Seaman, and Viljoen (2007)
Salvia africana-caerulea L.	Aerial parts	methanol:chloroform (1:1, v/v)	500 (H37Ra)		Kamatou et al. (2007)
Salvia albicaulis Benth.	Aerial parts	methanol:chloroform (1:1, v/v)	500 (H37Ra)		Kamatou et al. (2007)
Salvia aurita L.f. var. aurita	Aerial parts	methanol:chloroform (1:1, v/v)	500 (H37Ra)		Kamatou et al. (2007)
Salvia chamelaeagnea K. Bergius	Aerial parts	methanol:chloroform (1:1, v/v)	500 (H37Ra)		Kamatou et al. (2007)
Salvia disermas L.	Aerial parts	methanol:chloroform (1:1, v/v)	500 (H37Ra)		Kamatou et al. (2007)
Salvia dolomitica Codd.	Aerial parts	methanol:chloroform (1:1, v/v)	100 (H37Ra)		Kamatou et al. (2007)

Continued

TABLE 7.2 Medicinal Plants Tested for Their Antimycobacterial Activity—cont'd

Plant Name	Plant Part Used	Solvent System	Activity Against *M. tb* H37Rv (MIC in μg/mL)	Activity Against Other *Mycobacterium* spp. (MIC in μg/mL)	References
Salvia lanceolata Lam.	Aerial parts	methanol:chloroform (1:1, v/v)	500 (H37Ra)		Kamatou et al. (2007)
Salvia muirii L. Bol.	Aerial parts	methanol:chloroform (1:1, v/v)	500 (H37Ra)		Kamatou et al. (2007)
Salvia radula Benth.	Aerial parts	methanol:chloroform (1:1, v/v)	100 (H37Ra)		Kamatou et al. (2007)
Salvia repens Burch. ex. Benth.	Aerial parts	methanol:chloroform (1:1, v/v)	500 (H37Ra)		Kamatou et al. (2007)
Salvia runcinata L.f.	Aerial parts	methanol:chloroform (1:1, v/v)	500 (H37Ra)		Kamatou et al. (2007)
Salvia schlechteri Briq.	Aerial parts	methanol:chloroform (1:1, v/v)	500 (H37Ra)		Kamatou et al. (2007)
Salvia stenophylla Burch. ex. Benth.	Aerial parts	methanol:chloroform (1:1, v/v)	500 (H37Ra)		Kamatou et al. (2007)
Salvia verbenaca L.	Aerial parts	methanol:chloroform (1:1, v/v)	100 (H37Ra)		Kamatou et al. (2007)
Sclerocarya birrea (A. Rich.) Hocsht	Bark	Acetone	>100 (H37Ra)		Green et al. (2010)

Schotia brachypetala Sond.	Bark	Acetone	50 (H37Ra)		Green et al. (2010)
Securidaca longepedunculata Fresen	Roots	Acetone	>100 (H37Ra)		Green et al. (2010)
Solanum panduriforme E. may	Leaves	Water	>100 (H37Ra)		Green et al. (2010)
		Methanol	>100 (H37Ra)		Green et al. (2010)
Terminalia sericea Burch ex. D.C.	Bark	Acetone	25 (H37Ra)		Green et al. (2010)
Typha capensis (Rohrb.) N. E. Br.	Aerial parts	Methanol		312.50 (*M. audouinii*)	Nielsen et al. (2012)
Warburgia salutaris (Bertol.f.) Chiov	Leaves	Acetone	25 (H37Ra)		Green et al. (2010)
Ximenia caffra Sond.	Roots	Acetone	>100		Green et al. (2010)
Zantedeschia aethiopica (L.) Spreng.	Leaves and stems	Methanol		312.50 (*M. audouinii*)	Nielsen et al. (2012)
Ziziphus mucronata Willd.	Bark	Acetone	50 (H37Ra)		Green et al. (2010)

ATCC, American Type Culture Collection; *BCG*, bacillus Calmette-Guérin; *MIC*, minimum inhibitory concentration.

Aquatic extracts of medicinal plants from the Brazilian xeric shrub land have been screened for antibiofilm activities against *Staphylococcus epidermis*. They showed activity at a concentration range of 0.4–4.0 mg/mL (Trentin et al., 2013). *Azadirachta indica*, *Vitex negundo*, *Tridax procumbens*, and *Ocimum tenuiflorum* were tested against *Escherichia coli* biofilm. The results were reported as follows; the first two plant extracts indicated 100% biofilm inhibition at 100 μg/mL and the other two plant extracts indicated 60% biofilm inhibition (Namasivayam & Roy, 2013). Plant-derived natural compounds from Krameria, *Aesculus hippocastanum* and *Chelidonium majus* indicated antibiofilm activity against *Staphylococcus aureus* and *Staphylococcus epidermidis* at an effective dose (EC_{50}) of 24.5 and 15.2 μM for *S. aureus* and 8.6 and 7.6 μM for *S. epidermidis*.

Researchers have screened plants for antibiofilm activity: *Mentha piperita* (peppermint) had activity against *P. aeruginosa*, *Candida albicans*, and *Listeria monocytogenes* at an inhibition percentage of 38%, 28%, and 50%, respectively (Sandasi, Leonard, Vuuren, & Viljoen, 2011). *Rosmarinus officinalis* (rosemary) and *Melaleuca alternifolia* were tested for inhibitory properties against *Listeria monocytogenes*. The plants inhibited biofilm formation by 50% (Sandasi et al., 2011). The marula plant (*Sclerocarya birrea*) reduced biofilm formation of *P. aeruginosa* by 75% (Sarkar et al., 2014).

Inhibition of biofilm formation by *Mycobacterium* species using plants extracts has been reported by Abidi, Ahmed, Sherwani, Bibi, and Kazmi (2014). It was found that extract of *A. indica* (Neem) was efficient in reduction potential, which was 173% against *M. smegmatis* biofilm. Further analysis of biofilm inhibition with regards to a cell-based phenotypic screening of *Mycobacterium* has been done and led to the discovery of a small molecule, TCA1. It has bactericidal activity against both drug susceptible and resistant *Mtb* and can contribute to the inhibition of biofilm formation (Wang et al., 2013).

7.3.3 Immune Stimulatory Activity

Immunomodulators as a term have been used in literature to include botanical medicines believed to influence the immunity (Spelman et al., 2006). Due to the adverse effects of the first-line drugs mentioned before, many botanicals have been investigated for their potential as immunomodulators. Synthetic cytokines are not only used to alter the phenotype of the immune response but a combination of phytochemicals and cytokines can also be seen as a novel approach to therapeutics (Spelman et al., 2006). Engwerda, Andrew, Murphy, and Mynott (2001) described the potential of combination therapy by making use of bromelain, a mixture of cysteine proteases from the stem of pineapple plants and IL-12. Limited secretion of cytokines was shown by bromelain used individually but with the addition of IL-12 a synergistic outcome was achieved. Combination therapies could enhance not just the acquired immune responses of the host but also the innate immunity, which are both needed for a prolonged protective immune response for the control of *Mtb* infection.

Many herbal products have been shown to alter the immune response by showcasing immunomodulatory properties and so have many compounds derived from botanicals such as isoflavonoids, indoles, phytosterols, polysaccharides, sesquiterpenes, alkaloids, glucans, and tannins (Patwardhan & Gautam, 2005). Already many plants have been investigated for their potential as immunomodulators; for example, the study of Madan, Sharma, Inamdar, Singh Rao, and Singh (2008), where *Aloe vera* extracts were made and given to rats that, after 5 days, showed a marked increase in the white blood cells as well as an increase in their humoral immunity. Many in vitro and in vivo studies have shown the immunomodulatory effects of medicinal plants through modulation of cytokines as well as the process of phagocytosis. These all form an integral part of treating and eliminating *Mtb* in infected individuals.

Many herbal products have been shown to alter the host immune response by showcasing their immunomodulatory properties, which have a major role to play in combating infections. The classes of compounds from a natural origin that possess these properties include isoflavonoids, polysaccharides, sesquiterpenes, and alkaloids to name a few (Patwardhan & Gautam, 2005). Lead compounds further support the important role played by traditional medicines and ethnopharmacology in the production and development of new anti-TB drugs.

7.3.4 Hepatoprotective Activity

During TB infection, drug-induced liver toxicity is a major problem and may occur with all the current treatment regimens. The liver plays a central role in metabolism and detoxification of drugs, and in doing so, it is susceptible to injury. Isoniazid (6–9 months), rifampicin (4 months), isoniazid/rifampicin (4 months) as well as the twofold treatment with pyrazinamide with either ethambutol or a fluoroquinolone has shown hepatotoxicity (American Thoracic Society, 2000; Younossian, Rochat, Ketterer, Wacker, & Janssens, 2005). There are many factors that influence the severity of the toxic effect; these include, age, sex, cofactors, regimen, and the occurrence of HIV/AIDS, as well as Hepatitis B and C (Saukkonen et al., 2006). Oxidative stress and the subsequent increase in lipid peroxides are mechanisms of pathogenesis that can cause hepatic injury (Kumar, Modi, & Saxena, 2013) Isoniazid and rifampicin in part cause hepatotoxicity through the mechanism of oxidative stress and lipid peroxidation (Kumar et al., 2013). Countless infected individuals cease treatment due to the side-effects experienced with anti-TB drugs.

Experimental studies have shown that the hepatic injury due to anti-TB drugs can be lowered and eliminated by the use of herbal formulations (Tasaduq et al., 2003). Many plant-derived samples have been reported to show hepatoprotective abilities against anti-TB drug-induced hepatotoxicity such as the aqueous leaf extracts from the famous Neem tree, *A. indica* and the polyherbal formulation called Liv.52 (Kumar et al., 2013).

Another popular plant with hepatoprotective abilities is garlic (*Allium sativum*) that significantly lowers important liver transaminases alanine aminotransferase

(ALT) and aspartate aminotransferase (AST), which are biomarkers for liver injury, as well as the bilirubin levels. Garlic is well known for its therapeutic properties and these can all be ascribed to several known secondary metabolites such as thiosulfanates, terpenoids, and steroids to name a few (Kumar et al., 2013). A few modern drugs, such as tricholine citrate, trithioparamethoxy phenylpropene, phospholipids, and a combination of L-ornithine, L-aspartate and pancreatin, silymarin, and ursodeoxycholic acid, are known to exhibit hepatoprotective properties (Kumar et al., 2013). Although there are several drugs in the market that aid in liver difficulties, many have side-effects associated with them and no complete solution has yet been found to completely manage injuries to the liver without any side-effects. Many people also rely on herbal formulations as modern synthetic drugs come at a significant price.

Many plant species have been used traditionally for their abilities to provide hepatoprotection and for their high antioxidant contents. Many commercial products are also widely available with their claims against hepatotoxicity and with high antioxidant properties, for example, Liv.52. The main mechanism of herbal drugs in their abilities to provide hepatoprotection, stems from the inhibition of CYP P450 2E1 as well as their antioxidant activity (Jaeschke et al., 2002). Many drugs and treatments that are available for treating hepatotoxicity have severe side-effects. Herbal drugs and natural substitutes that are available may be able to provide the same efficacy but without the side-effects. Many plants have therefore been under investigation for their potential in providing hepatoprotection. Recent studies of plants to investigate their potential for hepatoprotection have yielded the isolation of 177 different phytoconstituents from plants belonging to 55 different families that have these properties. Hepatoprotectant plants contain many chemical constituents that display this activity such as phenols, coumarins, monoterpenes, glycosides, alkaloids, and xanthenes. It is thought that the underlying mechanism of these hepatoprotective plants is their antioxidant potential (Ahmed, Rao, Ahemad, & Ibrahim, 2012). According to Kumar, an increase in several liver enzyme activities and the subsequent reduction in lipid peroxide keeps the cellular integrity of the liver and also control the increase in the liver serum levels of ALT, AST, and alkaline phosphatase. Bilirubin, formerly known as hematoidin, is a breakdown product of heme found in the red blood cells. This product is excreted in both the bile and urine and elevated levels may indicate the incidence of certain diseases. The main function that has been attributed to bilirubin is the physiological role as a cellular antioxidant (Baranano, Rao, Ferris, & Snyder, 2002).

7.4 PRODUCT DEVELOPMENT AND FUTURE PROSPECTS

With the high plant diversity in Southern Africa, it is surprising that we as a country have not produced phytomedicines and natural products to treat TB. The process of transferring innovation and new technologies into bioprospecting medicinal products is a lengthy, expensive, and complicated process.

From the information in this chapter, it is evident that there are many candidate plants to be developed further into fully fledged products. Naturally, these products and the development process need to be regulated to incorporate all the parties involved. Firstly, the candidate plant/natural product needs to be scientifically validated for efficacy and safety. Secondly, most of the plants, if not all, have indigenous and/or traditional knowledge associated with them. For the process of bioprospecting from a natural source, these rightful knowledge holders need to be identified and an agreement needs to be established so that there will be equal and fair share of benefits, according to the "National Environmental Management Biodiversity Act, 2004 (Act No. 10 of 2004, NEMBA or Biodiversity Act) and the "Bioprospecting, Access and Benefit Sharing" (BABS) regulations of 2008 of South Africa. Thirdly, the cultivation of medicinal plants has to be established to limit the harvesting of wild material and decrease the burden on conservation of medicinal plants. And finally, the manufacturing and marketing of these products need to be done on an ethical basis, with good manufacturing processes. The development of a new product might seem impossible for some, if one considers all the hurdles, but it is possible in the current framework and status of the medical environment, especially with the trend to a greener and natural lifestyle.

7.5 CONCLUSION

Many medicinal plant extracts with various extracting solvents used, have shown effective activity against both *M. tuberculosis* and many other *Mycobacterium* spp. All of these studies have shown that a number of medicinal plants have variable activity against the *Mycobacterium* spp. Most of the medicinal plants tested have effective activity against the most important *Mycobacterium* spp. especially against *M. tuberculosis*. Further testing, which may include bioassay guided fractionation, may yield active components and phytochemicals of medicinal importance. Not only is the active killing of the *Mycobacterium* important but also other distinguishing factors that may assist infected individuals, such as hepatoprotection and immune stimulatory properties.

REFERENCES

Abidi, S. H., Ahmed, K., Sherwani, S. K., Bibi, N., & Kazmi, S. U. (2014). Detection of *Mycobacterium Smegmatis* biofilm and its control by natural agents. *International Journal of Current Microbiology and Applied Sciences*, *3*(4), 801–812.

Ahmed, F. M., Rao, S. A., Ahemad, R. S., & Ibrahim, M. (2012). Phytochemical studies and hepatoprotective activity of *Melia azedarach* Linn, against CCl_4 induced hepatotoxicity in rats. *Journal of Pharmacy Research*, *5*(5), 42–45.

Amada44. (2014). *Aspalathus linearis*. Available from: https://commons.wikimedia.org/wiki/File:Aspalathus_linearis_0222.jpg.

American Thoracic Society. (2000). Targeted tuberculin testing and treatment of latent tuberculosis infection. *American Journal of Respiratory and Critical Care Medicine*, *161*, 221–247.

Antas, P. R. Z., Ding, L., Hackman, J., Reeves-Hammock, L., Shintani, A. K., Schiffer, J., et al. (2006). Decreased CD4+ lymphocytes and innate response in adults with previous extrapulmonary tuberculosis. *Journal of Allergy and Clinical Immunology*, *117*, 916–923.

Armstrong, J. A., & Hart, P. D. (1971). Response of cultured macrophages to *Mycobacterium tuberculosis*, with observations on fusion of lysosomes with phagosomes. *Journal of Experimental Medicine*, *134*, 713–740.

Baranano, D. E., Rao, M., Ferris, C. D., & Snyder, S. H. (2002). Biliverdin reductase: A major physiologic cytoprotectant. *Proceedings of the National Academy of Sciences*, *99*(25), 16093–16098.

Barry, C. E., Boshoff, H. I. M., & Dowd, C. S. (2004). Prospects for clinical introduction of Nitroimidazole antibiotics for the treatment of tuberculosis. *Current Pharmaceutical Design*, *10*, 3239–3262.

Biswas, R., Agarwal, R. K., Bhilegaonkar, K. N., Kumar, A., Nambiar, P., Rawat, S., et al. (2011). Cloning and sequencing of biofilm-associated protein (bapA) gene and its occurrence in different serotypes of *Salmonella*. *Letters in Applied Microbiology*, *52*(2), 138–143.

Bloom, B. R. (1994). *Tuberculosis: Pathogenesis, protection and control*. Washington, DC: American Society of Microbiology.

Boom, W. H., Canaday, D. H., Fulton, S. A., Gehring, A. J., Rojas, R. E., & Torres, M. (2003). Human immunity to *M. tuberculosis*: T cell subsets and antigen processing. *Tuberculosis*, *83*, 98–106.

Boyd, A., & Chakrabarty, A. M. (1994). Role of aginate lyase in cell detachment of *Pseudomonas aeruginosa*. *Applied and Environmental Microbiology*, *60*(7), 2355–2359.

Branda, S. S., Vik, A., Friedman, L., & Kolter, R. (2005). Biofilm: The matrix revisited. *Trends in Microbiology*, *13*(1), 20–26.

Brennan, P. J. (2003). Structure, function, and biogenesis of the cell wall of *Mycobacterium tuberculosis*. *Tuberculosis*, *83*(1–3), 91–97.

Brosch, R., Gordon, S. V., Marmiesse, M., Brodin, P., Buchrieser, C., Eiglmeier, K., et al. (2002). A new evolutionary scenario for the *Mycobacterium tuberculosis* complex. *Proceedings of the National Academy of Sciences*, *99*, 3684–3689.

Brown, M. R. W., Allison, D. G., & Gilbert, P. (1988). Resistance of bacterial biofilms to antibiotics a growth-rate related effect? *Journal of Antimicrobial Chemotherapy*, *22*(6), 777–780.

Carr, J. (2006). *Mycobacterium_tuberculosis_8438_lores.jpg*. Available from: https://commons.wikimedia.org/wiki/File:Mycobacterium_tuberculosis_8438_lores.jpg.

CDC, Centre for Disease Control and Prevention. (2012). *Core curriculum on tuberculosis: What the clinician should know*. Available from: http://www.cdc.gov/diseasesconditions/.

Coronado-Aceves, E. W., Sanchez- Escalante, J. J., Lopez-Cervantes, J., Robles, Zepeda, R. E., Velasquez, C., Sanchez-Machado, D. I., et al. (2016). Antimycobacterial activity of medicinal plants used by the Mayo people of Sonora, Mexico. *Journal of Ethnopharmacology*, *190*, 106–115.

Costerton, J. W., Lewandowski, Z., Caldwell, D. E., Korber, D. R., & Lappin-Scott, H. M. (1995). Microbial biofilms. *Annual Review of Microbiology*, *49*, 711–745.

van Crevel, R., Ottenhoff, T. H. M., & van der Meer, J. W. M. (2002). Innate Immunity to *Mycobacterium tuberculosis*. *Clinical Microbiology Reviews*, *15*(2), 294–309.

Davey, M. E., & O'Toole, G. A. (2000). Microbial biofilms: From ecology to molecular genetics. *Microbiology and Molecular Biology Reviews*, *64*(4), 847–867.

Donlan, R. M., & Costerton, J. W. (2002). Biofilms: Survival mechanisms of clinically relevant microorganisms. *Clinical Microbiology Reviews*, *15*(2), 167–193.

Dzoyem, J. P., Aro, A. O., McGaw, L. J., & Eloff, J. N. (2016). Antimycobacterial activity against different pathogens and selectivity index of fourteen medicinal plants used in southern Africa to treat tuberculosis and respiratory ailments. *South African Journal of Botany*, *102*, 70–74.

Engwerda, C. R., Andrew, D., Murphy, M., & Mynott, T. L. (2001). Bromelain activates murine macrophages and natural killer cells *in vitro*. *Cellular Immunology*, *210*(1), 5–10.

ExpressPharma. (2014). *Developing a herbal-based drug for Hepatitus B*. Available from: http://www.pharma.financialexpress.com/sections/res/2059-developing-a-herbal-based-drug-for-hepatitis-B.

Fitzgerald, D. W., Sterling, T. R., & Haas, D. W. (2009). *Mycobacterium tuberculosis*. In G. L. Mandell, J. E. Bennett, & R. Dolan (Eds.), *Mandell, douglas, and bennett's principles and practice of infectious diseases* (7th ed.). Orlando, FL: Saunders Elsevier.

Flynn, J. L., & Chan, J. (2003). Immune evasion by *Mycobacterium tuberculosis*: Living with the enemy. *Current Opinion in Immunology*, *15*, 450–455.

Freebase Data Dumps. (2013). *Granuloma*. Available from: https://www.wikidata.org/wiki/Q1129338.

Ghadimi, D., De Vrese, M., Heller, K. J., & Schrezenmeir, J. (2010). Lactic acid bacteria enhance autophagic ability of mononuclear phagocytes by increasing Th1 autophagy-promoting cytokine (IFN-γ) and nitric oxide (NO) levels and reducing Th2 autophagy-restraining cytokines (IL-4 and IL-13) in response to *Mycobacterium tuberculosis* antigen. *International Immunopharmacology*, *10*, 694–706.

Green, E., Samie, A., Obi, C. L., Bessong, P. O., & Ndip, R. N. (2010). Inhibitory properties of selected South African medicinal plants against *Mycobacterium tuberculosis*. *Journal of Ethnopharmacology*, *130*(1), 151–157.

Grotzke, J. E., & Lewinsohn, D. M. (2005). Role of CD8$^+$ T lymphocytes in control of *Mycobacterium tuberculosis* infection. *World Health*, *7*, 776–788.

Gupta, A., Kaul, A., Tsolaki, A. G., Kishore, U., & Bhakta, S. (2012). *Mycobacterium tuberculosis*: Immune evasion, latency and reactivation. *Immunobiology*, *217*, 363–374.

Gurib-Fakim, A. (2006). Medicinal plants: Traditions of yesterday and drugs of tomorrow. *Molecular Aspects of Medicine*, *27*, 1–93.

Gutacker, M. M., Smoot, J. C., Migliaccio, C. A., Ricklefs, S. M., Hua, S., Cousins, D. V., et al. (2002). Genome-wide analysis of synonymous single nucleotide polymorphisms in *Mycobacterium tuberculosis* complex organisms: Resolution of genetic relationships among closely related microbial strains. *Genetics*, *162*, 1533–1542.

Häggström, M. (2014). Medical gallery of Mikael Häggström. *WikiJournal of Medicine*, *1*(2), ISSN: 2002–4436. http://dx.doi.org/10.15347/wjm/2014.008.

Himalaya wellness. (2014). Available from: http://www.himalayawellness.com.

Hughes, A. L., Friedman, R., & Murray, M. (2002). Genome-wide pattern of synonymous nucleotide substitution in two complete genomes of *Mycobacterium tuberculosis*. *Emerging Infectious Diseases*, *8*, 1342–1346.

Islam, M. S., Richards, J. P., & Ojha, A. K. (2012). Targeting drug tolerance in mycobacteria: a perspective from mycobacterial biofilms. *Expert review of anti-infective therapy*, *10*(9), 1055–1066.

Jaeschke, H., Gores, G. J., Cederbaum, A. I., Hinson, J. A., Pessayre, D., & Lemasters, J. J. (2002). Mechanism of hepatotoxicity. *Toxicological Sciences*, *65*(2), 166–176.

Kamatou, G. P. P., Van Vuuren, S. F., Van Heerden, F. R., Seaman, T., & Viljoen, A. M. (2007). Antibacterial and antimycobacterial activities of South African *Salvia* species and isolated compounds from *S. Chamelaeagnea*. *South African Journal of Botany*, *73*, 552–557.

Klein, L. L., Petukhova, V., Wan, B., Wang, Y., Santasiero, B. D., Lankin, D. C., et al. (2014). A novel indigoid anti-tuberculosis agent. *Bioorganic & Medicinal Chemistry Letters*, *24*, 268–270.

Korbel, D. S., Schneider, B. E., & Schaible, U. E. (2008). Innate immunity in tuberculosis: Myths and truth. *Microbes and Infection*, *10*(9), 995–1004.

Kuete, V., Ngameni, B. F., Simo, C. C., Tankeu, R. K., Ngadjui, B. T., et al. (2008). Antimicrobial activity of the crude extracts and compounds from *Ficus chlamydocarpa* and *Ficus cordata* (Moraceae). *Journal of Ethnopharmacology*, *120*, 17–24.

Kumar, A., Farhana, A., Guidry, L., Saini, V., Hondalus, M., & Steyn, A. J. C. (2011). Redox homeostasis in mycobacteria: The key to tuberculosis control? *Expert Reviews in Molecular Medicine*, *13*(1), 1–25.

Kumar, V., Modi, P. K., & Saxena, K. K. (2013). Exploration of hepatoprotective activity of aqueous extract of *Tinospora cordifolia* – an experimental study. *Asian Journal of Pharmaceutical and Clinical Research*, *6*(1), 87–91.

Lakshmanan, D., Werngren, J., Jose, L., Suja, K. P., Nair, M. S., Varma, R. L., et al. (2011). Ethyl p-methoxycinnamate isolated from a traditional anti-tuberculosis medicinal herb inhibits drug resistant strains of *Mycobacterium tuberculosis in vitro*. *Fitoterapia*, *82*(5), 757–761.

Lall, N., & Meyer, J. J. M. (1999). *In vitro* inhibition of drug-resistant and drug-sensitive strains of *Mycobacterium tuberculosis* by ethnobotanically selected South African plants. *Journal of Ethnopharmacology*, *66*(3), 347–354.

Lall, N., & Meyer, J. J. M. (2001). Inhibition of drug-sensitive and drug-resistant strains of *Mycobacterium tuberculosis* by diospyrin, isolated from *Euclea natalensis*. *Journal of Ethnopharmacology*, *78*(2), 213–216.

Leitao, S. G., Castro, O., Fonseca, E. N., Juliao, L. S., Tavares, E. S., Leo, R. R. T., et al. (2006). Screening of central and South American plant extracts for antimycobacterial activity by the alamar blue test. *Brazilian Journal of Pharmacognosy*, *16*(1), 6–11.

Lienhardt, C., Vernon, A., & Raviglione, M. C. (2010). New drugs and new regimens for the treatment of tuberculosis: Review of the drug development pipeline and implications for national programmes. *Current Opinion in Pulmonary Medicine*, *16*, 186–193.

Lurie, M. B. (1964). *Resistance to tuberculosis: Experimental studies in native and acquired defense mechanisms*. Cambridge, Mass: Harvard University Press.

Madan, J., Sharma, A. K., Inamdar, N., Singh Rao, H., & Singh, H. (2008). Immunomodulatory properties of *Aloe vera* gel in mice. *International Journal of Green Pharmacy*, *2*(3), 152–154.

Madariaga, M. G., Lalloo, U. G., & Swindells, S. (2008). Extensively drug-resistant tuberculosis. *The American Journal of Medicine*, *121*(10), 835–844.

Mahapatra, A., Mativandlela, S. P. N., Binneman, B., Fourie, P. B., Hamilton, C. J., Meyer, J. J. M., et al. (2007). Activity of 7-methyljuglone derivatives against *Mycobacterium tuberculosis* and as subversive substrates for mycothiol disulphide reductase. *Bioorganic and Medical Chemistry*, *15*(24), 7638–7646.

Ma, Z., Lienhardt, C., McIlleron, H., Nunn, A. J., & Wang, X. (2010). Series tuberculosis 5 global tuberculosis drug development pipeline: The need and the reality. *Online*, *6736*(10), 1–10.

Manjunatha, U., Boshoff, H. I., & Barry, C. E. (2009). The mechanism of action of PA-824: Novel insights from transcriptional profiling. *Communicative and Iintegrative Biology*, *2*(3), 215–218.

Mann, A., Amupitan, J. O., Oyewale, O. A., Okogun, J. I., Ibrahim, K., Oladosu, P., et al. (2008). Evaluation of in vitro antimycobacterial activity of Nigerian plants used for treatment of respiratory diseases. *African Journal of Biotechnology*, *7*(11), 1630–1636.

Mativandlela, S. P. N., Meyer, J. J. M., Hussein, A. A., Houghton, P. J., Hamilton, C. J., & Lall, N. (2008). Activity against *Mycobacterium smegmatis* and *Mycobacterium tuberculosis* by extract of South African medicinal plants. *Phytotherapy Research*, *22*, 841–845.

Matteelli, A., Carvalho, A. C. C., Dooley, K. E., & Kritski, A. (2010). TMC207: The first compound of a new class of potent anti-tuberculosis drugs. *Future Microbiology*, *5*(6), 849–858.

McGaw, L. J., Lall, N., Meyer, J. J. M., & Elof, J. N. (2008). The potential of South African plants against *Mycobacterium* infections. *Journal of Ethnopharmacology*, *119*, 482–500.

Microbiologybytes. (2014). *Mycobacterium tuberculosis*. Available from: http://www.microbiologybytes.com/video/Mtuberculosis.html.

Migliori, G. B., De Laco, G., Besozzi, G., Centis, R., & Cirillo, D. M. (2007). *First tuberculosis cases in Italy resistant to all tested drugs*. Available from: http://www.eurosurveillance.org.

Mills, B. (2008). *Stick model of a streptomycin molecule*. Available from: https://commons.wikimedia.org/wiki/File:Streptomycin-1ntb-xtal-3D-sticks.png.

Molina-Salinas, G. M., Ramos-Guerra, M. C., Vargas-Villarreal, J., Mata-Cardenas, B. D., Becerril-Montes, P., & Said-Fernandez, S. (2006). Bactericidal activity of organic extracts from *Flourensia cernua* DC against strains of *Mycobacterium tuberculosis*. *Archives of Medical Research, 37*(1), 45–49.

Molina-Salinas, G. M., Perez-Lopez, A., Becerril-Montes, P., Salazar-Aranda, R., Said-Fernández, S., & de Torres, N. W. (2007). Evaluation of the flora of Northern Mexico for in vitro antimicrobial and antituberculosis activity. *Journal of Ethnopharmacology, 109*, 435–441.

Namasivayam, S. K. R., & Roy, E. A. (2013). Antibiofilm effect of medicinal plant extracts against clinical isolate of biofilm of *Escherichia coli*. *International Journal of Pharmacy and Pharmaceutical Research, 5*(2), 486–489.

Nativa, Health, Science, Nature. (2014). Available from: http://www.nativa.co.za/products/category/linctagon/.

NIAID. (2007). Tuberculosis, which results from an infection with Mycobacterium tuberculosis, can usually be cured with a combination of first-line drugs taken for several months. Shown here are the four drugs in the standard regimen of first-line drugs. Available from: https://commons.wikimedia.org/wiki/File:First-Line_Treatment_of_Tuberculosis_(TB)_for_Drug-Sensitive_TB_(5102307249).jpg.

Nielsen, T. R. H., Kuete, V., Jager, A. K., Meyer, J. J. M., & Lall, N. (2012). Antimicrobial activity of selected South African medicinal plants. *BMC Complementary and Alternative Medicine, 12*, 74.

Ntutela, S., Smith, P., Matika, L., Mukinda, J., Arendse, H., Allie, N., et al. (2009). Efficacy of *Artemisia afra* phytotherapy in experimental tuberculosis. *Tuberculosis Journal, 89*(1), 33–40.

Ojha, A., Anand, M., Bhatt, A., Kremer, L., Jacobs, W. R., & Hatfull, G. F. (2005). GroEL1: A dedicated chaperone involved in mycolic acid biosynthesis during biofilm formation in mycobacteria. *Cell*, 1–13.

Oosthuizen, C. B. (2015). *Mycobacterium smegmatis biofilm*. University of Pretoria.

Orhan, D. D., Ozcelik, B., Hosbas, S., & Vural, M. (2012). Assessment of antioxidant, antibacterial, antimycobacterial and antifungal activities of some plants used as folk medicine in Turkey against dermatophytes and yeast like fungi. *Turkish Journal of Biology, 36*, 672–686.

Parsek, M. R., & Singh, P. K. (2003). Bacterial biofilms: An emerging link to disease pathogenesis. *Annual Review of Microbiology, 57*, 677–701.

Patwardhan, B., & Gautam, M. (2005). Botanical immunodrugs: Scope and opportunities. *Drug Discovery Today, 10*(7), 495–502.

Peganum. (2014). *Pelargonium sidoides*. Available from: https://commons.wikimedia.org/wiki/File:Pelargonium_sidoides_%2814251109977%29.jpg.

Ramos, D. F., Leitao, G. G., Costas, F., Abrue, L., Villarreal, J. V., Leitao, S. G., et al. (2008). Investigation of the antimycobacterial activity of 36 plant extracts from the brazilian Atlantic forest. *Brazilian Journal of Pharmaceutical Sciences, 44*(4), 669–674.

Recht, J., Martinez, A., Torello, S., & Kolter, R. (2000). Genetic analysis of sliding motility in *Mycobacterium smegmatis*. *Journal for Bacteriology, 182*(15), 4348–4351.

Reid, A. (2016a). *Hypoxis hemerocallidea*. University of Pretoria.

Reid, A. (2016b). *Euclea natalensis*. University of Pretoria.

Reid, A. (2016c). *(A) Cryptocarya latifolia, (B) Thymus vulgaris, (C) Ekebergia capensis*. University of Pretoria.

Rohde, K., Yates, R. M., Purdy, G. E., & Russell, D. G. (2007). *Mycobacterium tuberculosis* and the environment within the phagosome. *Immunological Reviews, 219*, 37–54.

Russell, D. G., Barry, C. E., & Flynn, J. L. (2010). Tuberculosis: What we don't know can, and does, hurt us. *Science, 328*, 852–855.

Sandasi, M., Leonard, C. M., Van Vuuren, S. F., & Viljoen, A. M. (2011). Peppermint (*Mentha piperita*) inhibits microbial biofilms *in vitro*. *South African Journal of Botany*, *77*(2011), 80–85.

Sarkar, R., Chaudhary, S. K., Sharma, A., Yadav, K. K., Nema, N. K., Sekhoacha, M., et al. (2014). Anti-biofilm activity of Marula – a study with the standardized bark extract. *Journal of Ethnophamacology*, *154*, 170–175.

Saukkonen, J. J., Cohn, D. L., Jasmer, R. M., Schenker, S., Jereb, J. A., Nolan, C. M., et al. (2006). An official ATS statement: Hepatotoxicity of antituberculosis therapy. *American Journal of Respiratory and Critical Care Medicine*, *174*(8), 935–952.

Schwander, S., & Dheda, K. (2011). Human lung immunity against *Mycobacterium tuberculosis*, insights into pathogenesis and protection. *American Journal of Respiratory Critical Care Medicine*, *183*, 696–707.

Seidel, V., & Taylor, P. W. (2004). *In vitro* activity of extracts and constituents of *Pelargonium* against rapidly growing mycobacteria. *International Journal of Antimicrobial Agents*, *23*(6), 613–619.

Shirtliff, M. E., Mader, J. T., & Camper, A. K. (2002). Molecular interactions in biofilms. *Chemistry & Biology*, *9*(8), 859–871.

SoulCity. (2015). *Literature review of TB in South Africa*. Available from: https://www.soulcity.org.za/projects/tuberculosis/research/literature-review-on-tuberculosis-in-south-africa.pdf.

Spelman, K., Burns, J. J., Nichols, D., Winters, N., Ottersberg, S., & Tenborg, M. (2006). Modulation of cytokine expression by traditional medicine: A review of herbal immunomodulators. *Alternative Medicine Review*, *11*(2), 128–150.

Street, R. A., & Prinsloo, G. (2013). Commercially important medicinal plants of South Africa: A review. *Journal of Chemistry*, *2013*, 1–16.

Suthar, A. B., Lawn, S. D., Del Amo, J., Getahun, H., Dye, C., Sculier, D., et al. (2012). Antiretroviral therapy for prevention of tuberculosis in adults with HIV: A systematic review and meta-analysis. *Plosmedicine.org*, *9*(7), 1–15.

TAG (Treatment Action Group). (2013). *2012 Report on tuberculosis research funding trends, 2005–2011*. Available from: http://www.treatmentactiongroup.org/tbrd2012.

Tasaduq, S. A., Singh, K., Sethi, S., Sharma, S. C., Bedi, K. L., Singh, J., et al. (2003). Hepatocurative and antioxidant profile of HP-1, a polyherbal phytomedicine. *Human and Experimental Toxicology*, *12*, 639–645.

Tasneen, R., Li, S., Peloquin, C. A., Taylor, D., Williams, K. N., Andries, K., et al. (2011). Sterilizing activity of novel TMC 207-and PA-824-containing regimens in a murine module of tuberculosis. *Antimicrobial Agents and Chemotherapy*, *55*(12), 5485–5492.

Tomlinson, C. (2011). *Streptomycin*. TBOnline. Available from: https://www.tbonline.info/posts/2011/9/1/streptomycin/.

Torrado, E., Robinson, R. T., & Cooper, A. M. (2011). Cellular response to mycobacteria: Balancing protection and pathology. *Trends in Immunology*, *32*(2), 66–72.

Tostmann, A., Boeree, M. J., Aarnoutse, R. E., de Lange, W. C. M., van der Ven, A. J. A.M., & Dekhuijzen, R. (2008). Antituberculosis drug-induced hepatotoxicity: Concise up-to-date review. *Journal of Gastroenterology and Hepatology*, *23*(2), 192–202.

Tosun, F., Kizilay, C. A., Sener, B., Vural, M., & Palittapongampim, P. (2004). Antimycobacterial screening of some Turkish plants. *Journal of Ethnopharmacology*, *95*, 273–275.

Trentin, D. S., Silva, D. B., Amaral, M. W., Zimmer, K. R., Silva, M. V., Lopes, N. O., et al. (2013). Tannins possessing bacteriostatic effect impair *Pseudomonas aeruginosa* adhesion and biofilm formation. *PLoS One*, *8*(6), 66257.

Udwadia, Z. F., Amale, R. A., Ajbani, K. K., & Rodrigues, C. (2012). Totally drug resistant tuberculosis in India. *Clinical Infectious Diseases*, *54*(4), 579–581.

Van Wyk, B. (2008). A broad review of commercially important southern African medicinal plants. *Africa, 119*, 342–355.

Verma, S., & Singh, S. P. (2008). Current and future status of herbal medicines. *Veterinary World, 1*(11), 347–350.

Wang, F., Sambandan, D., Halder, R., Wang, J., Batt, S. M., Weinrick, B., et al. (2013). Identification of a small molecule with activity against drug-resistant and persistent tuberculosis. *Proceedings of the National Academy of Sciences, 110*(270), 2510–2517.

Wauthoz, N., Balde, A., Balde, E. S., Van Damme, M., & Duez, P. (2007). Ethnopharmacology of *Mangifera indica* L. bark and pharmacological studies of its main C-Glucosylxanthone, Mangiferin. *International Journal of Biomedical and Pharmaceutical Sciences, 1*(2), 112–119.

Western Cape Government. (2012). *Tuberculosis*. Available from: http://www.westerncape.gov.za/eng/your_life/4502.

WGND, Working group on new TB drugs. (2014). Available from: http://www.newtbdrugs.org/pipeline/.

WHO, World Health Organization. (2016). *Tuberculosis report*. Available from: http://www.who.int/tb/publications/global_report/en/.

York, T. (2012). *An ethnopharmacological study of plants used for treating respiratory infections in rural Maputaland* (M.Sc. thesis). KwaZulu-Natal, South Africa: The University of Zululand.

Young, D. B., & Cole, S. T. (1993). Leprosy, tuberculosis, and the new genetics. *Journal of Bacteriology, 175*(1), 1–6.

Younossian, A. B., Rochat, T., Ketterer, J. P., Wacker, J., & Janssens, J. P. (2005). High hepatotoxicity of pyrazinamide and ethambutol for treatment of latent tuberculosis. *European Respiratory Journal, 26*, 462–464.

Zambrano, M. M., & Kolter, R. (2005). Mycobacterial biofilms: A greasy way to hold it together. *Cell, 123*(5), 762–764.

Chapter 8

Medicinal Plants Used in the Treatment of Superficial Skin Infections: From Traditional Medicine to Herbal Soap Formulations

Murunwa Madzinga, Quenton Kritzinger, Namrita Lall
University of Pretoria, Pretoria, South Africa

Chapter Outline

Medicinal Plants for Holistic Health and Well-Being. http://dx.doi.org/10.1016/B978-0-12-812475-8.00008-1

8.1 SUPERFICIAL SKIN INFECTIONS

Human skin is made up of the cellular epidermis, an underlying dermis and subcutis of connective tissues (Igarashi, Nishino, & Nayar, 2005, pp. 13–16; McGrath, Eady, & Pope, 2004). These layers consist of different types of cellular level elements resulting in different skin structures exhibiting different functions and various types of light propagation (Igarashi et al., 2005, pp. 13–16). The epidermis is the outermost layer of the skin and is composed of four sublayers: the innermost layer, the stratum corneum (horny cell layer); stratum granulosum (granular layer); stratum spinosum (prickle cell layer); and the stratum basale (basal cell layer) (Igarashi et al., 2005, pp. 13–16; McGrath et al., 2004). The soles of the feet and the hand palms have an extra layer referred to as stratum lucidum (clear layer), which is immediately after the stratum corneum (Igarashi et al., 2005, pp. 13–16). The dermis is the second layer of the skin, which is thicker than the epidermis and is made of collagen and elastin fibers (Igarashi et al., 2005, pp. 13–16; Venus, Waterman, & McNab, 2011). The dermis has a papillary layer (thin upper layer) and reticular layer (thicker lower layer) (Igarashi et al., 2005, pp. 13–16; Venus et al., 2011). The subcutis is the third and the last layer of the skin, which is not categorized as another layer. The constituents of this layer are mainly fats (Igarashi et al., 2005, pp. 13–16).

The skin is the largest organ of the human body, and it is constantly exposed to harmful chemical, physical, and biological pollutants. It provides primary defense against microbes as it acts as a major interface between environment and the body. The skin is colonized by a large number of microorganisms, most of which are harmless and even beneficial (Grice & Segre, 2011). Symbiotic microorganisms occupy a wide range of niches of the skin and provide protection against invasion by pathogenic microorganisms. The microbiota of the skin is dependent on the site of the skin. There is, however, a balance between the host and the microorganisms and disruptions in the balance can result in skin disorders or infections (Grice & Segre, 2011).

Skin infections are caused by a large number of microbes, including bacteria and fungi. The most common bacterial causal agents are *Staphylococcus aureus*, *Streptococcus pyogenes*, *Pseudomonas aeruginosa*, and the bacteroides group (Otang & Afolayan, 2015), with *S. aureus* being responsible for a number of cutaneous infections such as cellulites. The most common fungal agents responsible for skin infections are those associated with keratin, such as the members from the *Trichophyton* genus (Nweze, 2010; Otang & Afolayan, 2015).

Over the years, the occurrence of skin diseases has also increased in most parts of the world because of their association with the human immunodeficiency virus (HIV) and acquired immune deficiency syndrome (De Wet, Nciki, & van Vuuren, 2013). South Africa is one of the countries with the largest population of HIV-infected people, and it is also the country with the largest prevalence of HIV. More than 80% of people infected with HIV develop skin diseases at some stage of the HIV disease (Tschachler, Bergstresser, & Stingl, 1996).

In 2012, the World Health Organization (WHO) reported that burns have become one of the more prevalent problems of public health. In South Africa, burns are one of 15 major causes of death of children and young adults between the ages of 5 and 29, with over 19,000 fire-related deaths reported annually (De Wet et al., 2013). The victims of burns are susceptible to serious infection by *P. aeruginosa*. The hot and humid climatic conditions of most parts of South Africa also play a major role in the cases and occurrence of skin infections. Climatic factors are very problematic in sub-Saharan Africa where over 78% of the population has been said to have been infected with tinea capitis (De Wet et al., 2013).

In many countries a high incidence of skin infections is not regarded as a significant health problem because of the low level of severity and lethality. However, skin ailments present a major health concern because they occur worldwide and are often persistent and difficult to treat (Otang & Afolayan, 2015). They also have a significant impact on the quality of life. Skin diseases have been associated with increased stress, anxiety, and depression in people, and patients with severe cutaneous skin infections have increased suicidality diagnoses, especially patients of adolescence stage (De Wet et al., 2013; Smidt et al., 2010).

8.1.1 Common Superficial Fungal Infections (Mycoses)

Some in vivo studies done in the last few decades showed that the infecting dose for hairless skin is as little as six fungal conidia (Mendez-Tovar, 2010). Fungal infection is established as a multistep process. The fungus must first adhere to the substrate. Adherence is followed by germination of conidia, and the infection process ends with the invasion of the superficial skin layer, the stratum corneum. In vitro studies have shown that adherence of most dermatophytes to the skin takes at least 6 h, germination of most of the conidia takes at least 12 h, and the invasion of stratum corneum takes 72 h (Baldo et al., 2011; Mendez-Tovar, 2010). One of the organisms, with an exception of 6 h germination, is *Trichophyton mentagrophytes*, which starts germination at least 4 h postexposure (Baldo et al., 2011; Mendez-Tovar, 2010).

The adherence of dermatophytes is attributed to the possession of specific adhesins for some of the carbohydrates that are present on the skin (Mendez-Tovar, 2010). Once they adhere to a structure with keratin, the production and the release of enzymes such as keratinases, metalloproteases, and serine proteases take place. The released enzymes break bonds of the keratinized tissue, and some of these enzymes act as antigens that induce inflammation of varying degrees (Mendez-Tovar, 2010). Tissue damage is the combination of enzymatic action and a defense mechanism, which they activate.

For most dermatophytes a minimum infective dose must be contracted for the agent to establish infection (Moriello, 2003). The dose varies with the type of dermatophyte and also with the health status of the host. Some hosts may have predisposition factors such as stress and malnutrition for developing the infection (Chermette, Ferreiro, & Guillot, 2008; Moriello, 2003).

Fungal infections can be categorized as causing superficial, cutaneous, or subcutaneous infections (Garber, 2001). Subcutaneous fungal infections affect deeper layers of the skin and may also affect the blood stream. Superficial and cutaneous tissues are subjected to fungal infection when fungal organisms contaminate and colonize the epidermal surface of the skin and the hair follicles. Cutaneous infections affect the deep layers of the skin, and they are usually noninvasive. Dermatophytes affect the outermost layer of the skin thereby causing dermatophytosis (infection caused by dermatophytes) (Garber, 2001; Soares et al., 2013; Weitzman & Summerbell, 1995).

Dermatophytes are parasitic fungi dependent on keratin in the nails, skin, and hair (White, 2006, pp. 1–4). They break the keratin, and they are also capable of invading living tissues leading to certain injuries. Dermatophytes belong primarily to three genera according to the morphology and formation of their conidia, namely; *Trichophyton*, *Microsporum*, and *Epidermophyton* (Soares et al., 2013). They cause dermatophytoses such as tinea pedis (athlete's foot), tinea capitis (ringworm), and tinea unguium (nail fungal infections) (Stevens, 2001). *Candida albicans* is also a common microbe, which infects the skin, mouth, vagina, and the digestive tract. They have a high prevalence in African countries most probably because of the hot and humid environment (Nweze, 2010; Otang & Afolayan, 2015).

Dermatophytes can be zoophilic, geophilic, or anthropophilic depending on what is its reservoir (Outerbridge, 2006). Zoophilic dermatophytes are adapted to living on animals and rarely reside in the soil but have the ability to infect humans. The most common zoophilic dermatophyte that infects dogs and cats is *Microsporum canis*. Geophilic species reside in the soil and infect both humans and animals, with *Microsporum gypseum* commonly infecting dogs and cats. Anthropophilic dermatophytes are human pathogens but can infect animals having humans as reservoirs (Outerbridge, 2006).

Human infection occurs through the direct contact with infected individuals as well as contact of contaminated products or specimens such as hair, animal epidermal scales, and soil. Sharing personal belongings such as combs, shoes, and clothing also play a crucial role in the transmission of causal agents (Mendez-Tovar, 2010). Animal infection usually occurs through the contact of contaminated soil. This is mostly true for animals that spend most of their time outdoors (Chermette et al., 2008).

Some parts of the body are more susceptible than others and this is due to anatomical and environmental factors. The fungi under the nail area remain in contact with the skin for a long time, and this allows it to grow and invade the keratin layers eventually (Mendez-Tovar, 2010). The sweat, maceration, and alkaline pH between the fingers give the fungi a perfect environmental condition for development. Conidia on the scalp are protected by hair and remain in contact with the skin for quite a long time. The conditions on the groin area also favor the growth of dermatophytes as the joint of the scrotum and thigh provides the fungi with a moist, nutritious, and warm environment (Mendez-Tovar, 2010).

8.1.2 Common Bacterial Skin Infections

Most skin bacteria fall into four phyla, namely; Actinobacteria, Firmicutes (*Bacillus* genus), Bacteroides (such as *Bacteroides fragilis*, *B. vulgatus*, and *B. distasonis*), and the Proteobacteria (such as microbes of the *Escherichia* and *Helicobacter* genera). The Actinobacteria phylum includes most of the prevalent microbes in normal skin (Grice et al., 2009).

Bacterial skin infections are the 28th most common diagnoses of hospitalized patients (Stulberg, Penrod, & Blatny, 2002). The most common bacterial skin infections are cellulitis, impetigo, and folliculitis (Stulberg et al., 2002). Cellulitis is an erythematous infection of the dermis and subcutaneous tissues, and it is most common near the breaks of the skin. Impetigo is the infection that causes lifting of the stratum corneum resulting in a noticeable bullous (blister) effect, and it is most common in children of age two to five. Folliculitis affects the hair follicles, and it is characterized by the depth of the affected hair follicle (Stulberg et al., 2002).

The mechanism of infection development used by bacteria is the same as that of fungi (Section 8.1.1). The bacteria must first adhere to the host cells. Once adherence is accomplished, the bacteria must invade the tissue, and finally it must elaborate toxins. Two types of toxins are secreted, namely; the endotoxins and the exotoxins. Endotoxins are lipopolysaccharide chains and are usually found in the cell walls of Gram-negative bacteria. These are responsible for eliciting immune response, and their oversecretion usually leads to progressive skin infection resulting in necrotizing fasciitis (the extensive necrosis of subcutaneous tissues) (Ki & Rotstein, 2008; Luksich, Holds, & Hartstein, 2002). Exotoxins are actively produced proteins that are responsible for tissue damage and tissue dysfunction (Ki & Rotstein, 2008).

Tissue damage may be a result of enzymatic reactions, cellular dysregulation, or pore formation. Virulent *S. aureus* and *S. pyogenes* secrete a special group of exotoxins referred to as superantigens (Commons et al., 2014; Ki & Rotstein, 2008). They bind the conserved portions of T-cell receptors thereby activating T lymphocytes. The release of a large amount of cytokines leads to a grossly exaggerated inflammatory response, which may lead to severe tissue necrosis (Ki & Rotstein, 2008).

8.1.3 Prevalence of Skin Infections Worldwide

About 10% of hospitalized patients are affected by skin infections worldwide (Ki & Rotstein, 2008). Superficial mycoses are common worldwide and are believed to affect 20%–25% of the world's population, and the incidence continues to increase making them one of the most frequent forms of infection

(Ameen, 2010; Havlickova, Czaika, & Friedrich, 2008). The WHO estimates the world's occurrence of dermatophytosis at about 20% (Sharma, Kumawat, Sharma, Seth, & Chandra, 2015).

8.1.4 Prevalence of Skin Infections in Africa

The prevalence of skin infections in Africa varies among countries. The lowest prevalence in African countries was reported in Bamako, Mali (11.7%) and the highest prevalence in rural Ethiopia (48%) (Shrestha, Gurung, & Rosdahl, 2012). Research conducted in Nigeria on the prevalence of superficial fungal infections has shown that infection percentage range from 3.4% to 55% among the Nigerian regions (Oke, Onayemi, Olasode, Omisore, & Oninla, 2014).

8.1.5 Current Treatments of Skin Infections

Treatment therapy of fungal skin infections (dermatophytoses) involves the use of antifungal azoles, which are divided into two groups; the imidazoles and triazoles. Imidazoles are strictly used for superficial mycoses, whereas the triazoles have broader applications (Soares et al., 2013). Some of the antidermatophytic treatments involve the use of griseofulvin, terbinafine and also lufenuron (Moriello, 2003).

Griseofulvin is an orally administered antibiotic and has been used exclusively to control the development of dermatophytes because it presents only fungistatic action. This drug interferes with polymerization and nucleic acid synthesis of spindle microtubules causing abnormalities in cell division (Moriello, 2003; Soares et al., 2013). Terbinafine is an allylamine, lipophilic drug, which inhibits the fungal growth by inhibiting the biosynthesis of ergosterol through the inhibition of the fungal enzyme, squalene epoxidase (Castanon-Olivares, Manzano-Gayoso, Lopez-Martinez, De la Rosa-Velazquez, & Soto-Reyes-Solis, 2001; Moriello, 2003). It is an orally administered drug that reaches the stratum corneum and may persist there for weeks. Lufenuron is a benzoylphenylurea drug, which works by disrupting chitin synthesis. Chitin is an important component of the outer layer of cell wall of fungi (Moriello, 2003).

Cephalosporins are usually used to treat bacterial skin infections and are effective against most bacterial skin pathogens including the methicillin-resistant *S. aureus* (MRSA). Intravenous vancomycin is usually the first line of therapy against MRSA (Archer, 1998). Linezolid is also another antibiotic that has demonstrated activity against antibiotic-resistant positive organisms. Linezolid selectively binds to the 50S ribosomal unit and by so doing it prevents formation of the initiation complex, which prevents the cross-resistance with antimicrobial agents (Weigelt et al., 2005).

Antibiotics become ineffective over time as the microbes gain resistance against them (Dyer, Gerenraich, & Wadhams, 1998; Kownatzki, 2003). Cellular targets for antifungal drugs are limited as there is an existence of similarity between fungi and the host. Commonly used drugs achieve therapeutic effect by targeting biosynthetic pathways and therefore there is a target-specific resistance mechanism (Soares et al., 2013). The dermatophytic fungus achieves resistance by mutating the gene, which produces proteins targeted by the treating agents, for example, a terbinafine resistance has been reported, which involves the mutation of squalene epoxidase gene that results in the substitution of amino acid 393 from leucine to phenylalanine (Soares et al., 2013). By the year 1998, *S. aureus* had already gained resistance against all antibiotics except vancomycin (Levy, 1998).

Resistance may be due to acquired genes that destroy antibiotics before they have any effect due to the acquired efflux pumps that extrude antibacterial agents form the cell before they reach their target site (Tenover, 2006). Alternatively the bacteria may acquire or mutate an already existing gene so that the cell no longer has the binding site for a specific antibacterial agent (Tenover, 2006). In general, resistance may be through mutation and selection or by acquiring it from other bacteria (Tenover, 2006).

Plant extracts are more diluted than the pure compounds, which are isolated and used in most pharmaceuticals and cosmetics. The therapeutic compound in plant extracts is found in a pool of compounds and usually in a complex containing other compounds that mask its toxicity, and this results in plant extracts being less toxic than synthetic compounds and also isolated compounds (Aburjai & Natsheh, 2003). This therapeutic complex is environmental friendly as it is the same way that nature made it; it is therefore easily degraded and also possesses less to no toxicity. There is a high probability that a plant extract will contain additional active principles that may be chemically and therapeutically related to the primary constituent of interest (Aburjai & Natsheh, 2003).

8.2 SOUTH AFRICAN PLANTS USED IN THE TREATMENT OF BACTERIAL AND FUNGAL SKIN INFECTIONS

South Africa boasts a unique and diverse botanical heritage with ~30,000 plant species of which 3000 species are recorded and used as *materia medica* (Germishuizen & Meyer, 2003; Van Vuuren, 2008). In addition to the unique botanical diversity, South Africa has a huge cultural diversity with traditional healing being integrated into each ethnic group.

Medicinal plants contribute significantly to the primary health of the population of South Africa, and they have been used by people of rural communities for decades throughout the country for the treatment of different ailments including skin diseases (De Wet et al., 2013; Otang & Afolayan, 2015). Furthermore, the use of herbal medicine to treat animal diseases is widespread among the small-scale farmers of South Africa most probably owing to availability, low cost, and convenience (McGaw & Eloff, 2008). In traditional medicine, different methods exist regarding herbal preparation. For example, grinding the plant material to a fine powder and making a paste is the most common preparation for the treatment of skin infections and superficial burns (Coopoosamy & Naidoo, 2012).

Numerous reports, albeit in literature or through personal communication, exist of plants indigenous to South Africa that have been used to treat skin infections in humans and animals. For example, the two plants of the Asteraceae family, namely *Bidens pilosa* Linn. and *Bidens bipinnata* Linn. are used to dress wounds by Vhavenda people of the Limpopo province, north of South Africa (personal communication). The plants have also been reported to have antioxidant and antibacterial activity (Bartolome, Villasenor, & Yang, 2013). The fruits of *Solanum incanum* Linn. are also used by the same people for the treatment of ringworm, and the stems of *Synadenium cupulare* (Boiss.) LC Wheeler. are used for treatment of skin ailments of livestock (personal communication). In the following section of this chapter, selected South African plants used in the treatment of skin infections will be discussed in further detail.

8.2.1 *Elephantorrhiza elephantina* (Burch.) Skeels (Eland's Bean)

The Eland's bean (Fig. 8.4A) belongs to the Fabaceae family and has a number of unbranched, annual stems of about 1 m growing from an underground red rhizome. It has clusters of small, cream-colored flowers placed halfway up the aerial stem. It has pods of about 200 mm long as the characteristic feature of the plant. It is found across most of the South African grasslands (Van Wyk, Van Oudtshoorn, & Gericke, 1997).

Roots and rhizome are boiled in water for external use to treat acne. It is used in combination with *Pentanisia prunelloides* Walp. to treat eczema and acne (Mabona, Viljoen, Shikanga, Marston, & Van Vuuren, 2013; Van Wyk & Gericke, 2000). The root is grated and immersed in water for about 24 h and strained afterward and ready for external use. It is known to form a protective layer on the skin and the mucosa. This property has been linked to the enhancement of tissue regeneration (Van Wyk et al., 1997). Phytochemical screenings of this plant indicated the presence of catechins in rhizomes and tannins as the active ingredients (Maphosa, Masika, & Moyo, 2009; Van Wyk et al., 1997). The presence of phenolic compounds is reflected by the red color of the rhizome (Maphosa et al., 2009; Van Wyk et al., 1997). In 2009, Maphosa et al. reported that the plant causes a decrease in respiratory rate when ingested.

8.2.2 *Melianthus comosus* Vahl (Honey Flower)

Melianthus comosus (Fig. 8.4B), belonging to Melianthaceae family, is a multi-branched shrub growing up to 3 m in height (Van Wyk et al., 1997). Every part of this plant gives off a strong, unpleasant odor when disturbed or bruised. This is the character that gives it its Afrikaans common name "kruidjie-roer-my-nie," meaning herb-touch-me-not (Harris, 2004; Van Wyk et al., 1997). The leaves of this plant are clustered at the tips of the branches and have about five pairs of leaflets with toothed margins (Fig. 8.4C). The flowers are small and green with small, reddish petals that are borne in short clusters and followed with a four-winged bladdery capsule (Van Wyk et al., 1997). The plant is restricted to southern African parts with a wide distribution in the dry interior of South Africa (Van Wyk et al., 1997).

Leaf and root poultices and decoctions are topically used to treat septic wounds, ringworm, and also to reduce swellings (Mabona et al., 2013). The plant is also traditionally used to treat snakebite; this forms part of the Khoi healing culture (Van Wyk et al., 1997).

In 2000, Kelmanson et al. conducted a study on the Zulu medicinal plants. They studied the antibacterial activity of South African plant species using aqueous, methanolic, and ethyl acetate plant extracts. They found that *M. comosus* had the highest antibacterial activity against the Gram-negative bacteria, *Escherichia coli* and *P. aeruginosa.* They also found that methanolic extracts were more active than other extracts (Kelmanson, Jager, & van Staden, 2000).

The plant contains a number of toxic bufadienolides such as melianthusigenin, which have been reported to have caused human death (Van Wyk et al., 1997). The root contains considerable amounts of triterpenoids, oleanolic acid, and cinnamic acid derivatives. Not much is known about the healing properties of this plant, but it is believed to be due to the presence of triterpenoids as some triterpenoids are used in wound-healing therapy (Van Wyk et al., 1997).

8.2.3 *Dicoma anomala* Sond. (Fever Bush)

The fever bush (Fig. 8.4D) is a perennial herb, and it grows up to 0.3 m at an altitude of 152–2200 m. It grows on hillsides, stony places, and on flat grasslands (Netnou & Herman, 2003; Van Wyk & Gericke, 2000). It has erect stems thinly covered by hairs, and these stems emerge from woody rootstock that are also thinly covered by hairs (Balogun & Tom Ashafa, 2016). It is distributed throughout the African continent including Lesotho. It is found in all provinces of South Africa excluding the Eastern Cape and Western Cape (Balogun & Tom Ashafa, 2016; Netnou & Herman, 2003; Tselanyane, 2006). *Dicoma anomala* belongs to one of the largest plant families, the Asteraceae, which synthesizes secondary metabolites such as terpenoids, flavonoids, and polyacetylenes (Bremer, 1994, p. 277; Tselanyane, 2006).

The charred roots, stems, and the leaf paste of this plant are used traditionally as an antiseptic for wounds and also for treatment of ringworm by the Basotho people of Lesotho and by Basotho tribe of eastern Free State in treatment of dermatosis (Balogum & Tom Ashafa, 2016; Van Wyk & Gericke, 2000). The tuber of the plant is sometimes used in combination with *E. elephantina* to treat acne (Mabona et al., 2013). Mabona et al. (2013) reported noteworthy antimicrobial activities of the organic extract of this plant against *S. aureus*, and *T. mentagrophytes* (one of the causal agents of dermatophytosis) giving validation of its traditional use as a treatment of ringworm. Very little information has been reported on the pharmacology and the bioactive compounds of *D. anomala.* Similarly, little attention has been paid on the toxicity of this plant, but the continuous high-dose intake has been reported to alter hepatic and thyroid functions (Balogun & Tom Ashafa, 2016).

8.2.4 *Kigelia africana* (Lam.) Benth. (Sausage Tree)

The sausage tree (Fig. 8.4E) (Bignoniaceae family) is widely distributed in southern, central, and western Africa and strictly distributed in the northern and north-eastern parts of South Africa (Joffe, 2003; Van Wyk et al., 1997). This large, rounded tree has smooth gray bark and a thick trunk. The leaves have 7–11 leaflets with the terminal one being the largest. It has huge grayish-brown fruits like a sausage (of up to a meter) that hangs over a long fibrous stalk and have a number of seeds embedded in them (Fig. 8.4F) (Van Wyk et al., 1997). It not only has dark maroon flowers, which are very large and attractive, but also has an unpleasant smell (Owolabi, Omogbai, & Obasuyi, 2007; Van Wyk et al., 1997).

Traditional healers of South Africa have been using the sausage tree for treatment of various skin ailments including fungal infections (Owolabi et al., 2007). The ash of fruit of this plant has been used for treatment of eczema, and ringworm. Grace et al. (2002) found that the fruit has anti-*S. aureus* activity, which is implicated in atopic eczema.

Owolabi et al. (2007) conducted a study of the bark of the sausage tree using ethanolic extracts and aqueous extracts to evaluate the antibacterial and antifungal activity against *E. coli*, *P. aeruginosa*, *S. aureus*, and the yeast, *C. albicans*. The authors found that the ethanolic extracts had antifungal and antibacterial activity against *S. aureus* and *C. albicans*, and the water extracts showed no activity at all. It is likely that the antimicrobial activity is due to the presence of kigelin (Fig. 8.1) and related compounds, which are available in the fruits. The bark together with roots has been reported to contain dihydroisocoumarin kigelin as a major compound and is the main compound responsible for the beneficial effect of the external use of the bark (Van Wyk et al., 1997).

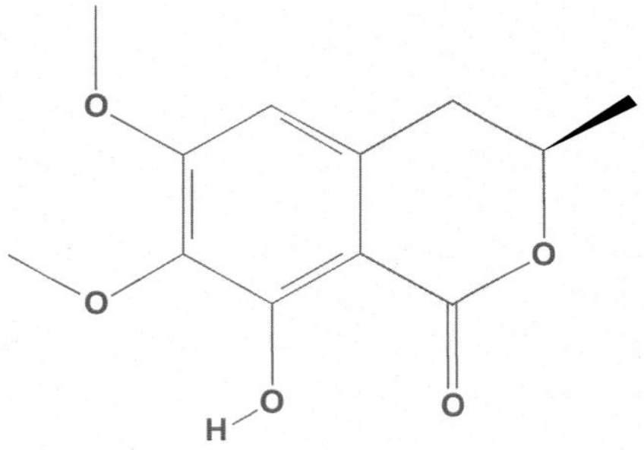

FIGURE 8.1 Chemical structure of Kiegelin, a compound found in the fruit of *Kigelia africana* that exhibits antimicrobial activity against *Escherichia coli*, *Pseudomonas aeruginosa*, *Staphylococcus aureus*, and *Candida albicans*.

8.2.5 *Ekebergia capensis* Sparrm. (Cape Ash)

Cape ash (Meliaceae) is a medium-sized tree, which grows to a height of about 10–12 m. It has a gray and rough bark, which peels off in thick flakes. It has white to pinkish flowers in spring, which end off as rounded, reddish, and fleshy fruits in summer (Fig. 8.4G). It is distributed across the eastern parts of South Africa (Ngeny, Magri, Mutai, Mwikwabe, & Bii, 2013; Van Wyk et al., 1997).

It has been reported that a powdered bark infusion is sometimes mixed with flour for treatment of abscesses, boils, and acne (Grierson & Afolayan, 1999; Van Wyk et al., 1997).

The antimicrobial activity of the bark of this plant was evaluated by Ngeny et al. (2013) using organic [hexane, dichloromethane (DCM), methanol] and water extracts. The extracts were tested against bacteria *P. aeruginosa* ATCC 27853, *S. aureus* ATCC 25923, *E. coli* ATCC 25922, *Klebsiella pneumoniae*, MRSA (clinical isolate), and the fungi *T. mentagrophytes*, *M. gypseum* (clinical isolate), *C. albicans* ATCC 90028, *Candida krusei* ATCC 6258, *Candida parapsilosis* ATCC 22019, and *Cryptococcus neoformans* ATCC 66031. The DCM extract was found to be active against *S. aureus*, whereas methanolic extracts had activity against MRSA, *P. aeruginosa*, *T. mentagrophytes*, and *M. gypseum*. The aqueous extracts were active against MRSA, *P. aeruginosa*, and *M. gypseum*. This validated the efficacy of this plant against most skin pathogens (Ngeny et al., 2013).

8.2.6 *Diospyros mespiliformis* Hochst. ex A.DC. (African Ebony, Jackleberry)

This tree, which belongs to the Ebenaceae family, grows very tall up to a height of 24 m. The bark is dark brown when the tree is still young and turns gray as it matures. The young leaves and twigs are covered in downy hair, and leaves are rarely lost from the young tree (Fig. 8.4H and I) (Mutshinyalo, 2007). The mature leaves are glossy, leathery, and dark green (Mutshinyalo, 2007; Orwa, Mutua, Kindt, Jamnadass, & Anthony, 2009). It has small inconspicuous flowers. It is usually found in the woodlands, savannah regions of South Africa, and most African countries, growing on termite mounds (Mutshinyalo, 2007; Orwa et al., 2009).

The root and the leaf decoctions are traditionally used for the treatment of scars, skin rashes, bruises, wounds, and ringworm (Van Wyk & Gericke, 2000; Van Wyk et al., 1997; Von Koenen, 2001, p. 113). The validation of its use in treatment of ringworm was provided by Adeniyi, Odelola, and Oso (1996) and by Mabona et al. (2013) who found the plant to have a minimum inhibition concentration of 0.10 and 0.50 mg/mL against *T. mentagrophytes* and *M. canis*, respectively.

8.2.7 *Aloe ferox* Mill. (Bitter Aloe)

This aloe (also referred to as *Aloe candelabrum* A. Berger) (Xanthorrhoeaceae, previously Aloaceae) is a polymorphic, robust plant with very dry leaves attached to a single stem and widely distributed along the eastern parts of South Africa (Chen, Van Wyk, Vermaak, & Viljoen, 2012; Van Wyk et al., 1997). It has dry fleshy leaves that are dull green to reddish and dark brown spines. The "ferox" in the name refers to the thorny sharp spines of the leaves (Chen et al., 2012). The bright red to orange flowers of this plant appears from May to August and appears to be arranged in erect, candle-shaped clusters (Van Wyk et al., 1997).

Aloe ferox (Fig. 8.4J and K) is used by the South African people especially the Khoisan for the treatment of eczema, skin irritations, bruises, acne, and burns (Van Wyk, 2008; Van Wyk et al., 1997). The inner leaf parenchyma has become popular in skin care products (Van Wyk, 2008). The gel of this plant is different from the gel of *Aloe vera* Linn. (True aloe), but there is limited information about the compounds of its gel (Chen et al., 2012). Because of the scarcity and importance of bitter aloe, *A. candelabrum* A. Berger has been included into the regime of the traditional medicine to achieve the same effects that are achieved by the cape aloes. *A. candelabrum* has the same morphology and chemical components as the cape aloes with differences lying in their gel having different phenolic compounds (Chen et al., 2012).

Barrantes and Guinea (2003) illustrated that cape aloes are rich in aloins (Fig. 8.2), which have the ability to inhibit collagenase and metalloprotease activity. This led to the investigation of the effects of whole leaf juice on wound healing and repair of the skin. The study found that whole leaf juice accelerates wound closure and inhibits microbial growth selectively (Barrantes & Guinea, 2003). There is minimal data concerning the toxicity of products containing the extracts of this plant (Chen et al., 2012). In 2015, Abosede et al. evaluated the cytotoxicity of this plant using the brine shrimp (*Artemia salina* L.) assay. It was found that the ethanolic, aqueous, and acetone root extracts together with ethanolic and aqueous leaf extracts were mildly toxic to the brine shrimp. The

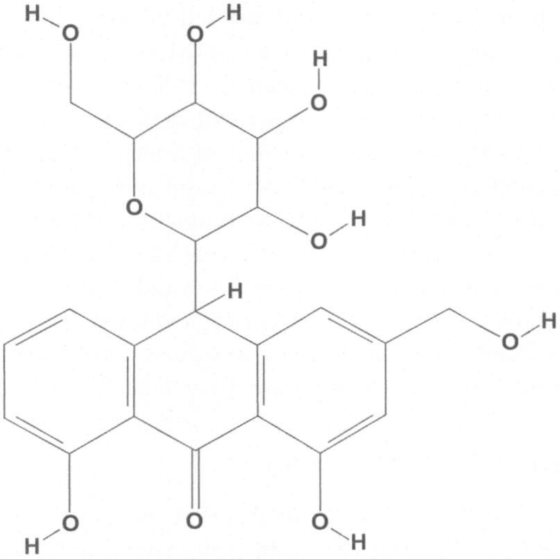

FIGURE 8.2 Chemical structure of aloin, a compound isolated from various Aloes with the capacity to inhibit metalloproteinases.

highest lethality recorded was that of aqueous leaf extract (4.7%), and lowest was that of acetone root extract (0.7%). However, in a study by Mwale and Masika (2012), as reported by Abosede et al. (2015), it was found that the aqueous extract exhibited no toxic effects on the rats at all grade tested. There have been reports of allergic and hypersensitive responses to some products. There has also been a report of the development of acute oliguric renal failure and liver dysfunction in a 47-year-old man from Soweto, South Africa, following ingestion of herbal tea containing cape aloe (Chen et al., 2012; Luyckx et al., 2002).

8.2.8 *Warburgia salutaris* (G. Bertol.) Chiov. (Pepper-Bark, Fever Tree)

The pepper-bark tree (Canellaceae) grows up to 10 m in height (Leonard & Viljoen, 2015). It has a rough, mottled bark with a reddish inner side (Van Wyk et al., 1997). The flowers are greenish-yellow and produced between the leaves. The fruits (Fig. 8.4L) are round and green and contain a number of seeds, which are flat (Van Wyk et al., 1997). *Warburgia salutaris* is restricted to the north-eastern parts of South Africa and has been exploited for its bark (Mbambezeli, 2004; Van Wyk et al., 1997). The epithet salutaris means salutary to health, reflecting its first medicinal use as tonic (Leonard & Viljoen, 2015).

The bark (Fig. 8.4M) of the pepper-bark tree has been used by the Venda people, north of Limpopo, for treatment of skin ailments of both animals and humans (personal communication—Makondelela, V). The powdered bark has been reported to be mixed with fats and applied topically to the infected area (Leonard & Viljoen, 2015). Mabona et al. (2013) found that the organic extracts of the leaves also provide the antimicrobial activities provided by the bark at a low concentration but enough to inhibit skin pathogens.

Tablets have also been produced from the leaves of this plant, such as those manufactured by Phyto Nova and used in treatment of oral thrush (Leonard & Viljoen, 2015; Van Wyk, 2008). It has been used as a natural antibiotic and thought to be effective against thrush (Van Wyk, 2008). The bark of this tree contains a large amount of drimane sesquiterpenoids such as polygodial (Fig. 8.3). Polygodial has been used to adjust antibiotics and antifungals having

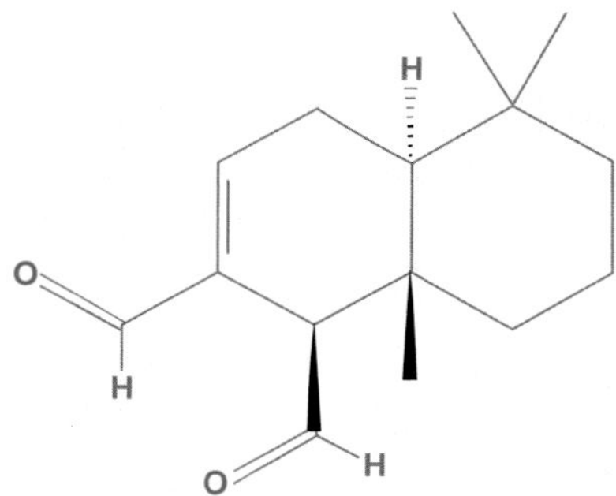

FIGURE 8.3 The chemical structure of polygodial, a compound isolated from *Warburgia salutaris*, which has been used to adjust antimicrobial compounds exhibiting poor permeability.

FIGURE 8.4 Photographs showing the morphological features of selected plants used in the treatment of skin infections: (A) *Elephantorrhiza elephantina**; (B and C) Leaves and fruits of *Melianthus comosus*∞; (D) Flowers and leaves of *Dicoma anomala**; (E and F) Tree and fruits of *Kigelia africana**; (G) Fruits of *Ekebergia capensis*∞; (H and I) Leaves of *Diospyros mespiliformis**∞; (J and K) *Aloe ferox*∞; (L and M) *Warburgia salutaris* fruits and bark∞. *Photo credits: *H.G.W.J. Schweickerdt Herbarium, University of Pretoria and ∞Wikimedia Commons.*

a poor permeability (Van Wyk, 2008; Van Wyk et al., 1997). Polygodial has been demonstrated by Kubo and Taniguchi (1988) to be more fungicidal than fungistatic (reported by Leonard & Viljoen, 2015). In 2010, Paiva et al. reported that the sesquiterpenoids are the compounds responsible for the anti-*Bacillus subtilis*, anti-*K. pneumoniae*, and anti-*S. aureus* activity of this plant.

The toxicity of this plant is poorly investigated, but the *Warburgia* genus has, however, been contraindicated in pregnancy, could lead to abortion (Leonard & Viljoen, 2015).

8.2.9 *Senecio serratuloides* DC. (Two-Day Plant)

The two-day plant is an herbaceous perennial plant having erect stems sprouting from a woody rootstock and growing up to a 1 m in height (Van Wyk et al., 1997). It has characteristic leaves with serrated margins, which grow up to 60 mm. Its small, yellow flowers are borne in sparse clusters toward the end of the branches. The plant is distributed in South African areas having summer rainfall (Van Wyk et al., 1997).

The lay people of Maputaland, the northern region of Kwazulu-Natal, are among the South African people who have been using medicinal plants for the relief of skin infections (De Wet et al., 2013). In 2013, De Wet et al. conducted an ethnobotanical survey in this area and reported that the lay people have over 40 plant species that they use for over 10 skin ailments including ringworm, acne, and rashes. The plants reported belonged to 35 families, with most of them residing within the Fabaceae family, which is known worldwide for its wound-healing effects (De Wet et al., 2013; Dey, Gupta, & De, 2012). The most frequently used plant was *Senecio serratuloides*, and this plant is a well-recorded South African plant used in speeding up wound healing as it is used as antiseptic (De Wet et al., 2013).

The leaves of this plant are usually applied directly to the skin, and sometimes powdered burnt leaves are used (Van Wyk et al., 1997). The leaves are macerated and used for bathing to treat skin ailments such as rash and eczema and also used for wound healing (Nciki, 2015). It is usually used in combination with *Drimia delagoensis* (Baker) Jessop (Nciki, 2015). There is minimal information regarding its efficacy, and the plant may also have hepatotoxic pyrrolizidine alkaloids, which have been linked to serious cell damage and induction of the formation of tumors in the liver and lungs when the plant is used regularly (Gould, 2015, p. 35; Van Wyk et al., 1997).

8.3 HERBAL SOAPS INFUSED WITH PLANTS EXTRACTS

Of the estimated number of medicinal plants used in South Africa only a few are available commercially as teas, tablets, ointments, and such, but not much as infusions in herbal soaps (Street & Prinsloo, 2013; Van Wyk, 2008). Soap acts as an emulsifier or a surfactant that softens the horny layer of the epidermis and also as a germicide that increases the permeability of the microbial envelope resulting in a disruption of the integrity of microbial cells (Esimone, Nworu, Ekong, & Okereke, 2007).

Most of the commercially available herbal soaps infused with South African plants are for treatment of uneven toned skin and acne but not for other skin infections. A soap of African extracts is one such soap used for uneven toned skin (http://www.africanextracts.com). It is infused with *rooibos*, *Aspalathus linearis* (Burm.f.) R. Dahlgren. Rondavel is an example of another soap, which is infused with honeybush, *Cyclopia genistoides* (L) R. Br.

(http://www.rondavelsoaps.co.za). In this section some of the *traditional* soaps infused with plant extracts from across the world are discussed.

Osedudu is a soft soap manufactured in Nigeria and is mostly used by the Yorubas of south western Nigeria (George et al., 2006). Some of the formulations are also in Nigerian open markets and are referred to as Black soaps. The soaps are used as toiletries and also employed as the basis for a number of herbal formulations intended for topical applications on ulcers, lesions, and skin infections (Moody, Adebiyi, & Adeniyi, 2004). Some of the plants infused in these soaps by traditional healers are *A. vera* and *Ageratum conyzoides* Linn. These two plants are incorporated for treatment of boils, ringworm, body odor, and wounds (Moody et al., 2004). In 2004, Moody et al. evaluated the antimicrobial activity of the Osedudu soaps by testing them against *P. aeruginosa*, *E. coli*, *S. aureus*, *C. albicans, Aspergillus clavatus*, and *Trichophyton* species. The authors found that the soaps infused with plant extracts lost activity against tested microbes as compared with soaps without plant extracts. This may be because of interactions between secondary metabolites of plants and other ingredients of the soap (Moody et al., 2004).

In eastern Nigeria, some of the plants with the ability to foam have been employed as soap for bathing and for treating skin and wound infections. One of these plants is *Cassia alata* Linn. *C. alata* contains secondary metabolites such as saponins, phenols, alkaloids, flavonoids, and anthraquinone glycosides, which are known to possess surface activity (Esimone et al., 2007). In 1981, Benjamin and Lamikanra (as reported by Esimone et al., 2007) found that *C. alata* have excellent wound-healing properties. Esimone et al. (2007) evaluated an herbal soap containing ethanolic extract of *C. alata* (95% ethanol) by testing the soap for the activity against *S. aureus*, *B. subtilis*, *E. coli*, *P. aeruginosa*, and *C. albicans*. The authors found that the soap had an excellent activity against all of the tested microorganism with an interesting activity against *S. aureus* with inhibition zone diameter of 18.5 mm.

Curcuma longa L. (turmeric) is a commonly used spice and household medicine in Songkhla, Thailand. It has 7% yellow oil, which is volatile and includes turmerone and zingiberene (Ungphaiboon et al., 2005). The rhizome of turmeric has been used by the old Thai to treat peptic ulcers and skin infections among others. Ungphaiboon et al. (2005) tested a clear herbal soap containing 0.5% of turmeric extract against *B. subtilis*, *S. aureus*, *C. albicans*, and *C. neoformans*. It was found that the soap infused with turmeric had inhibitory zone diameters (IZD) of 17, 14.2, 21.2, and 28.5 mm against *B. subtilis*, *S. aureus*, *C. albicans*, and *C. neoformans*, respectively, soon after the soap was prepared. The activity decreased as the storage time of the soap increased; after 12 months of storage the soap had an IZD of 15.3, 10, 19.3, and 19 mm against *B. subtilis*, *S. aureus*, *C. albicans*, and *C. neoformans*, respectively.

In 2013, Ibeneghu assessed 20 herbal soaps (constituents of the soaps are listed in Table 8.1) used for their antimicrobial activity against bacteria and yeast responsible for some skin infections. All of the tested soaps had activity against Gram-positive organisms (*Staphylococcus epidermidis*, *S. aureus*, and

TABLE 8.1 The Constituents of the Twenty Herbal Soaps Used in Nigeria

Soap Code	Constituents
A–K, S and T	Undisclosed
L, M, and N	*Aloe vera*, camwood extract, cocoa-pod ash solution, honey, palm kernel oil, palm-bunch ash solution, and shea butter
O	*Aloe vera*, avocado oil, honey, lime juice, and palm kernel oil
P	*Aloe vera*, honey, and palm kernel oil
Q	*Aloe vera*, camwood extract, cocoa-pod ash solution, honey, palm kernel oil, palm-bunch ash solution, shea butter, and lime juice
R	*Aloe vera*, camwood extract, cocoa-pod ash solution, honey, palm kernel oil, palm-bunch ash solution, shea butter, and lime juice

Adapted from Ibeneghu, O.A. (2013). The antimicrobial assessment of some Nigerian herbal soap. *African Journal of Traditional Complementary Medicine, 10*(6), 513–518.

B. subtilis) over Gram-negative (*E. coli*) and had no activity against the yeasts, *C. albicans*, and *Candida pseudotropicalis*.

8.4 CONCLUSION

Despite the well-documented ethnobotanical literature, there is limited scientific information available regarding the efficacy of herbal soaps in prevention and treatment of skin infections. More work beyond the laboratory validation of medicinal plants is still needed. The use of extracts may help reduce the infection cases, the spread, and the resistance of microbes as plant extracts usually contain additional active principles, which are chemically and therapeutically related to the primary constituent of interest resulting in low probability of resistance as there will be a number of actives. The use of leaf extracts are more sustainable in the long run than the extracts of other plant parts as the plant recovers more quickly from leaf loss than the loss or disruptions of the bark and the roots. The potential does indeed exist for the development of new and possible more effective drugs for the treatment of superficial skin infections.

REFERENCES

Abosede, W. O., Sunday, A., & Jide, A. A. (2015). Toxicological investigations of *Aloe ferox* Mill. extracts using Brine shrimp (*Artemia salina* L.) assay. *Pakistan Journal of Pharmaceutical Sciences*, *28*(2), 635–640.

Aburjai, T., & Natsheh, F. M. (2003). Plants used in cosmetics. *Phytotherapy Research*, *17*, 987–1000.

Adeniyi, B. A., Odelola, H. A., & Oso, B. A. (1996). Antimicrobial potential of *Diospyros mespiliformis* (Ebenaceae). *African Journal of Medicine and Medical Science*, *25*, 22–24.

Ameen, M. (2010). Epidemiology of superficial fungal infections. *Clinic in Dermatology, 28*, 197–201.

Archer, G. L. (1998). *Staphylococcus aureus*: A well-armed pathogen. *Clinical Infectious Diseases, 26*, 1179–1181.

Baldo, A., Monod, M., Mathy, A., Cambier, L., Bagut, E. T., Defaweux, V., et al. (2011). Mechanisms of skin adherence and invasion by dermatophytes. *Mycoses, 55*(3), 218–223.

Balogun, F. O., & Tom Ashafa, A. O. (2016). Acute and subchronic oral toxicity evaluation of aqueous root extract of dicoma anomala Sond. in Wistar rats. *Evidence-Based Complementary and Alternative Medicine, 2016*.

Barrantes, E., & Guinea, M. (2003). Inhibition of collagenase and metalloproteinases by aloins and aloe gel. *Life Sciences, 72*, 843–850.

Bartolome, A. P., Villasenor, I. M., & Yang, W. C. (2013). *Bidens pilosa* L. (Asteraceae): Botanical properties, traditional uses, phytochemistry and pharmacology. *Evidence-Based Complementary and Alternative Medicine, 2013*, 1–51.

Bremer, K. (1994). *Asteraceae: Cladistics and classification*. Portland: Edition Timber Press.

Castanon-Olivares, L. R., Manzano-Gayoso, P., Lopez-Martinez, R., De la Rosa-Velazquez, I. A., & Soto-Reyes-Solis, E. (2001). Effectiveness of terbinafine in the eradication of *Microsporum canis* from laboratory cats. *Mycoses, 44*, 95–97.

Chen, W., Van Wyk, B. E., Vermaak, I., & Viljoen, A. M. (2012). Cape aloes-a review of the phytochemistry, pharmacology and commercialisation of *Aloe ferox*. *Phytochemistry Letters, 5*, 1–12.

Chermette, R., Ferreiro, L., & Guillot, J. (2008). Dermatophytoses in animals. *Mycopathologia, 166*, 385–405.

Commons, R. J., Smeesters, P. R., Proft, T., Fraser, J. D., Robins-Browne, R., & Curtis, N. (2014). *Streptococcal superantigens*: Categorization and clinical associations. *Trends in Molecular Medicine, 20*(1), 48–62.

Coopoosamy, R. M., & Naidoo, K. K. (2012). An ethnobotanical study medicinal plants used by traditional healers in Durban, South Africa. *African Journal of Pharmacy and Pharmacology, 6*(11), 818–823.

De Wet, H., Nciki, S., & van Vuuren, S. F. (2013). Medicinal plants used for the treatment of various skin disorders by a rural community in northern Maputaland, South Africa. *Journal of Ethnobiology and Ethnomedicine, 9*(1), 51–60.

Dey, A., Gupta, B., & De, J. N. (2012). Traditional phytotherapy against skin diseases and in wound healing of the tribes of Purulia district, West Bengal, India. *Journal of Medicinal Plants Research, 33*, 4825–4831.

Dyer, D. L., Gerenraich, K. B., & Wadhams, P. S. (1998). Testing a new alcohol-free hand sanitizer to combat infection. *Official Journal of Association of Perioperative Registered Nurses, 68*(2), 239–251.

Esimone, C., Nworu, C., Ekong, U., & Okereke, B. (2007). Evaluation of the antiseptic properties of *Cassia alata*-based herbal soap. *The Internet Journal of Alternative Medicine, 6*(1), 1–5.

Garber, G. (2001). An overview of fungal infections. *Drugs, 61*(1), 1–12.

George, A. O., Ogunbiyi, A. O., & Daramola, O. O. M. (2006). Cutaneous adornment in the Yoruba of south-western Nigerian-past and present. *International Journal of Dermatology, 45*, 23–27.

Germishuizen, G., & Meyer, N. L. (Eds.). (2003). *Plants of southern Africa: An annotated checklist. Strelitzia 14* (p. iv). Pretoria, South Africa: National Botanical Institute.

Gould, A. N. (2015). *Senecio serratuloides var. in wound healing: Efficacy and mechanistic investigations in porcine wound model* (Ph.D. thesis). South Africa: Department of Health Sciences, University of Witwatersrand.

Grace, O. M., Light, M. E., Lindsey, K. L., Moholland, D. A., van Staden, J. V., & Jager, A. K. (2002). Antibacterial activity and isolation of antibacterial compounds from fruit of the traditional African medicinal plant, *Kigelia africana. South African Journal of Botany*, *68*, 220–222.

Grice, E. A., Kong, H. H., Conlan, S., Deming, C. B., Davis, J., Young, A. C., et al. (2009). Topographical and temporal diversity of the human skin microbiome. *Science*, *324*(5931), 1190–1192.

Grice, E. A., & Segre, J. A. (2011). The skin microbiome. *Nature Reviews: Microbiology*, *9*, 244–253.

Grierson, D. S., & Afolayan, A. J. (1999). An ethnobotanical study of plants used for the treatment of wounds in the Easter cape, South Africa. *Journal of Ethnopharmacology*, *67*, 327–332.

Harris, S. (2004). *Melianthus comosus*. Plantzafrica: South African National Biodiversity Institute, South Africa. http://www.plantzafrica.com/plantklm/melicomo.htm.

Havlickova, B., Czaika, V. A., & Friedrich, M. (2008). Epidemiological trends in skin mycoses worldwide. *Mycoses*, *52*(95), 2–15.

Ibeneghu, O. A. (2013). The antimicrobial assessment of some Nigerian herbal soap. *African Journal of Traditional Complementary Medicine*, *10*(6), 513–518.

Igarashi, T., Nishino, K., & Nayar, S. K. (2005). *The appearance of human skin*. New York, USA: Department of Computer Science, Columbia University.

Joffe, P. (2003). *Kigelia africana (Lam.) Benth*. Plantzafrica: South African National Biodiversity Institute, South Africa. http://www.plantzafrica.com/plantklm/kigeliaafric.htm.

Kelmanson, J. E., Jager, A. K., & van Staden, J. V. (2000). Zulu medicinal plants with antibacterial activity. *Journal of Ethnopharmacology*, *69*, 241–246.

Ki, V., & Rotstein, C. (2008). Bacterial skin and soft tissue infections in adults: A review of their epidemiology, pathogenesis, diagnosis, treatment and site of care. *Canadian Journal of Infectious Diseases and Medical Microbiology*, *19*(2), 173–184.

Kownatzki, E. (2003). Hand hygiene and skin health. *Journal of Hospital Infection*, *55*(4), 239–245.

Kubo, I., & Taniguchi, M. (1988). Polygodial, an antifungal potentiator. *Journal of Natural Products*, *51*(1), 22–29.

Leonard, C. M., & Viljoen, A. M. (2015). Warburgia: A comprehensive review of the botany, traditional uses and phytochemistry. *Journal of Ethnopharmacology*, *165*, 260–285.

Levy, S. B. (1998). The challenge of antibiotic resistance. *Scientific American*, *278*(3), 32–39.

Luksich, J. A., Holds, J. B., & Hartstein, M. E. (2002). Conservative management of necrotizing fasciitis of the eyelids. *Ophthalmology*, *109*(11), 2118–2122.

Luyckx, V. A., Ballanine, R., Claeys, M., Cuyckens, F., van den Heuvel, H., Cimanga, R. K., et al. (2002). Herbal remedy-associated acute renal failure secondary to Cape aloes. *American Journal of Kidney Diseases*, *39*(3), E13.

Mabona, U., Viljoen, A., Shikanga, E., Marston, A., & Van Vuuren, S. (2013). Antimicrobial activity of southern African medicinal plants with dermatological relevance: From an ethnopharmacological screening approach, to combination studies and the isolation of a bioactive compound. *Journal of Ethnopharmacology*, *148*, 45–55.

Maphosa, V., Masika, P. J., & Moyo, B. (2009). Toxicity evaluation of the aqueous extract of the rhizome of *Elephantorrhiza elaphantina* (Burch.) Skeels. (Fabaceae), in rats. *Food and Chemical Toxicology*, *48*(196), 196–201.

Mbambezeli, G. (2004). *Warburgia salutaris*. Plantzafrica: South African National Biodiversity Institute, South Africa. http://www.plantzafrica.com/plantwxyz/warburg.htm.

McGaw, L. J., & Eloff, J. N. (2008). Ethnoveterinary use of southern African plants and scientific evaluation of their medicinal properties. *Journal of Ethnopharmacology*, *119*, 559–574.

McGrath, J. A., Eady, R. A. J., & Pope, F. M. (2004). Anatomy and organization of human skin. In T. Burns, S. Breathnach, N. Cox, & C. Griffiths (Eds.), *Rook's textbook of dermatology* (pp. 3.1–3.84). Oxford: Blackwell Science Ltd.

Mendez-Tovar, L. J. (2010). Pathogenesis of dermatophytosis and tines versicolor. *Clinics in Dermatology*, *28*, 185–186.

Moody, J. O., Adebiyi, O. A., & Adeniyi, B. A. (2004). Do *Aloe vera* and *Ageratum conyzoides* enhance the anti-microbial activity of traditional medicinal soft soaps (Osedudu)? *Journal of Ethnopharmacology*, *92*, 57–60.

Moriello, K. A. (2003). Treatment of dermatophytes in dogs and cats: Review of published studies. *Veterinary Dermatology*, *15*, 99–107.

Mutshinyalo, T. (2007). *Diospyros mespiliformis*. Plantzafrica: South African National Biodiversity Institute, South Africa. http://www.plantzafrica.com/plantcd/diospyrosmespil.htm.

Mwale, M., & Masika, P. J. (2012). Toxicological studies on the leaf extract of *Aloe ferox* Mill. (Aloaceae). *Scientific Research and Essays*, *7*(15), 1605–1613.

Nciki, S. (2015). *Validating the traditional use of medicinal plants in Maputaland to treat skin diseases* (M.Sc. dissertation). Department of Health Sciences, University of Witwatersrand.

Netnou, N., & Herman, P. P. J. (2003). Dicoma. Strelitzia 14. In G. Germishuizen, & N. L. Meyer (Eds.), *Plants of Southern Africa: An annotated checklist* (pp. 209–210). Pretoria, South Africa: National Botanical Institute.

Ngeny, L. C., Magri, E., Mutai, C., Mwikwabe, N., & Bii, C. (2013). Antimicrobial properties and toxicity of *Hagenia abyssinica* (Bruce) J.F.Gmel, *Fuerstia africana* T.C.E. Fries, *Asparagus racemosus* (Willd.) and *Ekebergia capensis* Sparrm. *African Journal of Pharmacology and Therapeutics*, *2*(3), 76–82.

Nweze, E. I. (2010). Dermatophytosis in western Africa: A review. *Pakistan Journal of Biological Sciences*, *13*(13), 649–656.

Oke, O. O., Onayemi, O., Olasode, O. A., Omisore, A. G., & Oninla, O. A. (2014). The prevalence and pattern of superficial fungal infections among school children in Ile-Ife, South-Western Nigeria. *Dermatology Research and Practice*, *2014*.

Orwa, C., Mutua, A., Kindt, R., Jamnadass, R., & Anthony, S. (2009). *Agroforestree database: A tree reference and selection guide version 4.0*. Kenya: World Agroforestry Centre.

Otang, W. M., & Afolayan, A. J. (2015). Antimicrobial and antioxidant efficacy of *Citrus limon* L. peel extracts used for skin diseases by Xhosa tribe of Amathole district, Eastern Cape, South Africa. *South African Journal of Botany*, *102*, 46–49.

Outerbridge, C. A. (2006). Mycologic disorders of the skin. *Clinical Techniques in Small Animal Practice*, *21*, 128–134.

Owolabi, O. J., Omogbai, E. K. I., & Obasuyi, O. (2007). Antifungal and antibacterial activities of the ethanolic and aqueous extract of *Kigela africana* (Bignoniaceae) stem bark. *African Journal of Biotechnology*, *6*(14), 1677–1680.

Paiva, P. M. G., Gomes, F. S., Napoleao, T. H., Sa, R. A., Correia, M. T. S., & Coelho, L. C. B.B. (2010). Antimicrobial activity of secondary metabolites and lectins from plants. In A. Mendez-Villas (Ed.), *Current research, technology and education topics in applied microbiology and microbial biotechnology* (pp. 396–406). Brazil: Formatex Research Center.

Sharma, V., Kumawat, T. K., Sharma, A., Seth, R., & Chandra, S. (2015). Dermatophytosis: Diagnosis and its treat. *African Journal of Microbiology Research*, *9*(19), 1286–1293.

Shrestha, D. P., Gurung, D., & Rosdahl, I. (2012). Prevalence of skin diseases and impact on quality of life in hilly region of Nepal. *Journal of Institute of Medicine*, *3*(34), 44–49.

Smidt, A. C., Lai, J. S., Cella, D., Patel, S., Mancini, A. J., & Chamlin, S. L. (2010). Development and validation of skindex-teen, a quality-of-life instrument for adolescents with skin diseases. *Archives of Dermatology*, *146*(8), 865–869.

Soares, L. A., de Cassia-Orlandi-Sardi, J., Gullo, F. F., de Souza-Pitangui, N., Scorzoni, L., Leite, F. S., et al. (2013). Anti dermatophytic therapy-prospects for the discovery of new drugs from natural products. *Brazilian Journal of Microbiology*, *44*(4), 1035–1042.

Stevens, J. (2001). *Fungal skin infections*. School of Medicine, University of New Mexico.

Street, R. A., & Prinsloo, G. (2013). Commercially important medicinal plants of South Africa: A review. *Journal of Chemistry*, *2013*, 1–16.

Stulberg, D. L., Penrod, M. A., & Blatny, R. A. (2002). Common bacterial skin infection. *American Family Physician*, *66*(1), 119–124.

Tenover, F. C. (2006). Mechanism of antimicrobial resistance in bacteria. *American Journal of Infection Control*, *34*(5), S3–S4.

Tschachler, E., Bergstresser, P. R., & Stingl, G. (1996). HIV-related skin diseases. *Lancet*, *348*, 659–663.

Tselanyane, M. L. (2006). *The ethnobotany and ethnopharmacology of Dicoma anomala* (Ph.D. thesis). Department of Medicine, University of Cape Town.

Ungphaiboon, S., Supavita, T., Singchangchai, P., Sungkarak, S., Rattanasuwan, P., & Itharat, A. (2005). Study on antioxidant and antimicrobial activities of turmeric clear liquid soap for wound treatment of HIV patients. *Songklanakarin Journal of Science and Technology*, *27*(2), 570–578.

Van Vuuren, S. F. (2008). Antimicrobial activity of South African medicinal plants. *Journal of Ethnopharmacology*, *119*, 462–472.

Van Wyk, B. E. (2008). Abroad review of commercially important southern African medicinal plants. *Journal of Ethnopharmacology*, *199*, 342–355.

Van Wyk, B. E., & Gericke, N. (2000). *People's plants: A guide to useful plants of southern Africa* (pp. 33, 42–43, 247, 250–251). Pretoria: Briza Publications.

Van Wyk, B. E., Van Oudtshoorn, B., & Gericke, N. (1997). *Medicinal plants of South Africa* (pp. 40–41, 104–105, 114–117, 148–149, 160–161, 172–173, 238–239, 272–273). Pretoria: Briza publications.

Venus, M., Waterman, J., & McNab, I. (2011). Basic physiology of the skin. *Surgery*, *29*(10), 471–474.

Von Koenen, E. (2001). *Medicinal poisonous and edible plants in Namibia* (4th ed.). Windhoek, Namibia: Klaus-Hess publishers.

Weigelt, J., Itani, K., Stevens, D., Lau, W., Dryden, M., Knirsch, C., et al. (2005). Linezoid versus vancomycin in treatment of complicated skin and soft tissue infections. *American Society for Microbiology*, *49*(6), 2260–2266.

Weitzman, I., & Summerbell, R. C. (1995). The dermatophytes. *Clinical Microbiology Reviews*, *8*(2), 240–259.

White, S. (2006). *Fungal skin infections*. Davis, USA: University of California.

Chapter 9

Garlic (*Allium sativum*) and Its Associated Molecules, as Medicine

Carel B. Oosthuizen, Anna-Mari Reid, Namrita Lall
University of Pretoria, Pretoria, South Africa

Chapter Outline

9.1 INTRODUCTION: *ALLIUM SATIVUM* AND ITS HISTORICAL RELEVANCE

Originating from Middle Asia, brought by the Sumerians (2600–2100 BC), garlic was effectively known for its medicinal properties. In ancient times many cultures welcomed the medicinal effects attributed to garlic. These ranged from the ancient Egyptians using garlic as a nutritional supplement for use by the laborers (Petrovska & Cekovska, 2010; Rivlin, 2001); the ancient Indians using it for common weakness, skin disease to hemorrhoids as well as the ancient Greeks who used garlic for its medicinal relief of colic and sea sickness. It has also formed part of the diet of the military soldiers and was associated with strength and endurance (Petrovska & Cekovska, 2010; Rivlin, 2001). Garlic was brought all the way to Great Britain from the Mediterranean Sea and used for numerous purposes with Russia relabeling garlic as their Russian penicillin due to its natural antibiotic properties and its use during World War II to treat wounded soldiers (Petrovska & Cekovska, 2010).

Medicinal Plants for Holistic Health and Well-Being. http://dx.doi.org/10.1016/B978-0-12-812475-8.00009-3

Allium sativum, commonly known as garlic belongs to the Alliaceae family. The plant forms a bulb that is consumed as food and as a medicine. The flowers are hermaphrodite and are pollinated by bees and other insects. The plant prefers well-drained acidic soils, with medium light to perform at its best (Huxley, 1992). All the parts of the plant are edible, but the bulb is mostly consumed, raw or cooked for flavoring. *Allium* species are one of the world's oldest cultivated plants due to its long storage life. Garlic is originally native to Central Asia (Kazakhstan, Kyrgyzstan, Turkmenistan, Tajikistan, and Uzbekistan) and the north-eastern parts of Iran, although it is also cultivated widely around the world (Kew.org., 2016).

Garlic has a wide range of actions; not only has it been reported that it contains antibacterial, antiviral, antifungal, and antiprotozoal properties but it also has beneficial effects on the cardiovascular and immune systems (Harris, Cottrell, Plummer, & Lloyd, 2001). Garlic essential polysulfide oils are the active oils of garlic that contribute toward the aforementioned biological properties.

9.2 PRODUCT DEVELOPMENT

Many products of garlic and garlic-derived compounds have been commercialized. Garlic products exist in four different forms such as essential oils, garlic oil macerates, garlic powders, and garlic extracts. A standard intake of garlic has been established in clinical trials with the daily dose of dehydrated garlic powder set at 900 mg. Aged garlic is within a range of 1–7.2 g/day. However, at very high doses no severe toxic side effects were reported in any of the clinical trials undertaken (Amagase, Petesch, Matsuura, Kasuga, & Itakura, 2001).

9.3 RESEARCH FINDINGS OF GARLIC EXTRACTS

Garlic and garlic extracts contain numerous chemical compounds, many of which have shown significant biological activities. The preparation of these extracts is important. Fresh extract and aged extracts differ greatly in their chemical composition. During the aging process, certain enzymes break down chemicals into subsequent compounds. Intact garlic bulbs contain high amounts of γ-glutamylcysteines. These reserve compounds can be hydrolyzed and oxidized to form alliin, which accumulates naturally during storage of garlic bulbs at cool temperatures. After processing, such as cutting, crushing, chewing, or dehydration, the vacuolar enzyme, alliinase, rapidly lysis the cytosolic cysteine sulfoxides (alliin) to form cytotoxic and odoriferous alkyl alkane-thiosulfinates such as allicin. Allicin and other thiosulfinates instantly decompose to other compounds, such as diallyl sulfide (DAS), diallyl disulfide (DADS), diallyl trisulfide (DATS), dithiins, and ajoene. At the same time, γ-glutamylcysteines are converted to S-allyl cysteine (SAC) via a pathway other than the alliin/allicin pathway. SAC contributes heavily to the health benefits of garlic (Bayan, Koulivand, & Gorji, 2014).

SAC is found after hydrolysis of aqueous garlic extraction catalyzed by enzyme γ-glutamyl transpeptidase. Allicin (thiosulfonate) is found after lysis of alliin by enzyme alliinase. Allicin can undergo nonenzymatic decomposition into DAS, DADS, DATS, diallyl tetrasulfide, and allyl methyl trisulfide (Tocmo, Liang, Lin, & Huang, 2015). This is summarized in Fig. 9.1. Due to the presence of sulfur glycosides, garlic is well known for its distinctive smell when crushed or sliced. It is mostly these sulfur compounds (alliin, allicin, DAS, and ajoene) that are attributable to its medicinal properties. Garlic also contains important components such as water, cellulose, amino acids, lipids, etheric oil, a complex of fructosamine, steroid saponosides, organic acids, minerals, vitamins, and enzymes, to name but a few (Fig. 9.2).

FIGURE 9.1 Garlic is a versatile medicinal plant sold in many forms. (A) Freshly harvested garlic. (B) Garlic being sold in a market. (C) Products containing garlic as an ingredient (Barbieri, 2016; Fir0002/Flagstaffotos, 2016).

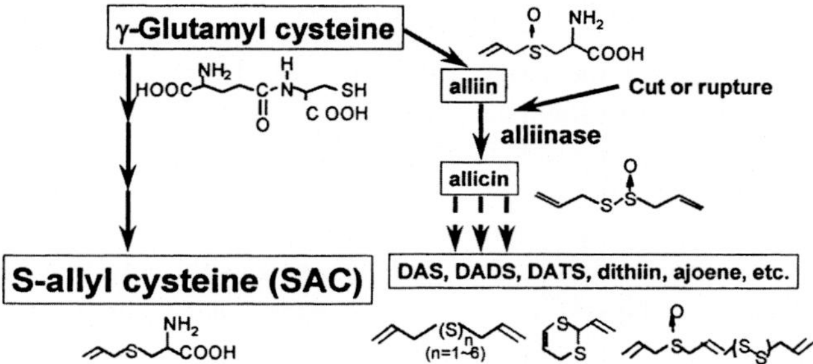

FIGURE 9.2 Conversion of garlic compounds (Amagase et al., 2001). *DADS*, diallyl disulfide; *DATS*, diallyl trisulfide; *DAS*, diallyl sulfide.

Other important compounds that are present in a garlic homogenate are 1-propenyl allyl thiosulfonate, allyl methyl thiosulfinate, (E, Z)-4, 5, 9-trithiadodeca 1, 6, 11-triene 9-oxide (Ajoene), and γ-L-glutamyl-*S*-alkyl-L-cysteine.

9.3.1 Biological Properties

The medicinal properties of garlic are well documented and well known throughout the world. Not only does it serve as a flavoring agent in many dishes but the bulb and extracts from it can also be used to treat multiple diseases and ailments. Table 9.1 summarizes the results from some studies done on the extract and associated compounds.

A study by Nencini et al. (2007), found that not only fresh homogenates of garlic were able to restore the nonenzymatic parameters studied in a hepatoprotective effect after injury induced by ethanol to control values but also improved the activity levels of the enzymes investigated. Several compounds found commonly in garlic were evaluated for the virucidal activity. Gebreyohannes and Gebreyohannes (2013) found that ajoene showed the most significant activity followed by allicin, allyl methyl thiosulfonate, and methyl allyl thiosulfinate. However, no virucidal activity was found for any of the polar fractions of alliin, deoxyalliin, DADS, or DATS. Rats with ethanol-induced gastric ulcers and gastric acid secretions were treated with raw and boiled garlic. The raw and boiled garlic were able to reduce histamine-stimulated acid secretion as well as secretion of basal acid. Raw garlic was able to show a decrease in the ulcer index when compared with the controls used. This further emphasizes the use of raw or boiled garlic for the treatment of gastric ulcers and gastric acid secretion (Amir, Dhaheri, Jaberi, Marzouqi, & Bastaki, 2011).

9.3.2 Nutritional Value

Investigations into the nutritional properties of garlic in its different forms and preparations have reported the presence of many important minerals. Bulbs that are completely ripe will produce crude proteins, oil, energy, fiber, ash, dimethyl sulfite as well as several essential oil and minerals such as K, P, Mg, Na, Ca, and Fe among others (Haciseferogulları, Ozcan, Demir, & Calisir, 2005). Suleria et al. (2015) also established that garlic bulbs consist of water (65%), carbohydrates (28%), organosulfur compounds (2.3%), proteins (2%), amino acids [free] (1.2%), and fiber (1.5%). Raw garlic contains water (58.58/100 g), carbohydrates (33.06/100 g), and proteins (6.36/100 g). Garlic powder consists mainly of carbohydrates and proteins as well as thiamin that has a high bioavailability due to the sulfur-containing components. Selenium is the main mineral found in garlic with levels that are 9 ppm higher than in any other plant (Suleria et al., 2015).

TABLE 9.1 Biological Activities of Garlic Extracts and/or Its Compounds

Biological Activity	Effect	Preparation	References
Extract			
Antibacterial	*Staphylococcus aureus*	Aqueous, ethanol, chloroform extract	EL-mahmood (2009)
	Escherichia coli, Salmonella typhi	Aqueous and ethanolic extract	Ankri and Mirelman (1999)
	Bacillus subtilis, Klebsiella pneumoniae	Aqueous, methanol and ethanol extract	Meriga, Mopuri, and MuraliKrishna (2012) and Pundir, Jain, and Sharma (2010)
	Helicobacter pylori	Extract	Lui et al. (2010)
	Salmonella enteritidis	Extract	Benkeblia (2004)
	Shigella spp, *Proteus mirabilis*	Extract	Eja et al. (2007)
	Actinobacillus pleuropneumoniae serotype 9	Extract	Becker et al. (2012)
	Streptococcus mutans	Extract	Loesche (1986)
Antifungal	*Candida albicans, Candida tropicalis, Blastoschizomyces capitatus*	Extract	Avato, Tursil, Vitali, Miccolis, and Candido (2000)
	Botrytis cinerea, Trichoderma harzianum	Extract	Lanzotti, Barile, Antignani, Bonanomi, and Scala (2012)
	Ascosphaera apis	Essential oil vapors	Kloucek et al. (2012)
	Paracoccidioides brasiliensis	Extract	Thomaz, Apitz-Castro, Marques, Travassos, and Taborda (2008)

Continued

TABLE 9.1 Biological Activities of Garlic Extracts and/or Its Compounds—cont'd

Biological Activity	Effect	Preparation	References
	Aspergillus niger	Extract	Yoshida et al. (1987)
	Dermatophytes, saprophytes	Ethanol extract	Shamim, Ahmed, and Azhar (2004)
	Cryptococcal spp.	Alcoholic extract	Khan and Katiyar (2000)
	B. cinerea, Mycosphaerella arachidicola, Physalospora piricola	Extract	Wang and Ng (2001)
Antiparasitic	*Trypanosoma* sp, *Entamoeba histolytica, Giardia lamblia*	Extract	Lun, Burri, Menzinger, and Kaminsky (1994)
	Trypanosoma cruzi	Extract	Gallwitz et al. (1999)
	Plasmodium spp, *Giardia* spp		
	Trypanosoma, Plasmodium, Giardia, and *Leishmania* spp.	Extract	Anthony, Fyfe, and Smith (2005)
	Hymenolepiasis, Giardiasis	Aqueous extract	Soffar and Mokhtar (1991)
	Haemonchus contortus	Ethanol, dichloromethane and water extract	Ahmed, Laing, and Nsahlai (2012)
Cardiovascular	Hypotensive via increasing nitric oxide synthesis	Extract	Al-Qattan et al. (2006)
	Hypotensive (endothelial dependent and independent)	Not applicable	Fallon et al. (1998)
	Induces vasodilation with H_2S	Extract	Ginter and Simko (2010)

	Angiotensin converting enzyme-inhibiting activity	Aqueous extract	Sener, Sakarcan, and Yegen (2007)
	Stimulation of nitric oxide generation in endothelial cells	Garlic-derived polysulfides	Ginter and Simko (2010)
	Bradycardia	Aqueous extract	Nwokocha, Ozolua, Owu, Nwokocha, and Ugwu (2011)
	Hepatopulmonary syndrome	Garlic powder	Thevenot et al. (2009)
	Decreases systolic blood pressure	Aged garlic	Harauma and Moriguchi (2006)
	Vasorelaxant	Not applicable	Zahid Ashraf et al. (2005)
	Coronary artery disease	Extract	Verma, Rajeevan, Jain, and Bordia (2005)
	Reduce myocardial infarction	Not mentioned	Yang, Chan, Hu, Walden, and Tomlinson (2011)
	Antithrombotic	Extract and derived	Choi and Park (2012)
	Antiatherosclerotic	Volatiles derived	Calvo-Gómez, Morales-Lopez, and Lopez (2004)
Hypolipidemia	Hypolipidemic effect	Capsule of garlic preparation	Duda, Suliburska, and Pupek-Musialik (2008)
	Hypolipidemic effect	Extract	Kuda, Iwai, and Yano (2004)
	Hypocholesterolemic		
	Hypotriacylglyceride		
	Hypoglycaemic	Extract	Sengupta, Ghosh, and Bhattacharjee (2004)
	Hypolipidemic effect	Organosulfur compound	Lii et al. (2012)
Immune system	Immunomodulation	Extract	Chandrashekar and Venkatesh (2012)
	Antioxidant properties	Organosulfur compounds in aged garlic	Cruz et al. (2007)

Continued

TABLE 9.1 Biological Activities of Garlic Extracts and/or Its Compounds—cont'd

Biological Activity	Effect	Preparation	References
Allicin			
Antimicrobial	*E. coli*	PF[a]	Ankri and Mirelman (1999)
	Cytomegalovirus	PF[a]	Ankri and Mirelman (1999)
	Influenza B virus	PF[a]	Ankri and Mirelman (1999)
	Herpes simplex virus type 2	PF[a]	Ankri and Mirelman (1999)
	Parainfluenza virus type 3	PF[a]	Ankri and Mirelman (1999)
	Vaccinia virus	PF[a]	Ankri and Mirelman (1999)
	Vesicular stomatitis virus	PF[a]	Ankri and Mirelman (1999)
	Human Rhinovirus type 2	PF[a]	Ankri and Mirelman (1999)
	Common cold virus	PF[a]	Josling (2001)
Antihypertensive	Lowered intraocular pressure	PF[a]	Chu, Ogidigben, Han, and Potter (1993)
	Lowered blood pressure	PF[a]	Elkayam et al. (2001)
Antiatherosclerosis	Antioxidant action, lipoprotein modification, and inhibition of low-density lipoprotein uptake and degradation by macrophages	PF[a]	Gonen et al. (2005)
Antifungal	*C. albicans*	PF[a]	Ankri and Mirelman (1999)
	Saccharomyces cerevisiae	PF[a]	Ogita, Fujita, Taniguchi, and Tanaka (2006)
	Aspergillus fumigatus	PF[a]	An et al. (2009)

Antithrombotic	Antiplatelet	PF[a]	Agarwal (1996)
Antiparasitic	*E. histolytica*	PF[a]	Ankri and Mirelman (1999)
	G. lamblia	PF[a]	Ankri and Mirelman (1999)
	Schistosoma mansoni	PF[a]	Lima et al. (2011)
	Trypanosoma brucei	PF[a]	Waag et al. (2010)
Antimalarial	*Plasmodium berghei*	PF[a]	Coppi, Cabinian, Mirelman, and Sinnis (2006)
	Plasmodium falciparum	PF[a]	Waag et al. (2010)
Hepatoprotection	Galactosamine/endotoxin challenged rats	PF[a]	Vimal and Devaki (2004)
Antitumor	B-16 melanoma and MCA-105	PF[a]	Patya et al. (2004)
Antihyperlipidemic	Antihyperlipidemic	PF[a]	Sela et al. (2004)
Anticancer	Cell proliferation of colon cancer cells	PF[a]	Bat-Chen, Golan, Peri, Ludmer, and Schwartz (2010)
Immune stimulatory	Effects on functions of peripheral blood cells	PF[a]	Salman, Bergman, Bessler, Punsky, and Djaldetti (1999)
Alliin			
Antifungal	Murine pulmonary aspergillosis	PF[a]	Appel et al. (2010)
Ajoene			
Antifungal	Mice infected with *P. brasiliensis*	PF[a]	Thomaz et al. (2008)
	A. niger	PF[a]	Yoshida et al. (1987)
	C. albicans	PF[a]	Yoshida et al. (1987)
Antiparasitic	*T. cruzi*	PF[a]	Gallwitz et al. (1999)
Antithrombotic	Antiplatelet	PF[a]	Apitz-Castro, Badimon, and Badimon (1994)

[a]*Pure form of the compound.*

9.3.3 Antibacterial

The two most reported compounds from garlic that have shown antibacterial activity are allicin and ajoene. These compounds have shown to inhibit the growth of Gram-positive and Gram-negative bacteria such as *Escherichia coli*. The antibacterial activity of garlic is attributed to the compound allicin. It has sulfhydryl-modifying activity and inhibits sulfhydryl enzymes. Cysteine and glutathione counteract the thiolation activity of allicin. Sulfur-containing compounds have been identified as quorum quenching molecules, inhibiting processes such as biofilm formation, persistence, and virulence. What is unclear is whether the chemical composition and the amount of the sulfur are important for this activity (Bayan et al., 2014; Jakobsen et al., 2012).

9.3.4 Cardiovascular

Garlic appears effective in reducing parameters associated with cardiovascular disease, even if more in-depth and appropriate studies are required. The beneficial effects of garlic supplementation in reducing blood pressure and offering cardio-protection seem to be due to its ability to counteract oxidative stress. This activity can be linked to the antioxidant activity related to garlic extract (Cruz et al., 2007). In ample studies in animal models as well as clinical human trials, garlic, garlic extracts, and compounds within have shown the ability to lower blood cholesterol and triglycerides, they can change blood lipoproteins and have the added effect of affecting coagulation parameters. In a review paper on garlic as preventative against atherosclerosis, the authors found that garlic may be valuable in the prevention as well as the treatment of this disease (Lau, Adetumbi, & Sanchez, 1983).

9.3.5 Immune Stimulatory Effect

Allicin as one of the main ingredients of garlic were examined by Sela et al. (2004) for their in vitro effects on T-cells. Allicin is known to inhibit T-cell migration through fibronectin by downregulating the reorganization of cortical actin and T-cell polarization and T-cell adhesion to fibronectin. Allicin was also able to inhibit T-cell adhesion to the endothelial cells and transendothelial migration. The proposed mechanism responsible for these inhibitory functions is the ability of allicin to downregulate phosphorylation of Pyk2, an intracellular component of the focal adhesion kinases, and the reduction of the expression of vascular cell adhesion protein 1 and fibronectin-specific alpha-4-beta-1-integrin very late antigen-4 (Caspasso, 2013; Sela et al., 2004). Several compounds isolated from garlic have been proven to modulate cytokine production and leukocyte proliferation, and therefore Hodge, Hodge, and Han (2002) investigated the possibility of garlic to assist in the treatment of inflammatory bowel disease. They found that low concentrations of garlic extract had the ability to inhibit the T-helper 1 (Th1)-response, interleukin 12 (IL-12), and the other inflammatory cytokines such as interleukin 2 (IL-2), interleukin 6 (IL-6), and interleukin 8 (IL-8). This together with the ability to upregulate the production of interleukin 10 (IL-10) may have the ability to naturally assist with the treatment of inflammation associated with inflammatory bowel disease.

9.3.6 Cancer Treatment

Many chemical constituents found in garlic have been tested for their ability to inhibit cancer, both in in vitro and in vivo models. It has been shown that many water- and lipid-soluble allyl sulfur compounds are able to induce prevention of chemically induced tumors by the active blocking of nitrosamine formation as well as metabolism thereof. Several publications reported on the therapeutical properties of garlic and its constituents, especially having an effect on altering an individual's risk of cancer and heart disease (Fenwick and Hanley, 1985; Milner, 1996, 1999; Orekhov & Grunwald, 1997; Yoshida et al., 1999). Preclinical investigations found that garlic and its constituents can not only inhibit the risk of cancer but also alter the formation and suppression of tumors. There have been reports of the effect of garlic and especially its sulfur constituents to subdue the occurrences of tumors in breast, colon, esophagus, lung, skin, and uterine cancers (Amagase & Milner, 1993; Hussain, Jannu, & Rao, 1990; Ip, Lisk, & Stoewsand, 1992; Lui, Lin, & Milner, 1992; Shukla, Singh, & Srivastava, 1999; Song & Milner, 1999; Sumiyoshi & Wargovich, 1990; Wargovich, Woods, Eng, Stephens, & Gray, 1988). Cancer protection may be in several forms but especially in the following manners (Milner, 2001):

- Blockage of *N*-nitroso compound formation and the suppression of several carcinogens
- Enhancing DNA repair
- Reducing cell proliferation
- Induction of apoptosis

A known compound from garlic, ajoene was topically applied to the tumors of 21 patients who had either nodular or superficial basal cell carcinoma (BCC). What Tilli et al. (2003) found was that the tumors in 17 of the patients reduced in size. The mechanism by which ajoene was able to reduce these tumor sizes was by means of a decrease in the apoptosis-suppressing protein (Caspasso, 2013; Tilli et al., 2003). Further analysis was done and it was found that ajoene induced apoptosis in a dose-dependent manner in BCC primary cultures. It was established that ajoene had the ability to reduce BCC tumor size mainly through the induction of the mitochondria-dependent route of apoptosis (Caspasso, 2013; Tilli et al., 2003). The ability of DATS to affect the viability of MCF-7 human breast cancer cells and nontumorigenic MCF-12a mammary epithelial cells was investigated by Das, Banik, and Ray (2007). This garlic constituent had the ability to affect the viability of these two cell lines by the decreasing a number of cells in the cell cycle phase of G (2)/M and also by inducing apoptosis. Apoptosis levels in the MCF-7 cells were significantly higher, simultaneously, with higher levels of cyclin B1. Das et al. (2007), noted that DATS had a novel method of treatment of breast cancer cells by the manner in which it induces apoptosis. Compounds derived from garlic were also used to treat glioblastoma cells, which produced reactive oxygen species that had the ability to induce apoptosis through phosphorylation of p38 mitogen-activated protein kinase and the activation of the c-Jun N-terminal kinases 1-pathway (Caspasso, 2013). Lamm and Riggs (2001) found that aged garlic extract may be an effective treatment solution for bladder cancer in the mouse bladder tumor line-2 murine bladder tumor model. Schafer and Kaschula (2014), provided a review that hypothesized that the chemopreventative mechanism of action of the garlic organosulfur compounds' is through the immunomodulatory and antiinflammatory properties it may possess.

In another study, garlic has proven to show effective anticarcinogenic activities. Garlic was shown to stop mitosis in all phases of cancer cells as well as effective activity against patients suffering from leukemia as shown in an in vivo study done by Hassan (2004).

9.3.7 Clinical Significance

The use of garlic and its extracts have been evaluated and verified through many different clinical studies, from antibacterial in the early stages to anticancer, and the ability to lower blood pressure and cholesterol. It is important to note that the human trials are based on preventative and/or therapeutic use. Statistical significance of the clinical use of garlic is still debatable as many studies have contradictory results and in some instances the size of the trial and geography do not permit absolute and concrete answers. Another issue that needs to be addressed is the clinical protocol that is being followed, with little standardization occurring throughout the different studies. These variations include preparation of the extract or form of the dosage, parameters assessed, the number of patients participating in the trial, and geographical and historical significance, to name but a few. The use and clinical assessment of garlic as a medicine have culminated in many studies for its consideration as diversely biologically active and as a natural medicine.

When considering infectious diseases, garlic has shown to inhibit an array of bacteria, fungi, viruses, and worms, supporting the historical use of this plant as a medicine. In one clinical study, the garlic preparations were compared to commonly used antibiotics (penicillin, streptomycin, chloramphenicol, erythromycin, and tetracyclines). Garlic was able to combat bacteria and fungi that have developed resistance to these antibiotics (Harris et al., 2001). Some of the infectious diseases that have been tested in human clinical studies and showed an overall lowering of bacterial/fungal/viral load includes the following: *E. coli*, *Klebsiella pneumonia*, Mycobacteria, *Salmonella* spp, *Staphylococcus* spp, *Candida albicans*, Herpes simplex, Human rhinovirus, and Vesicular stomatitis virus.

The effects of garlic on cancer have been evaluated in a few human clinical studies. Garlic has shown to increase the number of natural-killer cells, indirectly targeting cancerous cells (You et al., 2006). This result suggests, and is supported by other findings, that the anticancer activity related to garlic use is most likely due to indirect activity by enhancing the natural immune response (Khanum, Anilakumar, & Viswanathan, 2004). Other cancer-related activities have also been associated with some constituents of garlic. These were able to induce apoptosis, inhibition of the formation of nitrosamines and decreasing tumor growth (Belman, 1983).

The final clinical assessment that has been studied significantly is the effect of garlic on cardiovascular-related illnesses. This is related to the atherosclerosis inhibition. It is well known throughout folklore that garlic reduces the risk of heart diseases, cholesterol, and strokes (Lau et al., 1983). A direct correlation between garlic consumption and reduced risk of cardiovascular diseases has been found when combining many clinical trials (Rahman & Lowe, 2006). Garlic was shown to be able to reduce serum cholesterol levels in hypercholesterolemic individuals (Zhang et al., 2001). In addition to this, garlic extracts were able to inhibit platelet aggregation and lowering blood pressure (Ackermann et al., 2001).

Taking all this into consideration the clinical assessment of garlic as medicine shows enough merit to be justified as a medicine. It also substantiates and validates the historical use and beneficial effects related to Garlic. This plant and its constituents have had a long scientific background and will continue to be tested for many years to come.

9.4 CONCLUSION

As shown with all the studies done in the past 50 years, garlic is a multiactivity plant. Although a lot of work has been done, many clinical trials need to be undertaken to further prove or validate the many therapeutic in vitro and in vivo results found over many centuries. This may further establish and summarize the therapeutic uses that have been noted for garlic and all its derived compounds and preparations. Garlic and its derived compounds have the ability to supplement many first-line treatments for various diseases and disorders and may also substitute multiple drug intakes with a single intake of capsules or liquid medicines. Safety and quality checks need to be implemented to ensure that the various preparations of garlic-derived compounds and constituents are all standardized. Garlic as a plant serves as a sort of pioneer medicinal plant, which has been around for such a long time and studied intensely in the last century but still poses new questions that need answers. This indicates to the immense possibilities in the phytomedicine realm, and that under studied or new lead plants has the potential to be useful in more ways than we think.

REFERENCES

Ackermann, R. T., Mulrow, C. D., Ramirez, G., Gardner, C. D., Morbidoni, L., & Lawrence, V. A. (2001). Garlic shows promise for improving some cardiovascular risk factors. *Archives of Internal Medicine Journal*, *161*, 813–824.

Agarwal, K. C. (1996). Therapeutic actions of garlic constituents. *Medicinal Research Reviews*, *16*(1), 111–124.

Ahmed, M., Laing, M. D., & Nsahlai, I. V. (2012). In vitro anthelmintic activity of crude extracts of selected medicinal plants against *Haemonchus contortus* from sheep. *Journal of Helminthology*, *87*(2), 174–179.

Al-Qattan, K. K., Thomson, M., Al-Mutawa'a, S., Al-Hajeri, D., Drobiova, H., & Ali, M. (2006). Nitric oxide mediates the blood-pressure lowering effect of garlic in the rat two-kidney, one-clip model of hypertension. *Journal of Nutrition*, *136*, 774–776.

Amagase, H., & Milner, J. A. (1993). Impact of various sources of garlic and their constituents on 7, 12-dimethylbenz[a]anthracene binding to mammary cell DNA. *Carcinogenesis*, *14*(8), 1627–1631.

Amagase, H., Petesch, B. L., Matsuura, H., Kasuga, S., & Itakura, Y. (2001). Intake of garlic and its bioactive components. *The Journal of Nutrition*, *131*(3), 955–962.

Amir, N., Dhaheri, A. A., Jaberi, N. A., Marzouqi, F. A., & Bastaki, S. M. A. (2011). Comparative effect of garlic (*Allium sativum*), onion (*Allium cepa*), and black seed (*Nigella sativa*) on gastric acid secretion and gastric ulcer. *Research and Reports in Medicinal Chemistry*, (1), 3–9.

Ankri, S., & Mirelman, D. (1999). Antimicrobial properties of allicin from garlic. *Microbes and Infection*, *2*, 125–129.

An, M., Shen, H., Cao, Y., Zhang, J., Cai, Y., Wang, R., et al. (2009). Allicin enhances the oxidative damage effect of amphotericin B against *Candida albicans*. *International Journal of Antimicrobial Agents*, *33*(3), 258–263.

Anthony, J. P., Fyfe, L., & Smith, H. (2005). Plant active components – a resource for antiparasitic agents. *Trends in Parasitology*, *21*(10), 462–468.

Apitz-Castro, R., Badimon, J. J., & Badimon, L. (1994). A garlic derivative ajoene, inhibits platelet deposition on severely damaged vessel wall in an in vivo porcine experimental model. *Thrombosis Research*, *75*(3), 243–249.

Appel, E., Vallon-Eberhard, A., Rabinkov, A., Brenner, O., Shin, I., Sasson, K., et al. (2010). Therapy of murine pulmonary aspergillosis with antibody-alliinase conjugates and alliin. *Antimicrobial Agents and Chemotherapy*, *54*(2), 898–906.

Avato, P., Tursil, E., Vitali, C., Miccolis, V., & Candido, V. (2000). Allylsulfide constituents of garlic volatile oil as antimicrobial agents. *Phytomedicine*, *7*, 239–243.

Barbieri. (2016). https://commons.wikimedia.org/wiki/Allium_sativum#/media/File:Allium_sativum_-_Garlic_-_01.jpg.

Bat-Chen, W., Golan, T., Peri, I., Ludmer, Z., & Schwartz, B. (2010). Allicin purified from fresh garlic cloves induces apoptosis in colon cancer cells via Nrf2. *Nutrition and Cancer*, *62*(7), 947–957.

Bayan, L., Koulivand, P. H., & Gorji, A. (2014). Garlic: A review of potential therapeutic effects. *Avicenna Journal of Phytomedicine*, *4*(1), 1–14.

Becker, P. M., van Wikselaar, P. G., Mul, M. F., Pol, A., Engel, B., et al. (2012). *Actinobacillus pleuropneumoniae* is impaired by the garlic volatile allyl methyl sulfide (AMS) in vitro and in-feed garlic alleviates pleuropneumonia in a pig model. *Veterinary Microbiology*, *154*(3–4), 316–324.

Belman, S. (1983). Onion and garlic oils inhibit tumor promotion. *Carcinogenesis*, *4*, 1063–1065.

Benkeblia, N. (2004). Antimicrobial activity of essential oil extracts of various onions (*Allium cepa*) and garlic (*Allium sativum*). *Lebensmittel-Wissenschaft & Technologie*, *37*, 263–268.

Calvo-Gómez, O., Morales-Lopez, J., & Lopez, M. G. (2004). Solid-phase microextraction-gas chromatographic-mass spectrometric analysis of garlic oil obtained by hydrodistillation. *Journal of Chromatography A.*, *1*, 91–93.

Caspasso, A. (2013). Antioxidant action and therapeutic efficacy of *Allium sativum* L. *Molecules*, *18*(1), 690–700.

Chandrashekar, P. M., & Venkatesh, Y. P. (2012). Fructans from aged garlic extract produce a delayed immunoadjuvant response to ovalbumin antigen in BALB/c mice. *Immunopharmacology and Immunotoxicology*, *34*(1), 174–180.

Choi, Y. H., & Park, H. S. (2012). Apoptosis induction of U937 human leukemia cells by diallyl trisulfide induces through generation of reactive oxygen species. *Journal of Biomedical Science*, *19*(50).

Chu, T. C., Ogidigben, M., Han, J. C., & Potter, D. E. (1993). Allicin-induced hypotension in rabbit eyes. *Journal of Ocular Pharmacological*, *9*(3), 201–209.

Coppi, A., Cabinian, M., Mirelman, D., & Sinnis, P. (2006). Antimalarial activity of Allicin, a biologically active compound from garlic cloves. *Antimicrobial Agents and Chemotherapy*, *50*(5), 1731–1737.

Cruz, C., Correa-Rotter, R., Sanchez-Gonzales, D. J., Hernandez-Pando, R., Maldonado, P. D., & Martinez-Martinez, C. M. (2007). Renoprotective and response to ovalbumin antigen in BALB/c mice. *American Journal of Physiology Renal Physiology*, *34*, 174–180.

Das, A., Banik, N. L., & Ray, S. K. (2007). Garlic compounds generate reactive oxygen species leading to activation of stress kinases and cysteine proteases for apoptosis in human glioblastoma T98G and U87MG cells. *Cancer*, *110*, 1083–1095.

Duda, G., Suliburska, J., & Pupek-Musialik, D. (2008). Effects of short-term garlic supplementation on lipid metabolism and antioxidant status in hypertensive adults. *Pharmacological Reports*, *60*(2), 163–170.

Eja, M. E., Asikong, B. E., Abriba, C., Arikpo, G. E., Anwan, E. E., & Enyi-Idoh, K. H. (2007). A comparative assessment of the antimicrobial effects of garlic (*Allium sativum*) and antibiotics on diarrheagenic organisms. *Southeast Asian Journal of Tropical Medicine and Public Health*, *38*(2).

EL-mahmood, M. A. (2009). Efficacy of crude extracts of garlic (*Allium sativum* Linn) against nosocomial *Escherichia coli, Staphylococcus aureus, Streptococcus pneumoniea* and *Pseudomonas aeruginosa*. *Journal of Medicinal Plants Research*, *3*, 179–185.

Elkayam, A., Mirelman, D., Peleq, E., Wilchek, M., Miron, T., Rabinkov, A., et al. (2001). The effects of allicin and enalapril in fructose-induced hyperinsulinemic hyperlipidemic hypertensive rats. *American Journal of Hypertension*, *14*(4–1), 377–381.

Fallon, M. B., Abrams, G. A., Abdel-Razek, T. T., Dai, J., Chen, Y. F., Luo, B., et al. (1998). Garlic prevents hypoxic pulmonary hypertension in rats. *The American Journal of Physiology*, *275*(2), 283–287.

Fenwick, G. R., & Hanley, A. B. (1985). Allium species poisoning. *Veterinary Record*, *116*(1).

Fir0002. (2016). *flagstaffotos.com.au*. https://commons.wikimedia.org/wiki/Allium_sativum#/media/File:Garlic_plant.jpg.

Gallwitz, H., Bonse, S., Martinez-Cruz, A., Schlichting, I., Schumacher, K., & Krauth-Siegel, R. L. (1999). Ajoene is an inhibitor and subversive substrate of human glutathione reductase and *Trypanosoma cruzi* trypanothione reductase: Crystallographic, kinetic and spectroscopic studies. *Journal of Medicinal Chemistry*, *42*(3), 364–372.

Gebreyohannes, G., & Gebreyohannes, M. (2013). Medicinal values of garlic: A review. *International Journal of Medicine and Medicinal Sciences*, *5*(9), 401–408.

Ginter, E., & Simko, V. (2010). Garlic (*Allium sativum* L.) and cardiovascular diseases. *Bratislavske lekarske listy*, *111*(8), 452–456.

Gonen, A., Harats, D., Rabinkov, A., Miron, T., Mirelman, D., Wilchek, M., et al. (2005). The anti-atherogenic effect of allicin: Possible mode of action. *Pathobiology*, *72*(6), 325–334.

Haciseferogulları, H., Ozcan, M., Demir, F., & Calisir, S. (2005). Some nutritional and technological properties of garlic (*Allium sativum* L.). *Journal of Food Engineering*, *68*(4), 463–469.

Harauma, A., & Moriguchi, T. (2006). Aged garlic extract improves blood pressure in spontaneously hypertensive rats more safely than raw garlic. *Journal of Nutrition*, *136*, 769–773.

Harris, J. C., Cottrell, S. L., Plummer, S., & Lloyd, D. (2001). Antimicrobial properties of *Allium sativum* (garlic). *Applied Microbiology and Biotechnology*, *57*, 282–286.

Hassan, H. T. (2004). Ajoene (natural garlic compound): A new anti-leukaemia agent for AML therapy. *Leukemia Research*, *28*(7), 667–671.

Hodge, G., Hodge, S., & Han, P. (2002). *Allium sativum* (garlic) suppresses leukocyte inflammatory cytokine production in vitro: Potential therapeutic use in the treatment of inflammatory bowel disease. *Cytometry*, *48*(4), 209–215.

Hussain, S. P., Jannu, L. N., & Rao, A. R. (1990). Chemopreventative action of garlic on methylcholanthrene-induce carcinogenesis in the uterine cervix of mice. *Cancer Letters*, *49*(2), 175–180.

Huxley, A. J. (1992). *Acanthus montanus plant description and geographical distribution.* Available from: http://www.doacs.state.fl.us/pi/enpp/98-mar-apr.htm.

Ip, C., Lisk, D. J., & Stoewsand, G. S. (1992). Mammary cancer prevention by regular garlic and selenium-enriched garlic. *Nutrition and Cancer*, *17*, 279–286.

Jakobsen, T. H., van Gennip, M., Phipps, R. K., Shanmugham, M. S., Christensen, L. D., Alhede, M., et al. (2012). Ajoene, a Sulfur-rich molecule from garlic, inhibits genes controlled by quorum sensing. *Antimicrobial Agents and Chemotherapy*, *56*(5), 2314–2325.

Josling, P. (2001). Preventing the common cold with a garlic supplement: A double-blind placebo-controlled survey. *Advances in Therapy*, *18*(4), 189–193.

Kew.org. (2016). *Allium sativum (garlic).* Available from: http://www.kew.org/science-conservation/plants-fungi/allium-sativum-garlic.

Khan, Z. K., & Katiyar, R. (2000). Potent antifungal activity of garlic (*Allium sativum*) against murine disseminated cryptococcosis. *Pharmaceutical Biology*, *38*, 87–100.

Khanum, F., Anilakumar, K. R., & Viswanathan, K. R. (2004). Anticarcinogenic properties of garlic: A review. *Critical Reviews in Food Science and Nutrition*, *44*, 479–488.

Kloucek, P., Smid, J., Flesar, J., Havlik, J., Titera, D., Rada, V., et al. (2012). In vitro inhibitory activity of essential oil vapors against *Ascosphaera apis*, *7*(2), 253–256.

Kuda, T., Iwai, A., & Yano, T. (2004). Effect of red pepper *Capsicum annuum* var. *conoides* and garlic *Allium sativum* on plasma lipid levels and cecal microflora in mice fed beef tallow. *Food and Chemical Toxicology: An International Journal Published for the British Industrial Biological Research Association*, *42*(10), 1695–1700.

Lamm, D. L., & Riggs, D. R. (2001). Enhanced immunocompetence by garlic: Role in bladder cancer and other malignancies. *The Journal of Malignancies*, *131*(3), 1067–1070.

Lanzotti, V., Barile, E., Antignani, V., Bonanomi, G., & Scala, F. (2012). Antifingal saponins from bulbs of garlic, *Allium sativum* L. var. Voghiera. *Phytochemistry*, *78*, 126–134.

Lau, B. H. S., Adetumbi, M. A., & Sanchez, A. (1983). *Allium sativum* (garlic) and atherosclerosis: A review. *Nutrition Research*, *3*, 119–128.

Lii, C. K., Huang, C. Y., Chen, H. W., Chow, M. Y., Lin, Y. R., Huang, C. A., et al. (2012). Diallyl trisulfide suppresses the adipogenesis of 3T3-L1 preadipocytes through ERK activation. *Food and Chemical Toxicology: An International Journal Published for the British Industrial Biological Research Association*, *50*(3–4), 478–484.

Lima, C. M., Freitas, F. I., Morais, L. C., Cavalcanti, M. G., Silva, L. F., Padilha, R. J., et al. (2011). Ultrastructural study on the morphological changes to male worms of *Schistosoma mansoni* after in vitro exposure to allicin. *Revista da Sociedade Brasileira de Medicina Tropical*, *44*(33), 327–330.

Loesche, W. J. (1986). Role of *Streptococcus mutans* in human dental decay. *Microbiology Reviews*, *50*(4), 353–358.

Lui, J. Z., Lin, R. I., & Milner, J. A. (1992). Inhibition of 7, 12-dimethylbenz (a)anthracene-induced mammary tumors and DNA adducts by garlic powder. *Carcinogenesis*, *13*, 1847–1851.

Lui, S., Sun, Y., li, W., Yu, H., Li, X., Liu, Z., et al. (2010). The antibacterial mode of action of allitridi for its potential use as a therapeutic agent against *Heliobacter pylori* infection. *FEMS Microbioly Letters*, *303*, 183–189.

Lun, Z. R., Burri, C., Menzinger, M., & Kaminsky, R. (1994). Antiparasitic activity of diallyl trisulfide (Dasuansu) on human and animal pathogenic protozoa (*Trypanosoma* sp. *Entamoeba histolytica* and *Giardia lamblia*) in vitro. *Annales de la Societe belge de medicine tropicale*, *74*, 51–59.

Meriga, B., Mopuri, R., & MuraliKrishna, T. (2012). Insecticidal, antimicrobial and antioxidant activities of bulb extracts of *Allium sativum*. *Asian Pacific Journal of Tropical Medicine*, *5*, 391–395.

Milner, J. A. (1996). Garlic: Its anticarcinogenic and antitumorigenic properties. *Nutrition Reviews*, *54*(11), 82–86.

Milner, J. A. (1999). Functional foods and health promotion. *The Journal of Nutrition*, *129*(7), 1395–1397.

Milner, J. A. (2001). A historical perspective on garlic and cancer. *The Journal of Nutrition*, *131*(3), 1027–1031.

Nencini, C., Cavallo, F., Capasso, A., Franchi, G. G., Giorgio, G., & Micheli, L. (2007). Evaluation of antioxidative properties of *Allium* species growing in Italy. *Phytotherapy Research*, *21*(9), 874–878.

Nwokocha, C. R., Ozolua, R. I., Owu, D. U., Nwokocha, M. I., & Ugwu, A. C. (2011). Antihypertensive properties of *Allium sativum* (garlic) on normotensive and two kidney one clip hypertensive rats. *Nigerian Journal of Physiological Sciences: Official Publication of the Physiological Society of Nigeria*, *26*(2), 213–218.

Ogita, A., Fujita, K., Taniguchi, M., & Tanaka, T. (2006). Enhancement of the fungicidal activity of amphotericin B by allicin, an allyl-sulfur compound from garlic, against the yeast *Saccharomyces cerevisiae* as a model system. *Planta Medica*, *72*, 1247–1250.

Orekhov, A. N., & Grunwald, J. (1997). Effects of garlic on atherosclerosis. *Nutrition*, *13*(7–8), 656–663.

Patya, M., Zahalka, M. A., Vanichkin, A., Rabinkov, A., Miron, T., Mirelman, D., et al. (2004). Allicin stimulates lymphocytes and elicits an antitumor effect: A possible role of $p21^{ras}$. *International Immunology*, *16*(2), 275–281.

Petrovska, B. B., & Cekovska, S. (2010). Extracts from the history and medical properties of garlic. *Pharmacognosy Reviews*, *4*(7), 106–110.

Pundir, R. K., Jain, P., & Sharma, C. H. (2010). Antimicrobial activity of ethanolic extracts of *Syzygium aromaticum* and *Allium sativum* against food associated bacteria and fungi. *Ethnobotanical Leaflets*, *14*, 344–360.

Rahman, K., & Lowe, G. M. (2006). Garlic and cardiovascular disease: A critical review. *The Journal of Nutrition*, *6*, 736–740.

Rivlin, R. S. (2001). Historical perspective on the use of garlic. *Journal of Nutririon*, *131*(3), 951–954.

Salman, H., Bergman, M., Bessler, H., Punsky, I., & Djaldetti, M. (1999). Effect of a garlic derivative (alliin) on peripheral blood cell immune responses. *International Journal of Immunopharmacology*, *21*(9), 589–597.

Schafer, G., & Kaschula, C. H. (2014). The immunomodulation and anti-inflammatory effects of garlic organosulfur compounds in cancer chemoprevention. *Anti-Cancer Agents in Medicinal Chemistry*, *14*(2), 233–240.

Sela, U., Ganor, S., Hecht, I., Brill, A., Miron, T., Rabnikov, A., et al. (2004). Allicin inhibits SDF-1α-induced T cell interactions with fibronectin and endothelial cells by down-regulating cytoskeleton rearrangement, Pyk-2 phosphorylation and VLA-4 expression. *Nature Reviews. Immunology*, *111*(4), 391–399.

Sener, G., Sakarcan, A., & Yegen, B. C. (2007). Role of garlic in the prevention of ischemia-reperfusion injury. *Molecular Nutrition & Food Research*, *51*(11), 1345–1352.

Sengupta, A., Ghosh, S., & Bhattacharjee, S. (2004). Allium vegetables in cancer prevention: An overview. *Asian Pacific Journal of Cancer Prevention*, *5*(3), 237–245.

Shamim, S., Ahmed, S. W., & Azhar, I. (2004). Antifungal activity of allium, aloe, and solanum species. *Pharmaceutical Biology*, *42*, 491–498.

Shukla, Y., Singh, A., & Srivastava, B. (1999). Inhibition of carcinogen-induced activity of gamma-glutamyl transpeptidase by certain dietary constituents in mouse skin. *Biomedical and Environmental Sciences*, *12*(2), 110–115.

Soffar, S. A., & Mokhtar, G. M. (1991). Evaluation of the antiparasitic effect of aqueous garlic (*Allium sativum*) extract in hymenolepiasis nana and giardiasis. *Journal of the Egyptian Society of Parasitology*, *21*(2), 497–502.

Song, K., & Milner, J. A. (1999). Heating garlic inhibits its ability to suppress 7, 12-dimethylbenz (a) anthracene-induced DNA adduct formation in rat mammary tissue. *Journal of Nutrition*, *129*, 657–661.

Suleria, H. A. R., Butt, M. S., Khalid, N., Sultan, S., Raza, A., Aleem, M., et al. (2015). Garlic (*Allium sativum*): Diet based therapy of 21st century – a review. *Asian Pacific Journal of Tropical Disease*, *5*(4), 271–278.

Sumiyoshi, H., & Wargovich, M. J. (1990). Chemopreventation of 1, 2-dimethylhydrazine-induced colon cancer in mice by natural occurring organosulfur compounds. *Cancer Research*, *50*, 5084–5087.

Thevenot, T., Pastor, C. M., Cervoni, J. P., Jacquelinet, C., Nguyen-Khac, E., Richou, C., et al. (2009). Hepatopulmonary syndrome. *Gastroenterologie Clinique et biologique*, *33*(6–7), 565–579.

Thomaz, L., Apitz-Castro, R., Marques, A. F., Travassos, L. R., & Taborda, C. P. (2008). Experimental paracoccidioidomycosis: Alternative therapy with ajoene, compound from *Allium sativum*, associated with sulfamethoxazole/trimethoprim. *Medical Mycology*, *46*(2), 113–118.

Tilli, C. M. L.J., Stavast-Kooy, A. J. W., Vuerstaek, J. D. D., Thissen, M. R. T.M., Krekels, G. A. M., Ramaekers, F. C. S., et al. (2003). The garlic-derived organosulfur component ajoene decreases basal cell carcinoma tumor size by inducing apoptosis. *Archive of Dermatological Research*, *295*, 117–123.

Tocmo, R., Liang, D., Lin, Y., & Huang, D. (2015). Chemical and biochemical mechanisms underlying the cardioprotective roles of dietary organopolysulfides. *Frontiers in Nutrition*, *2*(1).

Verma, S. K., Rajeevan, V., Jain, P., & Bordia, A. (2005). Effect of garlic (*Allium sativum*) oil on exercise tolerance in patients with coronary artery disease. *Indian Journal of Physiology and Pharmacology*, *49*(1), 115–118.

Vimal, V., & Devaki, T. (2004). Hepatoprotective effect of allicin on tissue defense system in galactosamine/endotoxin challenged rats. *Journal of Ethnopharmacology*, *90*(1), 151–154.

Waag, T., Gelhaus, C., Rath, J., Stich, A., Leippe, M., & Schirmeister, T. (2010). Allicin and derivates are cysteine protease inhibitors with antiparasitic activity. *Bioorganic and Medicinal Chemistry Letters*, *20*(18), 5541–5543.

Wang, H. X., & Ng, T. B. (2001). Purification of allivin, a novel antifungal protein from bulbs of the round-cloved garlic. *Life Sciences*, *70*, 357–365.

Wargovich, M. J., Woods, C., Eng, V. W., Stephens, L. C., & Gray, K. (1988). Chemoprevention of N-nitrosomethylbenzylamine-induced esophageal cancer in rats by the naturally occurring thioether, diallyl sulfide. *Cancer Research*, *48*, 6872–6875.

Yang, Y., Chan, S. W., Hu, M., Walden, R., & Tomlinson, B. (2011). Effects of some common food constituents on cardiovascular disease. *ISRN Cardiology*, *2011*, 1–16.

Yoshida, S., Kasuga, S., Hayashi, N., Ushiroguchi, T., Matsuura, H., & Nakagawa, S. (1987). Antifungal activity of ajoene derived from garlic. *Applied and Environmental Microbiology*, *53*(3), 615–617.

Yoshida, H., Katsuzaki, H., Ohta, R., Ishikawa, K., Fukuda, H., Fujino, T., & Suzuki, A., et al. (1999). An organosulfur compound isolated from oil-macerated garlic extract, and its antimicrobial effect. *Bioscience, Biotechnology, and Biochemistry*, *63*(3), 588–590.

You, W. C., Brown, L. M., Zhang, L., Li, J. Y., Jin, M. L., Chang, Y. S., et al. (2006). Randomized double-blind factorial trial of three treatments to reduce the prevalence of precancerous gastric lesions. *Journal of the National Cancer Institute*, *98*(14), 974–983.

Zahid Ashraf, M., Hussein, M. E., & Fahim, M. (2005). Antiatherosclerotic effects of dietary supplementations of garlic and turmeric. Restoration of endothelial function in rats. *Life Sciences*, *77*(8), 837–857.

Zhang, X. H., Lowe, D., Giles, P., Fell, S., Board, A. R., Baughan, J. A., et al. (2001). A randomised trial of the effects of garlic oil upon coronary heart disease risk factors in trained male runners. *Blood Coagulationl Fibrinolysis*, *12*, 67–74.

Chapter 10

Maximizing Medicinal Plants: Steps to Realizing Their Full Potential

Bianca D. Fibrich, Namrita Lall
University of Pretoria, Pretoria, South Africa

Chapter Outline

10.1 MAXIMIZING THE VALUE OF MEDICINAL PLANTS

The value of plants in Western medicine is reflected in modern products implementing their actives. Although a popular form of medicine employed by a great deal of the third world population, medicinal plants and especially ethnomedicinal practices are, however, somewhat underappreciated and underestimated in modern times. To fully maximize the use of medicinal plants many aspects must be considered.

10.1.1 Educate the Global Population About the Use of Plants in Medicine

There exists a fallacy in current times surrounding the use of medicinal plants, comprises two myths; the first is that plants serve no medicinal value to

Medicinal Plants for Holistic Health and Well-Being. http://dx.doi.org/10.1016/B978-0-12-812475-8.00010-X

mankind, and the second is that plants are completely safe and offer no side effects. Although the greatest function of plants for mankind has been as food and sustenance, their ability to cure and treat certain conditions is not as widely accepted or appreciated. The global population needs to be educated to understand the history of plants as a medicine as well as understand the role of plants in the development of modern herbal drugs. Many consumers often consume prescription medication for the treatment of neurological or cardiac disorders without the knowledge that the active component is a plant-derived pharmaceutical. The role that phytochemicals play in the development of synthetic drugs also needs to be elaborated on. Although the application of plants is widely beneficial, the risks associated with toxic plants need to be addressed and the global population educated surrounding the limitations of plants.

10.1.2 The Use of Medicinal Plants Stemming From Ethnomedicinal Origin Needs to Be Fully Understood and Scientifically Validated

The role of ethnomedicine in modern drug discovery is just as unappreciated as the plants involved in their practice. More attention should be focused on the valuable information that is relayed over generations of traditional healers. These applications of medicinal plants should, however, not be considered to be written in stone but should importantly be considered when selecting plants for scientific investigation. Scientific validation of medicinal plants is of utmost importance for the treatment of conditions, especially concerning the safety of medicinal plant extracts as well as phytochemicals incorporated into the development of modern herbal drugs. Scientific testing and validation of the respective sample also ensures that the plant extract or isolated phytochemical can be used effectively for the treatment of a specific condition. Identifying the active component further allows for standardization of the medicine that does not occur in traditional medicinal practices where adulteration of species sharing physical characteristics can occur. Scientific testing and a condition-orientated approach also allow the best method of administration to be selected, and subsequently the herbal medicine can be processed to an appropriate and more palatable dosage form.

10.1.3 Bridge the Gap Between Modern Western Medicine and Traditional Medicine

The incorporation of traditional medicine into Western medicine and vice versa is an important step in bridging the gap between the two. This step is important for the development of a new platform where herbal medicines are used commonly as complementary or alternative medicine where appropriate with Western medicine. The use of herbal medicines may be as a supplement to the use of conventional Western medicine to enhance the efficacy or simply provide other desirable medicinal attributes to the treatment in question. Where appropriate the use of herbal medicine as an alternative to the conventional Western treatments, which unfortunately offer adverse side effects, should also be promoted. This should not, however, be encouraged in severe conditions or disease states where the benefits of Western medicine outweigh that of herbal medicine. Bridging the gap between these two crucial forms of medicine involves understanding that the goal of both is the health of the population and that each form of medicine offers its own respective advantages and disadvantages. In combining both we can reduce the disadvantages associated with each by replacing them with an alternative, which offers more advantage. Bridging the gap between these two forms of medicines also includes broadening the field of application. This, for example, includes permitting the use of herbal medicines as part of medical aid schemes, which conventionally only cover the Western practices. This will also improve the treatment of a patient not only using Western medicine as per prescription by a medical practitioner but also traditional medicine without informing the medical practitioner, allowing for many contraindications to take place in the course of treatment. Modern Western medicine and traditional medicine working together also holds the possibility of the development of new strategies that can be implemented in certain diseases and conditions.

10.1.4 Develop Sustainable Strategies for Commercialization

Another important aspect of commercialization of medicinal plants includes the development of a farming system for the medicinal plant at a large enough scale as well as sustainable harvesting practices. A great difficulty in large-scale farming of medicinal plants is the optimization of growing conditions because the production of a specific medicinal compound may be dependent on a specific set of circumstances. Finding a plant species that produces the compound of interest in the largest volume depends on age, part of the plant, harvesting season, and optimal storage conditions of harvested material. In instances where cultivation of the medicinal plants is implemented, a herbarium specimen should be made at the start of the cultivation practices; however, to ensure batch-to-batch consistency a sample should be taken and analyzed with each harvest to ensure quality control and monitor the effects of seasonal variation and age of the crop. Furthermore, these labor-intensive sections of commercialization should make use of unskilled labor and provide employees with training on farming of medicinal plants as well as the more skilled portions. The establishment of good agricultural practice principles related to cultivation and processing of medicinal plant material is important to ensure a sustainable supply of quality material.

10.2 CONCLUDING REMARKS

The value of natural remedies may be maximized directly as discussed above or indirectly through serving as a template from which synthetic molecules exhibiting similar features with diminished adverse effects may be developed. Having knowledge about biological pathways involving the synthesis of such valuable molecules also allows a better understanding that can be used to manipulate the process as desired. Studies investigating the activity of such molecules also provide insight into the mechanisms and interactions involved between the drug and its target that may be implemented in the development of alternative treatment regimes.

There are also some key issues involved in standardization of herbal extracts, which limit the value of these preparations. These include the fact that standardization involves the identification of either a specific pure compound or set of active components, which often does not include a range of other phytochemicals with additive value to the extract as a whole, and in certain instances, this reduced the broad application of the herb as reported in traditional literature to only serving as effective in particular applications. Maximizing the efficacy of a herbal medicine also includes diminishing toxicity through limiting the consumption of herbal material, which has been adulterated with toxic trace elements, other plants and interactions that may exist between the herbal material of interest and other consumed material.

Many of the conditions discussed in this book are without natural solution. The value of medicinal plants has been discussed and somewhat realized for many years; however, to fully realize the potential of this resource the value of traditional medicine needs to once again be taken seriously. A great deal of education surrounding the use of medicinal plants should be done to fully understand the value of medicinal plants from a science point of view. Once this can be done, the possibilities are endless.

Index

Note: 'Page numbers followed by "f" indicate figures, "t" indicate tables and "b" indicate boxes.'

B

C

G

H

T

U

V

W

Printed in the United States
By Bookmasters